SCHAUM'S
outlines

# Feedback and Control Systems

Second Edition

**Joseph J. DiStefano III, PhD**
*Departments of Computer Science and Medicine*
*University of California, Los Angeles*

**Allen R. Stubberud, PhD**
*Department of Electrical and Computer Engineering*
*University of California, Irvine*

**Ivan J. Williams, PhD**
*Space and Technology Group, TRW, Inc.*

Schaum's Outline Series

New York   Chicago   San Francisco   Athens   London   Madrid
Mexico City   Milan   New Delhi   Singapore   Sydney   Toronto

**JOSEPH J. DiSTEFANO III** received his MS in Control Systems and PhD in Biocybernetics from the University of California, Los Angeles (UCLA) in 1966. He is currently professor of Computer Science and Medicine, director of the Biocybernetics Research Laboratory, and chair of the Cybernetics Interdepartmental Program at UCLA. He is also on the Editorial boards of *Annals of Biomedical Engineering* and *Optimal Control Applications and Methods*, and is Editor and Founder of the *Modeling Methodology Forum* in the *American Journals of Physiology*. He is author of more than 100 research articles and books and is actively involved in systems modeling theory and software development as well as experimental laboratory research in physiology.

**ALLEN R. STUBBERUD** was awarded a BS degree from the University of Idaho, and the MS and PhD degrees from the University of California, Los Angeles (UCLA). He is presently professor of Electrical and Computer Engineering at the University of California, Irvine. Dr. Stubberud is the author of over 100 articles and books, and belongs to a number of professional and technical organizations, including the American Institute of Aeronautics and Astronautics (AIAA). He is a fellow of the Institute of Electrical and Electronics Engineers (IEEE), and the American Association of the Advancement of Sciences (AAAS).

**IVAN J. WILLIAMS** was awarded BS, MS, and PhD degrees by the University of California at Berkeley. He has instructed courses in control systems engineering at the University of California, Los Angeles (UCLA), and is presently a project manager at the Space and Technology Group of TRW, Inc.

2 3 4 5 6 7 8 9 0   CUS/CUS   1 0 9 8 7 6 5 4

ISBN       978-0-07-182948-9
MHID       0-07-182948-2

e-ISBN    978-0-07-183076-8   (basic e-book)
e-MHID     0-07-183076-6

e-ISBN    978-0-07-183077-5   (enhanced e-book)
e-MHID     0-07-183077-4

**Library of Congress Control Number: 2013946488**

# Preface

Feedback processes abound in nature and, over the last few decades, the word feedback, like *computer*, has found its way into our language far more pervasively than most others of technological origin. The conceptual framework for the theory of feedback and that of the discipline in which it is embedded—control systems engineering–have developed only since World War II. When our first edition was published, in 1967, the subject of linear continuous-time (or *analog*) control systems had already attained a high level of maturity, and it was (and remains) often designated *classical control* by the *conoscienti*. This was also the early development period for the digital computer and discrete-time data control processes and applications, during which courses and books in "sampled-data" control systems became more prevalent. Computer-controlled and *digital* control systems are now the terminology of choice for control systems that include digital computers or microprocessors.

In this second edition, as in the first, we present a concise, yet quite comprehensive, treatment of the fundamentals of feedback and control system theory and applications, for engineers, physical, biological and behavioral scientists, economists, mathematicians and students of these disciplines. Knowledge of basic calculus, and some physics are the only prerequisites. The necessary mathematical tools beyond calculus, and the physical and nonphysical principles and models used in applications, are developed throughout the text and in the numerous solved problems.

We have modernized the material in several significant ways in this new edition. We have first of all included discrete-time (digital) data signals, elements and control systems throughout the book, primarily in conjunction with treatments of their continuous-time (analog) counterparts, rather than in separate chapters or sections. In contrast, these subjects have for the most part been maintained pedagogically distinct in most other textbooks. Wherever possible, we have integrated these subjects, at the introductory level, in a *unified* exposition of continuous-time and discrete-time control system concepts. The emphasis remains on continuous-time and linear control systems, particularly in the solved problems, but we believe our approach takes much of the mystique out of the methodologic differences between the analog and digital control system worlds. In addition, we have updated and modernized the nomenclature, introduced state variable representations (models) and used them in a strengthened chapter introducing nonlinear control systems, as well as in a substantially modernized chapter introducing advanced control systems concepts. We have also solved numerous analog and digital control system analysis and design problems using special purpose computer software, illustrating the power and facility of these new tools.

The book is designed for use as a text in a formal course, as a supplement to other textbooks, as a reference or as a self-study manual. The quite comprehensive index and highly structured format should facilitate use by any type of readership. Each new topic is introduced either by section or by chapter, and each chapter concludes with numerous solved problems consisting of extensions and proofs of the theory, and applications from various fields.

Los Angeles, Irvine and
Redondo Beach, California
March, 1990

JOSEPH J. DISTEFANO, III
ALLEN R. STUBBERUD
IVAN J. WILLIAMS

# Contents

CONTENTS

CONTENTS

CONTENTS

# CONTENTS

# Chapter 1

## Introduction

### 1.1 CONTROL SYSTEMS: WHAT THEY ARE

In modern usage the word *system* has many meanings. So let us begin by defining what we mean when we use this word in this book, first abstractly then slightly more specifically in relation to scientific literature.

***Definition 1.1a***:      A **system** is an arrangement, set, or collection of things connected or related in such a manner as to form an entirety or whole.

***Definition 1.1b***:      A **system** is an arrangement of physical components connected or related in such a manner as to form and/or act as an entire unit.

The word **control** is usually taken to mean *regulate*, *direct*, or *command*. Combining the above definitions, we have

***Definition 1.2***:      A **control system** is an arrangement of physical components connected or related in such a manner as to command, direct, or regulate itself or another system.

In the most abstract sense it is possible to consider every physical object a control system. Everything alters its environment in some manner, if not actively then passively—like a mirror *directing* a beam of light shining on it at some acute angle. The mirror (Fig. 1-1) may be considered an elementary control system, controlling the beam of light according to the simple equation "the angle of reflection $\alpha$ equals the angle of incidence $\alpha$."

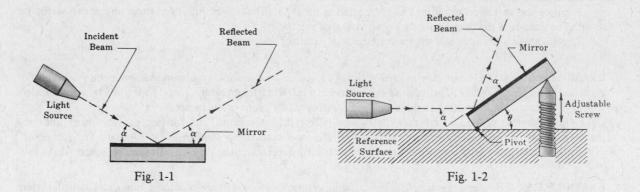

Fig. 1-1                Fig. 1-2

In engineering and science we usually restrict the meaning of control systems to apply to those systems whose major function is to *dynamically* or *actively* command, direct, or regulate. The system shown in Fig. 1-2, consisting of a mirror pivoted at one end and adjusted up and down with a screw at the other end, is properly termed a *control system*. The angle of reflected light is regulated by means of the screw.

It is important to note, however, that control systems of interest for analysis or design purposes include not only those manufactured by humans, but those that normally exist in nature, and control systems with both manufactured and natural components.

1

## 1.2  EXAMPLES OF CONTROL SYSTEMS

Control systems abound in our environment. But before exemplifying this, we define two terms: *input* and *output*, which help in identifying, delineating, or defining a control system.

**Definition 1.3:**    The **input** is the stimulus, excitation or command applied *to* a control system, typically from an external energy source, usually in order to produce a specified response *from* the control system.

**Definition 1.4:**    The **output** is the actual response obtained from a control system. It may or may not be equal to the specified response implied by the input.

Inputs and outputs can have many different forms. Inputs, for example, may be physical variables, or more abstract quantities such as *reference*, *setpoint*, or *desired* values for the output of the control system.

The purpose of the control system usually identifies or defines the output and input. If the output and input are given, it is possible to identify, delineate, or define the nature of the system components.

Control systems may have more than one input or output. Often all inputs and outputs are well defined by the system description. But sometimes they are not. For example, an atmospheric electrical storm may intermittently interfere with radio reception, producing an unwanted output from a loudspeaker in the form of static. This "noise" output is part of the total output as defined above, but for the purpose of simply identifying a system, spurious inputs producing undesirable outputs are not normally considered as inputs and outputs in the system description. However, it is usually necessary to carefully consider these extra inputs and outputs when the system is examined in detail.

The terms input and output also may be used in the description of any type of system, whether or not it is a control system, and a control system may be part of a larger system, in which case it is called a **subsystem** or **control subsystem**, and its inputs and outputs may then be internal variables of the larger system.

**EXAMPLE 1.1.**   An *electric switch* is a manufactured control system, controlling the flow of electricity. By definition, the apparatus or person flipping the switch is not a part of this control system.

Flipping the switch on or off may be considered as the input. That is, the input can be in one of two states, on or off. The output is the flow or nonflow (two states) of electricity.

The electric switch is one of the most rudimentary control systems.

**EXAMPLE 1.2.**   A *thermostatically controlled heater or furnace automatically regulating the temperature of a room or enclosure* is a control system. The input to this system is a reference temperature, usually specified by appropriately setting a thermostat. The output is the actual temperature of the room or enclosure.

When the thermostat detects that the output is less than the input, the furnace provides heat until the temperature of the enclosure becomes equal to the reference input. Then the furnace is automatically turned off. When the temperature falls somewhat below the reference temperature, the furnace is turned on again.

**EXAMPLE 1.3.**   The seemingly simple act of *pointing at an object with a finger* requires a biological control system consisting chiefly of the eyes, the arm, hand and finger, and the brain. The input is the precise direction of the object (moving or not) with respect to some reference, and the output is the actual pointed direction with respect to the same reference.

**EXAMPLE 1.4.**   A part of the human temperature control system is the *perspiration system*. When the temperature of the air exterior to the skin becomes too high the sweat glands secrete heavily, inducing cooling of the skin by evaporation. Secretions are reduced when the desired cooling effect is achieved, or when the air temperature falls sufficiently.

The input to this system may be "normal" or comfortable skin temperature, a "setpoint," or the air temperature, a physical variable. The output is the actual skin temperature.

**EXAMPLE 1.5.** The control system consisting of *a person driving an automobile* has components which are clearly both manufactured and biological. The driver wants to keep the automobile in the appropriate lane of the roadway. He or she accomplishes this by constantly watching the direction of the automobile with respect to the direction of the road. In this case, the direction or heading of the road, represented by the painted guide line or lines on either side of the lane may be considered as the input. The heading of the automobile is the output of the system. The driver controls this output by constantly measuring it with his or her eyes and brain, and correcting it with his or her hands on the steering wheel. The major components of this control system are the driver's hands, eyes and brain, and the vehicle.

## 1.3  OPEN-LOOP AND CLOSED-LOOP CONTROL SYSTEMS

Control systems are classified into two general categories: *open-loop* and *closed-loop* systems. The distinction is determined by the **control action**, that quantity responsible for activating the system to produce the output.

The term *control action* is classical in the control systems literature, but the word *action* in this expression does not always *directly* imply change, motion, or activity. For example, the control action in a system designed to have an object hit a target is usually the *distance* between the object and the target. Distance, as such, is not an action, but action (motion) is implied here, because the goal of such a control system is to reduce this distance to zero.

*Definition 1.5*:      An **open-loop** control system is one in which the control action is independent of the output.

*Definition 1.6*:      A **closed-loop** control system is one in which the control action is somehow dependent on the output.

Two outstanding features of open-loop control systems are:

1.  Their ability to perform accurately is determined by their calibration. To **calibrate** means to establish or reestablish the input-output relation to obtain a desired system accuracy.

2.  They are not usually troubled with problems of *instability*, a concept to be subsequently discussed in detail.

Closed-loop control systems are more commonly called *feedback* control systems, and are considered in more detail beginning in the next section.

To classify a control system as open-loop or closed-loop, we must distinguish clearly the components of the system from components that interact with but are not part of the system. For example, the driver in Example 1.5 was defined as part of that control system, but a human operator may or may not be a component of a system.

**EXAMPLE 1.6.** Most *automatic toasters* are open-loop systems because they are controlled by a timer. The time required to make "good toast" must be estimated by the user, who is not part of the system. Control over the quality of toast (the output) is removed once the time, which is both the input and the control action, has been set. The time is typically set by means of a calibrated dial or switch.

**EXAMPLE 1.7.** An *autopilot mechanism and the airplane it controls* is a closed-loop (feedback) control system. Its purpose is to maintain a specified airplane heading, despite atmospheric changes. It performs this task by continuously measuring the actual airplane heading, and automatically adjusting the airplane control surfaces (rudder, ailerons, etc.) so as to bring the actual airplane heading into correspondence with the specified heading. The human pilot or operator who presets the autopilot is not part of the control system.

## 1.4  FEEDBACK

Feedback is that characteristic of closed-loop control systems which distinguishes them from open-loop systems.

**Definition 1.7:**     **Feedback** is that property of a closed-loop system which permits the output (or some other controlled variable) to be compared with the input to the system (or an input to some other internally situated component or subsystem) so that the appropriate control action may be formed as some function of the output and input.

More generally, feedback is said to exist in a system when a *closed* sequence of cause-and-effect relations exists between system variables.

**EXAMPLE 1.8.**   The concept of feedback is clearly illustrated by the autopilot mechanism of Example 1.7. The input is the specified heading, which may be set on a dial or other instrument of the airplane control panel, and the output is the actual heading, as determined by automatic navigation instruments. A comparison device continuously monitors the input and output. When the two are in correspondence, control action is not required. When a difference exists between the input and output, the comparison device delivers a control action signal to the controller, the autopilot mechanism. The controller provides the appropriate signals to the control surfaces of the airplane to reduce the input-output difference. Feedback may be effected by mechanical or electrical connections from the navigation instruments, measuring the heading, to the comparison device. In practice, the comparison device may be integrated within the autopilot mechanism.

## 1.5  CHARACTERISTICS OF FEEDBACK

The presence of feedback typically imparts the following properties to a system.

1. Increased accuracy. For example, the ability to faithfully reproduce the input. This property is illustrated throughout the text.

2. Tendency toward oscillation or instability. This all-important characteristic is considered in detail in Chapters 5 and 9 through 19.

3. Reduced sensitivity of the ratio of output to input to variations in system parameters and other characteristics (Chapter 9).

4. Reduced effects of nonlinearities (Chapters 3 and 19).

5. Reduced effects of external disturbances or noise (Chapters 7, 9, and 10).

6. Increased bandwidth. The **bandwidth** of a system is a frequency response measure of how well the system responds to (or filters) variations (or frequencies) in the input signal (Chapters 6, 10, 12, and 15 through 18).

## 1.6  ANALOG AND DIGITAL CONTROL SYSTEMS

The signals in a control system, for example, the input and the output waveforms, are typically functions of some independent variable, usually time, denoted $t$.

**Definition 1.8:**     A signal dependent on a continuum of values of the independent variable $t$ is called a **continuous-time** signal or, more generally, a **continuous-data** signal or (less frequently) an **analog** signal.

**Definition 1.9:**     A signal defined at, or of interest at, only discrete (distinct) instants of the independent variable $t$ (upon which it depends) is called a **discrete-time**, a **discrete-data**, a **sampled-data**, or a **digital** signal.

We remark that *digital* is a somewhat more specialized term, particularly in other contexts. We use it as a synonym here because it is the convention in the control systems literature.

**EXAMPLE 1.9.** The continuous, sinusoidally varying voltage $v(t)$ or alternating current $i(t)$ available from an ordinary household electrical receptable is a continuous-time (analog) signal, because it is defined at *each and every instant* of time $t$ electrical power is available from that outlet.

**EXAMPLE 1.10.** If a lamp is connected to the receptacle in Example 1.9, and it is switched on and then immediately off every minute, the light from the lamp is a discrete-time signal, on only for an instant every minute.

**EXAMPLE 1.11.** The mean temperature $T$ in a room at precisely 8 A.M. (08 hours) each day is a discrete-time signal. This signal may be denoted in several ways, depending on the application; for example $T(8)$ for the temperature at 8 o'clock—rather than another time; $T(1), T(2), \ldots$ for the temperature at 8 o'clock on day 1, day 2, etc., or, equivalently, using a subscript notation, $T_1, T_2$, etc. Note that these discrete-time signals are *sampled* values of a continuous-time signal, the mean temperature of the room at all times, denoted $T(t)$.

**EXAMPLE 1.12.** The signals inside digital computers and microprocessors are inherently discrete-time, or discrete-data, or digital (or digitally coded) signals. At their most basic level, they are typically in the form of sequences of voltages, currents, light intensities, or other physical variables, at either of two constant levels, for example, $\pm 15$ V; light-on, light-off; etc. These *binary signals* are usually represented in alphanumeric form (numbers, letters, or other characters) at the inputs and outputs of such digital devices. On the other hand, the signals of analog computers and other analog devices are continuous-time.

Control systems can be classified according to the types of signals they process: continuous-time (analog), discrete-time (digital), or a combination of both (hybrid).

*Definition 1.10*:  **Continuous-time control systems**, also called **continuous-data control systems**, or **analog control systems**, contain or process only continuous-time (analog) signals and components.

*Definition 1.11*:  **Discrete-time control systems**, also called **discrete-data control systems**, or **sampled-data control systems**, have discrete-time signals or components at one or more points in the system.

We note that discrete-time control systems can have continuous-time as well as discrete-time signals; that is, they can be hybrid. The distinguishing factor is that a discrete-time or digital control system *must* include at least one discrete-data signal. Also, digital control systems, particularly of sampled-data type, often have both open-loop and closed-loop modes of operation.

**EXAMPLE 1.13.** A target tracking and following system, such as the one described in Example 1.3 (tracking and pointing at an object with a finger), is usually considered an analog or continuous-time control system, because the distance between the "tracker" (finger) and the target is a continuous function of time, and the objective of such a control system is to *continuously* follow the target. The system consisting of a person driving an automobile (Example 1.5) falls in the same category. Strictly speaking, however, tracking systems, both natural and manufactured, can have digital signals or components. For example, control signals from the brain are often treated as "pulsed" or discrete-time data in more detailed models which include the brain, and digital computers or microprocessors have replaced many analog components in vehicle control systems and tracking mechanisms.

**EXAMPLE 1.14.** A closer look at the thermostatically controlled heating system of Example 1.2 indicates that it is actually a sampled-data control system, with both digital and analog components and signals. If the desired room temperature is, say, 68°F (22°C) on the thermostat and the room temperature falls below, say, 66°F, the thermostat switching system closes the circuit to the furnace (an analog device), turning it on until the temperature of the room reaches, say, 70°F. Then the switching system automatically turns the furnace off until the room temperature again falls below 66°F. This control system is actually operating open-loop between switching instants, when the thermostat turns the furnace on or off, but overall operation is considered closed-loop. The thermostat receives a continuous-time signal at its input, the actual room temperature, and it delivers a discrete-time (binary) switching signal at its output, turning the furnace on or off. Actual room temperature thus varies continuously between 66° and 70°F, and *mean* temperature is controlled at about 68°F, the *setpoint* of the thermostat.

The terms discrete-time and discrete-data, sampled-data, and continuous-time and continuous-data are often abbreviated as *discrete*, *sampled*, and *continuous* in the remainder of the book, wherever the meaning is unambiguous. *Digital* or *analog* is also used in place of discrete (sampled) or continuous where appropriate and when the meaning is clear from the context.

## 1.7   THE CONTROL SYSTEMS ENGINEERING PROBLEM

Control systems engineering consists of *analysis* and *design* of control systems configurations.
**Analysis** is the investigation of the properties of an existing system. The **design** problem is the choice and arrangement of system components to perform a specific task.
Two methods exist for design:

1.   Design by analysis
2.   Design by synthesis

**Design by analysis** is accomplished by modifying the characteristics of an existing or standard system configuration, and **design by synthesis** by defining the form of the system directly from its specifications.

## 1.8   CONTROL SYSTEM MODELS OR REPRESENTATIONS

To solve a control systems problem, we must put the specifications or description of the system configuration and its components into a form amenable to analysis or design.
Three basic representations (models) of components and systems are used extensively in the study of control systems:

1.   Mathematical models, in the form of differential equations, difference equations, and/or other mathematical relations, for example, Laplace- and $z$-transforms
2.   Block diagrams
3.   Signal flow graphs

Mathematical models of control systems are developed in Chapters 3 and 4. Block diagrams and signal flow graphs are shorthand, graphical representations of either the schematic diagram of a system, or the set of mathematical equations characterizing its parts. Block diagrams are considered in detail in Chapters 2 and 7, and signal flow graphs in Chapter 8.

Mathematical models are needed when quantitative relationships are required, for example, to represent the detailed behavior of the output of a feedback system to a given input. Development of mathematical models is usually based on principles from the physical, biological, social, or information sciences, depending on the control system application area, and the complexity of such models varies widely. One class of models, commonly called *linear systems*, has found very broad application in control system science. Techniques for solving linear system models are well established and documented in the literature of applied mathematics and engineering, and the major focus of this book is linear feedback control systems, their analysis and their design. Continuous-time (continuous, analog) systems are emphasized, but discrete-time (discrete, digital) systems techniques are also developed throughout the text, in a unifying but not exhaustive manner. Techniques for analysis and design of *nonlinear* control systems are the subject of Chapter 19, by way of introduction to this more complex subject.

In order to communicate with as many readers as possible, the material in this book is developed from basic principles in the sciences and applied mathematics, and specific applications in various engineering and other disciplines are presented in the examples and in the solved problems at the end of each chapter.

# Solved Problems

## INPUT AND OUTPUT

**1.1.** Identify the input and output for the pivoted, adjustable mirror of Fig. 1-2.

The input is the angle of inclination of the mirror $\theta$, varied by turning the screw. The output is the angular position of the reflected beam $\theta + \alpha$ from the reference surface.

**1.2.** Identify a possible input and a possible output for a rotational generator of electricity.

The input may be the rotational speed of the prime mover (e.g., a steam turbine), in revolutions per minute. Assuming the generator has no load attached to its output terminals, the output may be the induced voltage at the output terminals.

Alternatively, the input can be expressed as angular momentum of the prime mover shaft, and the output in units of electrical power (watts) with a load attached to the generator.

**1.3.** Identify the input and output for an automatic washing machine.

Many washing machines operate in the following manner. After the clothes have been put into the machine, the soap or detergent, bleach, and water are entered in the proper amounts. The wash and spin cycle-time is then set on a timer and the washer is energized. When the cycle is completed, the machine shuts itself off.

If the proper amounts of detergent, bleach, and water, and the appropriate temperature of the water are predetermined or specified by the machine manufacturer, or automatically entered by the machine itself, then the input is the time (in minutes) for the wash and spin cycle. The timer is usually set by a human operator.

The output of a washing machine is more difficult to identify. Let us define *clean* as the absence of foreign substances from the items to be washed. Then we can identify the output as the percentage of cleanliness. At the start of a cycle the output is less than 100%, and at the end of a cycle the output is ideally equal to 100% (*clean* clothes are not always obtained).

For most coin-operated machines the cycle-time is preset, and the machine begins operating when the coin is entered. In this case, the percentage of cleanliness can be controlled by adjusting the amounts of detergent, bleach, water, and the temperature of the water. We may consider all of these quantities as inputs.

Other combinations of inputs and outputs are also possible.

**1.4.** Identify the organ-system components, and the input and output, and describe the operation of the biological control system consisting of a human being reaching for an object.

The basic components of this intentionally oversimplified control system description are the brain, arm and hand, and eyes.

The brain sends the required nervous system signal to the arm and hand to reach for the object. This signal is amplified in the muscles of the arm and hand, which serve as power actuators for the system. The eyes are employed as a sensing device, continuously "feeding back" the position of the hand to the brain.

Hand position is the output for the system. The input is object position.

The objective of the control system is to reduce the distance between hand position and object position to zero. Figure 1-3 is a schematic diagram. The dashed lines and arrows represent the direction of information flow,

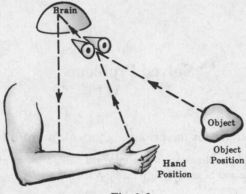

Fig. 1-3

## OPEN-LOOP AND CLOSED-LOOP SYSTEMS

**1.5.** Explain how a closed-loop automatic washing machine might operate.

Assume all quantities described as possible inputs in Problem 1.3, namely cycle-time, water volume, water temperature, amount of detergent, and amount of bleach, can be adjusted by devices such as valves and heaters.

A closed-loop automatic washer might continuously or periodically measure the percentage of cleanliness (output) of the items being washing, adjust the input quantities accordingly, and turn itself off when 100% cleanliness has been achieved.

**1.6.** How are the following open-loop systems calibrated: (*a*) automatic washing machine, (*b*) automatic toaster, (*c*) voltmeter?

(*a*)   Automatic washing machines are calibrated by estimating any combination of the following input quantities: (1) amount of detergent, (2) amount of bleach or other additives, (3) amount of water, (4) temperature of the water, (5) cycle-time.

On some washing machines one or more of these inputs is (are) predetermined. The remaining quantities must be estimated by the user and depend upon factors such as degree of hardness of the water, type of detergent, and type or strength of the bleach or other additives. Once this calibration has been determined for a specific type of wash (e.g., all white clothes, very dirty clothes), it does not normally have to be redetermined during the lifetime of the machine. If the machine breaks down and replacement parts are installed, recalibration may be necessary.

(*b*)   Although the timer dial for most automatic toasters is calibrated by the manufacturer (e.g., light-medium-dark), the amount of heat produced by the heating element may vary over a wide range. In addition, the efficiency of the heating element normally deteriorates with age. Hence the amount of time required for "good toast" must be estimated by the user, and this setting usually must be periodically readjusted. At first, the toast is usually too light or too dark. After several successively different estimates, the required toasting time for a desired quality of toast is obtained.

(*c*)   In general, a voltmeter is calibrated by comparing it with a known-voltage standard source, and appropriately marking the reading scale at specified intervals.

**1.7.** Identify the control action in the systems of Problems 1.1, 1.2, and 1.4.

For the mirror system of Problem 1.1 the control action is equal to the input, that is, the angle of inclination of the mirror $\theta$. For the generator of Problem 1.2 the control action is equal to the input, the rotational speed or angular momentum of the prime mover shaft. The control action of the human reaching system of Problem 1.4 is equal to the distance between hand and object position.

**1.8.** Which of the control systems in Problems 1.1, 1.2, and 1.4 are open-loop? Closed-loop?

Since the control action is equal to the input for the systems of Problems 1.1 and 1.2, no feedback exists and the systems are open-loop. The human reaching system of Problem 1.4 is closed-loop because the control action is dependent upon the output, hand position.

**1.9.** Identify the control action in Examples 1.1 through 1.5.

The control action for the electric switch of Example 1.1 is equal to the input, the on or off command. The control action for the heating system of Example 1.2 is equal to the difference between the reference and actual room temperatures. For the finger pointing system of Example 1.3, the control action is equal to the difference between the actual and pointed direction of the object. The perspiration system of Example 1.4 has its control action equal to the difference between the "normal" and actual skin surface temperature. The difference between the direction of the road and the heading of the automobile is the control action for the human driver and automobile system of Example 1.5.

**1.10.** Which of the control systems in Examples 1.1 through 1.5 are open-loop? Closed-loop?

The electric switch of Example 1.1 is open-loop because the control action is equal to the input, and therefore independent of the output. For the remaining Examples 1.2 through 1.5 the control action is clearly a function of the output. Hence they are closed-loop systems.

## FEEDBACK

**1.11.** Consider the voltage divider network of Fig. 1-4. The output is $v_2$ and the input is $v_1$.

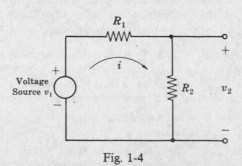

Fig. 1-4

(*a*) Write an equation for $v_2$ as a function of $v_1$, $R_1$, and $R_2$. That is, write an equation for $v_2$ which yields an open-loop system.

(*b*) Write an equation for $v_2$ in closed-loop form, that is, $v_2$ as a function of $v_1$, $v_2$, $R_1$, and $R_2$.

This problem illustrates how a passive network can be characterized as either an open-loop or a closed-loop system.

(*a*) From Ohm's law and Kirchhoff's voltage and current laws we have

$$v_2 = R_2 i \qquad i = \frac{v_1}{R_1 + R_2}$$

Therefore

$$v_2 = \left(\frac{R_2}{R_1 + R_2}\right) v_1 = f(v_1, R_1, R_2)$$

(*b*) Writing the current $i$ in a slightly different form, we have $i = (v_1 - v_2)/R_1$. Hence

$$v_2 = R_2 \left(\frac{v_1 - v_2}{R_1}\right) = \left(\frac{R_2}{R_1}\right) v_1 - \left(\frac{R_2}{R_1}\right) v_2 = f(v_1, v_2, R_1, R_2)$$

**1.12.** Explain how the classical economic concept known as the Law of Supply and Demand can be interpreted as a feedback control system. Choose the market price (selling price) of a particular item as the output of the system, and assume the objective of the system is to maintain price stability.

The Law can be stated in the following manner. The market *demand* for the item decreases as its price increases. The market *supply* usually increases as its price increases. The Law of Supply and Demand says that a stable market price is achieved if and only if the supply is equal to the demand.

The manner in which the price is regulated by the supply and the demand can be described with feedback control concepts. Let us choose the following four basic elements for our system: the Supplier, the Demander, the Pricer, and the Market where the item is bought and sold. (In reality, these elements generally represent very complicated processes.)

The input to our idealized economic system is *price stability* the "desired" output. A more convenient way to describe this input is *zero price fluctuation*. The output is the actual market price.

The system operates as follows: The Pricer receives a command (zero) for price stability. It estimates a price for the Market transaction with the help of information from its memory or records of past transactions. This price causes the Supplier to produce or supply a certain number of items, and the Demander to demand a number of items. The difference between the supply and the demand is the control action for this system. If the control action is nonzero, that is, if the supply is not equal to the demand, the Pricer initiates a change in the market price in a direction which makes the supply eventually equal to the demand. Hence both the Supplier and the Demander may be considered the feedback, since they determine the control action.

## MISCELLANEOUS PROBLEMS

**1.13.** (*a*) Explain the operation of ordinary traffic signals which control automobile traffic at roadway intersections. (*b*) Why are they open-loop control systems? (*c*) How can traffic be controlled more efficiently? (*d*) Why is the system of (*c*) closed-loop?

(*a*) Traffic lights control the flow of traffic by successively confronting the traffic in a particular direction (e.g., north-south) with a red (stop) and then a green (go) light. When one direction has the green signal, the cross traffic in the other direction (east-west) has the red. Most traffic signal red and green light intervals are predetermined by a calibrated timing mechanism.

(*b*) Control systems operated by preset timing mechanisms are open-loop. The control action is equal to the input, the red and green intervals.

(*c*) Besides preventing collisions, it is a function of traffic signals to generally control the *volume* of traffic. For the open-loop system described above, the volume of traffic does not influence the preset red and green timing intervals. In order to make traffic flow more smoothly, the green light timing interval must be made longer than the red in the direction containing the greater traffic volume. Often a traffic officer performs this task.

The ideal system would automatically measure the volume of traffic in all directions, using appropriate sensing devices, compare them, and use the difference to control the red and green time intervals, an ideal task for a computer.

(*d*) The system of (*c*) is closed-loop because the control action (the difference between the volume of traffic in each direction) is a function of the output (actual traffic volume flowing past the intersection in each direction).

**1.14.** (*a*) Describe, in a simplified way, the components and variables of the biological control system involved in walking in a prescribed direction. (*b*) Why is walking a closed-loop operation? (*c*) Under what conditions would the human walking apparatus become an open-loop system? A sampled-data system? Assume the person has normal vision.

(*a*) The major components involved in walking are the brain, eyes, and legs and feet. The input may be chosen as the desired walk direction, and the output the actual walk direction. The control action is determined by the eyes, which detect the difference between the input and output and send this information to the brain. The brain commands the legs and feet to walk in the prescribed direction.

(*b*) Walking is a closed-loop operation because the control action is a function of the output.

(c)  If the eyes are closed, the feedback loop is broken and the system becomes open-loop. If the eyes are opened and closed periodically, the system becomes a sampled-data one, and walking is usually more accurately controlled than with the eyes always closed.

**1.15.**  Devise a control system to fill a container with water after it is emptied through a stopcock at the bottom. The system must automatically shut off the water when the container is filled.

The simplified schematic diagram (Fig. 1-5) illustrates the principle of the ordinary toilet tank filling system.

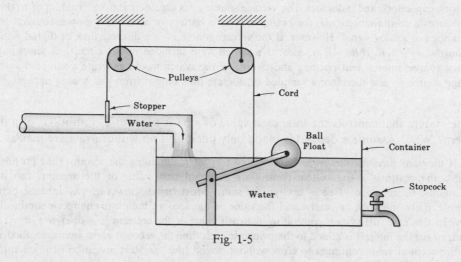

Fig. 1-5

The ball floats on the water. As the ball gets closer to the top of the container, the stopper decreases the flow of water. When the container becomes full, the stopper shuts off the flow of water.

**1.16.**  Devise a simple control system which automatically turns on a room lamp at dusk, and turns it off in daylight.

A simple system that accomplishes this task is shown in Fig. 1-6.

At dusk, the photocell, which functions as a light-sensitive switch, closes the lamp circuit, thereby lighting the room. The lamp stays lighted until daylight, at which time the photocell detects the bright outdoor light and opens the lamp circuit.

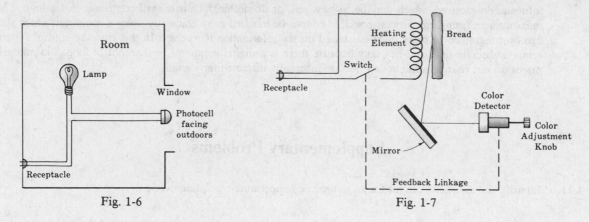

Fig. 1-6                                          Fig. 1-7

**1.17.**  Devise a closed-loop automatic toaster.

Assume each heating element supplies the same amount of heat to both sides of the bread, and toast quality can be determined by its color. A simplified schematic diagram of one possible way to apply the feedback principle to a toaster is shown in Fig. 1-7. Only one side of the toaster is illustrated.

The toaster is initially calibrated for a desired toast quality by means of the color adjustment knob. This setting never needs readjustment unless the toast quality criterion changes. When the switch is closed, the bread is toasted until the color detector "sees" the desired color. Then the switch is automatically opened by means of the feedback linkage, which may be electrical or mechanical.

**1.18.** Is the voltage divider network in Problem 1.11 an analog or digital device? Also, are the input and output analog or digital signals?

It is clearly an analog device, as are all electrical networks consisting only of passive elements such as resistors, capacitors, and inductors. The voltage source $v_1$ is considered an external input to this network. If it produces a continuous signal, for example, from a battery or alternating power source, the output is a continuous or analog signal. However, if the voltage source $v_1$ is a discrete-time or digital signal, then so is the output $v_2 = v_1 R_2 / (R_1 + R_2)$. Also, if a switch were included in the circuit, in series with an analog voltage source, intermittent opening and closing of the switch would generate a sampled waveform of the voltage source $v_1$ and therefore a sampled or discrete-time output from this analog network.

**1.19.** Is the system that controls the total cash value of a bank account a continuous or a discrete-time system? Why? Assume a deposit is made only once, and no withdrawals are made.

If the bank pays no interest and extracts no fees for maintaining the account (like putting your money "under the mattress"), the system controlling the total cash value of the account can be considered continuous, because the value is always the same. Most banks, however, pay interest periodically, for example, daily, monthly, or yearly, and the value of the account therefore changes periodically, *at discrete times*. In this case, the system controlling the cash value of the account is a *discrete system*. Assuming no withdrawals, the interest is added to the principle each time the account earns interest, called *compounding*, and the account value continues to grow without bound (the "greatest invention of mankind," a comment attributed to Einstein).

**1.20.** What *type* of control system, open-loop or closed-loop, continuous or discrete, is used by an ordinary stock market investor, whose objective is to profit from his or her investment.

Stock market investors typically follow the progress of their stocks, for example, their prices, periodically. They might check the bid and ask prices daily, with their broker or the daily newspaper, or more or less often, depending upon individual circumstances. In any case, they periodically *sample* the pricing signals and therefore the system is sampled-data, or discrete-time. However, stock prices normally rise and fall between sampling times and therefore the system operates open-loop during these periods. The feedback loop is closed only when the investor makes his or her periodic observations and acts upon the information received, which may be to buy, sell, or do nothing. Thus overall control is closed-loop. The measurement (sampling) process could, of course, be handled more efficiently using a computer, which also can be programed to make decisions based on the information it receives. In this case the control system remains discrete-time, but not only because there is a digital computer in the control loop. Bid and ask prices do not change continuously but are inherently discrete-time signals.

# Supplementary Problems

**1.21.** Identify the input and output for an automatic temperature-regulating oven.

**1.22.** Identify the input and output for an automatic refrigerator.

**1.23.** Identify an input and an output for an electric automatic coffeemaker. Is this system open-loop or closed-loop?

**1.24.** Devise a control system to automatically raise and lower a lift-bridge to permit ships to pass. No continuous human operator is permissible. The system must function entirely automatically.

**1.25.** Explain the operation and identify the pertinent quantities and components of an automatic, radar-controlled antiaircraft gun. Assume that no operator is required except to initially put the system into an operational mode.

**1.26.** How can the electrical network of Fig. 1-8 be given a *feedback* control system interpretation? Is this system analog or digital?

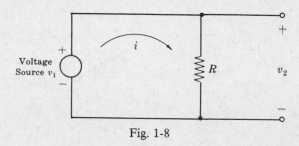

Fig. 1-8

**1.27.** Devise a control system for positioning the rudder of a ship from a control room located far from the rudder. The objective of the control system is to steer the ship in a desired heading.

**1.28.** What inputs in addition to the command for a desired heading would you expect to find acting on the system of Problem 1.27?

**1.29.** Can the application of "laissez faire capitalism" to an economic system be interpreted as a feedback control system? Why? How about "socialism" in its purest form? Why?

**1.30.** Does the operation of a stock exchange, for example, buying and selling equities, fit the model of the Law of Supply and Demand described in Problem 1.12? How?

**1.31.** Does a purely socialistic economic system fit the model of the Law of Supply and Demand described in Problem 1.12? Why (or why not)?

**1.32.** Which control systems in Problems 1.1 through 1.4 and 1.12 through 1.17 are digital or sampled-data and which are continuous or analog? Define the continuous signals and the discrete signals in each system.

**1.33.** Explain why economic control systems based on data obtained from typical accounting procedures are sampled-data control systems? Are they open-loop or closed-loop?

**1.34.** Is a rotating antenna radar system, which normally receives range and directional data once each revolution, an analog or a digital system?

**1.35.** What type of control system is involved in the treatment of a patient by a doctor, based on data obtained from laboratory analysis of a sample of the patient's blood?

# Answers to Some Supplementary Problems

**1.21.**    The input is the reference temperature. The output is the actual oven temperature.

**1.22.**    The input is the reference temperature. The output is the actual refrigerator temperature.

**1.23.**    One possible input for the automatic electric coffeemaker is the amount of coffee used. In addition, most coffeemakers have a dial which can be set for weak, medium, or strong coffee. This setting usually regulates a timing mechanism. The brewing time is therefore another possible input. The output of any coffeemaker can be chosen as coffee strength. The coffeemakers described above are open-loop.

# Chapter 2

# Control Systems Terminology

## 2.1 BLOCK DIAGRAMS: FUNDAMENTALS

A **block diagram** is a shorthand, pictorial representation of the cause-and-effect relationship between the input and output of a physical system. It provides a convenient and useful method for characterizing the functional relationships among the various components of a control system. System *components* are alternatively called *elements* of the system. The simplest form of the block diagram is the single *block*, with one input and one output, as shown in Fig. 2-1.

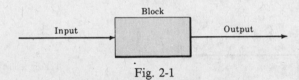

Fig. 2-1

The interior of the rectangle representing the block usually contains a description of or the name of the element, or the symbol for the mathematical operation to be performed on the input to yield the output. The *arrows* represent the direction of information or signal flow.

**EXAMPLE 2.1**

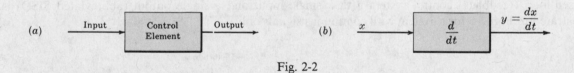

Fig. 2-2

The operations of addition and subtraction have a special representation. The block becomes a small circle, called a **summing point**, with the appropriate plus or minus sign associated with the arrows entering the circle. The output is the algebraic sum of the inputs. Any number of inputs may enter a summing point.

**EXAMPLE 2.2**

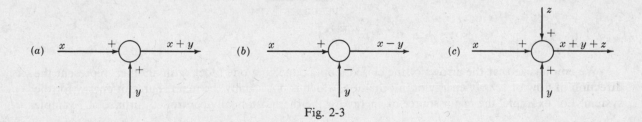

Fig. 2-3

15

Some authors put a cross in the circle: (Fig. 2-4)

Fig. 2-4

This notation is avoided here because it is sometimes confused with the multiplication operation.
In order to have the same signal or variable be an input to more than one block or summing point,
a **takeoff point** is used. This permits the signal to proceed unaltered along several different paths to
several destinations.

**EXAMPLE 2.3**

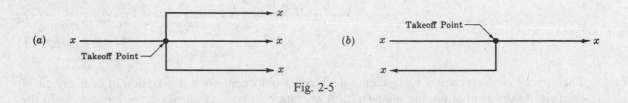

Fig. 2-5

## 2.2   BLOCK DIAGRAMS OF CONTINUOUS (ANALOG) FEEDBACK CONTROL SYSTEMS

The blocks representing the various components of a control system are connected in a fashion
which characterizes their functional relationships within the system. The basic configuration of a simple
closed-loop (feedback) control system with a single input and a single output (abbreviated SISO) is
illustrated in Fig. 2-6 for a system with continuous signals only.

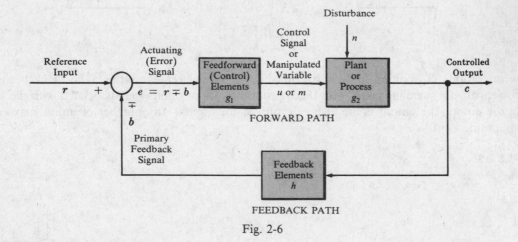

Fig. 2-6

We emphasize that the arrows of the closed loop, connecting one block with another, represent the
direction of flow of *control* energy or information, which is not usually the main source of energy for the
system. For example, the major source of energy for the thermostatically controlled furnace of Example

1.2 is often chemical, from burning fuel oil, coal, or gas. But this energy source would not appear in the closed control loop of the system.

## 2.3  TERMINOLOGY OF THE CLOSED-LOOP BLOCK DIAGRAM

It is important that the terms used in the closed-loop block diagram be clearly understood.

Lowercase letters are used to represent the input and output variables of each element as well as the symbols for the blocks $g_1$, $g_2$, and $h$. These quantities represent functions of time, unless otherwise specified.

**EXAMPLE 2.4.**  $r = r(t)$

In subsequent chapters, we use capital letters to denote Laplace transformed or $z$-transformed quantities, as functions of the complex variable $s$, or $z$, respectively, or Fourier transformed quantities (frequency functions), as functions of the pure imaginary variable $j\omega$. Functions of $s$ or $z$ are often abbreviated to the capital letter appearing alone. Frequency functions are never abbreviated.

**EXAMPLE 2.5.**  $R(s)$ may be abbreviated as $R$, or $F(z)$ as $F$. $R(j\omega)$ is never abbreviated.

The letters $r$, $c$, $e$, etc., were chosen to preserve the generic nature of the block diagram. This convention is now classical.

*Definition 2.1:*  The **plant** (or **process**, or **controlled system**) $g_2$ is the system, subsystem, process, or object controlled by the feedback control system.

*Definition 2.2:*  The **controlled output** $c$ is the output variable of the plant, under the control of the feedback control system.

*Definition 2.3:*  The **forward path** is the transmission path from the summing point to the controlled output $c$.

*Definition 2.4:*  The **feedforward (control) elements** $g_1$ are the components of the forward path that generate the control signal $u$ or $m$ applied to the plant. *Note:* Feedforward elements typically include controller(s), compensator(s) (or equalization elements), and/or amplifiers.

*Definition 2.5:*  The **control signal** $u$ (or **manipulated variable** $m$) is the output signal of the feedforward elements $g_1$ applied as input to the plant $g_2$.

*Definition 2.6:*  The **feedback path** is the transmission path from the controlled output $c$ back to the summing point.

*Definition 2.7:*  The **feedback elements** $h$ establish the functional relationship between the controlled output $c$ and the primary feedback signal $b$. *Note:* Feedback elements typically include sensors of the controlled output $c$, compensators, and/or controller elements.

*Definition 2.8:*  The **reference input** $r$ is an external signal applied to the feedback control system, usually at the first summing point, in order to command a specified action of the plant. It usually represents ideal (or desired) plant output behavior.

*Definition 2.9*:     The **primary feedback signal** $b$ is a function of the controlled output $c$, algebraically summed with the reference input $r$ to obtain the actuating (error) signal $e$, that is, $r \pm b = e$. *Note:* An *open-loop* system has no primary feedback signal.

*Definition 2.10*:    The **actuating** (or **error**) **signal** is the reference input signal $r$ plus or minus the primary feedback signal $b$. The *control action* is generated by the actuating (error) signal in a feedback control system (see Definitions 1.5 and 1.6). *Note:* In an *open-loop* system, which has no feedback, the actuating signal is equal to $r$.

*Definition 2.11*:    **Negative feedback** means the summing point is a subtractor, that is, $e = r - b$. **Positive feedback** means the summing point is an adder, that is, $e = r + b$.

## 2.4  BLOCK DIAGRAMS OF DISCRETE-TIME (SAMPLED-DATA, DIGITAL) COMPONENTS, CONTROL SYSTEMS, AND COMPUTER-CONTROLLED SYSTEMS

A *discrete-time (sampled-data or digital) control system* was defined in Definition 1.11 as one having discrete-time signals or components at one or more points in the system. We introduce several common discrete-time system components first, and then illustrate some of the ways they are interconnected in digital control systems. We remind the reader here that *discrete-time* is often abbreviated as *discrete* in this book, and *continuous-time* as *continuous*, wherever the meaning is unambiguous.

**EXAMPLE 2.6.**   A digital computer or microprocessor is a discrete-time (discrete or digital) device, a common component in digital control systems. The internal and external signals of a digital computer are typically discrete-time or digitally coded.

**EXAMPLE 2.7.**   A discrete system component (or components) with discrete-time input $u(t_k)$ and discrete-time output $y(t_k)$ signals, where $t_k$ are discrete instants of time, $k = 1, 2, \ldots$, etc., may be represented by a block diagram, as shown in Fig. 2-7.

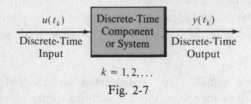

$$k = 1, 2, \ldots$$

Fig. 2-7

Many digital control systems contain both continuous and discrete components. One or more devices known as *samplers*, and others known as *holds*, are usually included in such systems.

*Definition 2.12*:    A **sampler** is a device that converts a continuous-time signal, say $u(t)$, into a discrete-time signal, denoted $u^*(t)$, consisting of a sequence of values of the signal at the instants $t_1, t_2, \ldots$, that is, $u(t_1), u(t_2), \ldots$, etc.

Ideal samplers are usually represented schematically by a switch, as shown in Fig. 2-8, where the switch is normally open except at the instants $t_1, t_2$, etc., when it is closed for an instant. The switch also may be represented as enclosed in a block, as shown in Fig. 2-9.

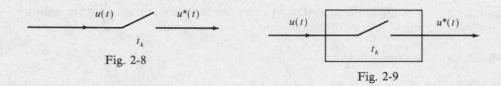

Fig. 2-8

Fig. 2-9

**EXAMPLE 2.8.**  The input signal of an ideal sampler and a few samples of the output signal are illustrated in Fig. 2-10. This type of signal is often called a *sampled-data signal*.

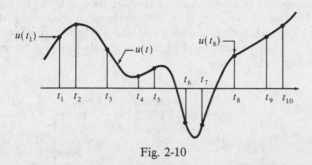

Fig. 2-10

Discrete-data signals $u(t_k)$ are often written more simply with the index $k$ as the only argument, that is, $u(k)$, and the sequence $u(t_1), u(t_2), \ldots$, etc., becomes $u(1), u(2), \ldots$, etc. This notation is introduced in Chapter 3. Although sampling rates are in general nonuniform, as in Example 2.8, uniform sampling is the rule in this book, that is, $t_{k+1} - t_k \equiv T$ for all $k$.

*Definition 2.13*:    A **hold**, or **data hold**, device is one that converts the discrete-time output of a sampler into a particular kind of continuous-time or analog signal.

**EXAMPLE 2.9.**  A **zero-order hold** (or **simple hold**) is one that maintains (i.e., holds) the value of $u(t_k)$ constant until the next sampling time $t_{k+1}$, as shown in Fig. 2-11. Note that the output $y_{H0}(t)$ of the zero-order hold is continuous, except at the sampling times. This type of signal is called a **piecewise-continuous** signal.

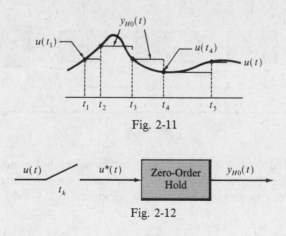

Fig. 2-11

Fig. 2-12

*Definition 2.14*:    An **analog-to-digital (A/D) converter** is a device that converts an analog or continuous signal into a discrete or digital signal.

***Definition 2.15*:**  A **digital-to-analog (D/A) converter is** a device that converts a discrete or digital signal into a continuous-time or analog signal.

**EXAMPLE 2.10.**  The sampler in Example 2.8 (Figs. 2-9 and 2-10) is an A/D converter.

**EXAMPLE 2.11.**  The zero-order hold in Example 2.9 (Figs. 2-11 and 2-12) is a D/A converter.

Samplers and zero-order holds are commonly used A/D and D/A converters, but they are not the only types available. Some D/A converters, in particular, are more complex.

**EXAMPLE 2.12.**  Digital computers or microprocessors are often used to control continuous plants or processes. A/D and D/A converters are typically required in such applications, to convert signals from the plant to digital signals, and to convert the digital signal from the computer into a control signal for the analog plant. The joint operation of these elements is usually synchronized by a clock and the resulting controller is sometimes called a *digital filter*, as illustrated in Fig. 2-13.

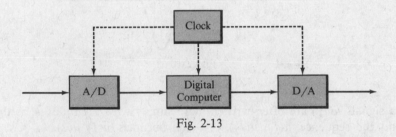

Fig. 2-13

***Definition 2.16*:**  A **computer-controlled system** includes a computer as the primary control element.

The most common computer-controlled systems have digital computers controlling analog or continuous processes. In this case, A/D and D/A converters are needed, as illustrated in Fig. 2-14.

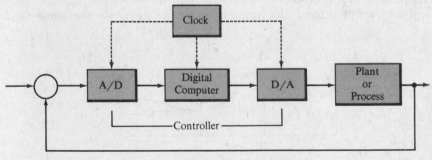

Fig. 2-14

The clock may be omitted from the diagram, as it synchronizes but is not an explicit part of signal flow in the control loop. Also, the summing junction and reference input are sometimes omitted from the diagram, because they may be implemented in the computer.

## 2.5  SUPPLEMENTARY TERMINOLOGY

Several other terms require definition and illustration at this time. Others are presented in subsequent chapters, as needed.

*Definition 2.17:*    A **transducer** is a device that converts one energy form into another.

For example, one of the most common transducers in control systems applications is the *potentiometer*, which converts mechanical position into an electrical voltage (Fig. 2-15).

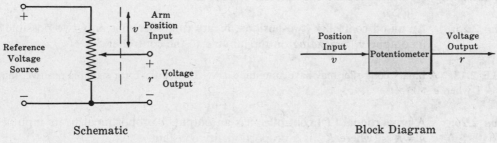

Schematic                                Block Diagram

Fig. 2-15

*Definition 2.18:*    The **command** $v$ is an input signal, usually equal to the reference input $r$. But when the energy form of the command $v$ is not the same as that of the primary feedback $b$, a transducer is required between the command $v$ and the reference input $r$ as shown in Fig. 2-16($a$).

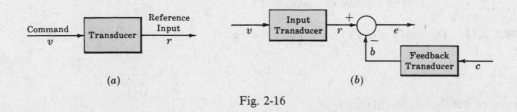

(a)                                    (b)

Fig. 2-16

*Definition 2.19:*    When the feedback element consists of a transducer, and a transducer is required at the input, that part of the control system illustrated in Fig. 2-16($b$) is called the **error detector**.

*Definition 2.20:*    A **stimulus**, or **test input**, is any externally (exogenously) introduced input signal affecting the controlled output $c$. *Note:* The reference input $r$ is an example of a stimulus, but it is not the only kind of stimulus.

*Definition 2.21:*    A **disturbance** $n$ (or **noise input**) is an undesired stimulus or input signal affecting the value of the controlled output $c$. It may enter the plant with $u$ or $m$, as shown in the block diagram of Fig. 2-6, or at the first summing point, or via another intermediate point.

*Definition 2.22:*    The **time response** of a system, subsystem, or element is the output as a function of time, usually following application of a prescribed input under specified operating conditions.

*Definition 2.23:*    A **multivariable system** is one with more than one input (**multiinput, MI-**), more than one output (**multioutput, -MO**), or both (**multiinput-multioutput, MIMO**).

*Definition 2.24*:      The term **controller** in a feedback control system is often associated with the elements of the forward path, between the actuating (error) signal $e$ and the control variable $u$. But it also sometimes includes the summing point, the feedback elements, or both, and some authors use the term controller and compensator synonymously. The context should eliminate ambiguity.

The following five definitions are examples of **control laws**, or **control algorithms**.

*Definition 2.25*:      An **on-off controller (two-position, binary controller)** has only two possible values at its output $u$, depending on the input $e$ to the controller.

**EXAMPLE 2.13.**   A binary controller may have an output $u = +1$ when the error signal is positive, that is, $e > 0$, and $u = -1$ when $e \leq 0$.

*Definition 2.26*:      A **proportional ($P$) controller** has an output $u$ proportional to its input $e$, that is, $u = K_P e$, where $K_P$ is a proportionality constant.

*Definition 2.27*:      A **derivative ($D$) controller** has an output proportional to the *derivative* of its input $e$, that is, $u = K_D \, de/dt$, where $K_D$ is a proportionality constant.

*Definition 2.28*:      An **integral ($I$) controller has an output $u$ proportional to the integral** of its input $e$, that is, $u = K_I \int e(t) \, dt$, where $K_I$ is a proportionality constant.

*Definition 2.29*:      **PD, PI, DI, and PID controllers** are combinations of proportional ($P$), derivative ($D$), and integral ($I$) controllers.

**EXAMPLE 2.14.**   The output $u$ of a PD controller has the form:

$$u_{\mathrm{PD}} = K_P e + K_D \frac{de}{dt}$$

The output of a PID controller has the form:

$$u_{\mathrm{PID}} = K_P e + K_D \frac{de}{dt} + K_I \int e(t) \, dt$$

## 2.6  SERVOMECHANISMS

The specialized feedback control system called a *servomechanism* deserves special attention, due to its prevalence in industrial applications and control systems literature.

*Definition 2.30*:      A **servomechanism** is a power-amplifying feedback control system in which the controlled variable $c$ is mechanical position, or a time derivative of position such as velocity or acceleration.

**EXAMPLE 2.15.**   An *automobile power-steering apparatus* is a servomechanism. The command input is the angular position of the steering wheel. A small rotational torque applied to the steering wheel is amplified hydraulically, resulting in a force adequate to modify the output, the angular position of the front wheels. The block diagram of such a system may be represented by Fig. 2-17. Negative feedback is necessary in order to return the control valve to the neutral position, reducing the torque from the hydraulic amplifier to zero when the desired wheel position has been achieved.

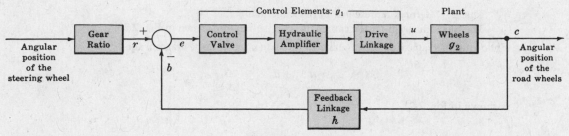

Fig. 2-17

## 2.7 REGULATORS

*Definition 2.31*:    A **regulator** or **regulating system** is a feedback control system in which the reference input or command is constant for long periods of time, often for the entire time interval during which the system is operational. Such an input is often called a **setpoint**.

A regulator differs from a servomechanism in that the primary function of a regulator is usually to maintain a constant controlled output, while that of a servomechanism is most often to cause the output of the system to follow a varying input.

# Solved Problems

## BLOCK DIAGRAMS

**2.1.**    Consider the following equations in which $x_1, x_2, \ldots, x_n$ are variables, and $a_1, a_2, \ldots, a_n$ are general coefficients or mathematical operators:

$$(a) \quad x_3 = a_1 x_1 + a_2 x_2 - 5$$
$$(b) \quad x_n = a_1 x_1 + a_2 x_2 + \cdots + a_{n-1} x_{n-1}$$

Draw a block diagram for each equation, identifying all blocks, inputs, and outputs.

($a$)    In the form the equation is written, $x_3$ is the output., The terms on the right-hand side of the equation are combined at a summing point, as shown in Fig. 2-18.

The $a_1 x_1$ term is represented by a single block, with $x_1$ as its input and $a_1 x_1$ as its output. Therefore the coefficient $a_1$ is put inside the block, as shown in Fig. 2-19. $a_1$ may represent any mathematical operation. For example, if $a_1$ were a constant, the block operation would be "multiply the input $x_1$ by the constant $a_1$." It is usually clear from the description or context of a problem what is meant by the symbol, operator, or description inside the block.

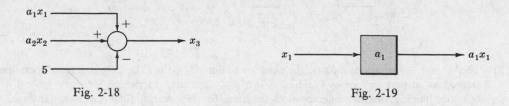

Fig. 2-18                                                    Fig. 2-19

The $a_2x_2$ term is represented in the same manner.

The block diagram for the entire equation is therefore shown in Fig. 2.20.

(*b*) Following the same line of reasoning as in part (*a*), the block diagram for

$$x_n = a_1x_1 + a_2x_2 + \cdots + a_{n-1}x_{n-1}$$

is shown in Fig. 2-21.

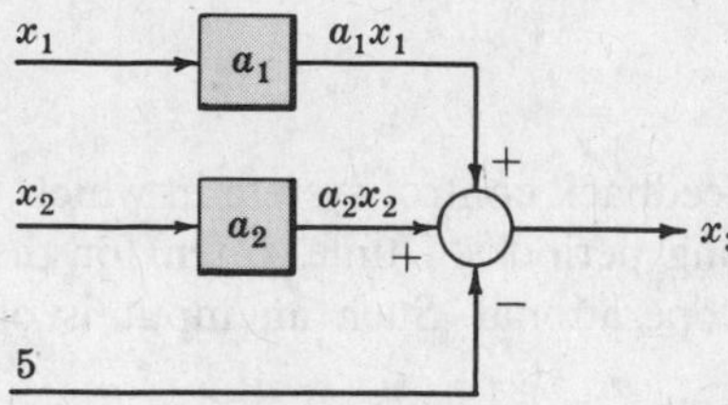


Fig. 2-20

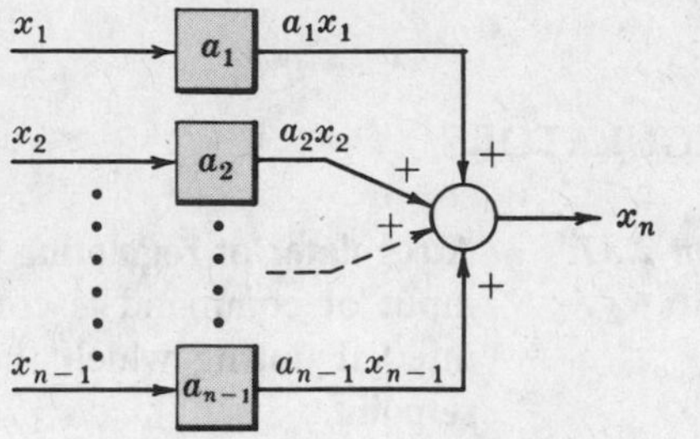


Fig. 2-21

**2.2.** Draw block diagrams for each of the following equations:

Mathcad

$$(a)\quad x_2 = a_1\left(\frac{dx_1}{dt}\right) \qquad (b)\quad x_3 = \frac{d^2x_2}{dt^2} + \frac{dx_1}{dt} - x_1 \qquad (c)\quad x_4 = \int x_3\,dt$$

(*a*) Two operations are specified by this equation, $a_1$ and differentiation $d/dt$. Therefore the block diagram contains two blocks, as shown in Fig. 2-22. Note the order of the blocks.

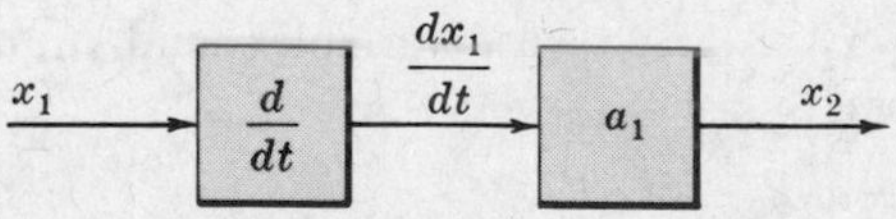


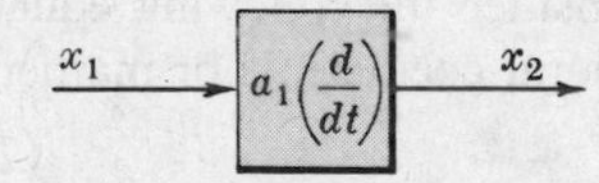


Fig. 2-23

Now, if $a_1$ were a constant, the $a_1$ block could be combined with the $d/dt$ block, as shown in Fig. 2-23, since no confusion about the order of the blocks would result. But, if $a_1$ were an unknown operator, the reversal of blocks $d/dt$ and $a_1$ would not necessarily result in an output equal to $x_2$, as shown in Fig. 2-24.

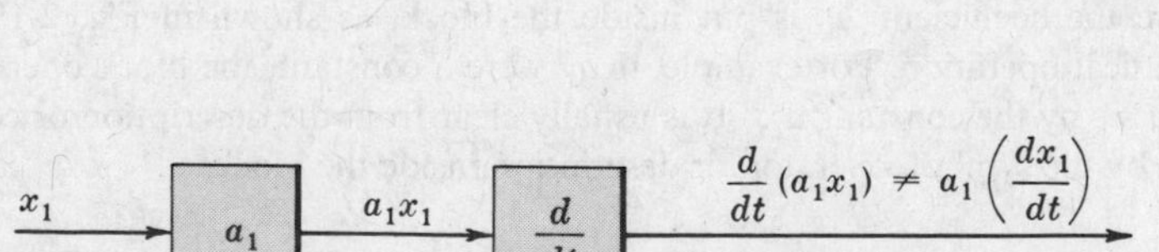


Fig. 2-24

(*b*) The + and − operations indicate the need for a summing point. The differentiation operation can be treated as in part (*a*), or by combining two first derivative operations into one second derivative operator block, giving two different block diagrams for the equation for $x_3$, as shown in Fig. 2-25.

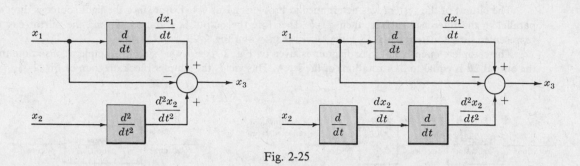

Fig. 2-25

(*c*)   The integration operation can be represented in block diagram form as Fig. 2-26.

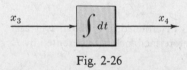

Fig. 2-26

**2.3.** Draw a block diagram for the pivoted, adjustable mirror mechanism of Section 1.1 with the output identified as in Problem 1.1. Assume that each 360° rotation of the screw raises or lowers the mirror $k$ degrees. Identify all the signals and components of the control system in the diagram.

The schematic diagram of the system is repeated in Fig. 2-27 for convenience.

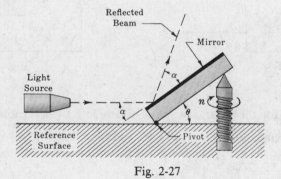

Fig. 2-27

Whereas the input was defined as $\theta$ in Problem 1.1, the specifications for this problem imply an input equal to the number of rotations of the screw. Let $n$ be the number of rotations of the screw such that $n = 0$ when $\theta = 0°$. Therefore $n$ and $\theta$ can be related by a block described by the constant $k$, since $\theta = kn$, as shown in Fig. 2-28.

Fig. 2-28                              Fig. 2-29

The output of the system was determined in Problem 1.1 as $\theta + \alpha$. But since the light source is directed parallel to the reference surface, then $\alpha = \theta$. Therefore the output is equal to $2\theta$, and the mirror can be represented by a constant equal to 2 in a block, as shown in Fig. 2-29.

The complete open-loop block diagram is given by Fig. 2-30. For this simple example we also note that the output $2\theta$ is equal to $2kn$ rotations of the screw. This yields the simpler block diagram of Fig. 2-31.

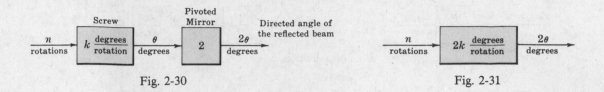

Fig. 2-30                                              Fig. 2-31

**2.4.** Draw an open-loop and a closed-loop block diagram for the voltage divider network of Problem 1.11.

The open-loop equation was determined in Problem 1.11 as $v_2 = (R_2/(R_1 + R_2))v_1$, where $v_1$ is the input and $v_2$ is the output. Therefore the block is represented by $R_2/(R_1 + R_2)$ (Fig. 2-32), and clearly the operation is multiplication.

The closed-loop equation is

$$v_2 = \left(\frac{R_2}{R_1}\right)v_1 - \left(\frac{R_2}{R_1}\right)v_2 = \left(\frac{R_2}{R_1}\right)(v_1 - v_2)$$

The actuating signal is $v_1 - v_2$. The closed-loop negative feedback block diagram is easily constructed with the only block represented by $R_2/R_1$, as shown in Fig. 2-33.

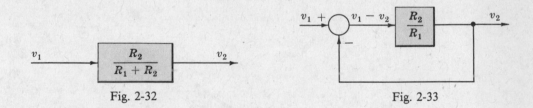

Fig. 2-32                                              Fig. 2-33

**2.5.** Draw a block diagram for the electric switch of Example 1.1 (see Problems 1.9 and 1.10).

Both the input and output are binary (two-state) variables. The switch is represented by a block, and the electrical power source the switch controls is not part of the control system. One possible open-loop block diagram is given by Fig. 2-34.

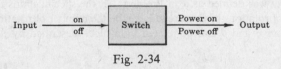

Fig. 2-34

For example, suppose the power source is an electrical current source. Then the block diagram for the switch might take the form of Fig. 2-35, where (again) the current source is not part of the control system, the input to the switch block is shown as a mechanical linkage to a simple "knife" switch, and the output is a nonzero current only when the switch is closed (on). Otherwise it is zero (off).

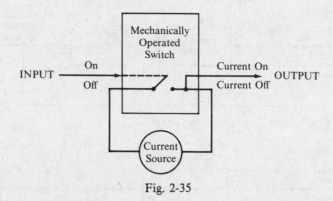

Fig. 2-35

**2.6.** Draw simple block diagrams for the control systems in Examples 1.2 through 1.5.

From Problem 1.10 we note that these systems are closed-loop, and from Problem 1.9 the actuating signal (control action) for the system in each example is equal to the input minus the output. Therefore negative feedback exists in each system.

For the thermostatically controlled furnace of Example 1.2, the thermostat can be chosen as the summing point, since this is the device that determines whether or not the furnace is turned on. The enclosure environment (outside) temperature may be treated as a **noise** input acting directly on the enclosure.

The eyes may be represented by a summing point in both the human pointing system of Example 1.3 and the driver-automobile system of Example 1.5. The eyes perform the function of monitoring the input and output.

For the perspiration system of Example 1.4, the summing point is not so easily defined. For the sake of simplicity let us call it the nervous system.

The block diagrams are easily constructed as shown below from the information given above and the list of components, inputs, and outputs given in the examples.

The arrows between components in the block diagrams of the biological systems in Examples 1.3 through 1.5 represent electrical, chemical, or mechanical signals controlled by the central nervous system.

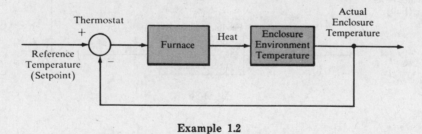

Example 1.2

Example 1.3

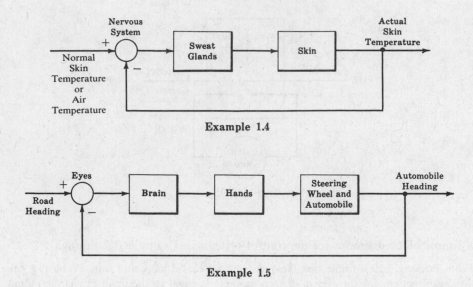

Example 1.4

Example 1.5

## BLOCK DIAGRAMS OF FEEDBACK CONTROL SYSTEMS

**2.7.** Draw a block diagram for the water-filling system described in Problem 1.15. Which component or components comprise the plant? The controller? The feedback?

The container is the plant because the water level *of the container* is being controlled (see Definition 2.1). The stopper valve may be chosen as the control element; and the ball-float, cord, and associated linkage as the feedback elements. The block diagram is given in Fig. 2-36.

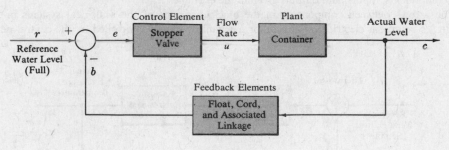

Fig. 2-36

The feedback is negative because the water flow rate to the container must decrease as the water level rises in the container.

**2.8.** Draw a simple block diagram for the feedback control system of Examples 1.7 and 1.8, the airplane with an autopilot.

The plant for this system is the airplane, including its control surfaces and navigational instruments. The controller is the autopilot mechanism, and the summing point is the comparison device. The feedback linkage may be simply represented by an arrow from the output to the summing point, as this linkage is not well defined in Example 1.8.

The autopilot provides control signals to operate the control surfaces (rudder, flaps, etc.). These signals may be denoted $u_1, u_2, \ldots$.

The simplest block diagram for this feedback system is given in Fig. 2-37.

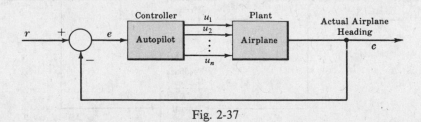

Fig. 2-37

## SERVOMECHANISMS

**2.9.** Draw a schematic and a block diagram from the following description of a *position servomechanism* whose function is to open and close a water valve.

At the input of the system there is a rotating-type potentiometer connected across a battery voltage source. Its movable (third) terminal is calibrated in terms of angular position (in radians). This output terminal is electrically connected to one terminal of a voltage amplifier called a *servoamplifier*. The servoamplifier supplies enough output power to operate an electric motor called a *servomotor*. The servomotor is mechanically linked with the water valve in a manner which permits the valve to be opened or closed by the motor.

Assume the loading effect of the valve on the motor is negligible; that is, it does not "resist" the motor. A 360° rotation of the motor shaft completely opens the valve. In addition, the movable terminal of a second potentiometer connected in parallel at its fixed terminals with the input potentiometer is mechanically connected to the motor shaft. It is electrically connected to the remaining input terminal of the servoamplifier. The potentiometer ratios are set so that they are equal when the valve is closed.

When a command is given to open the valve, the servomotor rotates in the appropriate direction. As the valve opens, the second potentiometer, called the *feedback potentiometer*, rotates in the same direction as the input potentiometer. It stops when the potentiometer ratios are again equal.

A schematic diagram (Fig. 2-38) is easily drawn from the preceding description. Mechanical connections are shown as dashed lines.

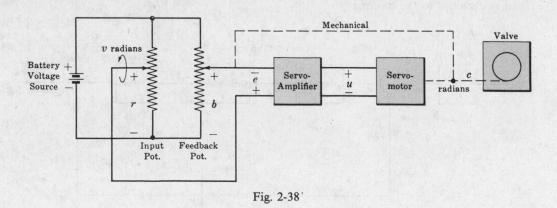

Fig. 2-38

The block diagram for this system (Fig. 2-39) is easily drawn from the schematic diagram.

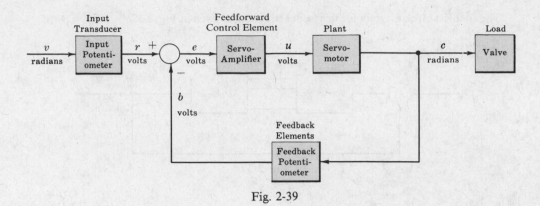

Fig. 2-39

**2.10.** Draw a block diagram for the elementary speed control system (velocity servomechanism) given in Fig. 2-40.

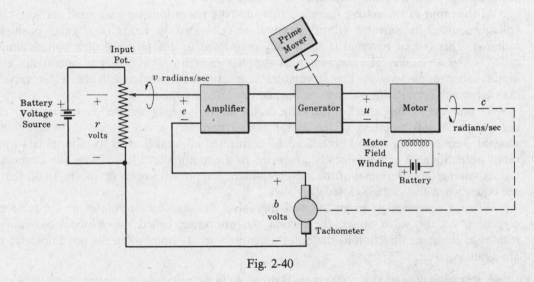

Fig. 2-40

The potentiometer is a rotating-type, calibrated in radians per seconds, and the prime-mover speed, motor field winding, and input potentiometer currents are constant functions of time. No load is attached to the motor shaft.

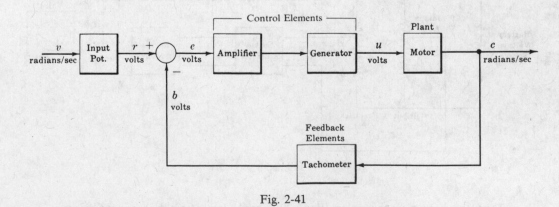

Fig. 2-41

The battery voltage sources for both the input potentiometer and motor field winding, and the prime-mover source for the generator are not part of the control loop of this servomechanism. The output of each of these sources is a constant function of time, and can be accounted for in the mathematical description of the input potentiometer, generator, and motor, respectively. Therefore the block diagram for this system is given in Fig. 2-41.

## MISCELLANEOUS PROBLEMS

**2.11.** Draw a block diagram for the photocell light switch system described in Problem 1.16. The light intensity in the room must be maintained at a level greater than or equal to a prespecified level.

One way of describing this system is with two inputs, one input chosen as minimum reference room-light intensity $r_1$, and the second as room sunlight intensity $r_2$. The output $c$ is actual room-light intensity.

The room is the plant. The manipulated variable (control signal) is the amount of light supplied to the room from both the lamp and the sun. The photocell and the lamp are the control elements because they control room-light intensity. Assume the minimum reference room-light intensity $r_1$ is equal to the intensity of room-light supplied by the lighted lamp alone. A block diagram for this system is given in Fig. 2-42.

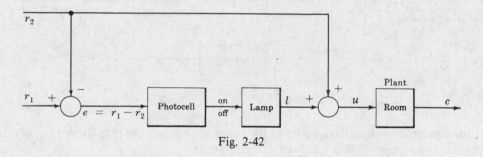

Fig. 2-42

The system is clearly open-loop. The actuating signal $e$ is independent of the output $c$, and is equal to the difference between the two inputs: $r_1 - r_2$. When $e \leq 0$, $l = 0$ (the light is off). When $e > 0$, $l = r_1$ (the light is on).

**2.12.** Draw a block diagram for the closed-loop traffic signal system described in Problem 1.13.

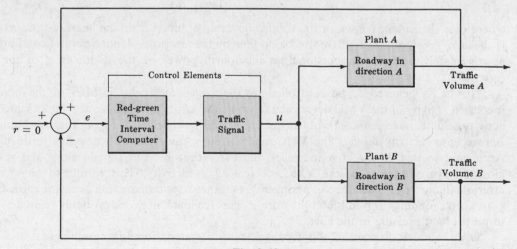

Fig. 2-43

This system has two outputs, the volume of traffic passing the intersection in one direction (the *A* direction), and the volume passing the intersection in the other direction (the *B* direction). The input is the command for equal traffic volumes in directions *A* and *B*; that is, the input is zero volume difference.

Suppose we call the mechanism for computing the appropriate red and green timing intervals the Red-Green Time Interval Computer. This device, in addition to the traffic signal, makes up the control elements. The plants are the roadway in direction *A* and the roadway in direction *B*. The block diagram of this traffic *regulator* is given in Fig. 2-43.

**2.13.** Draw a block diagram illustrating the economic Law of Supply and Demand, as described in Problem 1.12.

The block diagram is given by Fig. 2-44.

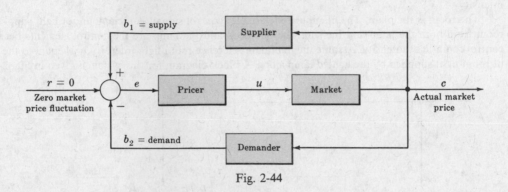

Fig. 2-44

**2.14.** The following very simplified model of the biological mechanism regulating human arterial blood pressure is an example of a feedback control system.

A well-regulated pressure must be maintained in the blood vessels (arteries, arterioles, and capillaries) supplying the tissues, so that blood flow is adequately maintained. This pressure is usually measured in the aorta (an artery) and is called the *blood pressure p*. It is not constant and normally has a range of 70–130 mm of mercury (mm Hg) in adults. Let us assume that *p* is equal to 100 mm Hg (on the average) in a normal individual.

A fundamental model of circulatory physiology is the following equation for arterial blood pressure:

$$p = Q\rho$$

where *Q* is the *cardiac output*, or the volume flow rate of blood from the heart to the aorta, and $\rho$ is the *peripheral resistance* offered to blood flow by the arterioles. Under normal conditions, $\rho$ is approximately inversely proportional to the fourth power of the diameter *d* of the vessels (arterioles).

Now *d* is believed to be controlled by the *vasomotor center* (VMC) of the brain, with increased activity of the VMC decreasing *d*, and vice versa. Although several factors affect VMC activity, the *baroreceptor cells* of the *arterial sinus* are believed to be the most important. Baroreceptor activity *inhibits* the VMC, and therefore functions in a negative feedback mode. According to this theory, if $\rho$ increases, the baroreceptors send signals along the vagus and glossopharyngeal nerves to the VMC, decreasing its activity. This results in an increase in arteriole diameter *d*, a decrease in peripheral resistance $\rho$, and (assuming constant cardiac output *Q*) a corresponding drop in blood pressure *p*. This feedback network probably regulates, at least in part, blood pressure in the aorta.

Draw a block diagram of this feedback control system, identifying all signals and components.

Let the aorta be the plant, represented by $Q$ (cardiac output); the VMC and arterioles may be chosen as the controller; the baroreceptors are the feedback elements. The input $p_0$ is the average normal (reference) blood pressure, 100 mm Hg. The output $p$ is the actual blood pressure. Since $\rho = k(1/d)^4$, where $k$ is a proportionality constant, the arterioles can be represented in the block by $k(\cdot)^4$. The block diagram is given in Fig. 2-45.

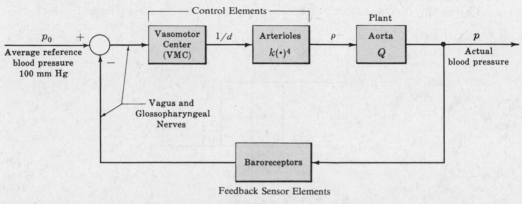

Fig. 2-45

**2.15.** The *thyroid* gland, an endocrine (internally secreting) gland located in the neck in the human, secretes *thyroxine* into the bloodstream. The bloodstream is the signal transmission system for most of the endocrine glands, just as conductive wires are the transmission system for the flow of electrical current, or pipes and tubes may be the transmission system for hydrodynamic fluid flow.

Like most human physiological processes, the production of thyroxine by the thyroid gland is automatically controlled. The amount of thyroxine in the bloodstream is regulated in part by a hormone secreted by the *anterior pituitary*, an endocrine gland suspended from the base of the brain. This "control" hormone is appropriately called *thyroid stimulating hormone* (TSH). In a simplified view of this control system, when the level of thyroxine in the circulatory system is higher than that required by the organism, TSH secretion is inhibited (reduced), causing a reduction in the activity of the thyroid. Hence less thyroxine is released by the thyroid.

Draw a block diagram of the simplified system described, identifying all components and signals.

Let the plant be the thyroid gland, with the controlled variable the level of thyroxine in the bloodstream. The pituitary gland is the controller, and the manipulated variable is the amount of TSH it secretes. The block diagram is given in Fig. 2-46.

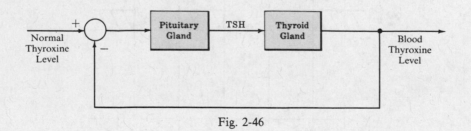

Fig. 2-46

We reemphasize that this is a very simplified view of this biological control system, as was that in the previous problem.

**2.16.** What type of **controller** is included in the more realistic thermostatically controlled heating system described in Example 1.14?

The thermostat-furnace controller has a binary output: furnace (full) on, or furnace off. Therefore it is an on-off controller. But it is not as simple as the sign-sensing binary controller of Example 2.13. The thermostat switch turns the furnace on when room temperature falls to 2° below its setpoint of 68°F (22°C), and turns it off when it rises to 2° above its setpoint.

Graphically, the characteristic curve of such a controller has the form given in Fig. 2-47.

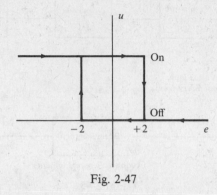

Fig. 2-47

This is called a **hysteresis** characteristic curve, because its output has a "memory"; that is, the switching points depend on whether the input $e$ is rising or falling when the controller switches states from on to off, or off to on.

**2.17.** Sketch the error, control, and controlled output signals as functions of time and discuss how the on-off controller of Problem 2.16 maintains the average room temperature specified by the setpoint (68°F) of the thermostat?

The signals $e(t)$, $u(t)$, and $c(t)$ typically have the form shown in Fig. 2-48, assuming the temperature was colder than 66°F at the start.

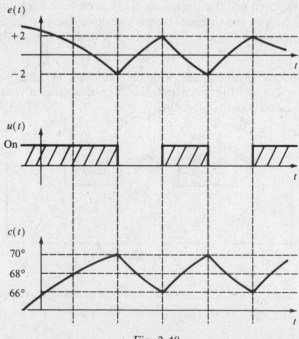

Fig. 2-48

The room temperature $c(t)$ is constantly varying. In each switching interval of the controller, it rises at an approximately constant rate, from 66° to 70°, or falls at an approximately constant rate, from 70° to 66°. The average temperature of the room is the mean value of this function $c(t)$, which is approximately 68°F.

**2.18.**   What major advantage does a computer-controlled system have over an analog system?

The controller (control law) in a computer-controlled system is typically implemented by means of software, rather than hardware. Therefore the class of control laws that can be implemented conveniently is substantially increased.

# Supplementary Problems

**2.19.**   The schematic diagram of a semiconductor voltage amplifier called an *emitter follower* is given in Fig. 2-49.

An equivalent circuit for this amplifier is shown in Fig. 2-50, where $r_p$ is the internal resistance of, and $\mu$ is a parameter of the particular semiconductor. Draw both an open-loop and a closed-loop block diagram for this circuit with an input $v_{\text{in}}$ and an output $v_{\text{out}}$.

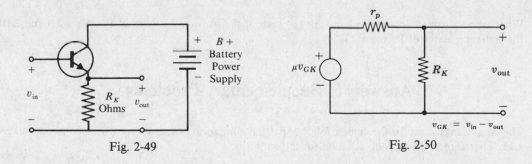

Fig. 2-49                                            Fig. 2-50

**2.20.**   Draw a block diagram for the human walking system of Problem 1.14.

**2.21.**   Draw a block diagram for the human reaching system described in Problem 1.4.

**2.22.**   Draw a block diagram for the automatic temperature-regulated oven of Problem 1.21.

**2.23.**   Draw a block diagram for the closed-loop automatic toaster of Problem 1.17.

**2.24.**   State the common dimensional units for the input and output of the following transducers: (*a*) accelerometer, (*b*) generator of electricity, (*c*) thermistor (temperature-sensitive resistor), (*d*) thermocouple.

**2.25.**   Which systems in Problems 2.1 through 2.8 and 2.11 through 2.21 are servomechanisms?

**2.26.**   The endocrine gland known as the *adrenal cortex* is located on top of each kidney (two parts). It secretes several hormones, one of which is *cortisol*. Cortisol plays an important part in regulating the metabolism of carbohydrates, proteins, and fats, particularly in times of stress. Cortisol production is controlled by adrenocorticotrophic hormone (ACTH) from the anterior pituitary gland. High blood cortisol inhibits ACTH production. Draw a block diagram of this simplified feedback control system.

**2.27.** Draw block diagrams for each of the following elements, first with voltage $v$ as input and current $i$ as output, and then vice versa: ($a$) resistance $R$, ($b$) capacitance $C$, ($c$) inductance $L$.

**2.28.** Draw block diagrams for each of the following mechanical systems, where force is the input and position the output: ($a$) a dashpot, ($b$) a spring, ($c$) a mass, ($d$) a mass, spring, and dashpot connected in series and fastened at one end (mass position is the output).

**2.29.** Draw a block diagram of a ($a$) parallel, ($b$) series $R$-$L$-$C$ network.

**2.30.** Which systems described in the problems of this chapter are regulators?

**2.31.** What type of sampled-data system described in this chapter might be used in implementing a device or algorithm for approximating the integral of a continuous function $u(t)$, using the well-known rectangular rule, or rectangular integration technique?

**2.32.** Draw a simple block diagram of a computer-controlled system in which a digital computer is used to control an analog plant or process, with the summing point and reference input implemented in software in the computer.

**2.33.** What type of controller is the stopper valve of the water-filling system of Problem 2.7?

**2.34.** What types of controllers are included in: ($a$) each of the servomechanisms of Problems 2.9 and 2.10, ($b$) the traffic regulator of Problem 2.12?

# Answers to Supplementary Problems

**2.19.** The equivalent circuit for the emitter follower has the same form as the voltage divider network of Problem 1.11. Therefore the open-loop equation for the output is

$$v_{\text{out}} = \frac{\mu R_K}{r_p + R_K}(v_{\text{in}} - v_{\text{out}}) = \left(\frac{\mu R_K}{r_p + (1 + \mu) R_K}\right) v_{\text{in}}$$

and the open-loop block diagram is given in fig. 2-51.

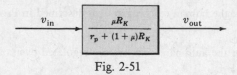

Fig. 2-51

The closed-loop output equation is simply

$$v_{\text{out}} = \frac{\mu R_K}{r_p + R_K}(v_{\text{in}} - v_{\text{out}})$$

and the closed-loop block diagram is given in Fig. 2-52.

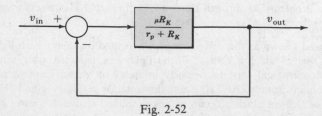

Fig. 2-52

**2.20.**

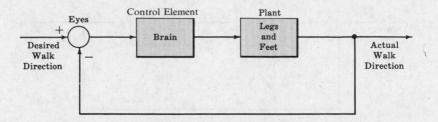

**2.21.**

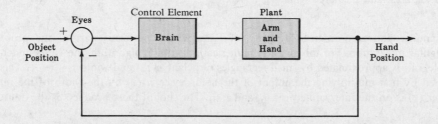

**2.22.**

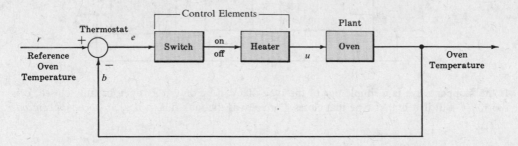

When $e > 0$ ($r > b$), the switch turns the heater on. When $e \leq 0$, the heater is turned off.

**2.23.**

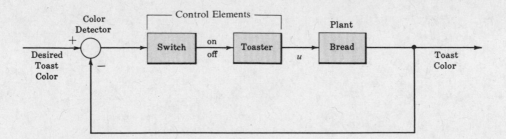

**2.24.** (*a*) The input to an accelerometer is acceleration. The output is displacement of a mass, voltage, or another quantity proportional to acceleration.

     (*b*) See Problem 1.2.

     (*c*) The input to a thermistor is temperature. The output is an electrical quantity measured in ohms, volts, or amperes.

     (*d*) The input to a thermocouple is a temperature difference. The output is a voltage.

**2.25.** The following problems describe servomechanisms: Examples 1.3 and 1.5 in Problem 2.6, and Problems 2.7, 2.8, 2.17, and 2.21.

**2.26.**

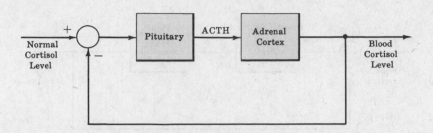

**2.30.** The systems of Examples 1.2 and 1.4 in Problem 2.6, and the systems of Problems 2.7, 2.8, 2.12, 2.13, 2.14, 2.15, 2.22, 2.23, and 2.26 are regulators.

**2.31.** The sampler and zero-order hold device of Example 2.9 performs part of the process required for rectangular integration. For this simplest numerical integration algorithm, the "area under the curve" (i.e., the integral) is approximated by small rectangles of height $u(t_k)$ and width $t_{k+1} - t_k$. This result could be obtained by first multiplying the output of the hold device $u^*(t)$ by the width of the interval $t_{k+1} - t_k$, when $u^*(t)$ is on the interval between $t_k$ and $t_{k+1}$. The sum of these products is the desired result.

**2.32.**

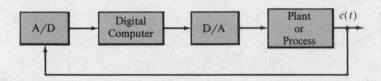

**2.33.** If the stopper valve is a simple one of the type that can be only fully open or fully closed, it is an *on-off controller*. But if it is that type that closes gradually as the tank fills, it is a *proportional controller*.

# Chapter 3

# Differential Equations, Difference Equations, and Linear Systems

## 3.1 SYSTEM EQUATIONS

A property common to all basic laws of physics is that certain fundamental quantities can be defined by numerical values. The physical laws define relationships between these fundamental quantities and are usually represented by equations.

**EXAMPLE 3.1.** The scalar version of Newton's second law states that, if a force of magnitude $f$ is applied to a mass of $M$ units, the acceleration $a$ of the mass is related to $f$ by the equation $f = Ma$.

**EXAMPLE 3.2.** Ohm's law states that, if a voltage of magnitude $v$ is applied across a resistor of $R$ units, the current $i$ through the resistor is related to $v$ by the equation $v = Ri$.

Many nonphysical laws can also be represented by equations.

**EXAMPLE 3.3.** The compound interest law states that, if an amount $P(0)$ is deposited for $n$ equal periods of time at an interest rate $I$ for each time period, the amount will grow to a value of $P(n) = P(0)(1 + I)^n$.

## 3.2 DIFFERENTIAL EQUATIONS AND DIFFERENCE EQUATIONS

Two classes of equations with broad application in the description of systems are differential equations and difference equations.

*Definition 3.1:*      A **differential equation** is any algebraic or transcendental equality which involves either differentials or derivatives.

Differential equations are useful for relating rates of change of variables and other parameters.

**EXAMPLE 3.4.** Newton's second law (Example 3.1) can be written alternatively as a relationship between force $f$, mass $M$, and the rate of change of the velocity $v$ of the mass with respect to time $t$, that is, $f = M(dv/dt)$.

**EXAMPLE 3.5.** Ohm's law (Example 3.2) can be written alternatively as a relationship between voltage $v$, resistance $R$, and the time rate of passage of charge through the resistor, that is, $v = R(dq/dt)$.

**EXAMPLE 3.6.** The diffusion equation in one dimension describes the relationship between the time rate of change of a quantity $T$ in an object (e.g., heat concentration in an iron bar) and the positional rate of change of $T$: $\partial T/\partial x = k(\partial T/\partial t)$, where $k$ is a proportionality constant, $x$ is a position variable, and $t$ is time.

*Definition 3.2:*      A **difference equation** is an algebraic or transcendental equality which involves more than one value of the dependent variable(s) corresponding to more than one value of at least one of the independent variable(s). The dependent variables do not involve either differentials or derivatives.

Difference equations are useful for relating the evolution of variables (or parameters) from one discrete instant of time (or other independent variable) to another.

39

**EXAMPLE 3.7.**   The compound interest law of Example 3.3 can be written alternatively as a difference equation relationship between $P(k)$, the amount of money after $k$ periods of time, and $P(k + 1)$, the amount of money after $k + 1$ periods of time, that is, $P(k + 1) = (1 + I)P(k)$.

## 3.3   PARTIAL AND ORDINARY DIFFERENTIAL EQUATIONS

*Definition 3.3*:        A **partial differential equation** is an equality involving one or more dependent and two or more independent variables, together with partial derivatives of the dependent with respect to the independent variables.

*Definition 3.4*:        An **ordinary (total) differential equation** is an equality involving one or more dependent variables, one independent variable, and one or more derivatives of the dependent variables with respect to the independent variable.

**EXAMPLE 3.8.**   The diffusion equation $\partial T/\partial x = k(\partial T/\partial t)$ is a partial differential equation. $T = T(x, t)$ is the dependent variable, which represents the concentration of some quantity at some position and some time in the object. The independent variable $x$ defines the position in the object, and the independent variable $t$ defines the time.

**EXAMPLE 3.9.**   Newton's second law (Example 3.4) is an ordinary differential equation: $f = M(dv/dt)$. The velocity $v = v(t)$ and the force $f = f(t)$ are dependent variables, and the time $t$ is the independent variable.

**EXAMPLE 3.10.**   Ohm's law (Example 3.5) is an ordinary differential equation: $v = R(dq/dt)$. The charge $q = q(t)$ and the voltage $v = v(t)$ are dependent variables, and the time $t$ is the independent variable.

**EXAMPLE 3.11.**   A differential equation of the form:

$$a_n \frac{d^n y}{dt^n} + a_{n-1} \frac{d^{n-1} y}{dt^{n-1}} + \cdots + a_1 \frac{dy}{dt} + a_0 y = u(t)$$

or, more compactly,

$$\sum_{i=0}^{n} a_i \frac{d^i y(t)}{dt^i} = u(t) \tag{3.1}$$

where $a_0, a_1, \ldots, a_n$ are constants, is an ordinary differential equation. $y(t)$ and $u(t)$ are dependent variables, and $t$ is the independent variable.

## 3.4   TIME VARIABILITY AND TIME INVARIANCE

In the remainder of this chapter, *time* is the only independent variable, unless otherwise specified. This variable is normally designated $t$, except that in difference equations the discrete variable $k$ is often used, as an abbreviation for the time instant $t_k$ (see Example 1.11 and Section 2.5); that is, $y(k)$ is used instead of $y(t_k)$, etc.

A **term** of a differential or difference equation consists of products and/or quotients of explicit functions of the independent variable, the dependent variables, and, for differential equations, derivatives of the dependent variables.

In the definitions of this and the next section, the term *equation* refers to either a differential equation or a difference equation.

*Definition 3.5*:        A **time-variable equation** is an equation in which one or more terms depend *explicitly* on the independent variable time.

*Definition 3.6*:        A **time-invariant equation** is an equation in which none of the terms depends *explicitly* on the independent variable time.

**EXAMPLE 3.12.**   The difference equation $ky(k+2) + y(k) = u(k)$, where $u$ and $y$ are dependent variables, is time-variable because the term $ky(k+2)$ depends explicitly on the coefficient $k$, which represents the time $t_k$.

**EXAMPLE 3.13.**   Any differential equation of the form:

$$\sum_{i=0}^{n} a_i \frac{d^i y}{dt^i} = \sum_{i=0}^{m} b_i \frac{d^i u}{dt^i} \tag{3.2}$$

where the coefficients $a_0, a_1, \ldots, a_n, b_0, b_1, \ldots, b_m$ are constants, is *time-invariant*. The equation depends *implicitly* on $t$, via the dependent variables $u$ and $y$ and their derivatives.

## 3.5   LINEAR AND NONLINEAR DIFFERENTIAL AND DIFFERENCE EQUATIONS

*Definition 3.7*:      A **linear term** is one which is first degree in the dependent variables and their derivatives.

*Definition 3.8*:      A **linear equation** is an equation consisting of a sum of linear terms. All others are **nonlinear equations**.

If any term of a differential equation contains higher powers, products, or transcendental functions of the dependent variables, it is nonlinear. Such terms include $(dy/dt)^3$, $u(dy/dt)$, and $\sin u$, respectively. For example, $(5/\cos t)(d^2 y/dt^2)$ is a term of first degree in the dependent variable $y$, and $2uy^3(dy/dt)$ is a term of fifth degree in the dependent variables $u$ and $y$.

**EXAMPLE 3.14.**   The ordinary differential equations $(dy/dt)^2 + y = 0$ and $d^2 y/dt^2 + \cos y = 0$ are nonlinear because $(dy/dt)^2$ is second degree in the first equation, and $\cos y$ in the second equation is *not* first degree, which is true of all transcendental functions.

**EXAMPLE 3.15.**   The difference equation $y(k+2) + u(k+1)y(k+1) + y(k) = u(k)$, in which $u$ and $y$ are dependent variables, is a nonlinear difference equation because $u(k+1)y(k+1)$ is second degree in $u$ and $y$. This type of nonlinear equation is sometimes called *bilinear* in $u$ and $y$.

**EXAMPLE 3.16.**   Any difference equation

$$\sum_{i=0}^{n} a_i(k) y(k+i) = \sum_{i=0}^{n} b_i(k) u(k+i) \tag{3.3}$$

in which the coefficients $a_i(k)$ and $b_i(k)$ depend only upon the independent variable $k$, is a linear difference equation.

**EXAMPLE 3.17.**   Any ordinary differential equation

$$\sum_{i=0}^{n} a_i(t) \frac{d^i y}{dt^i} = \sum_{i=0}^{m} b_i(t) \frac{d^i u}{dt^i} \tag{3.4}$$

where the coefficients $a_i(t)$ and $b_i(t)$ depend only upon the independent variable $t$, is a linear differential equation.

## 3.6   THE DIFFERENTIAL OPERATOR $D$ AND THE CHARACTERISTIC EQUATION

Consider the $n$th-order linear constant-coefficient differential equation

$$\frac{d^n y}{dt^n} + a_{n-1} \frac{d^{n-1} y}{dt^{n-1}} + \cdots + a_1 \frac{dy}{dt} + a_0 y = u \tag{3.5}$$

It is convenient to define a **differential operator**

$$D \equiv \frac{d}{dt}$$

and more generally an **$n$th-order differential operator**

$$D^n \equiv \frac{d^n}{dt^n}$$

The differential equation can now be written as

$$D^n y + a_{n-1} D^{n-1} y + \cdots + a_1 Dy + a_0 y = u$$

or

$$\left( D^n + a_{n-1} D^{n-1} + \cdots + a_1 D + a_0 \right) y = u$$

**Definition 3.9:**        The polynomial in $D$:

$$D^n + a_{n-1} D^{n-1} + \cdots + a_1 D + a_0 \qquad (3.6)$$

is called the **characteristic polynomial**.

**Definition 3.10:**        The equation

$$D^n + a_{n-1} D^{n-1} + \cdots + a_1 D + a_0 = 0 \qquad (3.7)$$

is called the **characteristic equation**.

The fundamental theorem of algebra states that the characteristic equation has exactly $n$ solutions $D = D_1, D = D_2, \ldots, D = D_n$. These $n$ solutions (also called **roots**) are not necessarily distinct.

**EXAMPLE 3.18.**   Consider the differential equation

$$\frac{d^2 y}{dt^2} + 3 \frac{dy}{dt} + 2 y = u$$

The characteristic polynomial is $D^2 + 3D + 2$. The characteristic equation is $D^2 + 3D + 2 = 0$, which has the two distinct roots: $D = -1$ and $D = -2$.

## 3.7   LINEAR INDEPENDENCE AND FUNDAMENTAL SETS

**Definition 3.11:**        A set of $n$ functions of time $f_1(t), f_2(t), \ldots, f_n(t)$ is called **linearly independent** if the only set of constants $c_1, c_2, \ldots, c_n$ for which

$$c_1 f_1(t) + c_2 f_2(t) + \cdots + c_n f_n(t) = 0$$

for all $t$ are the constants $c_1 = c_2 = \cdots = c_n = 0$.

**EXAMPLE 3.19.**   The functions $t$ and $t^2$ are linearly independent functions since

$$c_1 t + c_2 t^2 = t(c_1 + c_2 t) = 0$$

implies that $c_1 / c_2 = -t$. There are *no constants* that satisfy this relationship.

A *homogeneous* $n$th-order linear differential equation of the form:

$$\sum_{i=0}^{n} a_i \frac{d^i y}{dt^i} = 0$$

has at least one set of $n$ linearly independent solutions.

***Definition 3.12:***     Any set of *n linearly independent* solutions of a homogeneous *n*th-order linear differential equation is called a **fundamental set**.

There is no unique fundamental set. From a given fundamental set other fundamental sets can be generated by the following technique. Suppose that $y_1(t), y_2(t), \ldots, y_n(t)$ is a fundamental set for an *n*th-order linear differential equation. Then a set of *n* functions $z_1(t), z_2(t), \ldots, z_n(t)$ can be formed:

$$z_1(t) = \sum_{i=1}^{n} a_{1i} y_i(t), \, z_2(t) = \sum_{i=1}^{n} a_{2i} y_i(t), \ldots, z_n(t) = \sum_{i=1}^{n} a_{ni} y_i(t) \qquad (3.8)$$

where the $a_{ji}$ are a set of $n^2$ constants. Each $z_i(t)$ is a solution of the differential equation. This set of *n* solutions is a *fundamental set* if the determinant

$$\begin{vmatrix} a_{11} & a_{12} & \cdots & a_{1n} \\ a_{21} & a_{22} & \cdots & a_{2n} \\ \cdots\cdots\cdots\cdots\cdots\cdots \\ a_{n1} & a_{n2} & \cdots & a_{nn} \end{vmatrix} \neq 0$$

**EXAMPLE 3.20.**   The equation for simple harmonic motion, $d^2y/dt^2 + \omega^2 y = 0$, has as a fundamental set

$$y_1 = \sin \omega t \qquad y_2 = \cos \omega t$$

A second fundamental set is*

$$z_1 = \cos \omega t + j \sin \omega t = e^{j\omega t} \qquad z_2 = \cos \omega t - j \sin \omega t = e^{-j\omega t}$$

**Distinct Roots**

If the characteristic equation

$$\sum_{i=0}^{n} a_i D^i = 0$$

has distinct roots $D_1, D_2, \ldots, D_n$, then a fundamental set for the homogeneous equation

$$\sum_{i=0}^{n} a_i \frac{d^i y}{dt^i} = 0$$

is the set of functions $y_1 = e^{D_1 t}$, $y_2 = e^{D_2 t}, \ldots, y_n = e^{D_n t}$.

**EXAMPLE 3.21.**   The differential equation

$$\frac{d^2 y}{dt^2} + 3 \frac{dy}{dt} + 2y = 0$$

has the characteristic equation $D^2 + 3D + 2 = 0$ whose roots are $D = D_1 = -1$ and $D = D_2 = -2$. A fundamental set for this equation is $y_1 = e^{-t}$ and $y_2 = e^{-2t}$.

**Repeated Roots**

If the characteristic equation has repeated roots, then for each root $D_i$ of multiplicity $n_i$ (i.e., $n_i$ roots equal to $D_i$) there are $n_i$ elements of the fundamental set $e^{D_i t}, te^{D_i t}, \ldots, t^{n_i-1} e^{D_i t}$.

**EXAMPLE 3.22.**   The equation

$$\frac{d^2 y}{dt^2} + 2 \frac{dy}{dt} + y = 0$$

---

*The *complex exponential function* $e^w$, where $w = u + jv$ for real $u$ and $v$, and $j = \sqrt{-1}$, is defined in complex variable theory by $e^w \equiv e^u(\cos v + j \sin v)$. Therefore $e^{\pm j\omega t} = \cos \omega t \pm j \sin \omega t$.

with characteristic equation $D^2 + 2D + 1 = 0$, has the repeated root $D = -1$, and a fundamental set consisting of $e^{-t}$ and $te^{-t}$.

## 3.8   SOLUTION OF LINEAR CONSTANT-COEFFICIENT ORDINARY DIFFERENTIAL EQUATIONS

Consider the class of differential equations of the form:

$$\sum_{i=0}^{n} a_i \frac{d^i y}{dt^i} = \sum_{i=0}^{m} b_i \frac{d^i u}{dt^i} \tag{3.9}$$

where the coefficients $a_i$ and $b_i$ are constant, $u = u(t)$ (*the input*) *is a known time function*, and $y = y(t)$ (*the output*) *is the unknown solution of the equation*. If this equation describes a physical system, then generally $m \le n$, and $n$ is called the **order** of the differential equation. To completely specify the problem so that *a unique solution* $y(t)$ can be obtained, two additional items must be specified: (1) the interval of time over which a solution is desired and (2) a set of $n$ *initial conditions* for $y(t)$ and its first $n - 1$ derivatives. The time interval for the class of problems considered is defined by $0 \le t < +\infty$. This interval is used in the remainder of this book unless otherwise specified. The set of initial conditions is

$$y(0), \frac{dy}{dt}\bigg|_{t=0}, \ldots, \frac{d^{n-1}y}{dt^{n-1}}\bigg|_{t=0} \tag{3.10}$$

A problem defined over this interval and with these initial conditions is called an **initial value problem**.

The solution of a differential equation of this class can be divided into two parts, a *free response* and a *forced response*. The sum of these two responses constitutes the *total response*, or solution $y(t)$, of the equation.

## 3.9   THE FREE RESPONSE

The **free response** of a differential equation is the solution of the differential equation when the input $u(t)$ is identically zero.

If the input $u(t)$ is identically zero, then the differential equation has the form:

$$\sum_{i=0}^{n} a_i \frac{d^i y}{dt^i} = 0 \tag{3.11}$$

The solution $y(t)$ of such an equation depends only on the $n$ initial conditions in Equation (3.10).

**EXAMPLE 3.23.**   The solution of the homogeneous first-order differential equation $dy/dt + y = 0$ with initial condition $y(0) = c$, is $y(t) = ce^{-t}$. This can be verified by direct substitution. $ce^{-t}$ is the free response of any differential equation of the form $dy/dt + y = u$ with the initial condition $y(0) = c$.

The *free response* of a differential equation can always be written as a linear combination of the elements of a *fundamental set*. That is, if $y_1(t), y_2(t), \ldots, y_n(t)$ is a fundamental set, then *any* free response $y_a(t)$ of the differential equation can be represented as

$$y_a(t) = \sum_{i=1}^{n} c_i y_i(t) \tag{3.12}$$

where the *constants* $c_i$ are determined in terms of the initial conditions

$$y(0), \frac{dy}{dt}\bigg|_{t=0}, \ldots, \frac{d^{n-1}y}{dt^{n-1}}\bigg|_{t=0}$$

from the set of $n$ algebraic equations

$$y(0) = \sum_{i=1}^{n} c_i y_i(0), \frac{dy}{dt}\bigg|_{t=0} = \sum_{i=1}^{n} c_i \frac{dy_i}{dt}\bigg|_{t=0}, \ldots, \frac{d^{n-1}y}{dt^{n-1}}\bigg|_{t=0} = \sum_{i=1}^{n} c_i \frac{d^{n-1}y_i}{dt^{n-1}}\bigg|_{t=0} \tag{3.13}$$

The linear independence of the $y_i(t)$ guarantees that a solution to these equations can be obtained for $c_1, c_2, \ldots, c_n$.

**EXAMPLE 3.24.** The free response $y_a(t)$ of the differential equation

$$\frac{d^2 y}{dt^2} + 3\frac{dy}{dt} + 2y = u$$

with initial conditions $y(0) = 0$, $(dy/dt)|_{t=0} = 1$ is determined by letting

$$y_a(t) = c_1 e^{-t} + c_2 e^{-2t}$$

where $c_1$ and $c_2$ are unknown coefficients and $e^{-t}$ and $e^{-2t}$ are a fundamental set for the equation (Example 3.21). Since $y_a(t)$ must satisfy the initial conditions, that is,

$$y_a(0) = y(0) = 0 = c_1 + c_2 \qquad \frac{dy_a(t)}{dt}\bigg|_{t=0} = \frac{dy}{dt}\bigg|_{t=0} = 1 = -c_1 - 2c_2$$

then $c_1 = 1$ and $c_2 = -1$. The free response is therefore given by $y_a(t) = e^{-t} - e^{-2t}$.

## 3.10    THE FORCED RESPONSE

The **forced response** $y_b(t)$ of a differential equation is the solution of the differential equation when all the initial conditions

$$y(0), \frac{dy}{dt}\bigg|_{t=0}, \ldots, \frac{d^{n-1}y}{dt^{n-1}}\bigg|_{t=0}$$

are identically zero.

The implication of this definition is that the forced response depends only on the input $u(t)$. The *forced response* for a linear constant-coefficient ordinary differential equation can be written in terms of a *convolution integral* (see Example 3.38):

$$y_b(t) = \int_0^t w(t-\tau)\left[\sum_{i=0}^{m} b_i \frac{d^i u(\tau)}{d\tau^i}\right] d\tau \tag{3.14}$$

where $w(t-\tau)$ is the *weighting function (or kernel) of the differential equation*. This form of the convolution integral assumes that the weighting function describes a *causal* system (see Definition 3.22). This assumption is maintained below.

The weighting function of a linear constant-coefficient ordinary differential equation can be written as

$$w(t) = \sum_{i=1}^{n} c_i y_i(t) \qquad i \geq 0$$

$$= 0 \qquad\qquad t < 0 \tag{3.15}$$

where $c_1, \ldots, c_n$ are constants and the set of functions $y_1(t), y_2(t), \ldots, y_n(t)$ is a fundamental set of the differential equation. It should be noted that $w(t)$ *is a free response of the differential equation* and therefore requires $n$ initial conditions for complete specification. These conditions fix the values of the constants $c_1, c_2, \ldots, c_n$. The initial conditions which all weighting functions of linear differential equations must satisfy are

$$w(0) = 0, \frac{dw}{dt}\bigg|_{t=0} = 0, \ldots, \frac{d^{n-2}w}{dt^{n-2}}\bigg|_{t=0} = 0, \frac{d^{n-1}w}{dt^{n-1}}\bigg|_{t=0} = 1 \tag{3.16}$$

**EXAMPLE 3.25.**   The weighting function of the differential equation

$$\frac{d^2 y}{dt^2} + 3\frac{dy}{dt} + 2y = u$$

is a linear combination of $e^{-t}$ and $e^{-2t}$ (a fundamental set of the equation). That is,

$$w(t) = c_1 e^{-t} + c_2 e^{-2t}$$

$c_1$ and $c_2$ are determined from the two algebraic equations

$$w(0) = 0 = c_1 + c_2 \qquad \left. \frac{dw}{dt} \right|_{t=0} = 1 = -c_1 - 2c_2$$

The solution is $c_1 = 1$, $c_2 = -1$, and the weighting function is $w(t) = e^{-t} - e^{-2t}$.

**EXAMPLE 3.26.**   For the differential equation of Example 3.25, if $u(t) = 1$, then the forced response $y_b(t)$ of the equation is

$$y_b(t) = \int_0^t w(t-\tau) u(\tau)\, d\tau = \int_0^t \left[ e^{-(t-\tau)} - e^{-2(t-\tau)} \right] d\tau$$

$$= e^{-t} \int_0^t e^{\tau}\, d\tau - e^{-2t} \int_0^t e^{2\tau}\, d\tau = \frac{1}{2}(1 - 2e^{-t} + e^{-2t})$$

## 3.11   THE TOTAL RESPONSE

The **total response** of a linear constant-coefficient differential equation is the sum of the *free response* and the *forced response*.

**EXAMPLE 3.27.**   The total response $y(t)$ of the differential equation

$$\frac{d^2 y}{dt^2} + 3\frac{dy}{dt} + 2y = 1$$

with initial conditions $y(0) = 0$ and $(dy/dt)|_{t=0} = 1$ is the sum of the free response $y_a(t)$ determined in Example 3.24 and the forced response $y_b(t)$ determined in Example 3.26. Thus

$$y(t) = y_a(t) + y_b(t) = (e^{-t} - e^{-2t}) + \frac{1}{2}(1 - 2e^{-t} + e^{-2t}) = \frac{1}{2}(1 - e^{-2t})$$

## 3.12   THE STEADY STATE AND TRANSIENT RESPONSES

The *steady state response* and *transient response* are another pair of quantities whose sum is equal to the total response. These terms are often used for specifying control system performance. They are defined as follows.

*Definition 3.13:*   The **steady state response** is that part of the total response which *does not* approach zero as time approaches infinity.

*Definition 3.14:*   The **transient response** is that part of the total response which approaches zero as time approaches infinity.

**EXAMPLE 3.28.**   The total response for the differential equation in Example 3.27 was determined as $y = \frac{1}{2} - \frac{1}{2}e^{-t}$. Clearly, the steady state response is given by $y_{ss} = \frac{1}{2}$. Since $\lim_{t \to \infty} [-\frac{1}{2}e^{-t}] = 0$, the transient response is $y_T = -\frac{1}{2}e^{-t}$.

### 3.13   SINGULARITY FUNCTIONS: STEPS, RAMPS, AND IMPULSES

In the study of control systems and the equations which describe them, a particular family of functions called *singularity functions* is used extensively. Each member of this family is related to the others by one or more integrations or differentiations. The three most widely used singularity functions are the *unit step*, the *unit impulse*, and the *unit ramp*.

**Definition 3.15:**    A **unit step function** $\mathbf{1}(t - t_0)$ is defined by

$$\mathbf{1}(t - t_0) = \begin{cases} 1 & \text{for} \quad t > t_0 \\ 0 & \text{for} \quad t \le t_0 \end{cases} \tag{3.17}$$

The unit step function is illustrated in Fig. 3-1.

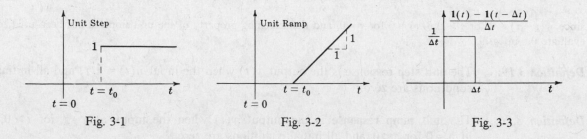

Fig. 3-1                              Fig. 3-2                              Fig. 3-3

**Definition 3.16:**    A **unit ramp function** is the integral of a unit step function

$$\int_{-\infty}^{t} \mathbf{1}(\tau - t_0)\, d\tau = \begin{cases} t - t_0 & \text{for} \quad t > t_0 \\ 0 & \text{for} \quad t \le t_0 \end{cases} \tag{3.18}$$

The unit ramp function is illustrated in Fig. 3-2.

**Definition 3.17:**    A **unit impulse function** $\delta(t)$ may be defined by

$$\delta(t) = \lim_{\substack{\Delta t \to 0 \\ \Delta t > 0}} \left[ \frac{\mathbf{1}(t) - \mathbf{1}(t - \Delta t)}{\Delta t} \right] \tag{3.19}*$$

where $\mathbf{1}(t)$ is the unit step function.

The pair $\begin{Bmatrix} \Delta t \to 0 \\ \Delta t > 0 \end{Bmatrix}$ may be abbreviated by $\Delta t \to 0^+$, meaning that $\Delta t$ approaches zero *from the right*. The quotient in brackets represents a rectangle of height $1/\Delta t$ and width $\Delta t$ as shown in Fig. 3-3. The limiting process produces a function whose height approaches infinity and width approaches zero. The area under the curve is equal to 1 for all values of $\Delta t$. That is,

$$\int_{-\infty}^{\infty} \delta(t)\, dt = 1$$

The unit impulse function has the following very important property:

**Screening Property:**    The integral of the product of a unit impulse function $\delta(t - t_0)$ and a function $f(t)$, continuous at $t = t_0$ over an interval which includes $t_0$, is equal to the function

---

*In a formal sense, Equation (*3.19*) defines the *one-sided derivative* of the unit step function. But neither the limit nor the derivative exist in the ordinary mathematical sense. However, Definition 3.17 is satisfactory for the purposes of this book, and many others.

$f(t)$ evaluated at $t_0$, that is,

$$\int_{-\infty}^{\infty} f(t)\, \delta(t - t_0)\, dt = f(t_0) \tag{3.20}$$

**Definition 3.18:**      The **unit impulse response** of a system is the output $y(t)$ of the system when the input $u(t) = \delta(t)$ and all initial conditions are zero.

**EXAMPLE 3.29.**    If the input-output relationship of a linear system is given by the convolution integral

$$y(t) = \int_0^t w(t - \tau)\, u(\tau)\, d\tau$$

then the unit impulse response $y_\delta(t)$ of the system is

$$y_\delta(t) = \int_0^t w(t - \tau)\, \delta(\tau)\, d\tau = \int_{-\infty}^{\infty} w(t - \tau)\, \delta(\tau)\, d\tau = w(t) \tag{3.21}$$

since $w(t - \tau) = 0$ for $\tau > t$, $\delta(\tau) = 0$ for $\tau < 0$, and the screening property of the unit impulse has been used to evaluate the integral.

**Definition 3.19:**      The **unit step response** is the output $y(t)$ when the input $u(t) = \mathbf{1}(t)$ and all initial conditions are zero.

**Definition 3.20:**      The **unit ramp response** is the output $y(t)$ when the input $u(t) = t$ for $t > 0$, $u(t) = 0$ for $t \le 0$, and all initial conditions are zero.

## 3.14   SECOND-ORDER SYSTEMS

In the study of control systems, linear constant-coefficient second-order differential equations of the form:

$$\frac{d^2 y}{dt^2} + 2\zeta\omega_n \frac{dy}{dt} + \omega_n^2 y = \omega_n^2 u \tag{3.22}$$

are important because higher-order systems can often be approximated by second-order systems. The constant $\zeta$ is called the **damping ratio**, and the constant $\omega_n$ is called the **undamped natural frequency** of the system. The forced response of this equation for inputs $u$ belonging to the class of singularity functions is of particular interest. That is, the *forced response* to a unit impulse, unit step, or unit ramp is the same as the *unit impulse response*, *unit step response*, or *unit ramp response* of a system represented by this equation.

Assuming that $0 \le \zeta \le 1$, the characteristic equation for Equation (3.22) is

$$D^2 + 2\zeta\omega_n D + \omega_n^2 = \left(D + \zeta\omega_n - j\omega_n\sqrt{1 - \zeta^2}\right)\left(D + \zeta\omega_n + j\omega_n\sqrt{1 - \zeta^2}\right) = 0$$

Hence the roots are

$$D_1 = -\zeta\omega_n + j\omega_n\sqrt{1 - \zeta^2} \equiv -\alpha + j\omega_d \qquad D_2 = -\zeta\omega_n - j\omega_n\sqrt{1 - \zeta^2} \equiv -\alpha - j\omega_d$$

where $\alpha \equiv \zeta\omega_n$ is called the **damping coefficient**, and $\omega_d \equiv \omega_n\sqrt{1 - \zeta^2}$ is called the **damped natural frequency**. $\alpha$ is the inverse of the **time constant** $\tau$ of the system, that is, $\tau = 1/\alpha$.

The weighting function of Equation (3.22) is $w(t) = (1/\omega_d)e^{-\alpha t}\sin\omega_d t$. The unit step response is given by

$$y_1(t) = \int_0^t w(t - \tau)\,\omega_n^2\, d\tau = 1 - \frac{\omega_n e^{-\alpha t}}{\omega_d}\sin(\omega_d t + \phi) \tag{3.23}$$

where $\phi \equiv \tan^{-1}(\omega_d/\alpha)$.

Figure 3-4 is a parametric representation of the unit step response. Note that the abscissa of this family of curves is normalized time $\omega_n t$, and the parameter defining each curve is the damping ratio $\zeta$.

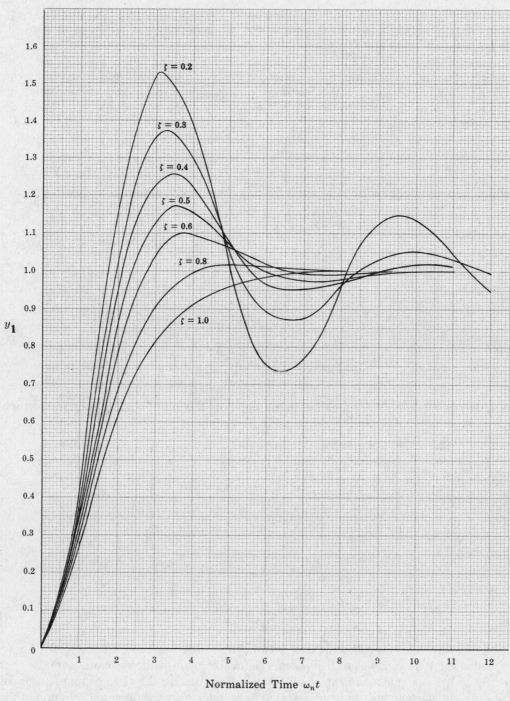

Fig. 3-4

## 3.15   STATE VARIABLE REPRESENTATION OF SYSTEMS DESCRIBED BY LINEAR DIFFERENTIAL EQUATIONS

In some problems of feedback and control, it is more convenient to describe a system by a set of first-order differential equations rather than by one or more $n$th-order differential equations. One reason is that quite general and powerful results from vector-matrix algebra can then be easily applied in deriving solutions for the differential equations.

**EXAMPLE 3.30.** Consider the differential equation form of Newton's second law, $f = M(d^2x/dt^2)$. It is clear from the meanings of velocity $v$ and acceleration $a$ that this second-order equation can be replaced by two first-order equations, $v = dx/dt$ and $f = M(dv/dt)$.

There are numerous ways to transform $n$th-order differential equations into $n$ first-order equations. One of these is quite prevalent in the literature, and straightforward, and we introduce only this transformation here, to illustrate the approach. Consider the $n$th-order, single-input linear constant-coefficient differential equation

$$\sum_{i=0}^{n} a_i \frac{d^i y}{dt^i} = u$$

This equation can always be replaced by the following $n$ first-order differential equations:

$$\frac{dx_1}{dt} = x_2$$

$$\frac{dx_2}{dt} = x_3$$

$$\vdots$$

$$\frac{dx_{n-1}}{dt} = x_n$$

$$\frac{dx_n}{dt} = -\frac{1}{a_n}\left[\sum_{i=0}^{n-1} a_i x_{i+1}\right] + \frac{1}{a_n}u \tag{3.24a}$$

where we have chosen $x_1 \equiv y$. Using *vector-matrix* notation, this set of equations can be written as

$$\begin{bmatrix} \dfrac{dx_1}{dt} \\ \dfrac{dx_2}{dt} \\ \vdots \\ \dfrac{dx_n}{dt} \end{bmatrix} = \begin{bmatrix} 0 & 1 & 0 & \cdots & 0 \\ 0 & 0 & 1 & \cdots & 0 \\ \vdots & \vdots & \vdots & \ddots & \vdots \\ -\dfrac{a_0}{a_n} & -\dfrac{a_1}{a_n} & -\dfrac{a_2}{a_n} & \cdots & -\dfrac{a_{n-1}}{a_n} \end{bmatrix} \begin{bmatrix} x_1 \\ x_2 \\ \vdots \\ x_n \end{bmatrix} + \begin{bmatrix} 0 \\ 0 \\ \vdots \\ 0 \\ \dfrac{1}{a_n} \end{bmatrix} u \tag{3.24b}$$

or, more compactly, as

$$\frac{d\mathbf{x}}{dt} = A\mathbf{x} + \mathbf{b}u \tag{3.24c}$$

In Equation $(3.24c)$ $\mathbf{x} \equiv \mathbf{x}(t)$ is called the **state vector**, with $n$ time functions $x_1(t), x_2(t), \ldots, x_n(t)$ as its elements, called the **state variables** of the system. The scalar input of the system is $u(t)$.

More generally, *multiinput-multioutput* (*MIMO*) *systems* described by one or more linear constant-coefficient differential equations can be represented by a vector-matrix differential equation of the form:

$$\begin{bmatrix} \dfrac{dx_1}{dt} \\ \dfrac{dx_2}{dt} \\ \vdots \\ \dfrac{dx_n}{dt} \end{bmatrix} = \begin{bmatrix} a_{11} & a_{12} & \cdots & a_{1n} \\ a_{21} & a_{22} & \cdots & a_{2n} \\ \vdots & & \ddots & \vdots \\ a_{n1} & a_{n2} & \cdots & a_{nn} \end{bmatrix} \begin{bmatrix} x_1 \\ x_2 \\ \vdots \\ x_n \end{bmatrix} + \begin{bmatrix} b_{11} & b_{12} & \cdots & b_{1r} \\ b_{21} & b_{22} & \cdots & b_{2r} \\ \vdots & & \ddots & \vdots \\ b_{n1} & b_{n2} & \cdots & b_{nr} \end{bmatrix} \begin{bmatrix} u_1 \\ u_2 \\ \vdots \\ u_r \end{bmatrix} \tag{3.25a}$$

or, more compactly, as

$$\frac{d\mathbf{x}}{dt} = A\mathbf{x} + B\mathbf{u} \qquad (3.25b)$$

In Equation ($3.25b$) $\mathbf{x}$ is defined as in Equation ($3.24c$), $A$ is the $n \times n$ matrix of constants $a_{ij}$, and $B$ is the $n \times r$ matrix of constants $b_{ij}$, each given in Equation ($3.25a$), and $\mathbf{u}$ is an $r$-vector of input functions.

### The Transition Matrix

The matrix equation

$$\frac{d\Phi}{dt} = A\Phi$$

where $\Phi$ is an $n \times n$ matrix of time functions, called the **transition matrix of the differential equation** ($3.24c$) or ($3.25b$), has a special role in the solution of vector-matrix differential equations like Equation ($3.25b$). If $I$ is the $n \times n$ *identity* or *unit* matrix, and $\Phi(0) = I$ is the initial condition of this homogeneous equation, the transition matrix has the special solution: $\Phi(t) = e^{At}$. In this case $e^{At}$ is an $n \times n$ matrix function defined by the infinite series:

$$e^{At} = I + At + \frac{A^2 t^2}{2!} + \frac{A^3 t^3}{3!} + \cdots$$

$\Phi$ also has the property, called the *transition property*, that for all $t_1$, $t_2$, and $t_3$: $\Phi(t_1 - t_2)\Phi(t_2 - t_3) = \Phi(t_1 - t_3)$.

To solve the differential equation ($3.24$) or ($3.25$), the time interval of interest must be specified, for example, $0 \le t < +\infty$, and an initial condition vector $\mathbf{x}(0)$ is also needed. In this case, the general solution of Equation (3.25) is

$$\mathbf{x}(t) = e^{At}\mathbf{x}(0) + \int_0^t e^{A(t-\tau)} B\mathbf{u}(\tau)\, d\tau \qquad (3.26)$$

The initial condition $\mathbf{x}(0)$ is sometimes referred to as the **state of the system at time t = 0**. From Equation ($3.26$) we see that knowledge of $\mathbf{x}(0)$, and the input $\mathbf{u}(t)$ on the interval $0 \le t < +\infty$, are adequate to completely determine the state variables for all time $t \ge 0$. Actually, knowledge of the state of the system at *any* time $t'$, $0 < t' < +\infty$, and knowledge of the input $u(t)$, $t' \le t < +\infty$, are adequate to completely define the state vector $\mathbf{x}(t)$ at all subsequent times $t \ge t'$.

## 3.16   SOLUTION OF LINEAR CONSTANT-COEFFICIENT DIFFERENCE EQUATIONS

Consider the class of difference equations

$$\sum_{i=0}^{n} a_i y(k+i) = \sum_{i=0}^{m} b_i u(k+i) \qquad (3.27)$$

where $k$ is the integer-valued discrete-time variable, the coefficients $a_i$ and $b_i$ are constant, $a_0$ and $a_n$ are nonzero, the input $u(k)$ is a known time sequence, and the output $y(k)$ is the unknown sequence solution of the equation. Since $y(k+n)$ is an explicit function of $y(k), y(k+1), \ldots, y(k+n-1)$, then $n$ is the **order of the difference equation**. To obtain a unique solution for $y(k)$, two additional items must be specified, the time sequence over which a solution is desired, and a set of $n$ initial conditions for $y(k)$. The time sequence for the class of problems treated in this book is the set of nonnegative integers, that is, $k = 0, 1, 2, \ldots$. The set of initial conditions is

$$y(0), y(1), \ldots, y(n-1) \qquad (3.28)$$

A problem defined over this time sequence and with these initial conditions is called an **initial value problem**.

Consider the $n$th-order linear constant-coefficient difference equation

$$y(k+n) + a_{n-1}y(k+n-1) + \cdots + a_1 y(k+1) + a_0 y(k) = u(k) \qquad (3.29)$$

It is convenient to define a **shift operator** $Z$ by the equation

$$Z[y(k)] \equiv y(k+1)$$

By repeated application of this operation, we obtain

$$Z^n[y(k)] = Z[Z[\ldots Z[y(k)]\ldots]] = y(k+n)$$

Similarly, a **unity operator** $I$ is defined by

$$I[y(k)] = y(k)$$

and $Z^0 \equiv I$. The operator $Z$ has the following important algebraic properties:

1. For constant $c$, $Z[cy(k)] = cZ[y(k)]$
2. $Z^m[y(k) + x(k)] = Z^m[y(k)] + Z^m[x(k)]$

The difference equation can thus be written as

$$Z^n[y(k)] + a_{n-1}Z^{n-1}[y(k)] + \cdots + a_1 Z[y(k)] + a_0 y(k) = u(k)$$

or

$$\left(Z^n + a_{n-1}Z^{n-1} + \cdots + a_1 Z + a_0\right)[y(k)] = u(k)$$

The equation

$$Z^n + a_{n-1}Z^{n-1} + \cdots + a_1 Z + a_0 = 0 \qquad (3.30)$$

is called the **characteristic equation** of the difference equation, and, by the fundamental theorem of algebra, it has exactly $n$ solutions: $Z = Z_1, Z = Z_2, \ldots, Z = Z_n$.

**EXAMPLE 3.31.** Consider the difference equation

$$y(k+2) + \frac{5}{6}y(k+1) + \frac{1}{6}y(k) = u(k)$$

The characteristic equation is $Z^2 + \frac{5}{6}Z + \frac{1}{6} = 0$ with two solutions, $Z = -\frac{1}{2}$ and $Z = -\frac{1}{3}$.

A homogeneous $n$th-order linear difference equation has at least one set of $n$ linearly independent solutions. Any such set is called a **fundamental set**. As with differential equations, fundamental sets are not unique.

If the characteristic equation has distinct roots $Z_1, Z_2, \ldots, Z_n$, a fundamental set for the homogeneous equation

$$\sum_{i=0}^{n} a_i y(k+i) = 0 \qquad (3.31)$$

is the set of functions $Z_1^k, Z_2^k, \ldots, Z_n^k$.

**EXAMPLE 3.32.** The difference equation

$$y(k+2) + \frac{5}{6}y(k+1) + \frac{1}{6}y(k) = 0$$

has the characteristic equation $Z^2 + \frac{5}{6}Z + \frac{1}{6} = 0$, with roots $Z = Z_1 = -\frac{1}{2}$ and $Z = Z_2 = -\frac{1}{3}$. A fundamental set of this equation is $y_1(k) = (-\frac{1}{2})^k$ and $y_2(k) = (-\frac{1}{3})^k$.

If the characteristic equation has repeated roots, then for each root $Z_i$ of multiplicity $n_i$, there are $n_i$ elements of the fundamental set $Z_i^k, kZ_i^k, \ldots, k^{n_i-2}Z_i^k, k^{n_i-1}Z_i^k$.

**EXAMPLE 3.33.** The equation $y(k+2) + y(k+1) + \frac{1}{4}y(k) = 0$ with the repeated root $Z = -\frac{1}{2}$ has a fundamental set consisting of $(-\frac{1}{2})^k$ and $k(-\frac{1}{2})^k$.

The free response of a difference equation of the form of Equation $(3.27)$ is the solution when the input sequence is identically zero. The equation then has the form of Equation $(3.31)$ and its solution depends only on the $n$ initial conditions $(3.28)$. If $y_1(k), y_2(k), \ldots, y_n(k)$ is a fundamental set, then any free response of the difference equation $(3.27)$ can be represented as

$$y_a(k) = \sum_{i=1}^{n} c_i y_i(k)$$

where the constants $c_i$ are determined in terms of the initial conditions $y_i(0)$ from the set of $n$ algebraic equations:

$$y(0) = \sum_{i=1}^{n} c_i y_i(0)$$

$$y(1) = \sum_{i=1}^{n} c_i y_i(1)$$

$$\vdots$$

$$y(n-1) = \sum_{i=1}^{n} c_i y_i(n-1) \tag{3.32}$$

The linear independence of the $y_i(k)$ guarantees a solution for $c_1, c_2, \ldots, c_n$.

**EXAMPLE 3.34.** The free response of the difference equation $y(k+2) + \frac{5}{6}y(k+1) + \frac{1}{6}y(k) = u(k)$ with initial conditions $y(0) = 0$ and $y(1) = 1$ is determined by letting

$$y_a(k) = c_1\left(-\frac{1}{2}\right)^k + c_2\left(-\frac{1}{3}\right)^k$$

where $c_1$ and $c_2$ are unknown coefficients and $(-\frac{1}{2})^k$ and $(-\frac{1}{3})^k$ are a fundamental set for the equation (Example 3.32). Since $y_a(k)$ must satisfy the initial conditions, that is,

$$y_a(0) = y(0) = 0 = c_1 + c_2$$

$$y_a(1) = y(1) = 1 = -\frac{1}{2}c_1 - \frac{1}{3}c_2$$

then $c_1 = -6$ and $c_2 = 6$. The free response is therefore given by $y_a(k) = -6(-\frac{1}{2})^k + 6(-\frac{1}{3})^k$.

The forced response $y_b(k)$ of a difference equation is its solution when all initial conditions $y(0), y(1), \ldots, y(n-1)$ are zero. It can be written in terms of a *convolution sum*:

$$y_b(k) = \sum_{j=0}^{k-1} w(k-j)\left[\sum_{i=0}^{m} b_i u(j+i)\right] \qquad k = 0, 1, \ldots, n \tag{3.33}$$

where $w(k-j)$ is the **weighting sequence of the difference equation**. Note that $y_b(0) = 0$ by definition of the forced response, and $w(k-j) = 0$ for $k < j$ (see Section 3.19). If $u(j) \equiv \delta(j) = 1$ for $j = 0$, and $\delta(j) = 0$ for $j \neq 0$, the special input called the **Kronecker delta sequence**, then the forced response $y_b(k) \equiv y_\delta(k)$ is called the **Kronecker delta response**.

The weighting sequence of a linear constant-coefficient difference equation can be written as

$$w(k-l) = \sum_{j=1}^{n} \frac{M_j(l)}{a_n M(l)} y_j(k) \tag{3.34}$$

where $y_1(k), y_2(k), \ldots, y_n(k)$ is a fundamental set of the difference equation, $M(l)$ is the **determinant**:

$$M(l) = \begin{vmatrix} y_1(l+1) & y_2(l+1) & \cdots & y_n(l+1) \\ y_1(l+2) & y_2(l+2) & \cdots & y_n(l+2) \\ \vdots & \vdots & \ddots & \vdots \\ y_1(l+n) & y_2(l+n) & \cdots & y_n(l+n) \end{vmatrix}$$

and $M_j(l)$ is the **cofactor** of the last element in the $j$th column of $M(l)$.

**EXAMPLE 3.35.** Consider the difference equation $y(k+2) + \frac{5}{6}y(k+1) + \frac{1}{6}y(k) = u(k)$. The weighting sequence is given by

$$w(k-l) = \frac{M_1(l)}{M(l)} y_1(k) + \frac{M_2(l)}{M(l)} y_2(k)$$

where $y_1(k) = (-\frac{1}{2})^k$, $y_2(k) = (-\frac{1}{3})^k$, $M_1(l) = -(-\frac{1}{3})^{l+1}$, $M_2(l) = (-\frac{1}{2})^{l+1}$, and

$$M(l) = \begin{vmatrix} \left(-\dfrac{1}{2}\right)^{l+1} & \left(-\dfrac{1}{3}\right)^{l+1} \\ \left(-\dfrac{1}{2}\right)^{l+2} & \left(-\dfrac{1}{3}\right)^{l+2} \end{vmatrix} = \frac{1}{36}\left(-\frac{1}{2}\right)^{l}\left(-\frac{1}{3}\right)^{l}$$

Therefore

$$w(k-l) = 12\left(-\frac{1}{2}\right)^{k-l} - 18\left(-\frac{1}{3}\right)^{k-l}$$

As for continuous systems, the **total response** of a difference equation is the sum of the free and forced responses of the equation. The **transient response** of a difference equation is that part of the total response which approaches zero as time approaches infinity. That part of the total response which does not approach zero is called the **steady state response**.

## 3.17 STATE VARIABLE REPRESENTATION OF SYSTEMS DESCRIBED BY LINEAR DIFFERENCE EQUATIONS

As with differential equations in Section 3.15, it is often useful to describe a system by a set of first-order difference equations, rather than by one or more $n$th-order difference equations.

**EXAMPLE 3.36.** The second-order difference equation

$$y(k+2) + \frac{5}{6}y(k+1) + \frac{1}{6}y(k) = u(k)$$

can be written as the two first-order equations:

$$x_1(k+1) = x_2(k)$$

$$x_2(k+1) = -\frac{5}{6}x_2(k) - \frac{1}{6}x_1(k) + u(k)$$

where we have chosen $x_1(k) \equiv y(k)$.

Consider the $n$th-order, single-input, linear constant-coefficient difference equation

$$\sum_{i=0}^{n} a_i y(k+i) = u(k)$$

This equation can always be replaced by the following $n$ first-order difference equations:

$$x_1(k+1) = x_2(k)$$
$$x_2(k+1) = x_3(k)$$
$$\vdots$$
$$x_{n-1}(k+1) = x_n(k)$$
$$x_n(k+1) = -\frac{1}{a_n}\left[\sum_{i=0}^{n-1} a_i x_{i+1}(k)\right] + \frac{1}{a_n}u(k) \qquad (3.35a)$$

where we have chosen $x_1(k) \equiv y(k)$. Using vector-matrix notation, this set of equations can be written as the *vector-matrix* difference equation

$$
\begin{bmatrix} x_1(k+1) \\ x_2(k+1) \\ \vdots \\ x_n(k+1) \end{bmatrix} = \begin{bmatrix} 0 & 1 & 0 & \cdots & 0 \\ 0 & 0 & 1 & & 0 \\ \cdot & & \cdot & & 0 \\ \cdot & & \cdot & & \\ \cdot & & \cdot & & 1 \\ -a_0/a_n & -a_1/a_n & \cdots & & -a_{n-1}/a_n \end{bmatrix} \begin{bmatrix} x_1(k) \\ x_2(k) \\ \vdots \\ x_n(k) \end{bmatrix} + \begin{bmatrix} 0 \\ 0 \\ \vdots \\ 0 \\ 1/a_n \end{bmatrix} u \quad (3.35b)
$$

or, more compactly, as

$$
\mathbf{x}(k+1) = A\mathbf{x}(k) + \mathbf{b}u \qquad (3.35c)
$$

In these equations, $\mathbf{x}(k)$ is an $n$-vector element of a time sequence called the **state vector**, made up of scalar elements $x_1(k), x_2(k), \ldots, x_n(k)$ called the **state variables** of the system at time $k$.

In general, *multiinput-multioutput* (MIMO) systems described by one or more linear constant-coefficient difference equations can be represented by

$$
\mathbf{x}(k+1) = A\mathbf{x}(k) + B\mathbf{u}(k) \qquad (3.36)
$$

where $\mathbf{x}(k)$ is the state vector of the system, as above, $A$ is an $n \times n$ matrix of constants $a_{ij}$, and $B$ is an $n \times r$ matrix of constants $b_{ij}$, each defined as in Equation (3.25a), and $\mathbf{u}(k)$ is an $r$-vector element of a (multiple) input sequence. Given a time sequence of interest $k = 0, 1, 2, \ldots$, and an initial condition vector $\mathbf{x}(0)$, the solution of Equation (3.36) can be written as

$$
\mathbf{x}(k) = A^k \mathbf{x}(0) + \sum_{j=0}^{k-1} A^{k-1-j} B\mathbf{u}(j) \qquad (3.37)
$$

Note that Equation (3.37) has a form similar to Equation (3.26). In general, however, $A^k$ need not have the properties of a transition matrix of a differential equation. But there is one very important case when $A^k$ does have such properties, that is, where $A^k$ *is* a transition matrix. This case provides the basis for *discretization* of differential equations, as illustrated next.

## Discretization of Differential Equations

Consider a *differential* system described by Equation (3.26). Suppose it is only necessary to have knowledge of the state variables at periodic time instants $t = 0, T, 2T, \ldots, kT, \ldots,$ . In this case, the following *sequence* of state vectors can be written as

$$
\mathbf{x}(T) = e^{AT}\mathbf{x}(0) + \int_0^T e^{A(T-\tau)} B\mathbf{u}(\tau)\, d\tau
$$

$$
\mathbf{x}(2T) = e^{AT}\mathbf{x}(T) + e^{AT} \int_T^{2T} e^{A(T-\tau)} B\mathbf{u}(\tau)\, d\tau
$$

$$
\vdots
$$

$$
\mathbf{x}(kT) = e^{AT}\mathbf{x}((k-1)T) + e^{A(k-1)T} \int_{(k-1)T}^{kT} e^{A(T-\tau)} B\mathbf{u}(\tau)\, d\tau
$$

If we suppress the parameter $T$, use the abbreviation $\mathbf{x}(k) \equiv \mathbf{x}(kT)$, and define a new *input sequence* by

$$
\mathbf{u}'(k) = e^{AkT} \int_{kT}^{(k+1)T} e^{A(T-\tau)} B\mathbf{u}(\tau)\, d\tau
$$

then the set of solution equations above can be replaced by the single *vector-matrix difference equation*

$$
\mathbf{x}(k+1) = e^{AT}\mathbf{x}(k) + \mathbf{u}'(k) \qquad (3.38)
$$

Note that $A' \equiv e^{AT}$ is a transition matrix in Equation (3.38).

### 3.18   LINEARITY AND SUPERPOSITION

The concept of linearity has been presented in Definition 3.8 as a property of differential and difference equations. In this section, linearity is discussed as a property of *general systems*, with one independent variable, time $t$. In Chapters 1 and 2, the concepts of system, input, and output were defined. The following definition of linearity is based on these earlier definitions.

**Definition 3.21:**   If all initial conditions in the system are zero, that is, if the system is completely at rest, then the system is a **linear system** if it has the following property:

(a)   If an input $u_1(t)$ produces an output $y_1(t)$, and

(b)   an input $u_2(t)$ produces an output $y_2(t)$,

(c)   then input $c_1u_1(t) + c_2u_2(t)$ produces an output $c_1y_1(t) + c_2y_2(t)$ for all pairs of inputs $u_1(t)$ and $u_2(t)$ and all pairs of constants $c_1$ and $c_2$.

Linear systems can often be represented by linear differential or difference equations.

**EXAMPLE 3.37.**   A system is *linear* if its input-output relationship can be described by the linear differential equation

$$\sum_{i=0}^{n} a_i(t)\frac{d^i y}{dt^i} = \sum_{i=0}^{m} b_i(t)\frac{d^i u}{dt^i} \tag{3.39}$$

where $y = y(t)$ is the system output and $u = u(t)$ is the system input.

**EXAMPLE 3.38.**   A system is linear if its input-output relationship can be described by the **convolution integral**

$$y(t) = \int_{-\infty}^{\infty} w(t,\tau)u(\tau)\,d\tau \tag{3.40}$$

where $w(t,\tau)$ is the **weighting function**, which embodies the internal physical properties of the system, $y(t)$ is the output, and $u(t)$ is the input.

The relationship between the systems of Examples 3.37 and 3.38 is discussed in Section 3.10. The concept of linearity is often expressed by the *principle of superposition*.

**Principle of Superposition:**   The response $y(t)$ of a linear system due to several inputs $u_1(t), u_2(t), \ldots, u_n(t)$ acting simultaneously is equal to the sum of the responses of each input acting alone, when all initial conditions in the system are zero. That is, if $y_i(t)$ is the response due to the input $u_i(t)$, then

$$y(t) = \sum_{i=1}^{n} y_i(t)$$

**EXAMPLE 3.39.**   A linear system is described by the linear algebraic equation

$$y(t) = 2u_1(t) + u_2(t)$$

where $u_1(t) = t$ and $u_2(t) = t^2$ are inputs, and $y(t)$ is the output. When $u_1(t) = t$ and $u_2(t) = 0$, then $y(t) = y_1(t) = 2t$. When $u_1(t) = 0$ and $u_2(t) = t^2$, then $y(t) = y_2(t) = t^2$. The total output resulting from $u_1(t) = t$ and $u_2(t) = t^2$ is then equal to

$$y(t) = y_1(t) + y_2(t) = 2t + t^2$$

The principle of superposition follows directly from the definition of linearity (Definition 3.21). Any system which satisfies the principle of superposition is linear.

## 3.19   CAUSALITY AND PHYSICALLY REALIZABLE SYSTEMS

The properties of a physical system restrict the form of its output. This restriction is embodied in the concept of *causality*.

**Definition 3.22:**    A system in which time is the independent variable is called **causal** if the output depends only on the present and past values of the input. That is, if $y(t)$ is the output, then $y(t)$ depends only on the input $u(\tau)$ for values of $\tau \le t$.

The implication of Definition 3.22 is that a *causal* system is one which cannot anticipate what its future input will be. Accordingly, causal systems are sometimes called **physically realizable** systems. An important consequence of causality (physical realizability) is that the weighting function $w(t, \tau)$ of a causal linear continuous system is identically zero for $\tau > t$; that is, future values of the input are weighted zero. For causal discrete systems, the weighting sequence $w(k - j) \equiv 0$ for $j > k$.

# Solved Problems

## SYSTEM EQUATIONS

**3.1.**    Faraday's law states that the voltage $v$ induced between the terminals of an inductor is equal to the time rate of change of flux linkages. (A flux linkage is defined as one line of magnetic flux linking one turn of the winding of the inductor.) Suppose it is experimentally determined that the number of flux linkages $\lambda$ is related to the current $i$ in the inductor as shown in Fig. 3-5. The curve is approximately a straight line for $-I_0 \le i \le I_0$. Determine a differential equation, valid for $-I_0 \le i \le I_0$, which relates the induced voltage $v$ and current $i$.

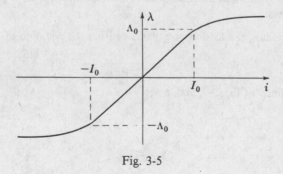

Fig. 3-5

Faraday's law can be written as $v = d\lambda/dt$. It is seen from the graph that

$$\lambda = \left( \frac{\Lambda_0}{I_0} \right) i = Li \qquad -I_0 \le i \le I_0$$

where $L \equiv \Lambda_0/I_0$ is called the *inductance* of the inductor. The equation relating $v$ and $i$ is obtained by substituting $Li$ for $\lambda$:

$$v = \frac{d\lambda}{dt} = \frac{d}{dt}(Li) = L\frac{di}{dt} \qquad \text{where} \quad -I_0 \le i \le I_0$$

**3.2.** Determine a differential equation relating the voltage $v(t)$ and the current $i(t)$ for $t \geq 0$ for the electrical network given in Fig. 3-6. Assume the capacitor is uncharged at $t = 0$, the current $i$ is zero at $t = 0$, and the switch $S$ closes at $t = 0$.

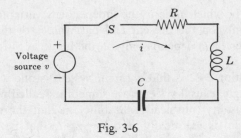

Fig. 3-6

By Kirchhoff's voltage law, the applied voltage $v(t)$ is equal to the sum of the voltage drops $v_R$, $v_L$, and $v_C$ across the resistor $R$, the inductor $L$, and the capacitor $C$, respectively. Thus

$$v = v_R + v_L + v_C = Ri + L\frac{di}{dt} + \frac{1}{C}\int_0^t i(\tau)\, d\tau$$

To eliminate the integral, both sides of the equation are differentiated with respect to time, resulting in the desired differential equation:

$$L\frac{d^2i}{dt^2} + R\frac{di}{dt} + \frac{i}{C} = \frac{dv}{dt}$$

**3.3.** Kepler's first two laws of planetary motion state that:

1. The orbit of a planet is an ellipse with the sun at a focus of the ellipse.

2. The radius vector drawn from the sun to a planet sweeps over equal areas in equal times.

Find a pair of differential equations that describes the motion of a planet about the sun, using Kepler's first two laws.

From Kepler's first law, the motion of a planet satisfies the equation of an ellipse:

$$r = \frac{p}{1 + e\cos\theta}$$

where $r$ and $\theta$ are defined in Fig. 3-7, and $p \equiv b^2/a = a(1 - e^2)$.

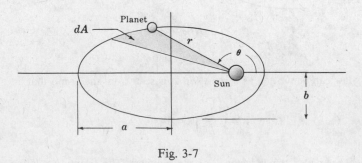

Fig. 3-7

In an infinitesimal time $dt$ the angle $\theta$ increases by an amount $d\theta$. The area swept out by $r$ over the period $dt$ is therefore equal to $dA = \frac{1}{2}r^2\, d\theta$. The rate at which the area is swept out by $r$ is constant

(Kepler's second law). Hence

$$\frac{dA}{dt} = \frac{1}{2}r^2\frac{d\theta}{dt} = \text{constant} \qquad \text{or} \qquad r^2\frac{d\theta}{dt} = k$$

The first differential equation is obtained by differentiating this result with respect to time:

$$2r\frac{dr}{dt}\frac{d\theta}{dt} + r^2\frac{d^2\theta}{dt^2} = 0 \qquad \text{or} \qquad 2\frac{dr}{dt}\frac{d\theta}{dt} + r\frac{d^2\theta}{dt^2} = 0$$

The second equation is obtained by differentiating the equation of the ellipse:

$$\frac{dr}{dt} = \left[\frac{pe\sin\theta}{(1 + e\cos\theta)^2}\right]\frac{d\theta}{dt}$$

Using the results that $d\theta/dt = k/r^2$ and $(1 + e\cos\theta) = p/r$, $dr/dt$ can be rewritten as

$$\frac{dr}{dt} = \frac{ek}{p}\sin\theta$$

Differentiating again and replacing $r^2(d\theta/dt)$ with $k$ yields

$$\frac{d^2r}{dt^2} = \left(\frac{e}{p}\right)\left(\frac{k^2}{r^2}\right)\cos\theta$$

But $\cos\theta = (1/e)[p/r - 1]$. Hence

$$\frac{d^2r}{dt^2} = \frac{k^2}{pr^2}\left[\frac{p}{r} - 1\right] = \frac{k^2}{r^3} - \frac{k^2}{pr^2}$$

Substituting $r(d\theta/dt)^2$ for $k^2/r^3$, we obtain the required second differential equation:

$$\frac{d^2r}{dt^2} - r\left(\frac{d\theta}{dt}\right)^2 + \frac{k^2}{pr^2} = 0 \qquad \text{or} \qquad \frac{d^2r}{dt^2} - r\left(\frac{d\theta}{dt}\right)^2 = -\frac{k^2}{pr^2}$$

**3.4.** A mathematical model for a feature of nervous system organization called *lateral inhibition* has been produced as a result of the work of several authors [2, 3, 4]. Lateral inhibitory phenomena can be simply described as inhibitory electrical interaction among laterally spaced, neighboring neurons (nerve cells). Each neuron in this model has a response $c$, measured by the frequency of discharge of pulses in its axon (the connection "cable" or "wire"). The response is determined by an excitation $r$ supplied by an external stimulus, and is diminished by whatever inhibitory influences are acting on the neurons as a result of the activity of neighboring neurons. In a system of $n$ neurons, the steady state response of the $k$th neuron is given by

$$c_k = r_k - \sum_{i=1}^{n} a_{k-i}c_i$$

where the constant $a_{k-i}$ is the inhibitory coefficient of the action of neuron $i$ on $k$. It depends only on the separation of the $k$th and $i$th neurons, and can be interpreted as a *spatial weighting function*. In addition, $a_m = a_{-m}$ (symmetrical spatial interaction).

(*a*)    If the effect of neuron $i$ on $k$ is not immediately felt, but exhibits a small time lag $\Delta t$, how should this model be modified?

(*b*)    If the input $r_k(t)$ is determined only by the output $c_k$, $\Delta t$ seconds prior to $t$ $[r_k(t) = c_k(t - \Delta t)]$, determine an approximate differential equation for the system of part (*a*).

(*a*)    The equation becomes

$$c_k(t) = r_k(t) - \sum_{i=1}^{n} a_{k-i}c_i(t - \Delta t)$$

(*b*)  Substituting $c_k(t - \Delta t)$ for $r_k(t)$,

$$c_k(t) - c_k(t - \Delta t) = -\sum_{i=1}^{n} a_{k-i} c_i(t - \Delta t)$$

Dividing both sides by $\Delta t$,

$$\frac{c_k(t) - c_k(t - \Delta t)}{\Delta t} = -\sum_{i=1}^{n} \left( \frac{a_{k-i}}{\Delta t} \right) c_i(t - \Delta t)$$

The left-hand side is approximately equal to $dc_k/dt$ for small $\Delta t$. If we additionally assume that $c_i(t - \Delta t) \cong c_i(t)$ for small $\Delta t$, then we get the approximate differential equation

$$\frac{dc_k}{dt} + \sum_{i=1}^{n} \left( \frac{a_{k-i}}{\Delta t} \right) c_i(t) = 0$$

**3.5.**  Determine a mathematical equation describing the sampled-data output of the ideal sampler described in Definition 2.12 and Example 2.8.

A convenient representation of the output of an ideal sampler is based on an extension of the concept of the unit impulse function $\delta(t)$ into an **impulse train**, defined for $t \geq 0$ as the function

$$m_{\mathrm{IT}}(t) = \delta(t) + \delta(t - t_1) + \delta(t - t_2) + \cdots = \sum_{k=0}^{\infty} \delta(t - t_k)$$

where $t_0 = 0$ and $t_{k+1} > t_k$. The sampled signal $u^*(t)$ is then given by

$$u^*(t) = u(t) m_{\mathrm{IT}}(t) = u(t) \sum_{k=0}^{\infty} \delta(t - t_k)$$

The utility of this representation is developed beginning in Chapter 4, following the introduction of transform methods.

**3.6.**  Show how the simple $R$-$C$ network given in Fig. 3-8 can be used to approximate the sample and (zero-order) hold function described in Example 2.9.

This system element operates as follows. When the sampling switch $S$ is closed, the capacitor $C$ is charged through the resistor $R$, and the voltage across $C$ approaches the input $u(t)$. When $S$ is opened, the capacitor cannot release its charge, because the current (charge) has nowhere to dissipate, so it *holds* its voltage until the next time $S$ is closed. If we describe the opening and closing of the switch by the simple

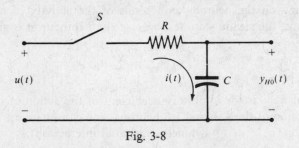

Fig. 3-8

function:

$$m_S(t) = \begin{cases} 1 & \text{if } S \text{ is closed} \\ 0 & \text{if } S \text{ is open} \end{cases}$$

we can say the current through $R$ and $C$ is *modulated* by $m_S(t)$. In these terms, we can write

$$i(t) = m_S(t) \left( \frac{u(t) - y_{H0}(t)}{R} \right)$$

and, since $i = C\,dy_{H0}/dt$, the differential equation for this circuit is

$$\frac{dy_{H0}}{dt} = \left(\frac{u - y_{H0}}{RC}\right) m_S(t)$$

We note that this is a *time-varying* differential equation, due to the multiplicative function $m_S(t)$ on the right-hand side. Also, as $RC$ becomes smaller, that is, $1/RC$ becomes larger, $dy_{H0}/dt$ becomes larger and the capacitor charges faster. Thus a smaller $RC$ in this circuit creates a better approximation of the sample and hold function.

**3.7.** If the sampler in the previous problem is ideal, and the sampling rate is uniform, with period $T$, what is the differential equation?

The ideal sampler impulse train modulating function $m_{IT}(t)$ was defined in Problem 3.5. Thus the differential equation of the sample and hold becomes

$$\frac{dy_{H0}}{dt} = \left(\frac{u - y_{H0}}{RC}\right) \sum_{k=0}^{\infty} \delta(t - kT)$$

In this idealization, impulses replace current pulses.

## CLASSIFICATIONS OF DIFFERENTIAL EQUATIONS

**3.8.** Classify the following differential equations according to whether they are ordinary or partial. Indicate the dependent and independent variables.

(a)    $\dfrac{dx}{dt} + \dfrac{dy}{dt} + x + y = 0$        $x = x(t)$        $y = y(t)$

(b)    $\dfrac{\partial f}{\partial x} + \dfrac{\partial f}{\partial y} + x + y = 0$        $f = f(x, y)$

(c)    $\dfrac{d}{dt}\left[\dfrac{\partial f}{\partial x}\right] = 0$        $f = x^2 + \dfrac{dx}{dt}$

(d)    $\dfrac{df}{dx} = x$        $f = y^2(x) + \dfrac{dy}{dx}$

(a)    Ordinary; independent variable $t$; dependent variables $x$ and $y$.

(b)    Partial; independent variables $x$ and $y$; dependent variable $f$.

(c)    Since $\partial f/\partial x = 2x$, then $(d/dt)[\partial f/\partial x] = 2(dx/dt) = 0$, which is an ordinary differential equation; independent variable $t$; dependent variable $x$.

(d)    $df/dx = 2y(dy/dx) + d^2y/dx^2 = x$, which is an ordinary differential equation; independent variable $x$; dependent variable $y$.

**3.9.** Classify the following linear differential equations according to whether they are time-variable or time-invariant. Indicate any time-variable terms.

(a)    $\dfrac{d^2y}{dt^2} + 2y = 0$        (c)    $\left(\dfrac{1}{t+1}\right)\dfrac{d^2y}{dt^2} + \left(\dfrac{1}{t+1}\right)y = 0$

(b)    $\dfrac{d}{dt}(t^2 y) = 0$        (d)    $\dfrac{d^2y}{dt^2} + (\cos t)y = 0$

(a)    Time-invariant.

(b)    $(d/dt)(t^2 y) = 2ty + t^2(dy/dt) = 0$. Dividing through by $t$, $t(dy/dt) + 2y = 0$ which is time-variable. The time-variable term is $t(dy/dt)$.

(c)    Multiplying through by $t + 1$, we obtain $d^2y/dt^2 + y = 0$ which is time-invariant.

(d)    Time-variable. The time-variable term is $(\cos t)y$.

**3.10.** Classify the following differential equations according to whether they are linear or nonlinear. Indicate the dependent and independent variables and any nonlinear terms.

$(a)$  $t\dfrac{dy}{dt} + y = 0$    $y = y(t)$        $(d)$  $(\cos t)\dfrac{d^2y}{dt^2} + (\sin 2t)y = 0$    $y = y(t)$

$(b)$  $y\dfrac{dy}{dt} + y = 0$    $y = y(t)$        $(e)$  $(\cos y)\dfrac{d^2y}{dt^2} + \sin 2y = 0$    $y = y(t)$

$(c)$  $\dfrac{dy}{dt} + y^2 = 0$    $y = y(t)$        $(f)$  $(\cos x)\dfrac{d^2y}{dt^2} + \sin 2x = 0$    $y = y(t),\ x = x(t)$

$(a)$  Linear; independent variable $t$; dependent variable $y$.

$(b)$  Nonlinear; independent variable $t$; dependent variable $y$; nonlinear term $y(dy/dt)$.

$(c)$  Nonlinear; independent variable $t$; dependent variable $y$; nonlinear term $y^2$.

$(d)$  Linear; independent variable $t$; dependent variable $y$.

$(e)$  Nonlinear; independent variable $t$; dependent variable $y$; nonlinear terms $(\cos y)\,d^2y/dt^2$ and $\sin 2y$.

$(f)$  Nonlinear; independent variable $t$; dependent variables $x$ and $y$; nonlinear terms $(\cos x)\,d^2y/dt^2$ and $\sin 2x$.

**3.11.** Why are all transcendental functions *not* of first degree?

Transcendental functions, such as the logarithmic, trigonometric, and hyperbolic functions and their corresponding inverses, are not of first degree because they are either defined by or can be written as infinite series. Hence their degree is in general equal to *infinity*. For example,

$$\sin x = \sum_{n=1}^{\infty} (-1)^{n-1} \frac{x^{2n-1}}{(2n-1)!} = x - \frac{x^3}{3!} + \frac{x^5}{5!} - \cdots$$

where the first term is first degree, the second is third degree, and so on.

## THE CHARACTERISTIC EQUATION

**3.12.** Find the characteristic polynomial and characteristic equation for each system:

$(a)$  $\dfrac{d^4y}{dt^4} + 9\dfrac{d^2y}{dt^2} + 7y = u$        $(b)$  $\dfrac{d^4y}{dt^4} + 9\dfrac{d^2y}{dt^2} + 7y = \sin u$

$(a)$  Putting $D^n \equiv d^n/dt^n$ for $n = 2$ and $n = 4$, the characteristic polynomial is $D^4 + 9D^2 + 7$; and the characteristic equation is $D^4 + 9D^2 + 7 = 0$.

$(b)$  Although the equation given in part $(b)$ is nonlinear by Definition 3.8 (the term $\sin u$ is not first degree in $u$), we can treat it as a linear equation if we arbitrarily put $\sin u = x$, and treat $x$ as a second dependent variable representing the input. In this case, part $(b)$ has the same answer as part $(a)$.

**3.13.** Determine the solution of the characteristic equation of the preceding problem.

Let $D^2 \equiv E$. Then $D^4 = E^2$, and the characteristic equation becomes quadratic:

$$E^2 + 9E + 7 = 0 \qquad E = -\frac{9 \pm \sqrt{53}}{2} \qquad \text{and} \qquad D = \pm\sqrt{\frac{-9 \pm \sqrt{53}}{2}}$$

## LINEAR INDEPENDENCE AND FUNDAMENTAL SETS

**3.14.** Show that a sufficient condition for a set of $n$ functions $f_1, f_2, \ldots, f_n$ to be linearly independent is that the determinant

$$\begin{vmatrix} f_1 & f_2 & \cdots & f_n \\ \dfrac{df_1}{dt} & \dfrac{df_2}{dt} & \cdots & \dfrac{df_n}{dt} \\ \cdots\cdots\cdots\cdots\cdots\cdots\cdots\cdots\cdots\cdots \\ \dfrac{d^{n-1}f_1}{dt^{n-1}} & \dfrac{d^{n-1}f_2}{dt^{n-1}} & \cdots & \dfrac{d^{n-1}f_n}{dt^{n-1}} \end{vmatrix}$$

be nonzero. This determinant is called the **Wronskian** of the functions $f_1, f_2, \ldots, f_n$.

Assuming the $f_i$ are differentiable at least $n-1$ times, let $n-1$ derivatives of

$$c_1 f_1 + c_2 f_2 + \cdots + c_n f_n = 0$$

be formed as follows, where the $c_i$ are unknown constants:

$$c_1 \frac{df_1}{dt} + c_2 \frac{df_2}{dt} + \cdots + c_n \frac{df_n}{dt} = 0$$

$$\cdots\cdots\cdots\cdots\cdots\cdots\cdots\cdots\cdots$$

$$c_1 \frac{d^{n-1}f_1}{dt^{n-1}} + c_2 \frac{d^{n-1}f_2}{dt^{n-1}} + \cdots + c_n \frac{d^{n-1}f_n}{dt^{n-1}} = 0$$

These equations may be considered as $n$ simultaneous linear homogeneous equations in the $n$ unknown constants $c_1, c_2, \ldots, c_n$, with coefficients given by the elements of the Wronskian. It is well known that these equations have a nonzero solution for $c_1, c_2, \ldots, c_n$ (i.e., not all $c_i$ are equal to zero) if and only if the determinant of the coefficients (the Wronskian) is equal to zero. Hence if the Wronskian is nonzero, then the only solution for $c_1, c_2, \ldots, c_n$ is the degenerate one, $c_1 = c_2 = \cdots = c_n = 0$. Clearly, this is equivalent to saying that if the Wronskian is nonzero the functions $f_1, f_2, \ldots, f_n$ are linearly independent, since the only solution to $c_1 f_1 + c_2 f_2 + \cdots + c_n f_n = 0$ is then $c_1 = c_2 = c_3 = \cdots = c_n = 0$. Hence a sufficient condition for the linear independence of $f_1, f_2, \ldots, f_n$ is that the Wronskian be nonzero. This condition is not *necessary*; that is, there exist sets of linearly independent functions for which the Wronskian *is* zero.

**3.15.** Show that the function $1, t, t^2$ are linearly independent.

The Wronskian of these three functions (see Problem 3.14) is

$$\begin{vmatrix} 1 & t & t^2 \\ 0 & 1 & 2t \\ 0 & 0 & 2 \end{vmatrix} = 2$$

Since the Wronskian is nonzero, the functions are linearly independent.

**3.16.** Determine a fundamental set for the differential equations:

(a) $\dfrac{d^3y}{dt^3} + 5\dfrac{d^2y}{dt^2} + 8\dfrac{dy}{dt} + 4y = u$    (b) $\dfrac{d^3y}{dt^3} + 4\dfrac{d^2y}{dt^2} + 6\dfrac{dy}{dt} + 4y = u$

(a) The characteristic polynomial is $D^3 + 5D^2 + 8D + 4$, which can be written in factored form as $(D+2)(D+2)(D+1)$. Corresponding to the root $D_1 = -1$ there is a solution $e^{-t}$, and corresponding to the repeated root $D_2 = D_3 = -2$ are the two solutions $e^{-2t}$ and $te^{-2t}$. The three solutions constitute a fundamental set.

(b) The characteristic polynomial is $D^3 + 4D^2 + 6D + 4$, which can be written in factored form as $(D + 1 + j)(D + 1 - j)(D + 2)$.

   A fundamental set is then $e^{(-1-j)t}$, $e^{(-1+j)t}$, and $e^{-2t}$.

**3.17.** For the differential equations of Problem 3.16, find fundamental sets different from those found in Problem 3.16.

(a) Choose any $3 \times 3$ nonzero determinant, say

$$\begin{vmatrix} 1 & 2 & -1 \\ -3 & 2 & 0 \\ 1 & 3 & -2 \end{vmatrix} = -5$$

Using the elements of the first row as coefficients $a_{1i}$ for the fundamental set $e^{-t}, e^{-2t}, te^{-2t}$ found in Problem 3.16, form

$$z_1 = e^{-t} + 2e^{-2t} - te^{-2t}$$

Using the second row, form

$$z_2 = -3e^{-t} + 2e^{-2t}$$

From the third row, form

$$z_3 = e^{-t} + 3e^{-2t} - 2te^{-2t}$$

The functions $z_1$, $z_2$, and $z_3$ constitute a fundamental set.

(b) For this equation generate the second fundamental set by letting

$$z_1 = e^{-2t}$$

$$z_2 = \frac{1}{2}e^{(-1+j)t} + \frac{1}{2}e^{(-1-j)t} = e^{-t}\left(\frac{e^{-jt} + e^{jt}}{2}\right)$$

$$= e^{-t}\left(\frac{\cos t - j\sin t + \cos t + j\sin t}{2}\right) = e^{-t}\cos t$$

$$z_3 = \frac{1}{2j}e^{(-1+j)t} - \frac{1}{2j}e^{(-1-j)t} = e^{-t}\left(\frac{e^{t} - e^{-jt}}{2j}\right)$$

$$= e^{-t}\left(\frac{\cos t + j\sin t - \cos t + j\sin t}{2j}\right) = e^{-t}\sin t$$

The coefficient determinant in this case is

$$\begin{vmatrix} 1 & 0 & 0 \\ 0 & \dfrac{1}{2} & \dfrac{1}{2} \\ 0 & \dfrac{1}{2j} & -\dfrac{1}{2j} \end{vmatrix} = -\frac{1}{2j}$$

## SOLUTION OF LINEAR CONSTANT-COEFFICIENT ORDINARY DIFFERENTIAL EQUATIONS

**3.18.** Show that any free response $y_a(t) = \sum_{k=1}^{n} c_k y_k(t)$ satisfies $\sum_{i=0}^{n} a_i (d^i y/dt^i) = 0$.

By the definition of a fundamental set, $y_k(t)$, $k = 1, 2, \ldots, n$, satisfies $\sum_{i=0}^{n} a_i (d^i y_k/dt^i) = 0$. Substituting $\sum_{k=1}^{n} c_k y_k(t)$ into this differential equation yields

$$\sum_{i=0}^{n} a_i \frac{d^i}{dt^i} \left[ \sum_{k=1}^{n} c_k y_k(t) \right] = \sum_{i=0}^{n} \sum_{k=1}^{n} a_i \frac{d^i}{dt^i} (c_k y_k(t)) = \sum_{k=1}^{n} c_k \left[ \sum_{i=0}^{n} a_i \frac{d^i y_k(t)}{dt^i} \right] = 0$$

The last equality is obtained because the term in the brackets is zero for all $k$.

**3.19.** Show that the forced response given by Equation (3.14)

$$y_b(t) = \int_0^t w(t - \tau) \left[ \sum_{i=0}^{m} b_i \frac{d^i u(\tau)}{d\tau^i} \right] d\tau$$

satisfies the differential equation

$$\sum_{i=0}^{n} a_i \frac{d^i y}{dt^i} = \sum_{i=0}^{m} b_i \frac{d^i u}{dt^i}$$

For simplification, let $r(t) \equiv \sum_{i=0}^{m} b_i (d^i u/dt^i)$. Then $y_b(t) = \int_0^t w(t - \tau) r(\tau) \, d\tau$ and

$$\frac{dy_b}{dt} = \int_0^t \frac{\partial w(t - \tau)}{\partial t} r(\tau) \, d\tau + w(t - \tau) r(\tau) \Big|_{\tau = t} = \int_0^t \frac{\partial w(t - \tau)}{\partial t} r(\tau) \, d\tau + 0 \cdot r(t)$$

Similarly,

$$\frac{d^2 y_b}{dt^2} = \int_0^t \frac{\partial^2 w(t - \tau)}{\partial t^2} r(\tau) \, d\tau, \ldots, \frac{d^{n-1} y_b}{dt^{n-1}} = \int_0^t \frac{\partial^{n-1} w(t - \tau)}{\partial t^{n-1}} r(\tau) \, d\tau$$

since, by Equation (3.16),

$$\frac{\partial^i w(t - \tau)}{\partial t^i} \Big|_{\tau = t} = \frac{d^i w(t)}{dt^i} \Big|_{t=0} = 0 \quad \text{for} \quad i = 0, 1, 2, \ldots, n - 2$$

The $n$th derivative is

$$\frac{d^n y_b}{dt^n} = \int_0^t \frac{\partial^n w(t - \tau)}{\partial t^n} r(\tau) \, d\tau + \frac{\partial^{n-1} w(t - \tau)}{\partial t^{n-1}} \Big|_{\tau = t} \cdot r(t) = \int_0^t \frac{\partial^n w(t - \tau)}{\partial t^n} r(\tau) \, d\tau + r(t)$$

since, by Equation (3.16),

$$\frac{\partial^{n-1} w(t - \tau)}{\partial t^{n-1}} \Big|_{\tau = t} = \frac{d^{n-1} w(t)}{dt^{n-1}} \Big|_{t=0} = 1$$

The summation of the $n$ derivatives is

$$\sum_{i=0}^{n} a_i \frac{d^i y_b}{dt^i} = \int_0^t \left[ \sum_{i=0}^{n} a_i \frac{\partial^i w(t - \tau)}{\partial t^i} \right] r(\tau) \, d\tau + r(t)$$

Finally, making the change of variables $t - \tau = \theta$ in the bracketed term yields

$$\sum_{i=0}^{n} a_i \frac{\partial^i w(\theta)}{\partial \theta^i} = \sum_{i=0}^{n} a_i \frac{d^i w(\theta)}{d\theta^i} = 0$$

because $w(\theta)$ is a free response (see Section 3.10 and Problem 3.18). Hence

$$\sum_{i=0}^{n} a_i \frac{d^i y_b}{dt^i} = r(t) \equiv \sum_{i=0}^{m} b_i \frac{d^i u}{dt^i}$$

**3.20.** Find the free response of the differential equation

$$\frac{d^3y}{dt^3} + 4\frac{d^2y}{dt^2} + 6\frac{dy}{dt} + 4y = u$$

with initial conditions $y(0) = 1$, $(dy/dt)|_{t=0} = 0$, and $(d^2y/dt^2)|_{t=0} = -1$.

From the results of Problems 3.16 and 3.17, a fundamental set for this equation is $e^{-2t}$, $e^{-t}\cos t$, $e^{-t}\sin t$. Hence the free response can be written as

$$y_a(t) = c_1 e^{-2t} + c_2 e^{-t}\cos t + c_3 e^{-t}\sin t$$

The initial conditions provide the following set of algebraic equations for $c_1, c_2, c_3$:

$$y_a(0) = c_1 + c_2 = 1 \qquad \frac{dy_a}{dt}\bigg|_{t=0} = -2c_1 - c_2 + c_3 = 0 \qquad \frac{d^2y_a}{dt^2}\bigg|_{t=0} = 4c_1 - 2c_3 = -1$$

from which $c_1 = \frac{1}{2}$, $c_2 = \frac{1}{2}$, $c_3 = \frac{3}{2}$. Therefore the free response is

$$y_a(t) = \frac{1}{2}e^{-2t} + \frac{1}{2}e^{-t}\cos t + \frac{3}{2}e^{-t}\sin t$$

**3.21.** Find the weighting function of the differential equation

$$\frac{d^2y}{dt^2} + 4\frac{dy}{dt} + 4y = 3\frac{du}{dt} + 2u$$

The characteristic equation is $D^2 + 4D + 4 = (D+2)^2 = 0$ with the repeated root $D = -2$. A fundamental set is therefore given by $e^{-2t}$, $te^{-2t}$, and the weighting function has the form

$$w(t) = c_1 e^{-2t} + c_2 te^{-2t}$$

with the initial conditions

$$w(0) = \left[c_1 e^{-2t} + c_2 te^{-2t}\right]\Big|_{t=0} = c_1 = 0 \qquad \frac{dw}{dt}\bigg|_{t=0} = \left[-2c_1 e^{-2t} + c_2 e^{-2t} - 2c_2 te^{-2t}\right]\Big|_{t=0} = c_2 = 1$$

Thus $w(t) = te^{-2t}$.

**3.22.** Find the forced response of the differential equation (Problem 3.21):

$$\frac{d^2y}{dt^2} + 4\frac{dy}{dt} + 4y = 3\frac{du}{dt} + 2u$$

where $u(t) = e^{-3t}$, $t \geq 0$.

The forced response is given by Equation (*3.14*) as

$$y_b(t) = \int_0^t w(t-\tau)\left[3\frac{du}{d\tau} + 2u\right]d\tau = 3\int_0^t w(t-\tau)\frac{du}{d\tau}d\tau + 2\int_0^t w(t-\tau)u\,d\tau$$

Integrating the first integral by parts,

$$\int_0^t w(t-\tau)\frac{du}{d\tau}d\tau = w(t-\tau)u(\tau)\big|_0^t - \int_0^t \frac{\partial w(t-\tau)}{\partial\tau}u\,d\tau$$

$$= w(0)u(t) - w(t)u(0) - \int_0^t \frac{\partial w(t-\tau)}{\partial\tau}u\,d\tau$$

But $w(0) = 0$; hence the forced response can be written as

$$y_b(t) = \int_0^t \left[-3\frac{\partial w(t-\tau)}{\partial\tau} + 2w(t-\tau)\right]u(\tau)\,d\tau - 3w(t)u(0)$$

From Problem 3.21, $w(t - \tau) = (t - \tau)e^{-2(t-\tau)}$; hence

$$\left[ -3\frac{\partial w(t - \tau)}{\partial \tau} + 2w(t - \tau) \right] = 3e^{-2(t-\tau)} - 4(t - \tau)e^{-2(t-\tau)}$$

and the forced response is

$$y_b(t) = 3e^{-2t}\int_0^t e^{2\tau}e^{-3\tau}\,d\tau - 4te^{-2t}\int_0^t e^{2\tau}e^{-3\tau}\,d\tau + 4e^{-2t}\int_0^t \tau e^{2\tau}e^{-3\tau}\,d\tau - 3te^{-2t}$$

$$= 7[e^{-2t} - e^{-3t} - te^{-2t}]$$

**3.23.** Find the output $y$ of a system described by the differential equation

$$\frac{d^2y}{dt^2} + 3\frac{dy}{dt} + 2y = 1 + t$$

with initial conditions $y(0) = 0$ and $(dy/dt)|_{t=0} = 1$.

Let $u_1 \equiv 1$, $u_2 \equiv t$. The response $y$ due to $u_1$ alone was determined in Example 3.27 as $y_1 = \frac{1}{2}(1 - e^{-2t})$. The free response $y_a$ for the differential equation was found in Example 3.24 to be $y_a = e^{-t} - e^{-2t}$. The forced response due to $u_2$ is given by Equation (3.14). Using the weighting function determined in Example 3.25, the forced response due to $u_2$ is

$$y_2 = \int_0^t w(t - \tau)u_2(\tau)\,d\tau = \int_0^t [e^{-(t-\tau)} - e^{-2(t-\tau)}]\tau\,d\tau$$

$$= e^{-t}\int_0^t \tau e^\tau\,d\tau - e^{-2t}\int_0^t \tau e^{2\tau}\,d\tau = \frac{1}{4}[4e^{-t} - e^{-2t} + 2t - 3]$$

Thus the forced response is

$$y_b = y_1 + y_2 = \frac{1}{4}[4e^{-t} - 3e^{-2t} + 2t - 1]$$

and the total response is

$$y = y_a + y_b = \frac{1}{4}[8e^{-t} - 7e^{-2t} + 2t - 1]$$

**3.24.** Find the transient and steady state responses of a system described by the differential equation

$$\frac{d^2y}{dt^2} + 3\frac{dy}{dt} + 2y = 1 + t$$

with the initial conditions $y(0) = 0$ and $(dy/dt)|_{t=0} = 1$.

The total response for this equation was determined in Problem 3.23 as

$$y = \frac{1}{4}[8e^{-t} - 7e^{-2t} + 2t - 1]$$

Since $\lim_{t \to \infty}[\frac{1}{4}(8e^{-t} - 7e^{-2t})] = 0$, the transient response is $y_T = \frac{1}{4}(8e^{-t} - 7e^{-2t})$. The steady state response is $y_{ss} = \frac{1}{4}(2t - 1)$.

## SINGULARITY FUNCTIONS

**3.25.** Evaluate: (a) $\int_5^8 t^2\delta(t - 6)\,dt$, (b) $\int_0^4 \sin t\delta(t - 7)\,dt$.

(a)   Using the screening property of the unit impulse function, $\int_5^8 t^2\delta(t - 6)\,dt = t^2|_{t=6} = 36$.

(b)   Since the interval of integration $0 \le t \le 4$ does not include the position of the unit impulse function $t = 7$, then $\int_0^4 \sin t\delta(t - 7)\,dt = 0$.

**3.26.** Show that the unit step response $y_1(t)$ of a causal linear system described by the convolution integral

$$y(t) = \int_0^t w(t-\tau)u(\tau)\,d\tau$$

is related to the unit impulse response $y_\delta(t)$ by the equation $y_1(t) = \int_0^t y_\delta(\tau)\,d\tau$.

The unit step response is given by $y_1(t) = \int_0^t w(t-\tau)u(\tau)\,d\tau$, where $\mathbf{1}(t)$ is a unit step function. In Example 3.29 it was shown that $y_\delta(t) = w(t)$. Hence

$$y_1(t) = \int_0^t y_\delta(t-\tau)u(\tau)\,d\tau = \int_0^t y_\delta(t-\tau)\,d\tau$$

Now make the change of variable $\theta = t - \tau$. Then $d\tau = -d\theta$, $\tau = 0$ implies $\theta = t$, $\tau = t$ implies $\theta = 0$, and the integral becomes

$$y_1(t) = -\int_t^0 y_\delta(\theta)\,d\theta = \int_0^t y_\delta(\theta)\,d\theta$$

**3.27.** Show that the unit ramp response $y_r(t)$ of a causal linear system described by the convolution integral (see Problem 3.26) is related to the unit impulse response $y_\delta(t)$ and the unit step response $y_1(t)$ by the equation

$$y_r(t) = \int_0^t y_1(\tau')\,d\tau' = \int_0^t \int_0^{\tau'} y_\delta(\theta)\,d\theta\,d\tau'$$

Proceeding as in Problem 3.26 with $w(t-\tau) = y_\delta(t-\tau)$ and $\tau$ changed to $t-\tau'$, we get

$$y_r(t) = \int_0^t y_\delta(t-\tau)\tau\,d\tau = \int_0^t (t-\tau')\,y_\delta(\tau')\,d\tau' = \int_0^t ty_\delta(\tau')\,d\tau' - \int_0^t \tau' y_\delta(\tau')\,d\tau'$$

From Problem 3.26, the first term can be written as $t\int_0^t y_\delta(\tau')\,d\tau' = ty_1(t)$. The second term can be integrated by parts, yielding

$$\int_0^t \tau' y_\delta(\tau')\,d\tau' = \tau' y_1(\tau')\big|_0^t - \int_0^t y_1(\tau')\,d\tau'$$

where $dy_1(\tau') = y_\delta(\tau')\,d\tau'$. Therefore

$$y_r(t) = ty_1(t) - ty_1(t) + \int_0^t y_1(\tau')\,d\tau' = \int_0^t y_1(\tau')\,d\tau'$$

Again using the result of Problem 3.26, we obtain the required equation.

## SECOND-ORDER SYSTEMS

**3.28.** Show that the weighting function of the second-order differential equation

$$\frac{d^2y}{dt^2} + 2\zeta\omega_n \frac{dy}{dt} + \omega_n^2 y = \omega_n^2 u$$

is given by $w(t) = (1/\omega_d)e^{-\alpha t}\sin\omega_d t$, where $\alpha \equiv \zeta\omega_n$, $\omega_d \equiv \omega_n\sqrt{1-\zeta^2}$, $0 \le \zeta \le 1$.

The characteristic equation

$$D^2 + 2\zeta\omega_n D + \omega_n^2 = 0$$

has the roots

$$D_1 = -\zeta\omega_n + j\omega_n\sqrt{1-\zeta^2} = -\alpha + j\omega_d$$

$$D_2 = -\zeta\omega_n - j\omega_n\sqrt{1-\zeta^2} = -\alpha - j\omega_d$$

One fundamental set is $y_1 = e^{-\alpha t}e^{j\omega_d t}$, $y_2 = e^{-\alpha t}e^{-j\omega_d t}$; and the weighting function can be written as

$$w(t) = c_1 e^{-\alpha t}e^{-j\omega_d t} + c_2 e^{-\alpha t}e^{j\omega_d t}$$

where $c_1$ and $c_2$ are, as yet, unknown coefficients. $w(t)$ can be rewritten as

$$w(t) = e^{-\alpha t} [ c_1 \cos \omega_d t - jc_1 \sin \omega_d t + c_2 \cos \omega_d t + jc_2 \sin \omega_d t ]$$

$$= (c_1 + c_2) e^{-\alpha t} \cos \omega_d t + j(c_2 - c_1) e^{-\alpha t} \sin \omega_d t$$

$$= Ae^{-\alpha t} \cos \omega_d t + Be^{-\alpha t} \sin \omega_d t$$

where $A \equiv c_1 + c_2$ and $B \equiv j(c_2 - c_1)$ are unknown coefficients determined from the initial conditions given by Equation (3.16). That is,

$$w(0) = \left[ Ae^{-\alpha t} \cos \omega_d t + Be^{-\alpha t} \sin \omega_d t \right]\Big|_{t=0} = A = 0$$

and

$$\frac{dw}{dt}\Big|_{t=0} = Be^{-\alpha t} [ \omega_d \cos \omega_d t - \alpha \sin \omega_d t ]\Big|_{t=0} = B\omega_d = 1$$

Hence

$$w(t) = \frac{1}{\omega_d} e^{-\alpha t} \sin \omega_d t$$

**3.29.** Determine the damping ratio $\zeta$, undamped natural frequency $\omega_n$, damped natural frequency $\omega_d$, damping coefficient $\alpha$, and time constant $\tau$ for the following second-order system:

$$2\frac{d^2 y}{dt^2} + 4\frac{dy}{dt} + 8y = 8u$$

Dividing both sides of the equation by 2, $d^2 y/dt^2 + 2(dy/dt) + 4y = 4u$. Comparing the coefficients of this equation with those of Equation (3.22), we obtain $2\zeta\omega_n = 2$ and $\omega_n^2 = 4$ with the solutions $\omega_n = 2$ and $\zeta = \frac{1}{2} = 0.5$. Now $\omega_d = \omega_n\sqrt{1 - \zeta^2} = \sqrt{3}$, $\alpha = \zeta\omega_n = 1$, and $\tau = 1/\alpha = 1$.

**3.30.** The **overshoot** of a second-order system in response to a unit step input is the difference between the maximum value attained by the output and the steady state solution. Determine the overshoot for the system of Problem 3.29 using the normalized family of curves given in Section 3.14.

Since the damping ratio of this system is $\zeta = 0.5$, the normalized curve corresponding to $\zeta = 0.5$ is used. This curve has its maximum value (peak) at $\omega_n t = 3.4$. From Problem 3.29, $\omega_n = 2$; hence the time $t_p$ at which the peak occurs is $t_p = 3.4/\omega_n = 3.4/2 = 1.7$ sec. The value attained at this time is 1.17, and the overshoot is $1.17 - 1.00 = 0.17$.

## STATE VARIABLE REPRESENTATION OF SYSTEMS DESCRIBED BY LINEAR DIFFERENTIAL AND DIFFERENCE EQUATIONS

**3.31.** Put the differential equation

$$\frac{d^2 y}{dt^2} = u$$

with initial conditions $y(0) = 1$ and $(dy/dt)|_{t=0} = -1$, into state variable form. Then develop a solution for the resulting vector-matrix equation in the form of Equation (3.26) and, from this specify the free response and the forced response. Also, for $u(t) = 1$, specify the transient and steady state responses.

Letting $x_1 \equiv y$ and $dx_1/dt = x_2$, the state variable representation is $dx_1/dt = x_2$ with $x_1(0) = 1$, and $dx_2/dt = u$ with $x_2(0) = -1$. The matrices $A$ and $B$ in the general equation form (3.25) are

$$A = \begin{bmatrix} 0 & 1 \\ 0 & 0 \end{bmatrix} \qquad \mathbf{b} = \begin{bmatrix} 0 \\ 1 \end{bmatrix}$$

Since $A^k = 0$ for $k \geq 2$, the transition matrix is

$$e^{At} = I + At = \begin{bmatrix} 1 & t \\ 0 & 1 \end{bmatrix}$$

and the solution of the state variable equation can be written as

$$\begin{bmatrix} x_1(t) \\ x_2(t) \end{bmatrix} = \begin{bmatrix} 1 & t \\ 0 & 1 \end{bmatrix} \begin{bmatrix} 1 \\ -1 \end{bmatrix} + \int_0^t \begin{bmatrix} 1 & (t-\tau) \\ 0 & 1 \end{bmatrix} \begin{bmatrix} 0 \\ u(\tau) \end{bmatrix} d\tau$$

or, after multiplying the matrices in each term,

$$x_1(t) = 1 - t + \int_0^t (t - \tau) u(\tau) \, d\tau$$

$$x_2(t) = -1 + \int_0^t u(\tau) \, d\tau$$

The *free responses* are

$$x_{1a}(t) = 1 - t$$
$$x_{2a}(t) = -1$$

and the *forced responses* are

$$x_{1b}(t) = \int_0^t (t - \tau) u(\tau) \, d\tau$$

$$x_{2b}(t) = \int_0^t u(\tau) \, d\tau$$

For $u(t) = 1$, $x_1(t) = 1 - t + t^2/2$ and $x_2(t) = -1 + t$. The *transient responses* are $x_{1T}(t) = 0$ and $x_{2T}(t) = 0$ and the *steady state responses* are $x_{1ss}(t) = 1 - t + t^2/2$ and $y_{2ss}(t) = -1 + t$.

**3.32.** Show that the weighting sequence of the difference equation $(3.29)$ has the form of Equation $(3.34)$.

The technique used to solve this problem is called *variation of parameters*. It is assumed that the forced response of Equation $(3.29)$ has the form:

$$y_b(k) = \sum_{j=1}^n c_j(k) y_j(k)$$

where $y_1(k), \dots, y_n(k)$ is a fundamental set of solutions and $c_1(k), \dots, c_n(k)$ is a set of unknown time-variable parameters to be determined. Since $y_b(0) = 0$ for any forced response of a difference equation, then $c_1(0) = 0, \dots, c_n(0) = 0$. The parameter $c_j(k+1)$ is written as $c_j(k+1) = c_j(k) + \Delta c_j(k)$. Thus

$$y_b(k+1) = \sum_{j=1}^n c_j(k) y_j(k+1) + \left[ \sum_{j=1}^n \Delta c_j(k) y_j(k+1) \right]$$

The increments $\Delta c_1(k), \dots, \Delta c_n(k)$ are chosen such that the term in the brackets is zero. This process is then repeated for $y_b(k+2)$ so that

$$y_b(k+2) = \sum_{j=1}^n c_j(k) y_j(k+2) + \left[ \sum_{j=1}^n \Delta c_j(k) y_j(k+2) \right]$$

Again the bracketed term is made zero by choice of the increments $\Delta c_1(k), \dots, \Delta c_n(k)$. Similar expressions are generated for $y_b(k+3), y_b(k+4), \dots, y_b(k+n-1)$. Finally,

$$y_b(k+n) = \sum_{j=1}^n c_j(k) y_j(k+n) + \left[ \sum_{j=1}^n \Delta c_j(k) y_j(k+n) \right]$$

In this last expression, the bracketed term is not set to zero. Now the summation in Equation $(3.29)$ is

$$\sum_{i=0}^n a_i y_b(k+i) = \sum_{j=1}^n c_j(k) \sum_{i=0}^n a_i y_j(k+i) + a_n \sum_{j=1}^n \Delta c_j(k) y_j(k+n) = u(k)$$

Since each element of the fundamental set is a free response, then

$$\sum_{i=0}^n a_i y_j(k+i) = 0$$

for each $j$. A set of $n$ linear algebraic equations in $n$ unknowns has thus been generated:

$$\sum_{j=1}^{n} \Delta c_j(k) y_j(k+1) = 0$$

$$\sum_{j=1}^{n} \Delta c_j(k) y_j(k+2) = 0$$

$$\vdots$$

$$\sum_{j=1}^{n} \Delta c_j(k) y_j(k+n) = \frac{u(k)}{a_n}$$

Now $\Delta c_j(k)$ can be written as

$$\Delta c_j(k) = \frac{M_j(k)}{M(k)} \frac{u(k)}{a_n}$$

where $M(k)$ is the determinant

$$M(k) = \begin{vmatrix} y_1(k+1) & y_2(k+1) & \cdots & y_n(k+1) \\ y_1(k+2) & y_2(k+2) & \cdots & y_n(k+2) \\ \vdots & \vdots & \ddots & \vdots \\ y_1(k+n) & y_2(k+n) & \cdots & y_n(k+n) \end{vmatrix}$$

$M_j(k)$ is the cofactor of the last element in the $j$th column of this determinant. The parameters $c_1(k), \ldots, c_n(k)$ are thus given by

$$c_j(k) = \sum_{l=0}^{k-1} \Delta c_j(l) = \sum_{l=0}^{k-1} \frac{M_j(l)}{M(l)} \frac{u(l)}{a_n}$$

The forced response then becomes

$$y_b(k) = \sum_{j=1}^{n} \sum_{l=0}^{k-1} \frac{M_j(l)}{M(l)} \frac{u(l)}{a_n} y_j(k)$$

$$= \sum_{l=0}^{k-1} \left[ \sum_{j=1}^{n} \frac{M_j(l)}{a_n M(l)} y_j(k) \right] u(l)$$

This last equation is in the form of a convolution sum with weighting sequence

$$w(k-l) = \sum_{j=1}^{n} \frac{M_j(l)}{a_n M(l)} y_j(k)$$

## LINEARITY AND SUPERPOSITION

**3.33.** Using the definition of linearity, Definition 3.21, show that any differential equation of the form:

$$\sum_{i=0}^{n} a_i(t) \frac{d^i y}{dt^i} = u$$

where $y$ is the output and $u$ is the input, is linear.

Let $u_1$ and $u_2$ be two arbitrary inputs, and let $y_1$ and $y_2$ be the corresponding outputs. Then, with all initial conditions equal to zero,

$$\sum_{i=0}^{n} a_i(t) \frac{d^i y_1}{dt^i} = u_1 \qquad \text{and} \qquad \sum_{i=0}^{n} a_i(t) \frac{d^i y_2}{dt^i} = u_2$$

Now form

$$c_1 u_1 + c_2 u_2 = c_1 \left[ \sum_{i=0}^{n} a_i(t) \frac{d^i y_1}{dt^i} \right] + c_2 \left[ \sum_{i=0}^{n} a_i(t) \frac{d^i y_2}{dt^i} \right]$$

$$= \sum_{i=0}^{n} a_i(t) \frac{d^i(c_1 y_1)}{dt^i} + \sum_{i=0}^{n} a_i(t) \frac{d^i(c_2 y_2)}{dt^i}$$

$$= \sum_{i=0}^{n} a_i(t) \frac{d^i}{dt^i}(c_1 y_1 + c_2 y)$$

Since this equation holds for all $c_1$ and $c_2$, the equation is linear.

**3.34.** Show that a system described by the convolution integral

$$y(t) = \int_{-\infty}^{\infty} w(t, \tau) u(\tau) \, d\tau$$

is linear, $y$ is the output and $u$ the input.

Let $u_1$ and $u_2$ be two arbitrary inputs and let

$$y_1 = \int_{-\infty}^{\infty} w(t, \tau) u_1(\tau) \, d\tau \qquad y_2 = \int_{-\infty}^{\infty} w(t, \tau) u_2(\tau) \, d\tau$$

Now let $c_1 u_1 + c_2 u_2$ be a third input and form

$$\int_{-\infty}^{\infty} w(t, \tau) \left[ c_1 u_1(\tau) + c_2 u_2(\tau) \right] d\tau = c_1 \int_{-\infty}^{\infty} w(t, \tau) u_1(\tau) \, d\tau + c_2 \int_{-\infty}^{\infty} w(t, \tau) u_2(\tau) \, d\tau$$

$$= c_1 y_1 + c_2 y_2$$

Since this relationship holds for all $c_1$ and $c_2$, the convolution integral is a linear operation (or transformation).

**3.35.** Use the Principle of Superposition to determine the output $y$ of Fig. 3-9.

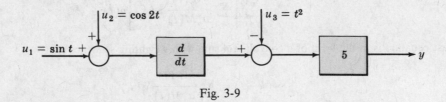

Fig. 3-9

For $u_2 = u_3 = 0$, $y_1 = 5(d/dt)(\sin t) = 5 \cos t$. For $u_1 = u_3 = 0$, $y_2 = 5(d/dt)(\cos 2t) = -10 \sin 2t$. For $u_1 = u_2 = 0$, $y_3 = -5t^2$. Therefore

$$y = y_1 + y_2 + y_3 = 5(\cos t - 2 \sin 2t - t^2)$$

**3.36.** A linear system is described by the weighting function

$$w(t, \tau) = e^{-|t - \tau|} \qquad \text{for all} \quad t, \tau$$

Suppose the system is excited by an input

$$u(t) = t \qquad \text{for all} \quad t$$

Find the output $y(t)$.

The output is given by the convolution integral (Example 3.38):

$$y(t) = \int_{-\infty}^{\infty} e^{-|t-\tau|}\tau\, d\tau = \int_{-\infty}^{t} e^{-(t-\tau)}\tau\, d\tau + \int_{t}^{\infty} e^{(t-\tau)}\tau\, d\tau$$

$$= e^{-t}\int_{-\infty}^{t} e^{\tau}\tau\, d\tau + e^{t}\int_{t}^{\infty} e^{-\tau}\tau\, d\tau$$

$$= e^{-t}\left[e^{\tau}(\tau-1)\big|_{-\infty}^{t}\right] + e^{t}\left[e^{-\tau}(-\tau-1)\big|_{t}^{\infty}\right] = 2t$$

## CAUSALITY

**3.37.** Two systems are defined by the relationships between their inputs and outputs as follows:

System 1: The input is $u(t)$ and at the same instant of time the output is $y(t) = u(t+T)$, $T > 0$.
System 2: The input is $u(t)$ and at the same instant of time the output is $y(t) = u(t-T)$, $T > 0$.

Are either of these systems causal?

In System 1, the output depends only on the input $T$ seconds in the future. Thus it is not causal. An operation of this type is called **prediction**.
In System 2, the output depends only on the input $T$ seconds in the past. Thus it is causal. An operation of this type is called a **time delay**.

# Supplementary Problems

**3.38.** Which of the following terms are first degree in the dependent variable $y = y(t)$? (a) $t^2 y$, (b) $\tan y$, (c) $\cos t$, (d) $e^{-y}$, (e) $te^{-t}$.

**3.39.** Show that a system defined by the equation $y = mu + b$, where $y$ is the output, $u$ is the input, and $m$ and $b$ are nonzero constants, is nonlinear according to Definition 3.21.

**3.40.** Show that any differential equation of the form

$$\sum_{i=0}^{n} a_i(t)\frac{d^i y}{dt^i} = \sum_{i=0}^{m} b_i(t)\frac{d^i u}{dt^i}$$

satisfies Definition 3.21. (See Example 3.37 and Problem 3.33).

**3.41.** Show that the functions $\cos t$ and $\sin t$ are linearly independent.

**3.42.** Show that the functions $\sin nt$ and $\sin kt$, where $n$ and $k$ are integers, are linearly independent if $n \neq k$.

**3.43.** Show that the functions $t$ and $t^2$ constitute a fundamental set for the differential equation

$$t^2\frac{d^2 y}{dt^2} - 2t\frac{dy}{dt} + 2y = 0$$

**3.44.** Find a fundamental set for

$$\frac{d^3 y}{dt^3} + 6\frac{d^2 y}{dt^2} + 21\frac{dy}{dt} + 26y = u$$

# Chapter 4

## The Laplace Transform and the *z*-Transform

### 4.1  INTRODUCTION

Several techniques used in solving engineering problems are based on the replacement of functions of a real variable (usually time or distance) by certain frequency-dependent representations, or by functions of a complex variable dependent upon frequency. A typical example is the use of Fourier series to solve certain electrical problems. One such problem consists of finding the current in some part of a linear electrical network in which the input voltage is a periodic or repeating waveform. The periodic voltage may be replaced by its Fourier series representation, and the current produced by each term of the series can then be determined. The total current is the sum of the individual currents (superposition). This technique often results in a substantial savings in computational effort.

Two very important transformation techniques for linear control system analysis are presented in this chapter: the *Laplace transform* and the *z-transform*. The Laplace transform relates time functions to frequency-dependent functions of a complex variable. The *z*-transform relates time sequences to a different, but related, type of frequency-dependent function. Applications of these mathematical transformations to solving linear constant-coefficient differential and difference equations are also discussed here. Together these methods provide the basis for the analysis and design techniques developed in subsequent chapters.

### 4.2  THE LAPLACE TRANSFORM

The Laplace transform is defined in the following manner:

***Definition 4.1:***   Let $f(t)$ be a real function of a real variable $t$ defined for $t > 0$. Then

$$\mathscr{L}[f(t)] \equiv F(s) \equiv \lim_{\substack{T \to \infty \\ \epsilon \to 0}} \int_{\epsilon}^{T} f(t) e^{-st}\, dt = \int_{0^+}^{\infty} f(t) e^{-st}\, dt \qquad 0 < \epsilon < T$$

is called the **Laplace transform** of $f(t)$. $s$ is a complex variable defined by $s \equiv \sigma + j\omega$, where $\sigma$ and $\omega$ are real variables* and $j = \sqrt{-1}$.

Note that the lower limit on the integral is $t = \epsilon > 0$. This definition of the lower limit is sometimes useful in dealing with functions that are discontinuous at $t = 0$. When *explicit* use is made of this limit, it will be abbreviated $t = \lim_{\epsilon \to 0} \epsilon \equiv 0^+$, as shown above in the integral on the right.

The real variable $t$ always denotes *time*.

***Definition 4.2:***   If $f(t)$ is defined and single-valued for $t > 0$ and $F(\sigma)$ is absolutely convergent for some real number $\sigma_0$, that is,

$$\int_{0^+}^{\infty} |f(t)| e^{-\sigma_0 t}\, dt = \lim_{\substack{T \to \infty \\ \epsilon \to 0}} \int_{\epsilon}^{T} |f(t)| e^{-\sigma_0 t}\, dt < +\infty \qquad 0 < \epsilon < T$$

then $f(t)$ is **Laplace transformable** for $\mathrm{Re}(s) > \sigma_0$.

---

*The real part $\sigma$ of a complex variable $s$ is often written as $\mathrm{Re}(s)$ (the real part of $s$) and the imaginary part $\omega$ as $\mathrm{Im}(s)$ (the imaginary part of $s$). Parentheses are placed around $s$ only when there is a possibility of confusion.

**EXAMPLE 4.1.**   The function $e^{-t}$ is Laplace transformable since

$$\int_{0^+}^{\infty} |e^{-t}| e^{-\sigma_0 t}\, dt = \int_{0^+}^{\infty} e^{-(1+\sigma_0)t}\, dt = \frac{1}{-(1+\sigma_0)} e^{-(1+\sigma_0)t}\Big|_{0^+}^{\infty} = \frac{1}{1+\sigma_0} < +\infty$$

if $1 + \sigma_0 > 0$ or $\sigma_0 > -1$.

**EXAMPLE 4.2.**   The Laplace transform of $e^{-t}$ is

$$\mathscr{L}[e^{-t}] = \int_{0^+}^{\infty} e^{-t} e^{-st}\, dt = \frac{-1}{(s+1)} e^{-(s+1)t}\Big|_{0^+}^{\infty} = \frac{1}{s+1} \qquad \text{for} \quad \operatorname{Re}(s) > -1$$

## 4.3   THE INVERSE LAPLACE TRANSFORM

The Laplace transform transforms a problem from the real variable time domain into the complex variable $s$-domain. After a solution of the transformed problem has been obtained in terms of $s$, it is necessary to "invert" this transform to obtain the time domain solution. The transformation from the $s$-domain into the $t$-domain is called the *inverse Laplace transform*.

**Definition 4.3:**   Let $F(s)$ be the Laplace transform of a function $f(t)$, $t > 0$. The contour integral

$$\mathscr{L}^{-1}[F(s)] \equiv f(t) = \frac{1}{2\pi j} \int_{c-j\infty}^{c+j\infty} F(s) e^{st}\, ds$$

where $j = \sqrt{-1}$ and $c > \sigma_0$ ($\sigma_0$ as given in Definition 4.2), is called the **inverse Laplace transform** of $F(s)$.

It is seldom necessary in practice to perform the contour integration defined in Definition 4.3. For applications of the Laplace transform in this book, it is never necessary. A simple technique for evaluating the inverse transform for most control system problems is presented in Section 4.8.

## 4.4   SOME PROPERTIES OF THE LAPLACE TRANSFORM AND ITS INVERSE

The Laplace transform and its inverse have several important properties which can be used advantageously in the solution of linear constant-coefficient differential equations. They are:

1. The Laplace transform is a *linear transformation* between functions defined in the $t$-domain and functions defined in the $s$-domain. That is, if $F_1(s)$ and $F_2(s)$ are the Laplace transforms of $f_1(t)$ and $f_2(t)$, respectively, then $a_1 F_1(s) + a_2 F_2(s)$ is the Laplace transform of $a_1 f_1(t) + a_2 f_2(t)$, where $a_1$ and $a_2$ are arbitrary constants.

2. The inverse Laplace transform is a *linear transformation* between functions defined in the $s$-domain and functions defined in the $t$-domain. That is, if $f_1(t)$ and $f_2(t)$ are the inverse Laplace transforms of $F_1(s)$ and $F_2(s)$, respectively, then $b_1 f_1(t) + b_2 f_2(t)$ is the inverse Laplace transform of $b_1 F_1(s) + b_2 F_2(s)$, where $b_1$ and $b_2$ are arbitrary constants.

3. The Laplace transform of the *derivative $df/dt$* of a function $f(t)$ whose Laplace transform is $F(s)$ is

$$\mathscr{L}\left[\frac{df}{dt}\right] = sF(s) - f(0^+)$$

where $f(0^+)$ is the initial value of $f(t)$, evaluated as the one-sided limit of $f(t)$ as $t$ approaches zero from positive values.

4.  The Laplace transform of the *integral* $\int_0^t f(\tau)\, d\tau$ of a function $f(t)$ whose Laplace transform is $F(s)$ is

$$\mathscr{L}\left[\int_0^t f(\tau)\, d\tau\right] = \frac{F(s)}{s}$$

5.  The initial value $f(0^+)$ of the function $f(t)$ whose Laplace transform is $F(s)$ is

$$f(0^+) = \lim_{t \to 0} f(t) = \lim_{s \to \infty} sF(s) \qquad t > 0$$

This relation is called the *Initial Value Theorem*.

6.  The final value $f(\infty)$ of the function $f(t)$ whose Laplace transform is $F(s)$ is

$$f(\infty) = \lim_{t \to \infty} f(t) = \lim_{s \to 0} sF(s)$$

if $\lim_{t \to \infty} f(t)$ exists. This relation is called the *Final Value Theorem*.

7.  The Laplace transform of a function $f(t/a)$ (*Time Scaling*) is

$$\mathscr{L}\left[f\left(\frac{t}{a}\right)\right] = aF(as)$$

where $F(s) = \mathscr{L}[f(t)]$.

8.  The inverse Laplace transform of the function $F(s/a)$ (*Frequency Scaling*) is

$$\mathscr{L}^{-1}\left[F\left(\frac{s}{a}\right)\right] = af(at)$$

where $\mathscr{L}^{-1}[F(s)] = f(t)$.

9.  The Laplace transform of the function $f(t - T)$ (*Time Delay*), where $T > 0$ and $f(t - T) = 0$ for $t \leq T$, is

$$\mathscr{L}[f(t - T)] = e^{-sT}F(s)$$

where $F(s) = \mathscr{L}[f(t)]$.

10.  The Laplace transform of the function $e^{-at}f(t)$ is given by

$$\mathscr{L}[e^{-at}f(t)] = F(s + a)$$

where $F(s) = \mathscr{L}[f(t)]$ (*Complex Translation*).

11.  The Laplace transform of the *product of two functions* $f_1(t)$ and $f_2(t)$ is given by the *complex convolution integral*

$$\mathscr{L}[f_1(t)f_2(t)] = \frac{1}{2\pi j} \int_{c-j\infty}^{c+j\infty} F_1(\omega)F_2(s - \omega)\, d\omega$$

where $F_1(s) = \mathscr{L}[f_1(t)]$, $F_2(s) = \mathscr{L}[f_2(t)]$.

12.  The inverse Laplace transform of the *product of the two transforms* $F_1(s)$ and $F_2(s)$ is given by the *convolution integrals*

$$\mathscr{L}^{-1}[F_1(s)F_2(s)] = \int_{0^+}^t f_1(\tau)f_2(t - \tau)\, d\tau = \int_{0^+}^t f_2(\tau)f_1(t - \tau)\, d\tau$$

where $\mathscr{L}^{-1}[F_1(s)] = f_1(t)$, $\mathscr{L}^{-1}[F_2(s)] = f_2(t)$.

**EXAMPLE 4.3.**  The Laplace transforms of the functions $e^{-t}$ and $e^{-2t}$ are $\mathscr{L}[e^{-t}] = 1/(s + 1)$, $\mathscr{L}[e^{-2t}] = 1/(s + 2)$. Then, by Property 1,

$$\mathscr{L}[3e^{-t} - e^{-2t}] = 3\mathscr{L}[e^{-t}] - \mathscr{L}[e^{-2t}] = \frac{3}{s + 1} - \frac{1}{s + 2} = \frac{2s + 5}{s^2 + 3s + 2}$$

**EXAMPLE 4.4.** The inverse Laplace transforms of the functions $1/(s+1)$ and $1/(s+3)$ are

$$\mathcal{L}^{-1}\left[\frac{1}{s+1}\right] = e^{-t} \qquad \mathcal{L}^{-1}\left[\frac{1}{s+3}\right] = e^{-3t}$$

Then, by Property 2,

$$\mathcal{L}^{-1}\left[\frac{2}{s+1} - \frac{4}{s+3}\right] = 2\mathcal{L}^{-1}\left[\frac{1}{s+1}\right] - 4\mathcal{L}^{-1}\left[\frac{1}{s+3}\right] = 2e^{-t} - 4e^{-3t}$$

**EXAMPLE 4.5.** The Laplace transform of $(d/dt)(e^{-t})$ can be determined by application of Property 3. Since $\mathcal{L}[e^{-t}] = 1/(s+1)$ and $\lim_{t \to 0} e^{-t} = 1$, then

$$\mathcal{L}\left[\frac{d}{dt}(e^{-t})\right] = s\left(\frac{1}{s+1}\right) - 1 = \frac{-1}{s+1}$$

**EXAMPLE 4.6.** The Laplace transform of $\int_0^t e^{-\tau}\,d\tau$ can be determined by application of Property 4. Since $\mathcal{L}[e^{-t}] = 1/(s+1)$, then

$$\mathcal{L}\left[\int_0^t e^{-\tau}\,d\tau\right] = \frac{1}{s}\left(\frac{1}{s+1}\right) = \frac{1}{s(s+1)}$$

**EXAMPLE 4.7.** The Laplace transform of $e^{-3t}$ is $\mathcal{L}[e^{-3t}] = 1/(s+3)$. The initial value of $e^{-3t}$ can be determined by the Initial Value Theorem as

$$\lim_{t \to 0} e^{-3t} = \lim_{s \to \infty} s\left(\frac{1}{s+3}\right) = 1$$

**EXAMPLE 4.8.** The Laplace transform of the function $(1 - e^{-t})$ is $1/s(s+1)$. The final value of this function can be determined from the Final Value Theorem as

$$\lim_{t \to \infty} (1 - e^{-t}) = \lim_{s \to 0} \frac{s}{s(s+1)} = 1$$

**EXAMPLE 4.9.** The Laplace transform of $e^{-t}$ is $1/(s+1)$. The Laplace transform of $e^{-3t}$ can be determined by application of Property 7 (Time Scaling), where $a = \frac{1}{3}$:

$$\mathcal{L}[e^{-3t}] = \frac{1}{3}\left[\frac{1}{\left(\frac{1}{3}s+1\right)}\right] = \frac{1}{s+3}$$

**EXAMPLE 4.10.** The inverse transform of $1/(s+1)$ is $e^{-t}$. The inverse transform of $1/(\frac{1}{3}s+1)$ can be determined by application of Property 8 (Frequency Scaling):

$$\mathcal{L}^{-1}\left[\frac{1}{\frac{1}{3}s+1}\right] = 3e^{-3t}$$

**EXAMPLE 4.11.** The Laplace transform of the function $e^{-t}$ is $1/(s+1)$. The Laplace transform of the function defined as

$$f(t) = \begin{cases} e^{-(t-2)} & t > 2 \\ 0 & t \le 2 \end{cases}$$

can be determined by Property 9, with $T = 2$:

$$\mathcal{L}[f(t)] = e^{-2s} \cdot \mathcal{L}[e^{-t}] = \frac{e^{-2s}}{s+1}$$

**EXAMPLE 4.12.** The Laplace transform of $\cos t$ is $s/(s^2+1)$. The Laplace transform of $e^{-2t}\cos t$ can be determined from Property 10 with $a = 2$:

$$\mathcal{L}[e^{-2t}\cos t] = \frac{s+2}{(s+2)^2+1} = \frac{s+2}{s^2+4s+5}$$

**EXAMPLE 4.13.** The Laplace transform of the product $e^{-2t}\cos t$ can be determined by application of Property 11 (Complex Convolution). That is, since $\mathscr{L}[e^{-2t}] = 1/(s+2)$ and $\mathscr{L}[\cos t] = s/(s^2+1)$, then

$$\mathscr{L}[e^{-2t}\cos t] = \frac{1}{2\pi j}\int_{c-j\infty}^{c+j\infty}\left(\frac{\omega}{\omega^2+1}\right)\left(\frac{1}{s-\omega+2}\right)d\omega = \frac{s+2}{s^2+4s+5}$$

The details of this contour integration are not carried out here because they are too complicated (see, e.g., Reference [1]) and unnecessary. The Laplace transform of $e^{-2t}\cos t$ was very simply determined in Example 4.12 using Property 10. There are, however, many instances in more advanced treatments of automatic control in which complex convolution can be used effectively.

**EXAMPLE 4.14.** The inverse Laplace transform of the function $F(s) = s/(s+1)(s^2+1)$ can be determined by application of Property 12. Since $\mathscr{L}^{-1}[1/(s+1)] = e^{-t}$ and $\mathscr{L}^{-1}[s/(s^2+1)] = \cos t$, then

$$\mathscr{L}^{-1}\left[\left(\frac{1}{s+1}\right)\left(\frac{s}{s^2+1}\right)\right] = \int_{0^+}^{t} e^{-(t-\tau)}\cos\tau\,d\tau = e^{-t}\int_{0^+}^{t} e^{\tau}\cos\tau\,d\tau = \tfrac{1}{2}(\cos t + \sin t - e^{-t})$$

## 4.5   SHORT TABLE OF LAPLACE TRANSFORMS

Table 4.1 is a short table of Laplace transforms. It is not complete, but when used in conjunction with the properties of the Laplace transform described in Section 4.4 and the partial fraction expansion techniques described in Section 4.7, it is adequate to handle all of the problems in this book. A more complete table of Laplace transform pairs is found in Appendix A.

**TABLE 4.1**

| Time Function | | Laplace Transform |
|---|---|---|
| Unit Impulse | $\delta(t)$ | $1$ |
| Unit Step | $\mathbf{1}(t)$ | $\dfrac{1}{s}$ |
| Unit Ramp | $t$ | $\dfrac{1}{s^2}$ |
| Polynomial | $t^n$ | $\dfrac{n!}{s^{n+1}}$ |
| Exponential | $e^{-at}$ | $\dfrac{1}{s+a}$ |
| Sine Wave | $\sin\omega t$ | $\dfrac{\omega}{s^2+\omega^2}$ |
| Cosine Wave | $\cos\omega t$ | $\dfrac{s}{s^2+\omega^2}$ |
| Damped Sine Wave | $e^{-at}\sin\omega t$ | $\dfrac{\omega}{(s+a)^2+\omega^2}$ |
| Damped Cosine Wave | $e^{-at}\cos\omega t$ | $\dfrac{s+a}{(s+a)^2+\omega^2}$ |

Table 4.1 can be used to find both Laplace transforms and inverse Laplace transforms. To find the Laplace transform of a time function which can be represented by some combination of the elementary functions given in Table 4.1, the appropriate transforms are chosen from the table and are combined using the properties in Section 4.4.

**EXAMPLE 4.15.**  The Laplace transform of the function $f(t) = e^{-4t} + \sin(t-2) + t^2 e^{-2t}$ is determined as follows. The Laplace transforms of $e^{-4t}$, $\sin t$, and $t^2$ are given in the table as

$$\mathscr{L}[e^{-4t}] = \frac{1}{s+4} \qquad \mathscr{L}[\sin t] = \frac{1}{s^2+1} \qquad \mathscr{L}[t^2] = \frac{2}{s^3}$$

Application of Properties 9 and 10, respectively, yields

$$\mathscr{L}[\sin(t-2)] = \frac{e^{-2s}}{s^2+1} \qquad \mathscr{L}[t^2 e^{-2t}] = \frac{2}{(s+2)^3}$$

Then Property 1 (Linearity) gives

$$\mathscr{L}[f(t)] = \frac{1}{s+4} + \frac{e^{-2s}}{s^2+1} + \frac{2}{(s+2)^3}$$

To find the inverse of the transform of a combination of those in Table 4.1, the corresponding time functions (inverse transforms) are determined from the table and combined appropriately using the properties in Section 4.4.

**EXAMPLE 4.16.**  The inverse Laplace transform of $F(s) = [(s+2)/s^2 + 4] \cdot e^{-s}$ can be determined as follows. $F(s)$ is first rewritten as

$$F(s) = \frac{se^{-s}}{s^2+4} + \frac{2e^{-s}}{s^2+4}$$

Now

$$\mathscr{L}^{-1}\left[\frac{s}{s^2+4}\right] = \cos 2t \qquad \mathscr{L}^{-1}\left[\frac{2}{s^2+4}\right] = \sin 2t$$

Application of Property 9 for $t > 1$ yields

$$\mathscr{L}^{-1}\left[\frac{se^{-s}}{s^2+4}\right] = \cos 2(t-1) \qquad \mathscr{L}^{-1}\left[\frac{2e^{-s}}{s^2+4}\right] = \sin 2(t-1)$$

Then Property 2 (Linearity) gives

$$\begin{aligned}
\mathscr{L}^{-1}[F(s)] &= \cos 2(t-1) + \sin 2(t-1) \qquad & t > 1 \\
&= 0 \qquad & t \le 1
\end{aligned}$$

## 4.6  APPLICATION OF LAPLACE TRANSFORMS TO THE SOLUTION OF LINEAR CONSTANT-COEFFICIENT DIFFERENTIAL EQUATIONS

The application of Laplace transforms to the solution of linear constant-coefficient differential equations is of major importance in linear control system problems. Two classes of equations of general interest are treated in this section. The first of these has the form:

$$\sum_{i=0}^{n} a_i \frac{d^i y}{dt^i} = u \qquad (4.1)$$

where $y$ is the output, $u$ is the input, the coefficients $a_0, a_1, \ldots, a_{n-1}$, are constants, and $a_n = 1$. The initial conditions for this equation are written as

$$\left.\frac{d^k y}{dt^k}\right|_{t=0^+} \equiv y_0^k \qquad k = 0, 1, \ldots, n-1$$

where $y_0^k$ are constants. The Laplace transform of Equation ($4.1$) is given by

$$\sum_{i=0}^{n}\left[a_i\left(s^i Y(s) - \sum_{k=0}^{i-1} s^{i-1-k} y_0^k\right)\right] = U(s) \qquad (4.2)$$

and the transform of the output is

$$Y(s) = \frac{U(s)}{\displaystyle\sum_{i=0}^{n} a_i s^i} + \frac{\displaystyle\sum_{i=0}^{n}\sum_{k=0}^{i-1} a_i s^{i-1-k} y_0^k}{\displaystyle\sum_{i=0}^{n} a_i s^i} \tag{4.3}$$

Note that the right side of Equation $(4.3)$ is the sum of two terms: a term dependent only on the input transform, and a term dependent only on the initial conditions. In addition, note that the denominator of both terms in Equation $(4.3)$, that is,

$$\sum_{i=0}^{n} a_i s^i = s^n + a_{n-1} s^{n-1} + \cdots + a_1 s + a_0$$

is the *characteristic polynomial* of Equation $(4.1)$ (see Section 3.6).

The time solution $y(t)$ of Equation $(4.1)$ is the inverse Laplace transform of $Y(s)$, that is,

$$y(t) = \mathscr{L}^{-1}\left[ \frac{U(s)}{\displaystyle\sum_{i=0}^{n} a_i s^i} \right] + \mathscr{L}^{-1}\left[ \frac{\displaystyle\sum_{i=0}^{n}\sum_{k=0}^{i-1} a_i s^{i-1-k} y_0^k}{\displaystyle\sum_{i=0}^{n} a_i s^i} \right] \tag{4.4}$$

The first term on the right is the *forced response* and the second term is the *free response* of the system represented by Equation $(4.1)$.

Direct substitution into Equations $(4.2)$, $(4.3)$, and $(4.4)$ yields the transform of the differential equation, the solution transform $Y(s)$, or the time solution $y(t)$, respectively. But it is often easier to directly apply the properties of Section 4.4 to determine these quantities, especially when the order of the differential equation is low.

**EXAMPLE 4.17.**   The Laplace transform of the differential equation

$$\frac{d^2 y}{dt^2} + 3\frac{dy}{dt} + 2y = \mathbf{1}(t) = \text{unit step}$$

with initial conditions $y(0^+) = -1$ and $(dy/dt)|_{t=0^+} = 2$ can be written directly from Equation $(4.2)$ by first identifying $n$, $a_i$, and $y_0^k$: $n = 2$, $y_0^0 = -1$, $y_0^1 = 2$, $a_0 = 2$, $a_1 = 3$, $a_2 = 1$. Substitution of these values into Equation $(4.2)$ yields

$$2Y + 3(sY + 1) + 1(s^2 Y + s - 2) = \frac{1}{s} \qquad \text{or} \qquad (s^2 + 3s + 2)Y = \frac{-(s^2 + s - 1)}{s}$$

It should be noted that when $i = 0$ in Equation $(4.2)$, the summation interior to the brackets is, by definition,

$$\left.\sum_{k=0}^{i-1}\right|_{i=0} = \sum_{k=0}^{k=-1} = 0$$

The Laplace transform of the differential equation can also be determined in the following manner. The transform of $d^2 y/dt^2$ is given by

$$\mathscr{L}\left[\frac{d^2 y}{dt^2}\right] = s^2 Y(s) - sy(0^+) - \left.\frac{dy}{dt}\right|_{t=0^+}$$

This equation is a direct consequence of Property 3, Section 4.4 (see Problem 4.17). With this information the transform of the differential equation can be determined by applying Property 1 (Linearity) of Section 4.4; that is,

$$\mathscr{L}\left[\frac{d^2 y}{dt^2} + 3\frac{dy}{dt} + 2y\right] = \mathscr{L}\left[\frac{d^2 y}{dt^2}\right] + \mathscr{L}\left[3\frac{dy}{dt}\right] + \mathscr{L}[2y] = (s^2 + 3s + 2)Y + s + 1 = \mathscr{L}[\mathbf{1}(t)] = \frac{1}{s}$$

The output transform $Y(s)$ is determined by rearranging the previous equation and is

$$Y(s) = \frac{-(s^2 + s - 1)}{s(s^2 + 3s + 2)}$$

The output time solution $y(t)$ is the inverse transform of $Y(s)$. A method for determining the inverse transform of functions like $Y(s)$ above is presented in Sections 4.7 and 4.8.

Now consider constant-coefficient equations of the form:

$$\sum_{i=0}^{n} a_i \frac{d^i y}{dt^i} = \sum_{i=0}^{m} b_i \frac{d^i u}{dt^i} \qquad (4.5)$$

where $y$ is the output, $u$ is the input, $a_n = 1$, and $m \le n$. The Laplace transform of Equation $(4.5)$ is given by

$$\sum_{i=0}^{n} \left[ a_i \left( s^i Y(s) - \sum_{k=0}^{i-1} s^{i-1-k} y_0^k \right) \right] = \sum_{i=0}^{m} \left[ b_i \left( s^i U(s) - \sum_{k=0}^{i-1} s^{i-1-k} u_0^k \right) \right] \qquad (4.6)$$

where $u_0^k = (d^k u/dt^k)|_{t=0^+}$. The output transform $Y(s)$ is

$$Y(s) = \left[ \frac{\sum_{i=0}^{m} b_i s^i}{\sum_{i=0}^{n} a_i s^i} \right] U(s) - \frac{\sum_{i=0}^{m} \sum_{k=0}^{i-1} b_i s^{i-1-k} u_0^k}{\sum_{i=0}^{n} a_i s^i} + \frac{\sum_{i=0}^{n} \sum_{k=0}^{i-1} a_i s^{i-1-k} y_0^k}{\sum_{i=0}^{n} a_i s^i} \qquad (4.7)$$

The time solution $y(t)$ is the inverse Laplace transform of $Y(s)$:

$$y(t) = \mathcal{L}^{-1} \left[ \frac{\sum_{i=0}^{m} b_i s^i}{\sum_{i=0}^{n} a_i s^i} U(s) - \frac{\sum_{i=0}^{m} \sum_{k=0}^{i-1} b_i s^{i-1-k} u_0^k}{\sum_{i=0}^{n} a_i s^i} \right] + \mathcal{L}^{-1} \left[ \frac{\sum_{i=0}^{n} \sum_{k=0}^{i-1} a_i s^{i-1-k} y_0^k}{\sum_{i=0}^{n} a_i s^i} \right] \qquad (4.8)$$

The first term on the right is the *forced response*, and the second term is the *free response* of a system represented by Equation $(4.5)$.

Note that the Laplace transform $Y(s)$ of the output $y(t)$ consists of ratios of polynomials in the complex variable $s$. Such ratios are generally called **rational (algebraic) functions**. If all initial conditions in Eq. $(4.8)$ are zero and $U(s) = 1$, $(4.8)$ gives the *unit impulse response*. The denominator of each term in $(4.8)$ is the *characteristic polynomial* of the system.

For problems in which initial conditions are not specified on $y(t)$ but on some other parameter of the system (such as the initial voltage across a capacitor not appearing at the output), $y_0^k$, $k = 0, 1, \ldots,$ $n - 1$, must be derived using the available information. For systems represented in the form of Equation $(4.5)$, that is, including derivative terms in $u$, computation of $y_0^k$ will also depend on $u_0^k$. Problem 4.38 illustrates these points.

The restriction $n \ge m$ in Equation $(4.5)$ is based on the fact that real systems have a *smoothing* effect on their input. By a smoothing effect, it is meant that variations in the input are made less pronounced (at least no more pronounced) by the action of the system on the input. Since a differentiator generates the slope of a time function, it accentuates the variations of the function. An integrator, on the other hand, sums the area under the curve of a time function over an interval of time and thus averages (smooths) the variations of the function.

In Equation $(4.5)$, the output $y$ is related to the input $u$ by an operation which includes $m$ differentiations and $n$ integrations of the input. Hence, in order that there be a smoothing effect (at least no accentuation of the variations) between the input and the output, there must be more (at least as many) integrations than differentiations; that is, $n \ge m$.

**EXAMPLE 4.18.**  A certain system is described by the differential equation

$$\frac{d^2y}{dt^2} = \frac{du}{dt}, \qquad y(0^+) = \frac{dy}{dt}\bigg|_{t=0^+} = 0$$

where the input $u$ is graphed in Fig. 4-1. The corresponding functions $du/dt$ and

$$y(t) = \int_{0^+}^{t}\int_{0^+}^{\theta} \frac{du}{d\alpha}\, d\alpha\, d\theta = \int_{0^+}^{t} u(\theta)\, d\theta$$

are also shown. Note from these graphs that differentiation of $u$ accentuates the variations in $u$ while integration smooths them.

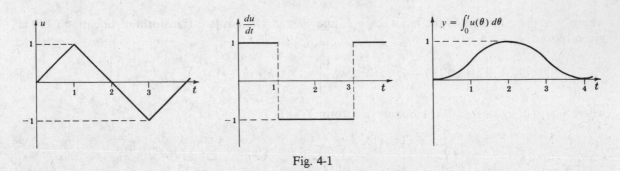

Fig. 4-1

**EXAMPLE 4.19.**  Consider a system described by the differential equation

$$\frac{d^2y}{dt^2} + 3\frac{dy}{dt} + 2y = \frac{du}{dt} + 3u$$

with initial conditions $y_0^0 = 1$, $y_0^1 = 0$. If the input is given by $u(t) = e^{-4t}$, then the Laplace transform of the output $y(t)$ can be obtained by direct application of Equation $(4.7)$ by first identifying $m$, $n$, $a_i$, $b_i$ and $u_0^0$: $n = 2$, $a_0 = 2$, $a_1 = 3$, $a_2 = 1$, $m = 1$, $u_0^0 = \lim_{t \to 0} e^{-4t} = 1$, $b_0 = 3$, $b_1 = 1$. Substitution of these values into Equation $(4.7)$ yields

$$Y(s) = \left(\frac{s+3}{s^2+3s+2}\right)\left(\frac{1}{s+4}\right) + \frac{s+3}{s^2+3s+2} - \frac{1}{s^2+3s+2}$$

This transform can also be obtained by direct application of Properties 1 and 3 of Section 4.4 to the differential equation, as was done in Example 4.17.

The linear constant-coefficient vector-matrix differential equations discussed in Section 3.15 also can be solved by Laplace transform techniques, as illustrated in the following example.

**EXAMPLE 4.20.**  Consider the vector-matrix differential equation of Problem 3.31:

$$\frac{d\mathbf{x}}{dt} = A\mathbf{x} + \mathbf{b}u$$

where

$$\mathbf{x}(t) = \begin{bmatrix} x_1(t) \\ x_2(t) \end{bmatrix} \qquad A = \begin{bmatrix} 0 & 1 \\ 0 & 0 \end{bmatrix} \qquad \mathbf{b} = \begin{bmatrix} 0 \\ 1 \end{bmatrix} \qquad \mathbf{x}(0) = \begin{bmatrix} 1 \\ -1 \end{bmatrix}$$

and with $u = \mathbf{1}(t)$, the unit step function. The Laplace transform of the vector-matrix form of this equation is

$$s\mathbf{X}(s) - \mathbf{x}(0) = A\mathbf{X}(s) + \frac{1}{s}\mathbf{b}$$

where $\mathbf{X}(s)$ is the vector Laplace transform whose components are the Laplace transforms of the components of $\mathbf{x}(t)$. This can be rewritten as

$$[sI - A]\mathbf{X}(s) = \mathbf{x}(0) + \frac{1}{s}\mathbf{b}$$

where $I$ is the *identity* or *unit* matrix. The Laplace transform of the solution vector $\mathbf{x}(t)$ can thus be written as

$$\mathbf{X}(s) = [sI - A]^{-1}\mathbf{x}(0) + \frac{1}{s}[sI - A]^{-1}\mathbf{b}$$

where $[\cdot]^{-1}$ represents the *inverse* of the matrix. Since

$$sI - A = \begin{bmatrix} s & -1 \\ 0 & s \end{bmatrix}$$

then

$$[sI - A]^{-1} = \frac{1}{s^2}\begin{bmatrix} s & 1 \\ 0 & s \end{bmatrix}$$

Substituting for $[sI - A]^{-1}$, $\mathbf{x}(0)$, and $\mathbf{b}$ gives

$$\mathbf{X}(s) = \begin{bmatrix} \dfrac{s-1}{s^2} \\[2mm] -\dfrac{1}{s} \end{bmatrix} + \begin{bmatrix} \dfrac{1}{s^3} \\[2mm] \dfrac{1}{s^2} \end{bmatrix}$$

where the first term is the Laplace transform of the *free response*, and the second term is the Laplace transform of the *forced response*. Using Table 4.1, the Laplace transform of these vectors can be inverted term by term, providing the solution vector:

$$\mathbf{x}(t) = \begin{bmatrix} \mathbf{1}(t) - t + t^2/2 \\ -\mathbf{1}(t) + t \end{bmatrix}$$

## 4.7  PARTIAL FRACTION EXPANSIONS

In Section 4.6 it was shown that the Laplace transforms encountered in the solution of linear constant-coefficient differential equations are rational functions of $s$ (i.e., ratios of polynomials in $s$). In this section an important representation of rational functions, the partial fraction expansion, is presented. It will be shown in the next section that this representation greatly simplifies the inversion of the Laplace transform of a rational function.

Consider the rational function

$$F(s) = \frac{\displaystyle\sum_{i=0}^{m} b_i s^i}{\displaystyle\sum_{i=0}^{n} a_i s^i} \tag{4.9}$$

where $a_n = 1$ and $n \geq m$. By the fundamental theorem of algebra, the denominator polynomial equation

$$\sum_{i=0}^{n} a_i s^i = 0$$

has $n$ roots. Some of these roots may be repeated.

**EXAMPLE 4.21.**   The polynomial $s^3 + 5s^2 + 8s + 4$ has three roots: $-2, -2, -1$. $-2$ is a repeated root.

Suppose the denominator polynomial equation above has $n_1$ roots equal to $-p_1$, $n_2$ roots equal to $-p_2, \ldots, n_r$ roots equal to $-p_r$, where $\sum_{i=1}^{r} n_i = n$. Then

$$\sum_{i=0}^{n} a_i s^i = \prod_{i=1}^{r} (s + p_i)^{n_i}$$

The rational function $F(s)$ can then be written as

$$F(s) = \frac{\sum_{i=0}^{m} b_i s^i}{\prod_{i=1}^{r} (s + p_i)^{n_i}}$$

The **partial fraction expansion** representation of the rational function $F(s)$ is

$$F(s) = b_n + \sum_{i=1}^{r} \sum_{k=1}^{n_i} \frac{c_{ik}}{(s + p_i)^k} \qquad (4.10a)$$

where $b_n = 0$ unless $m = n$. The coefficients $c_{ik}$ are given by

$$c_{ik} = \frac{1}{(n_i - k)!} \frac{d^{n_i - k}}{ds^{n_i - k}} \left[ (s + p_i)^{n_i} F(s) \right] \Big|_{s = -p_i} \qquad (4.10b)$$

The particular coefficients $c_{i1}$, $i = 1, 2, \ldots, r$, are called the **residues** of $F(s)$ at $-p_i$, $i = 1, 2, \ldots, r$. If none of the roots are repeated, then

$$F(s) = b_n + \sum_{i=1}^{n} \frac{c_{i1}}{s + p_i} \qquad (4.11a)$$

where

$$c_{i1} = (s + p_i) F(s) \big|_{s = -p_i} \qquad (4.11b)$$

**EXAMPLE 4.22.** Consider the rational function

$$F(s) = \frac{s^2 + 2s + 2}{s^2 + 3s + 2} = \frac{s^2 + 2s + 2}{(s + 1)(s + 2)}$$

The partial fraction expansion of $F(s)$ is

$$F(s) = b_2 + \frac{c_{11}}{s + 1} + \frac{c_{21}}{s + 2}$$

The numerator coefficient of $s^2$ is $b_2 = 1$. The coefficients $c_{11}$ and $c_{21}$ are determined from Equation $(4.11b)$ as

$$c_{11} = (s + 1) F(s) \big|_{s = -1} = \frac{s^2 + 2s + 2}{s + 2} \Big|_{s = -1} = 1$$

$$c_{21} = (s + 2) F(s) \big|_{s = -2} = \frac{s^2 + 2s + 2}{s + 1} \Big|_{s = -2} = -2$$

Hence

$$F(s) = 1 + \frac{1}{s + 1} - \frac{2}{s + 2}$$

**EXAMPLE 4.23.** Consider the rational function

$$F(s) = \frac{1}{(s + 1)^2 (s + 2)}$$

The partial fraction expansion of $F(s)$ is

$$F(s) = b_3 + \frac{c_{11}}{s + 1} + \frac{c_{12}}{(s + 1)^2} + \frac{c_{21}}{s + 2}$$

The coefficients $b_3, c_{11}, c_{12}, c_{21}$ are given by

$$b_3 = 0$$

$$c_{11} = \frac{d}{ds}(s+1)^2 F(s)\bigg|_{s=-1} = \frac{d}{ds}\frac{1}{s+2}\bigg|_{s=-1} = -1$$

$$c_{12} = (s+1)^2 F(s)\big|_{s=-1} = \frac{1}{s+2}\bigg|_{s=-1} = 1$$

$$c_{21} = (s+2) F(s)\big|_{s=-2} = 1$$

Thus
$$F(s) = -\frac{1}{s+1} + \frac{1}{(s+1)^2} + \frac{1}{s+2}$$

## 4.8  INVERSE TRANSFORMS USING PARTIAL FRACTION EXPANSIONS

In Section 4.6 it was shown that the solution to a linear constant-coefficient ordinary differential equation can be determined by finding the inverse Laplace transform of a rational function. The general form of this operation can be written using Equation (*4.10*) as

$$\mathscr{L}^{-1}\left[\frac{\displaystyle\sum_{i=0}^{m} b_i s^i}{\displaystyle\sum_{i=0}^{n} a_i s^i}\right] = \mathscr{L}^{-1}\left[b_n + \sum_{i=0}^{r}\sum_{k=1}^{n_i}\frac{c_{ik}}{(s+p_i)^k}\right] = b_n\delta(t) + \sum_{i=1}^{r}\sum_{k=1}^{n_i}\frac{c_{ik}}{(k-1)!}t^{k-1}e^{-p_i t} \quad (4.12)$$

where $\delta(t)$ is the unit impulse function and $b_n = 0$ unless $m = n$. We remark that the rightmost term in Equation (*4.12*) is the general form of the *unit impulse response* for Equation (*4.5*).

**EXAMPLE 4.24.**    The inverse Laplace transform of the function

$$F(s) = \frac{s^2 + 2s + 2}{(s+1)(s+2)}$$

is given by

$$\mathscr{L}^{-1}\left[\frac{s^2+2s+2}{(s+1)(s+2)}\right] = \mathscr{L}^{-1}\left[1 + \frac{1}{s+1} - \frac{2}{s+2}\right] = \mathscr{L}^{-1}[1] + \mathscr{L}^{-1}\left[\frac{1}{s+1}\right] - \mathscr{L}^{-1}\left[\frac{2}{s+2}\right] = \delta(t) + e^{-t} - 2e^{-2t}$$

which is the unit impulse response for the differential equation:

$$\frac{d^2 y}{dt^2} + 3\frac{dy}{dt} + 2y = \frac{d^2 u}{dt^2} + 2\frac{du}{dt} + 2u$$

**EXAMPLE 4.25.**    The inverse Laplace transform of the function

$$F(s) = \frac{1}{(s+1)^2(s+2)}$$

is given by

$$\mathscr{L}^{-1}\left[\frac{1}{(s+1)^2(s+2)}\right] = \mathscr{L}^{-1}\left[-\frac{1}{s+1} + \frac{1}{(s+1)^2} + \frac{1}{s+2}\right]$$

$$= -\mathscr{L}^{-1}\left[\frac{1}{s+1}\right] + \mathscr{L}^{-1}\left[\frac{1}{(s+1)^2}\right] + \mathscr{L}^{-1}\left[\frac{1}{s+2}\right] = -e^{-t} + te^{-t} + e^{-2t}$$

### 4.9   THE $z$-TRANSFORM

The $z$-transform is used to describe signals and components in discrete-time control systems. It is defined as follows:

**Definition 4.4:**     Let $\{f(k)\}$ denote a real-valued sequence $f(0), f(1), f(2), \ldots$, or equivalently, $f(k)$ for $k = 0, 1, 2, \ldots$. Then

$$\mathscr{Z}\{f(k)\} \equiv F(z) = \sum_{k=0}^{\infty} f(k)z^{-k}$$

is called the **$z$-transform** of $\{f(k)\}$. $z$ is a *complex variable defined* by $z \equiv \mu + j\nu$, where $\mu$ and $\nu$ are real variables and $j = \sqrt{-1}$.

**Remark 1:**     The $k$th term of the series in this definition is always the $k$th element of the sequence being $z$-transformed times $z^{-k}$.

**Remark 2:**     Often $\{f(k)\}$ is defined over equally spaced times: $0, T, 2T, \ldots, kT, \ldots$, where $T$ is a fixed time interval. The resulting sequence is thus sometimes written as $\{f(kT)\}$, or $f(kT)$, $k = 0, 1, 2, \ldots$, and $\mathscr{Z}\{f(kT)\} = \sum_{k=0}^{\infty} f(kT)z^{-k}$, but the dependence on $T$ is usually suppressed. We use the variable arguments $k$ and $kT$ interchangeably for time sequences, when there is no ambiguity.

**Remark 3:**     The $z$-transform is defined differently by some authors, as the transformation $z \equiv e^{sT}$, which amounts to a simple exponential change of variables between the complex variable $z = \mu + j\nu$ and the complex variable $s = \sigma + j\omega$ in the Laplace transform domain, where $T$ is the sampling period of the discrete-time system. This definition implies a sequence $\{f(k)\}$, or $\{f(kT)\}$, obtained by ideal sampling (sometimes called *impulse sampling*) of a continuous signal $f(t)$ at uniformly spaced times $kT$, $k = 1, 2, \ldots$. Then $s = \ln z/T$, and our definition above, that is, $F(z) = \sum_{k=0}^{\infty} f(kT)z^{-k}$, follows directly from the result of Problem 4.39. Additional relationships between continuous and discrete-time systems, particularly for systems with both types of elements, are developed further beginning in Chapter 6.

**EXAMPLE 4.26.**   The series $F(z) = 1 + z^{-1} + z^{-2} + \cdots + z^{-k} + \cdots$, is the $z$-transform of the sequence $f(k) = 1$, $k = 0, 1, 2, \ldots$.

If the rate of increase in the terms of the sequence $\{f(k)\}$ is no greater than that of some geometric series as $k$ approaches infinity, then $\{f(k)\}$ is said to be of **exponential order**. In this case, there exists a real number $r$ such that

$$F(z) = \sum_{k=0}^{\infty} f(k)z^{-k}$$

converges for $|z| > r$. $r$ is called the **radius of convergence** of the series. If $r$ is finite, the sequence $\{f(k)\}$ is called **$z$-transformable**.

**EXAMPLE 4.27.**   The series in Example 4.26 is convergent for $|z| > 1$ and can be written in closed form as the function

$$F(z) = \frac{1}{1 - z^{-1}} \qquad \text{for} \quad |z| > 1$$

If $F(z)$ exists for $|z| > r$, then the integral and derivative of $F(z)$ can be evaluated by operating term by term on the defining series. In addition, if

$$F_1(z) = \sum_{k=0}^{\infty} f_1(k)z^{-k} \qquad \text{for} \quad |z| > r_1$$

and

$$F_2(z) = \sum_{k=0}^{\infty} f_2(k)z^{-k} \qquad \text{for} \quad |z| > r_2$$

then

$$F_1(z)F_2(z) = \sum_{k=0}^{\infty} \left( \sum_{i=0}^{k} f_1(k-i)f_2(i) \right) z^{-k} = \sum_{k=0}^{\infty} \left( \sum_{i=0}^{k} f_2(k-i)f_1(i) \right) z^{-k}$$

The term $\sum_{i=0}^{k} f_1(k-i)f_2(i)$ is called the **convolution sum** of the sequences $\{f_1(k)\}$ and $\{f_2(k)\}$, where the radius of convergence is the larger of the two radii of convergence of $F_1(z)$ and $F_2(z)$.

**EXAMPLE 4.28.**   The derivative of the series in Example 4.26 is

$$\frac{dF}{dz} = -z^{-2} - 2z^{-3} - \cdots - kz^{-(k+1)} - \cdots$$

The indefinite integral is

$$\int F(z)\,dz = z + \ln z - z^{-1} + \cdots$$

**EXAMPLE 4.29.**   The $z$-transform of the sequence $f_2(k) = 2^k$, $k = 0,1,2,\ldots$, is

$$F_2(z) = 1 + 2z^{-1} + 4z^{-2} + \cdots$$

for $|z| > 2$. Let $F_1(z)$ be the $z$-transform in Example 4.26. Then

$$F_1(z)F_2(z) = \sum_{k=0}^{\infty} \left( \sum_{i=0}^{k} 1^{k-1}2^i \right) z^{-k} = \sum_{k=0}^{\infty} (2^{k+1} - 1) z^{-k} \qquad \text{for} \quad |z| > 2$$

The $z$-transform of the sequence $f(k) = A^k$, $k = 0,1,2,\ldots$, where $A$ is any finite complex number, is

$$\mathcal{Z}\{A^k\} = 1 + Az^{-1} + A^2 z^{-2} + \cdots$$

$$= \frac{1}{1 - Az^{-1}} = \frac{z}{z - A}$$

where the radius of convergence $r = |A|$. By suitable choice of $A$, the most common types of sequences can be defined and their $z$-transforms generated from this relationship.

**EXAMPLE 4.30.**   For $A = e^{\alpha T}$, the sequence $\{A^k\}$ is the sampled exponential $1, e^{\alpha T}, e^{2\alpha T}, \ldots$, and the $z$-transform of this sequence is

$$\mathcal{Z}\{e^{\alpha k T}\} = \frac{1}{1 - e^{\alpha T}z^{-1}}$$

with radius of convergence $r = |e^{\alpha T}|$.

The $z$-transform has an inverse very similar to that of the Laplace transform.

**Definition 4.5:**     Let $C$ be a circle centered at the origin of the $z$-plane and with radius greater than the radius of convergence of the $z$-transform $F(z)$. Then

$$\mathcal{Z}^{-1}[F(z)] \equiv \{f(k)\} = \frac{1}{2\pi j} \int_c F(z) z^{k-1}\,dz$$

is the **inverse of the $z$-transform** $F(z)$.

In practice, it is seldom necessary to perform the contour integration in Definition 4.5. For applications of $z$-transforms in this book, it is never necessary. The properties and techniques in the remainder of this section are adequate to evaluate the inverse transform for most discrete-time control system problems.

Following are some additional **properties of the $z$-transform and its inverse** which can be used advantageously in discrete-time control system problems.

1.   The $z$-transform and its inverse are *linear transformations* between the time domain and the $z$-domain. Therefore, if $\{f_1(k)\}$ and $F_1(z)$ are a transform pair and if $\{f_2(k)\}$ and $F_2(z)$ are a

transform pair, then $\{a_1 f_1(k) + a_2 f_2(k)\}$ and $a_1 F_1(z) + a_2 F_2(z)$ are a transform pair for any $a_1$ and $a_2$.

2.  If $F(z)$ is the $z$-transform of the sequence $f(0), f(1), f(2), \ldots$, then

$$z^n F(z) - z^n f(0) - z^{n-1} f(1) - \cdots - z f(n-1)$$

is the $z$-transform of the sequence $f(n), f(n+1), f(n+2), \ldots$, for $n > 1$. Note that the $k$th element of this sequence is $f(n+k)$.

3.  The initial term $f(0)$ of the sequence $\{f(k)\}$ whose $z$-transform is $F(z)$ is

$$f(0) = \lim_{z \to \infty} (1 - z^{-1}) F(z) = F(\infty)$$

This relation is called the **Initial Value Theorem**

4.  Let the sequence $\{f(k)\}$ have the $z$-transform $F(z)$, with radius of convergence $\leq 1$. Then the final value $f(\infty)$ of the sequence is given by

$$f(\infty) = \lim_{z \to 1} (1 - z^{-1}) F(z)$$

if the limit exists. This relation is called the **Final Value Theorem**.

5.  The inverse $z$-transform of the function $F(z/a)$ (**Frequency Scaling**) is

$$\mathcal{Z}^{-1}\left[ F\left( \frac{z}{a} \right) \right] = a^k f(k) \qquad k = 0, 1, 2, \ldots$$

where $\mathcal{Z}^{-1}[F(z)] = \{f(k)\}$.

6.  If $F(z)$ is the $z$-transform of the sequence $f(0), f(1), f(2), \ldots$, then $z^{-1} F(z)$ is the $z$-transform of the time-shifted sequence $f(-1), f(0), f(1), \ldots$, where $f(-1) \equiv 0$. This relationship is called the **Shift Theorem**.

**EXAMPLE 4.31.**  The $z$-transforms of the sequences $\{(\frac{1}{3})^k\}$ and $\{(\frac{1}{2})^k\}$ are $\mathcal{Z}\{(\frac{1}{3})^k\} = z/(z - \frac{1}{3})$, and $\mathcal{Z}\{(\frac{1}{2})^k\} = z/(z - \frac{1}{2})$. Then, by Property 1,

$$\mathcal{Z}\left\{ 3\left( \frac{1}{2} \right)^k - \left( \frac{1}{3} \right)^k \right\} = \frac{3z}{z - \frac{1}{2}} - \frac{z}{z - \frac{1}{3}}$$

$$= \frac{2z^2 - \dfrac{z}{2}}{z^2 - \dfrac{5z}{6} + \dfrac{1}{6}}$$

**EXAMPLE 4.32.**  The inverse $z$-transforms of the functions $z/(z + \frac{1}{2})$ and $z/(z - \frac{1}{4})$ are

$$\mathcal{Z}^{-1}\left[ \frac{z}{z + \frac{1}{2}} \right] = \left\{ \left( -\frac{1}{2} \right)^k \right\}, \qquad \mathcal{Z}^{-1}\left[ \frac{z}{z - \frac{1}{4}} \right] = \left\{ \left( \frac{1}{4} \right)^k \right\}$$

Then, by Property 1,

$$\mathcal{Z}^{-1}\left[ 2\frac{z}{z + \frac{1}{2}} - 4\frac{z}{z - \frac{1}{4}} \right] = 2\,\mathcal{Z}^{-1}\left[ \frac{z}{z + \frac{1}{2}} \right] - 4\,\mathcal{Z}^{-1}\left[ \frac{z}{z - \frac{1}{4}} \right] = \left\{ 2\left( -\frac{1}{2} \right)^k - 4\left( \frac{1}{4} \right)^k \right\}$$

**EXAMPLE 4.33.**  The $z$-transform of the sequence $1, \frac{1}{2}, \frac{1}{4}, \ldots, (\frac{1}{2})^k, \ldots$ is $z/(z - \frac{1}{2})$. Then, by Property 2, the $z$-transform of the sequence $\frac{1}{4}, \frac{1}{8}, \ldots, (\frac{1}{2})^{k+2}, \ldots$ is

$$z^2\left( \frac{z}{z - \frac{1}{2}} \right) - z^2 - \frac{z}{2} = \frac{1}{4}\frac{z}{z - \frac{1}{2}}$$

**EXAMPLE 4.34.**  The $z$-transform of $\{(\frac{1}{4})^k\}$ is $z/(z - \frac{1}{4})$. The initial value of $\{(\frac{1}{4})^k\}$ can be determined by the Initial Value Theorem as

$$\lim_{k \to 0} \left\{ \left( \tfrac{1}{4} \right)^k \right\} = \lim_{z \to \infty} (1 - z^{-1})\left( \frac{z}{z - \frac{1}{4}} \right) = 1$$

**EXAMPLE 4.35.** The $z$-transform of the sequence $\{1 - (\frac{1}{4})^k\}$ is $\frac{3}{4}z/(z^2 - \frac{5z}{4} + \frac{1}{4})$. The final value of this sequence can be determined from the Final Value Theorem as

$$\lim_{k \to \infty} \left\{ 1 - \left(\frac{1}{4}\right)^k \right\} = \lim_{z \to 1} (1 - z^{-1}) \left( \frac{\frac{3}{4}z}{z^2 - \frac{5z}{4} + \frac{1}{4}} \right) = 1$$

**EXAMPLE 4.36.** The inverse $z$-transform of $z/(z - \frac{1}{4})$ is $\{(\frac{1}{4})^k\}$. The inverse transform of $(\frac{z}{2})/(\frac{z}{2} - \frac{1}{4})$ is $\{2^k (\frac{1}{4})^k\} = \{(\frac{1}{2})^k\}$.

For the types of control problems considered in this book, the resulting $z$-transforms are rational algebraic functions of $z$, as illustrated below, and there are two practical methods for inverting them. The first is a numerical technique, generating a power series expansion by long division.

Suppose the $z$-transform has the form:

$$F(z) = \frac{b_n z^n + b_{n-1} z^{n-1} + \cdots + b_1 z + b_0}{a_n z^n + a_{n-1} z^{n-1} + \cdots + a_1 z + a_0}$$

It is easily rewritten in powers of $z^{-1}$ as

$$F(z) = \frac{b_n + b_{n-1} z^{-1} + \cdots + b_0 z^{-n}}{a_n + a_{n-1} z^{-1} + \cdots + a_0 z^{-n}}$$

by multiplying each term by $z^{-n}$. Then, by long division, the denominator is divided into the numerator, yielding a polynomial in $z^{-1}$ of the form:

$$F(z) = \frac{b_n}{a_n} + \frac{1}{a_n} \left( b_{n-1} - \frac{b_n a_{n-1}}{a_n} \right) z^{-1} + \cdots$$

**EXAMPLE 4.37.** The $z$-transform $z/(z - \frac{1}{2})$ is rewritten as $1/(1 - z^{-1}/2)$ which, by long division, has the form:

$$\frac{1}{1 - z^{-1}/2} = 1 + \left(\frac{1}{2}\right) z^{-1} + \left(\frac{1}{2}\right)^2 z^{-2} + \cdots$$

For the second inversion method, $F(z)$ is first expanded into a special partial fraction form and each term is inverted using the properties previously discussed.

Table 4.2 is a short table of $z$-transform pairs. When used in conjunction with the properties of the $z$-transform described earlier, and the partial fraction expansion techniques described in Section 4.7, it

**Table 4.2**

| $k$ th Term of the Time Sequence | $z$-Transform |
|---|---|
| 1 at $k$, 0 elsewhere (Kronecker delta sequence) | $z^{-k}$ |
| 1 (unit step sequence) | $\dfrac{z}{z-1}$ |
| $k$ (unit ramp sequence) | $\dfrac{z}{(z-1)^2}$ |
| $A^k$ (for complex numbers $A$) | $\dfrac{z}{z-A}$ |
| $kA^k$ | $\dfrac{Az}{(z-A)^2}$ |
| $\dfrac{(k+1)(k+2) \cdots (k+n-1)}{(n-1)!} A^k$ | $\dfrac{z^n}{(z-A)^n}$ |

is adequate to handle all the problems in this book. A more complete table of z-transform pairs is given in Appendix B.

The final transform pair in Table 4.2 can be used to generate many other useful transforms by proper choice of $A$ and use of Property 1.

The following examples illustrate how z-transforms can be inverted using the partial fraction expansion method.

**EXAMPLE 4.38.** To invert the z-transform $F(z) = 1/(z+1)(z+2)$, we form the partial fraction expansion of $F(z)/z$:

$$\frac{F(z)}{z} = \frac{1}{z(z+1)(z+2)} = \frac{\frac{1}{2}}{z} + \frac{-1}{z+1} + \frac{\frac{1}{2}}{z+2}$$

Then

$$F(z) = \frac{1}{2} - \frac{z}{z+1} + \frac{1}{2}\frac{z}{z+2}$$

which can be inverted term by term as

$$f(0) = 0$$

$$f(k) = -(-1)^k + \frac{1}{2}(-2)^k \quad \text{for all} \quad k \geq 1$$

**EXAMPLE 4.39.** To invert $F(z) = 1/(z+1)^2(z+2)$, we take the partial fraction expansion of $F(z)/z$:

$$\frac{F(z)}{z} = \frac{\frac{1}{2}}{z} + \frac{0}{z+1} + \frac{-1}{(z+1)^2} + \frac{-\frac{1}{2}}{z+2}$$

Then

$$F(z) = \frac{1}{2} - \frac{z}{(z+1)^2} - \frac{1}{2}\frac{z}{z+2}$$

$$f(k) = -k(-1)^k - \frac{1}{2}(-2)^k \quad \text{for all} \quad k \geq 1 \text{ and } f(0) = 0$$

**EXAMPLE 4.40.** Using the last transform pair in Table 4.2, the z-transform of the sequence $\{k^2/2\}$ can be generated by noting the following transform pairs:

$$\left\{\frac{(k+1)(k+2)}{2!}\right\} \leftrightarrow \frac{z^3}{(z-1)^3}$$

$$\{k\} \leftrightarrow \frac{z}{(z-1)^2}$$

$$\{1\} \leftrightarrow \frac{z}{z-1}$$

Since

$$\frac{(k+1)(k+2)}{2!} = \frac{k^2}{2} + \frac{3}{2}k + 1$$

then, by Property 1,

$$\mathcal{Z}\left\{\frac{k^2}{2}\right\} = \frac{z^3}{(z-1)^3} - \frac{3}{2}\frac{z}{(z-1)^2} - \frac{z}{z-1} = \frac{z(z+1)/2}{(z-1)^3}$$

Linear $n$th-order constant-coefficient difference equations can be solved using z-transform methods by a procedure virtually identical to that used to solve differential equations by Laplace transform methods. This is illustrated step by step in the following example.

**EXAMPLE 4.41.** The difference equation

$$x(k+2) + \frac{5}{6}x(k+1) + \frac{1}{6}x(k) = 1$$

with initial conditions $x(0) = 0$ and $x(1) = 1$ is $z$-transformed by applying Properties 1 and 2. By Property 1 (Linearity):

$$\mathcal{Z}\left\{x(k+2) + \frac{5}{6}x(k+1) + \frac{1}{6}x(k)\right\} = \mathcal{Z}\{x(k+2)\} + \frac{5}{6}\mathcal{Z}\{x(k+1)\} + \frac{1}{6}\mathcal{Z}\{x(k)\} = \mathcal{Z}\{1\}$$

By Property 2, if $\mathcal{Z}[x(k)] \equiv X(z)$, then

$$\mathcal{Z}\{x(k+1)\} = zX(z) - zx(0) = zX(z)$$

$$\mathcal{Z}\{x(k+2)\} = z^2X(z) - z^2x(0) - zx(1) = z^2X(z) - z$$

From Table 4.2, the $z$-transform of the unit step sequence is

$$\mathcal{Z}\{1\} = \frac{z}{z-1}$$

Direct substitution of these expressions into the transformed equation then gives

$$\left(z^2 + \frac{5}{6}z + \frac{1}{6}\right)X(z) - z = \frac{z}{z-1}$$

Thus the $z$-transform $X(z)$ of the solution sequence $x(k)$ is

$$X(z) = \frac{z}{z^2 + \frac{5}{6}z + \frac{1}{6}} + \frac{z}{(z-1)\left(z^2 + \frac{5}{6}z + \frac{1}{6}\right)} = X_a(z) + X_b(z)$$

Note that the first term $X_a(z)$ results from the initial conditions and the second term $X_b(z)$ results from the input sequence. Thus the inverse of the first term is the *free response*, and the inverse of the second term is the *forced response*. The first term can be inverted by forming the partial fraction expansion

$$\frac{X_a(z)}{z} = \frac{1}{z^2 + \frac{5}{6}z + \frac{1}{6}} = -\frac{6}{z + \frac{1}{2}} + \frac{6}{z + \frac{1}{3}}$$

From this,

$$X_a(z) = -6\frac{z}{z + \frac{1}{2}} + 6\frac{z}{z + \frac{1}{3}}$$

and from Table 4.2, the inverse of $X_a(z)$ (the free response) is

$$x_a(k) = -6\left(-\frac{1}{2}\right)^k + 6\left(-\frac{1}{3}\right)^k \qquad k = 0, 1, 2, \ldots$$

Similarly, to find the forced response, the following partial fraction expansion is formed:

$$\frac{X_b(z)}{z} = \frac{1}{(z-1)\left(z + \frac{1}{2}\right)\left(z + \frac{1}{3}\right)}$$

$$= \frac{\frac{1}{2}}{z-1} + \frac{4}{z + \frac{1}{2}} + \frac{-\frac{9}{2}}{z + \frac{1}{3}}$$

Thus

$$X_b(z) = \frac{\frac{1}{2}z}{z-1} + \frac{4z}{z + \frac{1}{2}} - \frac{\frac{9}{2}z}{z + \frac{1}{3}}$$

Then, from Table 4.2, the inverse of $X_b(z)$ (the forced response) is

$$x_b(k) = \frac{1}{2} + 4\left(-\frac{1}{2}\right)^k - \frac{9}{2}\left(-\frac{1}{3}\right)^k \qquad k = 0, 1, 2, \ldots$$

The total response $x(k)$ is

$$x(k) \equiv x_a(k) + x_b(k) = \frac{1}{2} - 2\left(-\frac{1}{2}\right)^k + \frac{3}{2}\left(-\frac{1}{3}\right)^k \qquad k = 0,1,2,\ldots$$

Linear constant-coefficient vector-matrix difference equations presented in Section 3.17 can also be solved by $z$-transform techniques, as illustrated in the following example.

**EXAMPLE 4.42.**   Consider the difference equation of Example 4.41 written in state variable form (see Example 3.36):

$$x_1(k+1) = x_2(k)$$

$$x_2(k+1) = -\frac{5}{6}x_2(k) - \frac{1}{6}x_1(k) + 1$$

with initial conditions $x_1(0) = 0$ and $x_2(0) = 1$. In vector-matrix form, these two equations are written as

$$\mathbf{x}(k+1) = A\mathbf{x}(k) + \mathbf{b}u(k)$$

where

$$A = \begin{bmatrix} 0 & 1 \\ -\frac{1}{6} & -\frac{5}{6} \end{bmatrix} \qquad \mathbf{b} = \begin{bmatrix} 0 \\ 1 \end{bmatrix} \qquad \mathbf{x}(k) = \begin{bmatrix} x_1(k) \\ x_2(k) \end{bmatrix} \qquad \mathbf{x}(0) = \begin{bmatrix} 0 \\ 1 \end{bmatrix}$$

$u(k) = 1$. The $z$-transform of the vector-matrix form of the equation is

$$z\mathbf{X}(z) - z\mathbf{x}(0) = A\mathbf{X}(z) + \frac{z}{z-1}\mathbf{b}$$

where $\mathbf{X}(z)$ is a vector-valued $z$-transform whose components are the $z$-transforms of the corresponding components of the state vector $\mathbf{x}(k)$. This transformed equation can be rewritten as

$$(zI - A)\mathbf{X}(z) = z\mathbf{x}(0) + \frac{z}{z-1}\mathbf{b}$$

where $I$ is the identity or unit matrix. The $z$-transform of the solution vector $\mathbf{x}(k)$ is

$$\mathbf{X}(z) = z(zI - A)^{-1}\mathbf{x}(0) + \frac{z}{z-1}(zI - A)^{-1}\mathbf{b}$$

where $(\cdot)^{-1}$ represents the *inverse* of the matrix. Since

$$zI - A = \begin{bmatrix} z & -1 \\ \frac{1}{6} & z + \frac{5}{6} \end{bmatrix}$$

then

$$(zI - A)^{-1} = \frac{1}{z^2 + \frac{5}{6}z + \frac{1}{6}} \begin{bmatrix} z + \frac{5}{6} & 1 \\ -\frac{1}{6} & z \end{bmatrix}$$

Substituting for $(zI - A)^{-1}$, $\mathbf{x}(0)$, and $\mathbf{b}$ yields

$$\mathbf{X}(z) = \begin{bmatrix} \dfrac{z}{z^2 + \frac{5}{6}z + \frac{1}{6}} \\[3mm] \dfrac{z^2}{z^2 + \frac{5}{6}z + \frac{1}{6}} \end{bmatrix} + \begin{bmatrix} \dfrac{z}{(z-1)\left(z^2 + \frac{5}{6}z + \frac{1}{6}\right)} \\[3mm] \dfrac{z^2}{(z-1)\left(z^2 + \frac{5}{6}z + \frac{1}{6}\right)} \end{bmatrix}$$

where the first term is the $z$-transform of the free response and the second of the forced response. Using the partial fraction expansion method and Table 4.2, the inverse of this $z$-transform is

$$\mathbf{x}(k) = \begin{bmatrix} \frac{1}{2} - 2\left(-\frac{1}{2}\right)^k + \frac{3}{2}\left(-\frac{1}{3}\right)^k \\[2mm] \frac{1}{2} + \left(-\frac{1}{2}\right)^k - \frac{1}{2}\left(-\frac{1}{3}\right)^k \end{bmatrix} \qquad k = 0,1,2,\ldots$$

## 4.10 DETERMINING ROOTS OF POLYNOMIALS

The results of Sections 4.7, 4.8, and 4.9 indicate that finding the solution of linear constant-coefficient differential and difference equations by transform techniques generally requires the determination of the roots of polynomial equations of the form:

$$Q_n(s) = \sum_{i=0}^{n} a_i s^i = 0$$

where $a_n = 1$, $a_0, a_1, \ldots, a_{n-1}$, are real constants and $s$ is replaced by $z$ for $z$-transform polynomials.

The roots of a second-order polynomial equation $s^2 + a_1 s + a_0 = 0$ can be obtained directly from the quadratic formula and are given by

$$s_1 = \frac{-a_1 + \sqrt{a_1^2 - 4a_0}}{2} \qquad s_2 = \frac{-a_1 - \sqrt{a_1^2 - 4a_0}}{2}$$

But for higher-order polynomials such analytical expressions do not, in general, exist. The expressions that do exist are very complicated. Fortunately, numerical techniques exist for determining these roots.

To aid in the use of these numerical techniques, the following general properties of $Q_n(s)$ are given:

1. If a repeated root of multiplicity $n_i$ is counted as $n_i$ roots, then $Q_n(s) = 0$ has exactly $n$ roots (Fundamental theorem of algebra).

2. If $Q_n(s)$ is divided by the factor $s + p$ until a constant remainder is obtained, the remainder is $Q_n(-p)$.

3. $s + p$ is a factor of $Q_n(s)$ if and only if $Q_n(-p) = 0$ [$-p$ is a root of $Q_n(s) = 0$].

4. If $\sigma + j\omega$ ($\sigma, \omega$ real) is a root of $Q_n(s) = 0$, then $\sigma - j\omega$ is also a root of $Q_n(s) = 0$.

5. If $n$ is odd, $Q_n(s) = 0$ has at least one real root.

6. The number of positive real roots of $Q_n(s) = 0$ cannot exceed the number of variations in sign of the coefficients in the polynomial $Q_n(s)$, and the number of negative roots cannot exceed the number of variations in sign of the coefficients of $Q_n(-s)$ (Descartes' rule of signs).

Of the techniques available for iteratively determining the roots of a polynomial equation (or equivalently the factors of the polynomial), some can determine only real roots and others both real and complex roots. Both types are presented below.

### Horner's Method

This method can be used to determine the *real roots* of the polynomial equation $Q_n(s) = 0$. The steps to be followed are:

1. Evaluate $Q_n(s)$ for real integer values of $s$, $s = 0, \pm 1, \pm 2, \ldots$, until for two consecutive integer values such as $k_0$ and $k_0 + 1$, $Q_n(k_0)$ and $Q_n(k_0 + 1)$ have opposite signs. A real root then lies between $k_0$ and $k_0 + 1$. Assume this root is positive without loss of generality. A first approximation of the root is taken to be $k_0$. Corrections to this approximation are obtained in the remaining steps.

2. Determine a sequence of polynomials $Q_n^l(s)$ using the recursive relationship

$$Q_n^{l+1}(s) = Q_n^l\left(\frac{k_l}{10^l} + s\right) = \sum_{i=0}^{n} a_i^{l+1} s^i \qquad l = 0, 1, 2, \ldots \qquad (4.13)$$

where $Q_n^0(s) = Q_n(s)$, and the values $k_l$, $l = 1, 2, \ldots$, are generated in Step 3.

3. Determine the integer $k_l$ at each iteration by evaluating $Q_n^l(s)$ for real values of $s$ given by $s = k/10^l$, $k = 0, 1, 2, \ldots, 9$. For two consecutive values of $k$, say $k_l$ and $k_{l+1}$, the values $Q_n(k_l/10^l)$ and $Q_n(k_{l+1}/10^l)$ have opposite signs.

4.  Repeat until the desired accuracy of the root has been achieved. The approximation of the real root after the $N$th iteration is given by

$$s_N = \sum_{l=0}^{N} \frac{k_l}{10^l} \qquad (4.14)$$

Each iteration increases the accuracy of the approximation by one decimal place.

## Newton's Method

This method can determine *real roots* of the polynomial equation $Q_n(s) = 0$. The steps to be followed are:

1.  Obtain a first approximation $s_0$ of a root by making an "educated" guess, or by a technique such as the one in Step 1 of Horner's method.

2.  Generate a sequence of improved approximations until the desired accuracy is achieved by the recursive relationship

$$s_{l+1} = s_l - \left. \frac{Q_n(s)}{\dfrac{d}{ds}[Q_n(s)]} \right|_{s=s_l}$$

which can be rewritten as

$$s_{l+1} = \frac{\displaystyle\sum_{i=0}^{n}(i-1)a_i s_l^i}{\displaystyle\sum_{i=1}^{n} i a_i s_l^{i-1}} \qquad (4.15)$$

where $l = 0, 1, 2, \ldots$.

This method does not provide a measure of the accuracy of the approximation. Indeed, there is no guarantee that the approximations converge to the correct value.

## Lin-Bairstow Method

This method can determine both *real and complex roots* of the polynomial equation $Q_n(s) = 0$. More exactly, this method determines quadratic factors of $Q_n(s)$ from which two roots can be determined by the quadratic formula. The roots can, of course, be either real or complex. The steps to be followed are:

1.  Obtain a first approximation of a quadratic factor

$$s^2 + \alpha_1 s + \alpha_0$$

of $Q_n(s) = \sum_{i=0}^{n} a_i s^i$ by some method, perhaps an "educated" guess. Corrections to this approximation are obtained in the remaining steps.

2.  Generate a set of constants $b_{n-2}, b_{n-3}, \ldots, b_0, b_{-1}, b_{-2}$ from the recursive relationship

$$b_{i-2} = a_i - \alpha_1 b_{i-1} - \alpha_0 b_i$$

where $b_n = b_{n-1} = 0$, and $i = n, n-1, \ldots, 1, 0$.

3.  Generate a set of constants $c_{n-2}, c_{n-3}, \ldots, c_1, c_0$ from the recursive relationship

$$c_{i-1} = b_{i-1} - \alpha_1 c_i - \alpha_0 c_{i+1}$$

where $c_n = c_{n-1} = 0$, and $i = n, n-1, \ldots, 1$.

4.  Solve the two simultaneous equations

$$c_0 \, \Delta\alpha_1 + c_1 \, \Delta\alpha_0 = b_{-1}$$

$$(-\alpha_1 c_0 - \alpha_0 c_1) \, \Delta\alpha_1 + c_0 \, \Delta\alpha_0 = b_{-2}$$

for $\Delta\alpha_1$ and $\Delta\alpha_0$. The new approximation of the quadratic factor is

$$s^2 + (\alpha_1 + \Delta\alpha_1)s + (\alpha_0 + \Delta\alpha_0)$$

5.  Repeat Steps 1 through 4 for the quadratic factor obtained in Step 4, until successive approximations are sufficiently close.

This method does not provide a measure of the accuracy of the approximation. Indeed, there is no guarantee that the approximations converge to the correct value.

### Root-Locus Method

This method can be used to determine both real and complex roots of the polynomial equation $Q_n(s) = 0$. The technique is discussed in Chapter 13.

## 4.11  COMPLEX PLANE: POLE-ZERO MAPS

The rational functions $F(s)$ for continuous systems can be rewritten as

$$F(s) = \frac{b_m \sum\limits_{i=0}^{m} (b_i/b_m)s^i}{\sum\limits_{i=0}^{n} a_i s^i} = \frac{b_m \prod\limits_{i=1}^{m} (s + z_i)}{\prod\limits_{i=0}^{n} (s + p_i)}$$

where the terms $s + z_i$ are factors of the numerator polynomial and the terms $s + p_i$ are factors of the denominator polynomial, with $a_n \equiv 1$. If $s$ is replaced by $z$, $F(z)$ represents a system function for discrete-time systems.

***Definition 4.6:***    Those values of the complex variable $s$ for which $|F(s)|$ [absolute value of $F(s)$] is zero are called the **zeros** of $F(s)$.

***Definition 4.7:***    Those values of the complex variable $s$ for which $|F(s)|$ is infinite are called the **poles** of $F(s)$.

**EXAMPLE 4.43.**    Let $F(s)$ be given by

$$F(s) = \frac{2s^2 - 2s - 4}{s^3 + 5s^2 + 8s + 6}$$

which can be rewritten as

$$F(s) = \frac{2(s+1)(s-2)}{(s+3)(s+1+j)(s+1-j)}$$

$F(s)$ has *finite zeros* at $s = -1$ and $s = 2$, and a *zero* at $s = \infty$. $F(s)$ has *finite poles* at $s = -3$, $s = -1-j$, and $s = -1 + j$.

**Poles** and **zeros** are complex numbers determined by two real variables, one representing the real part and the other the imaginary part of the complex number. A pole or zero can therefore be represented as a point in rectangular coordinates. The *abscissa* of this point represents the real part and the *ordinate* the imaginary part. In the $s$-plane, the abscissa is called the **σ-axis** and the ordinate the **jω-axis**. In the $z$-plane, the abscissa is called the **μ-axis** and the ordinate the **jν-axis**. The planes defined by these coordinate systems are generally called the **complex plane** ($s$-plane or $z$-plane). That half of the complex plane in which $\mathrm{Re}(s) < 0$ or $\mathrm{Re}(z) < 0$ is called the **left half of the $s$-plane** or **$z$-plane** (LHP), and that half in which $\mathrm{Re}(s) > 0$ or $\mathrm{Re}(z) > 0$ is called the **right half of the $s$-plane** or **$z$-plane** (RHP). That portion of the $z$-plane in which $|z| < 1$ is called (the interior of) the **unit circle** in the $z$-plane.

The location of a pole in the complex plane is denoted symbolically by a cross ($\times$), and the location of a zero by a small circle ($\circ$). The $s$-plane including the locations of the finite poles and zeros of $F(s)$ is called the **pole-zero map** of $F(s)$. A similar comment holds for the $z$-plane.

**EXAMPLE 4.44.** The rational function

$$F(s) = \frac{(s+1)(s-2)}{(s+3)(s+1+j)(s+1-j)}$$

has finite poles $s = -3$, $s = -1 - j$, and $s = -1 + j$, and finite zeros $s = -1$ and $s = 2$. The pole-zero map of $F(s)$ is shown in Fig. 4-2.

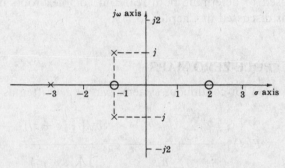

Fig. 4-2

## 4.12   GRAPHICAL EVALUATION OF RESIDUES*

Let $F(s)$ be a rational function written in its factored form:

$$F(s) = \frac{b_m \displaystyle\prod_{i=1}^{m} (s + z_i)}{\displaystyle\prod_{i=1}^{n} (s + p_i)}$$

Since $F(s)$ is a complex function, it can be written in *polar form* as

$$F(s) = |F(s)| e^{j\phi} = |F(s)| \underline{/\phi}$$

where $|F(s)|$ is the absolute value of $F(s)$ and $\phi \equiv \arg F(s) = \tan^{-1}[\mathrm{Im}\, F(s) / \mathrm{Re}\, F(s)]$.

$F(s)$ can further be written in terms of the polar forms of the factors $s + z_i$ and $s + p_i$ as

$$F(s) = \frac{b_m \displaystyle\prod_{i=1}^{m} |s + z_i|}{\displaystyle\prod_{i=1}^{n} |s + p_i|} \underline{\left/ \left[ \sum_{i=1}^{m} \phi_{iz} - \sum_{i=1}^{n} \phi_{ip} \right] \right.}$$

where $s + z_i = |s + z_i| \underline{/\phi_{iz}}$ and $s + p_i = |s + p_i| \underline{/\phi_{ip}}$.

---

*While $s$ is used to represent the complex variable in this section, it is not intended to represent the Laplace variable only but rather to be a general complex variable and the discussion is applicable to both the Laplace and $z$-transforms.

Each complex number $s$, $z_i$, $p_i$, $s + z_i$, and $s + p_i$ can be represented by a vector in the $s$-plane. If $p$ is a general complex number, then the vector representing $p$ has magnitude $|p|$ and direction defined by the angle

$$\phi = \tan^{-1}\left[\frac{\text{Im } p}{\text{Re } p}\right]$$

measured counterclockwise from the positive $\sigma$-axis.

A typical pole $-p_i$ and zero $-z_i$ are shown in Fig. 4-3, along with a general complex variable $s$. The *sum vectors* $s + z_i$ and $s + p_i$ are also shown. Note that the vector $s + z_i$ is a vector which starts at the zero $-z_i$ and terminates at $s$, and $s + p_i$ starts at the pole $-p_i$ and terminates at $s$.

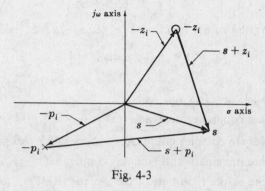

Fig. 4-3

For distinct poles of the rational function $F(s)$, the *residue* $c_{k1} \equiv c_k$ of the pole $-p_k$ is given by

$$c_k = (s + p_k)F(s)\big|_{s = -p_k} = \left.\frac{b_m(s + p_k)\prod\limits_{i=1}^{m}(s + z_i)}{\prod\limits_{i=1}^{n}(s + p_i)}\right|_{s = -p_k}$$

These residues can be determined by the following graphical procedure:

1.  Plot the pole-zero map of $(s + p_k)F(s)$.

2.  Draw vectors on this map starting at the poles and zeros of $(s + p_k)F(s)$, and terminating at $-p_k$. Measure the magnitude (in the scale of the pole-zero map) of these vectors and the angles of the vectors measured from the positive real axis in the counterclockwise direction.

3.  Obtain the magnitude $|c_k|$ of the residue $c_k$ as the product of $b_m$ and the magnitudes of the vectors from the zeros to $-p_k$, divided by the product of the magnitudes of the vectors from the poles to $-p_k$.

4.  Determine the angle $\phi_k$ of the residue $c_k$ as the sum of the angles of the vectors from the zeros to $-p_k$, minus the sum of the angles of the vectors from the poles to $-p_k$. This is true for positive $b_m$. If $b_m$ is negative, then add 180° to this angle.

The residue $c_k$ is given in polar form by

$$c_k = |c_k|e^{j\phi_k} = |c_k|\underline{/\phi_k}$$

or in rectangular form by

$$c_k = |c_k|\cos\phi_k + j|c_k|\sin\phi_k$$

This graphical technique is not directly applicable for evaluating residues of multiple poles.

### 4.13  SECOND-ORDER SYSTEMS

As indicated in Section 3.14, many control systems can be described or approximated by the general second-order differential equation

$$\frac{d^2y}{dt^2} + 2\zeta\omega_n\frac{dy}{dt} + \omega_n^2 y = \omega_n^2 u$$

The positive coefficient $\omega_n$ is called the **undamped natural frequency** and the coefficient $\zeta$ is the **damping ratio** of the system.

The Laplace transform of $y(t)$, when the initial conditions are zero, is

$$Y(s) = \left[\frac{\omega_n^2}{s^2 + 2\zeta\omega_n s + \omega_n^2}\right]U(s)$$

where $U(s) = \mathscr{L}[u(t)]$. The poles of the function $Y(s)/U(s) = \omega_n^2/(s^2 + 2\zeta\omega_n s + \omega_n^2)$ are

$$s = -\zeta\omega_n \pm \omega_n\sqrt{\zeta^2 - 1}$$

Note that:

1.  If $\zeta > 1$, both poles are negative and real.
2.  If $\zeta = 1$, the poles are equal, negative, and real ($s = -\omega_n$).
3.  If $0 < \zeta < 1$, the poles are complex conjugates with negative real parts ($s = -\zeta\omega_n \pm j\omega_n\sqrt{1 - \zeta^2}$).
4.  If $\zeta = 0$, the poles are imaginary and complex conjugate ($s = \pm j\omega_n$).
5.  If $\zeta < 0$, the poles are in the right half of the $s$-plane (RHP).

Of particular interest in this book is Case 3, representing an **underdamped second-order system**. The poles are complex conjugates with negative real parts and are located at

$$s = -\zeta\omega_n \pm j\omega_n\sqrt{1 - \zeta^2}$$

or at

$$s = -\alpha \pm j\omega_d$$

where $1/\alpha \equiv 1/\zeta\omega_n$ is called the **time constant** of the system and $\omega_d \equiv \omega_n\sqrt{1 - \zeta^2}$ is called the **damped natural frequency** of the system. For fixed $\omega_n$, Fig. 4-4 shows the locus of these poles as a function of $\zeta$, $0 < \zeta < 1$. The locus is a semicircle of radius $\omega_n$. The angle $\theta$ is related to the damping ratio by $\theta = \cos^{-1}\zeta$.

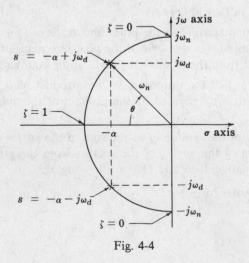

Fig. 4-4

A similar description for second-order systems described by difference equations does not exist in such a simple and useful form.

# Solved Problems

## LAPLACE TRANSFORMS FROM THE DEFINITION

**4.1.** Show that the unit step function $\mathbf{1}(t)$ is Laplace transformable and determine its Laplace transform.

Direct substitution into the equation of Definition 4.2 yields

$$\int_{0^+}^{\infty} |\mathbf{1}(t)| e^{-\sigma_0 t}\, dt = \int_{0^+}^{\infty} e^{-\sigma_0 t}\, dt = -\frac{1}{\sigma_0} e^{-\sigma_0 t}\Big|_{0^+}^{\infty} = \frac{1}{\sigma_0} < +\infty$$

for $\sigma_0 > 0$. The Laplace transform is given by Definition 4.1:

$$\mathscr{L}[\mathbf{1}(t)] = \int_{0^+}^{\infty} \mathbf{1}(t) e^{-st}\, dt = -\frac{1}{s} e^{-st}\Big|_{0^+}^{\infty} = \frac{1}{s} \quad \text{for} \quad \operatorname{Re} s > 0$$

**4.2.** Show that the unit ramp function $t$ is Laplace transformable and determine its Laplace transform.

Direct substitution into the equation of Definition 4.2 yields

$$\int_{0^+}^{\infty} |t| e^{-\sigma_0 t}\, dt = \frac{e^{-\sigma_0 t}}{\sigma_0^2}(-\sigma_0 t - 1)\Big|_{0^+}^{\infty} = \frac{1}{\sigma_0^2} < +\infty$$

for $\sigma_0 > 0$. The Laplace transform is given by Definition 4.1:

$$\mathscr{L}[t] = \int_{0^+}^{\infty} t e^{-st}\, dt = \frac{e^{-st}}{s^2}(-st - 1)\Big|_{0^+}^{\infty} = \frac{1}{s^2} \quad \text{for} \quad \operatorname{Re} s > 0$$

**4.3.** Show that the sine function $\sin t$ is Laplace transformable and determine its Laplace transform.

The integral $\int_{0^+}^{\infty} |\sin t| e^{-\sigma_0 t}\, dt$ can be evaluated by writing the integral over the positive half cycles of $\sin t$ as

$$\int_{n\pi}^{(n+1)\pi} \sin t\, e^{-\sigma_0 t}\, dt = \frac{e^{-\sigma_0 n\pi}}{\sigma_0^2 + 1}\left[e^{-\sigma_0 \pi} + 1\right]$$

for $n$ even, and over negative half-cycles of $\sin t$ as

$$-\int_{n\pi}^{(n+1)\pi} \sin t\, e^{-\sigma_0 t}\, dt = \frac{e^{-\sigma_0 n\pi}}{\sigma_0^2 + 1}\left[e^{-\sigma_0 \pi} + 1\right]$$

for $n$ odd. Then

$$\int_{0^+}^{\infty} |\sin t| e^{-\sigma_0 t}\, dt = \frac{e^{-\sigma_0 \pi} + 1}{\sigma_0^2 + 1} \sum_{n=0}^{\infty} e^{-\sigma_0 n\pi}$$

The summation converges for $e^{-\sigma_0 \pi} < 1$ or $\sigma_0 > 0$ and can be written in closed form as

$$\sum_{n=0}^{\infty} e^{-\sigma_0 n\pi} = \frac{1}{1 - e^{-\sigma_0 \pi}}$$

Then

$$\int_{0^+}^{\infty} |\sin t| e^{-\sigma_0 t}\, dt = \left[\frac{1 + e^{-\sigma_0 \pi}}{1 - e^{-\sigma_0 \pi}}\right]\left(\frac{1}{\sigma_0^2 + 1}\right) < +\infty \quad \text{for} \quad \sigma_0 > 0$$

Finally, $\mathscr{L}[\sin t] = \int_{0^+}^{\infty} \sin t\, e^{-st}\, dt = \frac{e^{-st}(-s \sin t - \cos t)}{s^2 + 1}\Big|_{0^+}^{\infty} = \frac{1}{s^2 + 1} \quad \text{for} \quad \operatorname{Re} s > 0$

**4.4.** Show that the Laplace transform of the unit impulse function is given by $\mathscr{L}[\delta(t)] = 1$.

Direct substitution of Equation (3.19) into the equation of Definition 4.1 yields

$$\int_{0^+}^{\infty} \delta(t) e^{-st} \, dt = \int_{0^+}^{\infty} \lim_{\Delta t \to 0} \left[ \frac{\mathbf{1}(t) - \mathbf{1}(t - \Delta t)}{\Delta t} \right] e^{-st} \, dt$$

$$= \lim_{\Delta t \to 0} \left[ \int_{0^+}^{\infty} \frac{\mathbf{1}(t)}{\Delta t} e^{-st} \, dt - \int_{0^+}^{\infty} \frac{\mathbf{1}(t - \Delta t)}{\Delta t} e^{-st} \, dt \right] = \lim_{\Delta t \to 0} \frac{1}{\Delta t} \left[ \frac{1}{s} - \frac{e^{-\Delta t s}}{s} \right]$$

where the Laplace transform of $\mathbf{1}(t)$ is $1/s$, as shown in Problem 4.1, and the second term is obtained using Property 9. Now

$$e^{-\Delta t s} = 1 - \Delta t s + \frac{(\Delta t s)^2}{2!} - \frac{(\Delta t s)^3}{3!} + \cdots$$

(see Reference [1]). Thus

$$\mathscr{L}[\delta(t)] = \lim_{\Delta t \to 0} \frac{1}{\Delta t} \left[ \frac{1}{s} - \frac{e^{-\Delta t s}}{s} \right] = \lim_{\Delta t \to 0} \frac{1}{\Delta t} \left[ \Delta t - \frac{(\Delta t)^2 s}{2!} + \frac{(\Delta t)^3 s^2}{3!} - \cdots \right] = 1$$

## PROPERTIES OF THE LAPLACE TRANSFORM AND ITS INVERSE

**4.5.** Show that $\mathscr{L}[a_1 f_1(t) + a_2 f_2(t)] = a_1 F_1(s) + a_2 F_2(s)$, where $F_1(s) = \mathscr{L}[f_1(t)]$ and $F_2(s) = \mathscr{L}[f_2(t)]$ (Property 1).

By definition,

$$\mathscr{L}[a_1 f_1(t) + a_2 f_2(t)] = \int_{0^+}^{\infty} [a_1 f_1(t) + a_2 f_2(t)] e^{-st} \, dt$$

$$= \int_{0^+}^{\infty} a_1 f_1(t) e^{-st} \, dt + \int_{0^+}^{\infty} a_2 f_2(t) e^{-st} \, dt$$

$$= a_1 \int_{0^+}^{\infty} f_1(t) e^{-st} \, dt + a_2 \int_{0^+}^{\infty} f_2(t) e^{-st} \, dt$$

$$= a_1 \mathscr{L}[f_1(t)] + a_2 \mathscr{L}[f_2(t)] = a_1 F_1(s) + a_2 F_2(s)$$

**4.6.** Show that $\mathscr{L}^{-1}[a_1 F_1(s) + a_2 F_2(s)] = a_1 f_1(t) + a_2 f_2(t)$, where $\mathscr{L}^{-1}[F_1(s)] = f_1(t)$ and $\mathscr{L}^{-1}[F_2(s)] = f_2(t)$ (Property 2).

By definition,

$$\mathscr{L}^{-1}[a_1 F_1(s) + a_2 F_2(s)] = \frac{1}{2\pi j} \int_{c-j\infty}^{c+j\infty} [a_1 F_1(s) + a_2 F_2(s)] e^{st} \, ds$$

$$= \frac{1}{2\pi j} \int_{c-j\infty}^{c+j\infty} a_1 F_1(s) e^{st} \, ds + \frac{1}{2\pi j} \int_{c-j\infty}^{c+j\infty} a_2 F_2(s) e^{st} \, ds$$

$$= a_1 \left[ \frac{1}{2\pi j} \int_{c-j\infty}^{c+j\infty} F_1(s) e^{st} \, ds \right] + a_2 \left[ \frac{1}{2\pi j} \int_{c-j\infty}^{c+j\infty} F_2(s) e^{st} \, ds \right]$$

$$= a_1 \mathscr{L}^{-1}[F_1(s)] + a_2 \mathscr{L}^{-1}[F_2(s)] = a_1 f_1(t) + a_2 f_2(t)$$

**4.7.** Show that the Laplace transform of the derivative $df/dt$ of a function $f(t)$ is given by $\mathscr{L}[df/dt] = sF(s) - f(0^+)$, where $F(s) = \mathscr{L}[f(t)]$ (Property 3).

By definition,

$$\mathscr{L}\left[ \frac{df}{dt} \right] = \lim_{\substack{T \to \infty \\ \epsilon \to 0}} \int_{\epsilon}^{T} \frac{df}{dt} e^{-st} \, dt$$

Integrating by parts,

$$\lim_{\substack{T \to \infty \\ \epsilon \to 0}} \int_\epsilon^T \frac{df}{dt} e^{-st}\, dt = \lim_{\substack{T \to \infty \\ \epsilon \to 0}} \left[ f(t)\, e^{-st}\Big|_\epsilon^T + s\int_\epsilon^T f(t)\, e^{-st}\, dt \right] = -f(0^+) + sF(s)$$

where $\lim_{\epsilon \to 0} f(\epsilon) = f(0^+)$.

**4.8.** Show that

$$\mathscr{L}\left[ \int_0^t f(\tau)\, d\tau \right] = \frac{F(s)}{s}$$

where $F(s) = \mathscr{L}[f(t)]$ (Property 4).

By definition and a change in the order of integrations, we have

$$\mathscr{L}\left[ \int_0^t f(\tau)\, d\tau \right] = \int_{0^+}^\infty \int_0^t f(\tau)\, d\tau\, e^{-st}\, dt = \int_{0^+}^\infty f(\tau) \int_\tau^\infty e^{-st}\, dt\, d\tau$$

$$= \int_{0^+}^\infty f(\tau) \left[ -\frac{1}{s} e^{-st}\Big|_\tau^\infty \right] d\tau = \int_{0^+}^\infty f(\tau) \frac{e^{-s\tau}}{s}\, d\tau = \frac{F(s)}{s}$$

**4.9.** Show that $f(0^+) \equiv \lim_{t \to 0} f(t) = \lim_{s \to \infty} sF(s)$, where $F(s) = \mathscr{L}[f(t)]$ (Property 5).

From Problem 4.7,

$$\mathscr{L}\left[ \frac{df}{dt} \right] = sF(s) - f(0^+) = \lim_{\substack{T \to \infty \\ \epsilon \to 0}} \int_\epsilon^T \frac{df}{dt} e^{-st}\, dt$$

Now let $s \to \infty$, that is,

$$\lim_{s \to \infty} \left[ sF(s) - f(0^+) \right] = \lim_{s \to \infty} \left[ \lim_{\substack{T \to \infty \\ \epsilon \to 0}} \int_\epsilon^T \frac{df}{dt} e^{-st}\, dt \right]$$

Since the limiting processes can be interchanged, we have

$$\lim_{s \to \infty} \left[ \lim_{\substack{T \to \infty \\ \epsilon \to 0}} \int_\epsilon^T \frac{df}{dt} e^{-st}\, dt \right] = \lim_{\substack{T \to \infty \\ \epsilon \to 0}} \int_\epsilon^T \frac{df}{dt} \left( \lim_{s \to \infty} e^{-st} \right) dt$$

But $\lim_{s \to \infty} e^{-st} = 0$. Hence the right side of the equation is zero and $\lim_{s \to \infty} sF(s) = f(0^+)$.

**4.10.** Show that if $\lim_{t \to \infty} f(t)$ exists then $f(\infty) \equiv \lim_{t \to \infty} f(t) = \lim_{s \to 0} sF(s)$, where $F(s) = \mathscr{L}[f(t)]$ (Property 6).

From Problem 4.7,

$$\mathscr{L}\left[ \frac{df}{dt} \right] = sF(s) - f(0^+) = \lim_{\substack{T \to \infty \\ \epsilon \to 0}} \int_\epsilon^T \frac{df}{dt} e^{-st}\, dt$$

Now let $s \to 0$, that is,

$$\lim_{s \to 0} \left[ sF(s) - f(0^+) \right] = \lim_{s \to 0} \left[ \lim_{\substack{T \to \infty \\ \epsilon \to 0}} \int_\epsilon^T \frac{df}{dt} e^{-st}\, dt \right]$$

Since the limiting processes can be interchanged, we have

$$\lim_{s \to 0} \left[ \lim_{\substack{T \to \infty \\ \epsilon \to 0}} \int_\epsilon^T \frac{df}{dt} e^{-st}\, dt \right] = \lim_{\substack{T \to \infty \\ \epsilon \to 0}} \int_\epsilon^T \frac{df}{dt} \left( \lim_{s \to 0} e^{-st} \right) dt = \lim_{\substack{T \to \infty \\ \epsilon \to 0}} \int_\epsilon^T \frac{df}{dt}\, dt = f(\infty) - f(0^+)$$

Adding $f(0^+)$ to both sides of the last equation yields $\lim_{s \to 0} sF(s) = f(\infty)$ if $f(\infty) = \lim_{t \to \infty} f(t)$ exists.

**4.11.** Show that $\mathscr{L}[f(t/a)] = aF(as)$, where $F(s) = \mathscr{L}[f(t)]$ (Property 7).

By definition, $\mathscr{L}[f(t/a)] = \int_{0^+}^{\infty} f(t/a) e^{-st} \, dt$. Making the change of variable $\tau = t/a$,

$$\mathscr{L}\left[f\left(\frac{t}{a}\right)\right] = a\int_{0^+}^{\infty} f(\tau) e^{-(as)\tau} \, d\tau = aF(as)$$

**4.12.** Show that $\mathscr{L}^{-1}[F(s/a)] = af(at)$, where $f(t) = \mathscr{L}^{-1}[F(s)]$ (Property 8).

By definition,

$$\mathscr{L}^{-1}\left[F\left(\frac{s}{a}\right)\right] = \frac{1}{2\pi j}\int_{c-j\infty}^{c+j\infty} F\left(\frac{s}{a}\right) e^{st} \, ds$$

Making the change of variable $\omega = s/a$,

$$\mathscr{L}^{-1}\left[F\left(\frac{s}{a}\right)\right] = \frac{a}{2\pi j}\int_{c-j\infty}^{c+j\infty} F(\omega) e^{\omega(at)} \, d\omega = af(at)$$

**4.13.** Show that $\mathscr{L}[f(t-T)] = e^{-sT}F(s)$, where $f(t-T) = 0$ for $t \le T$ and $F(s) = \mathscr{L}[f(t)]$ (Property 9).

By definition,

$$\mathscr{L}[f(t-T)] = \int_{0^+}^{\infty} f(t-T) e^{-st} \, dt = \int_{T}^{\infty} f(t-T) e^{-st} \, dt$$

Making the change of variable $\theta = t - T$,

$$\mathscr{L}[f(t-T)] = \int_{0^+}^{\infty} f(\theta) e^{-s\theta} e^{-sT} \, d\theta = e^{-sT}F(s)$$

**4.14.** Show that $\mathscr{L}[e^{-at}f(t)] = F(s+a)$, where $F(s) = \mathscr{L}[f(t)]$ (Property 10).

By definition,

$$\mathscr{L}[e^{-at}f(t)] = \int_{0^+}^{\infty} e^{-at}f(t) e^{-st} \, dt = \int_{0^+}^{\infty} f(t) e^{-(s+a)t} \, dt = F(s+a)$$

**4.15.** Show that

$$\mathscr{L}[f_1(t)f_2(t)] = \frac{1}{2\pi j}\int_{c-j\infty}^{c+j\infty} F_1(\omega) F_2(s-\omega) \, d\omega$$

where $F_1(s) = \mathscr{L}[f_1(t)]$ and $F_2(s) = \mathscr{L}[f_2(t)]$ (Property 11).

By definition,

$$\mathscr{L}[f_1(t)f_2(t)] = \int_{0^+}^{\infty} f_1(t)f_2(t) e^{-st} \, dt$$

But

$$f_1(t) = \frac{1}{2\pi j}\int_{c-j\infty}^{c+j\infty} F_1(\omega) e^{\omega t} \, d\omega$$

Hence

$$\mathscr{L}[f_1(t)f_2(t)] = \frac{1}{2\pi j}\int_{0^+}^{\infty}\int_{c-j\infty}^{c+j\infty} F_1(\omega) e^{\omega t} \, d\omega \, f_2(t) e^{-st} \, dt$$

Interchanging the order of integrations yields

$$\mathscr{L}[f_1(t)f_2(t)] = \frac{1}{2\pi j}\int_{c-j\infty}^{c+j\infty} F_1(\omega) \int_{0^+}^{\infty} f_2(t) e^{-(s-\omega)t} \, dt \, d\omega$$

Since $\int_{0^+}^{\infty} f_2(t) e^{-(s-\omega)t}\, dt = F_2(s - \omega)$,

$$\mathscr{L}[f_1(t) f_2(t)] = \frac{1}{2\pi j} \int_{c-j\infty}^{c+j\infty} F_1(\omega) F_2(s - \omega)\, d\omega$$

**4.16.** Show that

$$\mathscr{L}^{-1}[F_1(s) F_2(s)] = \int_{0^+}^{t} f_1(\tau) f_2(t - \tau)\, d\tau$$

where $f_1(t) = \mathscr{L}^{-1}[F_1(s)]$ and $f_2(t) = \mathscr{L}^{-1}[F_2(s)]$ (Property 12).

By definition,

$$\mathscr{L}^{-1}[F_1(s) F_2(s)] = \frac{1}{2\pi j} \int_{c-j\infty}^{c+j\infty} F_1(s) F_2(s) e^{st}\, ds$$

But $F_1(s) = \int_{0^+}^{\infty} f_1(\tau) e^{-s\tau}\, d\tau$. Hence

$$\mathscr{L}^{-1}[F_1(s) F_2(s)] = \frac{1}{2\pi j} \int_{c-j\infty}^{c+j\infty} \int_{0^+}^{\infty} f_1(\tau) e^{-s\tau}\, d\tau\, F_2(s) e^{st}\, ds$$

Interchanging the order of integrations yields

$$\mathscr{L}^{-1}[F_1(s) F_2(s)] = \int_{0^+}^{\infty} \frac{1}{2\pi j} \int_{c-j\infty}^{c+j\infty} F_2(s) e^{s(t-\tau)}\, ds\, f_1(\tau)\, d\tau$$

Since

$$\frac{1}{2\pi j} \int_{c-j\infty}^{c+j\infty} F_2(s) e^{s(t-\tau)}\, ds = f_2(t - \tau)$$

then

$$\mathscr{L}^{-1}[F_1(s) F_2(s)] = \int_{0^+}^{\infty} f_1(\tau) f_2(t - \tau)\, d\tau = \int_{0^+}^{t} f_1(\tau) f_2(t - \tau)\, d\tau$$

where the second equality is true since $f_2(t - \tau) = 0$ for $\tau \geq t$.

**4.17.** Show that

$$\mathscr{L}\left[\frac{d^i y}{dt^i}\right] = s^i Y(s) - \sum_{k=0}^{i-1} s^{i-1-k} y_0^k$$

for $i > 0$, where $Y(s) = \mathscr{L}[y(t)]$ and $y_0^k = (d^k y/dt^k)|_{t=0^+}$

This result can be shown by mathematical induction. For $i = 1$,

$$\mathscr{L}\left[\frac{dy}{dt}\right] = sY(s) - y(0^+) = sY(s) - y_0^0$$

as shown in Problem 4.7. Now assume the result holds for $i = n - 1$, that is,

$$\mathscr{L}\left[\frac{d^{n-1} y}{dt^{n-1}}\right] = s^{n-1} Y(s) - \sum_{k=0}^{n-2} s^{n-2-k} y_0^k$$

Then $\mathscr{L}[d^n y/dt^n]$ can be written as

$$\mathscr{L}\left[\frac{d^n y}{dt^n}\right] = \mathscr{L}\left[\frac{d}{dt}\left(\frac{d^{n-1} y}{dt^{n-1}}\right)\right] = s\mathscr{L}\left[\frac{d^{n-1} y}{dt^{n-1}}\right] - \frac{d^{n-1} y}{dt^{n-1}}\bigg|_{t=0^+}$$

$$= s\left(s^{n-1} Y(s) - \sum_{k=0}^{n-2} s^{n-2-k} y_0^k\right) - y_0^{n-1} = s^n Y(s) - \sum_{k=0}^{n-1} s^{n-1-k} y_0^k$$

For the special case $n = 2$, we have $\mathscr{L}[d^2 y/dt^2] = s^2 Y(s) - s y_0^0 - y_0^1$.

**LAPLACE TRANSFORMS AND THEIR INVERSE FROM THE TABLE OF TRANSFORM PAIRS**

**4.18.** Find the Laplace transform of $f(t) = 2e^{-t}\cos 10t - t^4 + 6e^{-(t-10)}$ for $t > 0$.

From the table of transform pairs,

$$\mathscr{L}[e^{-t}\cos 10t] = \frac{s+1}{(s+1)^2 + 10^2} \qquad \mathscr{L}[t^4] = \frac{4!}{s^5} \qquad \mathscr{L}[e^{-t}] = \frac{1}{s+1}$$

Using Property 9, $\mathscr{L}[e^{-(t-10)}] = e^{-10s}/(s+1)$. Using Property 1,

$$\mathscr{L}[f(t)] = 2\mathscr{L}[e^{-t}\cos 10t] - \mathscr{L}[t^4] + 6\mathscr{L}[e^{-(t-10)}] = \frac{2(s+1)}{s^2 + 2s + 101} - \frac{24}{s^5} + \frac{6e^{-10s}}{s+1}$$

**4.19.** Find the inverse Laplace transform of

$$F(s) = \frac{2e^{-0.5s}}{s^2 - 6s + 13} - \frac{s-1}{s^2 - 2s + 2}$$

for $t > 0$.

$$\frac{2}{s^2 - 6s + 13} = \frac{2}{(s-3)^2 + 2^2} \qquad \frac{s-1}{s^2 - 2s + 2} = \frac{s-1}{(s-1)^2 + 1}$$

The inverse transforms are determined directly from Table 4.1 as

$$\mathscr{L}^{-1}\left[\frac{1}{(s-3)^2 + 2^2}\right] = e^{3t}\sin 2t \qquad \mathscr{L}^{-1}\left[\frac{s-1}{(s-1)^2 + 1}\right] = e^t\cos t$$

Using Property 9, *then* Property 2, results in

$$f(t) = \begin{cases} -e^t\cos t & 0 < t \le 0.5 \\ e^{3(t-0.5)}\sin 2(t - 0.5) - e^t\cos t & t > 0.5 \end{cases}$$

**LAPLACE TRANSFORMS OF LINEAR CONSTANT-COEFFICIENT DIFFERENTIAL EQUATIONS**

**4.20.** Determine the output transform $Y(s)$ for the differential equation

Mathcad

$$\frac{d^3y}{dt^3} + 3\frac{d^2y}{dt^2} - \frac{dy}{dt} + 6y = \frac{d^2u}{dt^2} - u$$

where $y = $ output, $u = $ input, and initial conditions are

$$y(0^+) = \left.\frac{dy}{dt}\right|_{t=0^+} = 0 \qquad \left.\frac{d^2y}{dt^2}\right|_{t=0^+} = 1$$

Using Property 3 or the result of Problem 4.17, the Laplace transforms of the terms of the equation are given as

$$\mathscr{L}\left[\frac{d^3y}{dt^3}\right] = s^3Y(s) - s^2y(0^+) - s\left.\frac{dy}{dt}\right|_{t=0^+} - \left.\frac{d^2y}{dt^2}\right|_{t=0^+} = s^3Y(s) - 1$$

$$\mathscr{L}\left[\frac{d^2y}{dt^2}\right] = s^2Y(s) - sy(0^+) - \left.\frac{dy}{dt}\right|_{t=0^+} = s^2Y(s)$$

$$\mathscr{L}\left[\frac{dy}{dt}\right] = sY(s) - y(0^+) = sY(s) \qquad \mathscr{L}\left[\frac{d^2u}{dt^2}\right] = s^2U(s) - su(0^+) - \left.\frac{du}{dt}\right|_{t=0^+}$$

where $Y(s) = \mathscr{L}[y(t)]$ and $U(s) = \mathscr{L}[u(t)]$. The Laplace transform of the given equation can now be

written as

$$\mathscr{L}\left[\frac{d^3y}{dt^3}\right] + 3\mathscr{L}\left[\frac{d^2y}{dt^2}\right] - \mathscr{L}\left[\frac{dy}{dt}\right] + 6\mathscr{L}[y]$$

$$= s^3 Y(s) - 1 + 3s^2 Y(s) - sY(s) + 6Y(s)$$

$$= \mathscr{L}\left[\frac{d^2u}{dt^2}\right] - \mathscr{L}[u] = s^2 U(s) - su(0^+) - \frac{du}{dt}\bigg|_{t=0^+} - U(s)$$

Solving for $Y(s)$, we obtain

$$Y(s) = \frac{(s^2-1)U(s)}{s^3 + 3s^2 - s + 6} - \frac{su(0^+) + \frac{du}{dt}\bigg|_{t=0^+}}{s^3 + 3s^2 - s + 6} + \frac{1}{s^3 + 3s^2 - s + 6}$$

**4.21.** What part of the solution of Problem 4.20 is the transform of the free response? The forced response?

The transform of the free response $Y_a(s)$ is that part of the output transform $Y(s)$ which does not depend on the input $u(t)$, its derivatives or its transform; that is,

$$Y_a(s) = \frac{1}{s^3 + 3s^2 - s + 6}$$

The transform of the forced response $Y_b(s)$ is that part of $Y(s)$ which depends on $u(t)$, its derivative and its transform; that is,

$$Y_b(s) = \frac{(s^2-1)U(s)}{s^2 + 3s^2 - s + 6} - \frac{su(0^+) + \frac{du}{dt}\bigg|_{t=0^+}}{s^3 + 3s^2 - s + 6}$$

**4.22.** What is the characteristic polynomial for the differential equation of Problems 4.20 and 4.21?

The characteristic polynomial is the denominator polynomial which is common to the transforms of the free and forced responses (see Problem 4.21), that is, the polynomial $s^3 + 3s^2 - s + 6$.

**4.23.** Determine the output transform $Y(s)$ of the system of Problem 4.20 for an input $u(t) = 5 \sin t$.

From Table 4.1, $U(s) \equiv \mathscr{L}[u(t)] = \mathscr{L}[5 \sin t] = 5/(s^2 + 1)$.
The initial values of $u(t)$ and $du/dt$ are $u(0^+) = \lim_{t \to 0} 5 \sin t = 0$, $(du/dt)|_{t=0^+} = \lim_{t \to 0} 5 \cos t = 5$.
Substituting these values into the output transform $Y(s)$ given in Problem 4.20,

$$Y(s) = \frac{s^2 - 9}{(s^3 + 3s^2 - s + 6)(s^2 + 1)}$$

## PARTIAL FRACTION EXPANSIONS

**4.24.** A rational function $F(s)$ can be represented by

$$F(s) = \frac{\displaystyle\sum_{i=0}^{n} b_i s^i}{\displaystyle\prod_{i=1}^{r} (s + p_i)^{n_i}} = b_n + \sum_{i=1}^{r} \sum_{k=1}^{n_i} \frac{c_{ik}}{(s + p_i)^k} \qquad (4.10a)$$

where the second form is the partial fraction expansion of $F(s)$. Show that the constants $c_{ik}$ are given by

$$c_{ik} = \frac{1}{(n_i - k)!} \frac{d^{n_i - k}}{ds^{n_i - k}}\left[(s + p_i)^{n_i} F(s)\right]\bigg|_{s = -p_i} \qquad (4.10b)$$

Let $(s + p_j)$ be the factor of interest and form

$$(s + p_j)^{n_j} F(s) = (s + p_j)^{n_j} b_n + \sum_{i=1}^{r} \sum_{k=1}^{n_i} \frac{(s + p_j)^{n_j} c_{ik}}{(s + p_i)^k}$$

This can be rewritten as

$$(s + p_j)^{n_j} F(s) = (s + p_j)^{n_j} b_n + \sum_{i=1}^{j-1} \sum_{k=1}^{n_i} \frac{(s + p_j)^{n_j} c_{ik}}{(s + p_i)^k}$$

$$+ \sum_{i=j+1}^{r} \sum_{k=1}^{n_i} \frac{(s + p_j)^{n_j} c_{ik}}{(s + p_i)^k} + \sum_{k=1}^{n_j} (s + p_j)^{n_j - k} c_{jk}$$

Now form

$$\frac{d^{n_j - l}}{ds^{n_j - l}} \left[ (s + p_j)^{n_j} F(s) \right] \Bigg|_{s = -p_j}$$

Note that the first three terms on the right-hand side of $(s + p_j)^{n_j} F(s)$ will have a factor $s + p_j$ in the numerator even after being differentiated $n_j - l$ times $(l = 1, 2, \ldots, n_j)$ and thus these three terms become zero when evaluated at $s = -p_j$. Therefore

$$\frac{d^{n_j - l}}{ds^{n_j - l}} \left[ (s + p_j)^{n_j} F(s) \right] \Bigg|_{s = -p_j} = \frac{d^{n_j - l}}{ds^{n_j - l}} \left[ \sum_{k=1}^{n_j} (s + p_j)^{n_j - k} c_{jk} \right] \Bigg|_{s = -p_j}$$

$$= \sum_{k=1}^{l} (n_j - k)(n_j - k - 1) \cdots (l - k + 1)(s + p_j)^{(-k + l)} c_{jk} \Bigg|_{s = -p_j}$$

Except for that term in the summation for which $k = l$, all the other terms are zero since they contain factors $s + p_j$. Then

$$\frac{d^{n_j - l}}{ds^{n_j - l}} \left[ (s + p_j)^{n_j} F(s) \right] \Bigg|_{s = -p_j} = (n_j - l)(n_j - l - 1) \cdots (1) c_{jl}$$

or

$$c_{jl} = \frac{1}{(n_j - l)!} \frac{d^{n_j - l}}{ds^{n_j - l}} \left[ (s + p_j)^{n_j} F(s) \right] \Bigg|_{s = -p_j}$$

**4.25.** Expand $Y(s)$ of Example 4.17 in a partial fraction expansion.

$Y(s)$ can be rewritten with the denominator polynomial in factored form as

$$Y(s) = \frac{-(s^2 + s - 1)}{s(s + 1)(s + 2)}$$

The partial fraction expansion of $Y(s)$ is [see Equation (4.11)]

$$Y(s) = b_3 + \frac{c_{11}}{s} + \frac{c_{21}}{s + 1} + \frac{c_{31}}{s + 2}$$

where $b_3 = 0$,

$$c_{11} = \frac{-(s^2 + s - 1)}{(s + 1)(s + 2)} \Bigg|_{s=0} = \frac{1}{2} \qquad c_{21} = \frac{-(s^2 + s - 1)}{s(s + 2)} \Bigg|_{s=-1} = -1 \qquad c_{31} = \frac{-(s^2 + s - 1)}{s(s + 1)} \Bigg|_{s=-2} = -\frac{1}{2}$$

Thus

$$Y(s) = \frac{1}{2s} - \frac{1}{s + 1} - \frac{1}{2(s + 2)}$$

**4.26.** Expand $Y(s)$ of Example 4.19 in a partial fraction expansion.

$Y(s)$ can be rewritten with the denominator polynomial in factored form as

$$Y(s) = \frac{s^2 + 9s + 19}{(s + 1)(s + 2)(s + 4)}$$

The partial fraction expansion of $Y(s)$ is [see Equation $(4.11)$]

$$Y(s) = b_3 + \frac{c_{11}}{s+1} + \frac{c_{21}}{s+2} + \frac{c_{31}}{s+4}$$

where $b_3 = 0$,

$$c_{11} = \left.\frac{s^2 + 9s + 19}{(s+2)(s+4)}\right|_{s=-1} = \frac{11}{3} \qquad c_{21} = \left.\frac{s^2 + 9s + 19}{(s+1)(s+4)}\right|_{s=-2} = -\frac{5}{2}$$

$$c_{31} = \left.\frac{s^2 + 9s + 19}{(s+1)(s+2)}\right|_{s=-4} = -\frac{1}{6}$$

Thus

$$Y(s) = \frac{11}{3(s+1)} - \frac{5}{2(s+2)} - \frac{1}{6(s+4)}$$

## INVERSE LAPLACE TRANSFORMS USING PARTIAL FRACTION EXPANSIONS

**4.27.** Determine $y(t)$ for the system of Example 4.17.

From the result of Problem 4.25, the transform of $y(t)$ can be written as

$$\mathscr{L}[y(t)] \equiv Y(s) = \frac{1}{2s} - \frac{1}{s+1} - \frac{1}{2(s+2)}$$

Therefore

$$y(t) = \frac{1}{2}\mathscr{L}^{-1}\left[\frac{1}{s}\right] - \mathscr{L}^{-1}\left[\frac{1}{s+1}\right] - \frac{1}{2}\mathscr{L}^{-1}\left[\frac{1}{s+2}\right] = \frac{1}{2}[1 - 2e^{-t} - e^{-2t}] \qquad t > 0$$

**4.28.** Determine $y(t)$ for the system of Example 4.19.

From the result of Problem 4.26, the transform of $y(t)$ can be written as

$$\mathscr{L}[y(t)] = Y(s) = \frac{11}{3(s+1)} - \frac{5}{2(s+2)} - \frac{1}{6(s+4)}$$

Therefore

$$y(t) = \frac{11}{3}e^{-t} - \frac{5}{2}e^{-2t} - \frac{1}{6}e^{-4t}$$

## ROOTS OF POLYNOMIALS

**4.29.** Find an approximation of a real root of the polynomial equation

$$Q_3(s) = s^3 - 3s^2 + 4s - 5 = 0$$

to an accuracy of three significant figures using *Horner's method*.

By Descartes' rule of signs, $Q_3(s)$ has three variations in the signs of its coefficients (1 to $-3$, $-3$ to 4, and 4 to $-5$). Thus there may be three positive real roots. $Q_3(-s) = -s^3 - 3s^2 - 4s - 5$ has no sign changes; therefore $Q_3(s)$ has no negative real roots and only real values of $s$ greater than zero need be considered.

*Step 1*—We have $Q_3(0) = -5$, $Q_3(1) = -3$, $Q_3(2) = -1$, $Q_3(3) = 7$. Therefore $k_0 = 2$ and the first approximation is $s_0 = k_0 = 2$.

*Step 2*—Determine $Q_3^1(s)$ as

$$Q_3^1(s) = Q_3^0(2+s) = (2+s)^3 - 3(2+s)^2 + 4(2+s) - 5 = s^3 + 3s^2 + 4s - 1$$

*Step 3*—$Q_3^1(0) = -1$, $Q_3^1(\frac{1}{10}) = -0.569$, $Q_3^1(\frac{2}{10}) = -0.072$, $Q_3^1(\frac{3}{10}) = 0.497$. Hence $k_1 = 0.2$ and $s_1 = k_0 + k_1 = 2.2$.

Now repeat Step 2 to determine $Q_3^2(s)$:

$$Q_3^2(s) = Q_3^1(0.2+s) = (0.2+s)^3 + 3(0.2+s)^2 + 4(0.2+s) - 1 = s^3 + 3.6s^2 + 5.32s - 0.072$$

Repeating Step 3: $Q_3^2(0) = -0.072$, $Q_3^2(1/100) = -0.018$, $Q_3^2(2/100) = 0.036$. Hence $k_2 = 0.01$ and $s_2 = k_0 + k_1 + k_2 = 2.21$ which is an approximation of the root accurate to three significant figures.

**4.30.** Find an approximation of a real root of the polynomial equation given in Problem 4.29 using *Newton's method*. Perform four iterations and compare the result with the solution of Problem 4.29.

The sequence of approximations is defined by letting $n = 3$, $a_3 = 1$, $a_2 = -3$, $a_1 = 4$, and $a_0 = -5$ in the recursive relationship of Newton's method [Equation (*4.15*)]. The result is

$$s_{l+1} = \frac{2s_l^3 - 3s_l^2 + 5}{3s_l^2 - 6s_l + 4} \qquad l = 0, 1, 2, \ldots$$

Let the first guess be $s_0 = 0$. Then

$$s_1 = \frac{5}{4} = 1.25$$

$$s_3 = \frac{2(3.55)^3 - 3(3.55)^2 + 5}{3(3.55)^2 - 6(3.55) + 4} = 2.76$$

$$s_2 = \frac{2(1.25)^3 - 3(1.25)^2 + 5}{3(1.25)^2 - 6(1.25) + 4} = 3.55$$

$$s_4 = \frac{2(2.76)^3 - 3(2.76)^2 + 5}{3(2.76)^2 - 6(2.76) + 4} = 2.35$$

The next iteration yields $s_5 = 2.22$ and the sequence is converging.

**4.31.** Find an approximation of a quadratic factor of the polynomial

$$Q_3(s) = s^3 - 3s^2 + 4s - 5$$

of Problems 4.29 and 4.30, using the *Lin-Bairstow method*. Perform two iterations.

*Step 1*—Choose as a first approximation the factor $s^2 - s + 2$.
The constants needed in Step 2 are $\alpha_1 = -1$, $\alpha_0 = 2$, $n = 3$, $a_3 = 1$, $a_2 = -3$, $a_1 = 4$, $a_0 = -5$.
*Step 2*—From the recursive relationship

$$b_{i-2} = a_i - \alpha_1 b_{i-1} - \alpha_0 b_i$$

$i = n, n-1, \ldots, 1, 0$, the following constants are formed:

$$b_1 = a_3 = 1 \qquad\qquad b_0 = a_2 + b_1 = -2$$
$$b_{-1} = a_1 + b_0 - 2b_1 = 0 \qquad b_{-2} = a_0 + b_{-1} - 2b_0 = -1$$

*Step 3*—From the recursive relationship

$$c_{i-1} = b_{i-1} - \alpha_1 c_i - \alpha_0 c_{i+1}$$

$i = n, n-1, \ldots, 1$, the following constants are determined:

$$c_1 = b_1 = 1 \qquad c_0 = b_0 + c_1 = -1$$

*Step 4*—The simultaneous equations

$$c_0 \Delta\alpha_1 + c_1 \Delta\alpha_0 = b_{-1}$$
$$(-\alpha_1 c_0 - \alpha_0 c_1) \Delta\alpha_1 + c_0 \Delta\alpha_0 = b_{-2}$$

can now be written as

$$-\Delta\alpha_1 + \Delta\alpha_0 = 0$$
$$-3\Delta\alpha_1 - \Delta\alpha_0 = -1$$

whose solution is $\Delta\alpha_1 = \frac{1}{4}$, $\Delta\alpha_0 = \frac{1}{4}$, and the new approximation of the quadratic factor is

$$s^2 - 0.75s + 2.25$$

If Steps 1 through 4 are repeated for $\alpha_1 = -0.75$, $\alpha_0 = 2.25$, the second iteration produces

$$s^2 - 0.7861s + 2.2583$$

## POLE-ZERO MAPS

**4.32.** Determine all of the poles and zeros of $F(s) = (s^2 - 16)/(s^5 - 7s^4 - 30s^3)$.

The finite poles of $F(s)$ are the roots of the denominator polynomial equation

$$s^5 - 7s^4 - 30s^3 = s^3(s + 3)(s - 10) = 0$$

Therefore $s = 0$, $s = -3$, and $s = 10$ are the finite poles of $F(s)$. $s = 0$ is a triple root of the equation and is called a **triple pole** of $F(s)$. These are the only values of $s$ for which $|F(s)|$ is infinite and are all the poles of $F(s)$. The finite zeros of $F(s)$ are the roots of the numerator polynomial equation

$$s^2 - 16 = (s - 4)(s + 4) = 0$$

Therefore $s = 4$ and $s = -4$ are the *finite zeros* of $F(s)$. As $|s| \to \infty$, $F(s) \cong 1/s^3 \to 0$. Then $F(s)$ has a triple zero at $s = \infty$.

**4.33.** Draw a pole-zero map for the function of Problem 4.32.

From the solution of Problem 4.32, $F(s)$ has *finite zeros* at $s = 4$ and $s = -4$, and *finite poles* at $s = 0$ (a triple pole), $s = -3$ and $s = 10$. The pole-zero map is shown in Fig. 4-5.

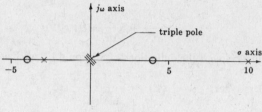

Fig. 4-5

**4.34.** Using the graphical technique, evaluate the residues of the function

$$F(s) = \frac{20}{(s + 10)(s + 1 + j)(s + 1 - j)}$$

The pole-zero map of $F(s)$ is shown in Fig. 4-6.

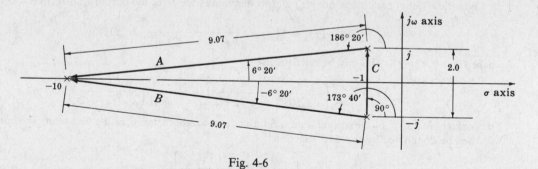

Fig. 4-6

Included in this pole-zero map are the vector displacements between the poles. For example, $A$ is the vector displacement of the pole $s = -10$ relative to the pole $s = -1 + j$. Clearly then, $-A$ is the vector displacement of the pole $s = -1 + j$ relative to the pole $s = -10$.

The magnitude of the residue at the pole $s = -10$ is

$$|c_1| = \frac{20}{|A|\,|B|} = \frac{20}{(9.07)(9.07)} = 0.243$$

The angle $\phi_1$ of the residue at $s = -10$ is the negative of the sum of the angles of $A$ and $B$, that is, $\phi_1 = -[186°20' + 173°40'] = -360°$. Hence $c_1 = 0.243$.

The magnitude of the residue at the pole $s = -1 + j$ is

$$|c_2| = \frac{20}{|-A||C|} = \frac{20}{(9.07)(2)} = 1.102$$

The angle $\phi_2$ of the residue at the pole $s = -1 + j$ is the negative of the sums of the angles of $-A$ and $C$:
$\phi_2 = -[6°20' + 90°] = -96°20'$. Hence $c_2 = 1.102\underline{/-96°20'} = -0.128 - j1.095$.

The magnitude of the residue at the pole $s = -1 - j$ is

$$|c_3| = \frac{20}{|-B||-C|} = \frac{20}{(9.07)(2)} = 1.102$$

The angle $\phi_3$ of the residue at the pole $s = -1 - j$ is the negative of the sum of the angles of $-B$ and $-C$:
$\phi_3 = -[-90° - 6°20'] = 96°20'$. Hence $c_3 = 1.102\underline{/96°20'} = -0.128 + j1.095$.

Note that the residues $c_2$ and $c_3$ of the complex conjugate poles are also complex conjugates. This is always true for the residues of complex conjugate poles.

## SECOND-ORDER SYSTEMS

**4.35.** Determine (*a*) the undamped natural frequency $\omega_n$, (*b*) the damping ratio $\zeta$, (*c*) the time constant $\tau$, (*d*) the damped natural frequency $\omega_d$, (*e*) characteristic equation for the second-order system given by

$$\frac{d^2y}{dt^2} + 5\frac{dy}{dt} + 9y = 9u$$

Comparing this equation with the definitions of Section 4.13, we have

(*a*)  $\omega_n^2 = 9$ or $\omega_n = 3$ rad/sec     (*c*)  $\tau = \frac{1}{\zeta\omega_n} = \frac{2}{5}$ sec     (*e*)  $s^2 + 5s + 9 = 0$

(*b*)  $2\zeta\omega_n = 5$ or $\zeta = \frac{5}{2\omega_n} = \frac{5}{6}$     (*d*)  $\omega_d = \omega_n\sqrt{1-\zeta^2} = 1.66$ rad/sec

**4.36.** How and why can the following system be approximated by a second-order system?

$$\frac{d^3y}{dt^3} + 12\frac{d^2y}{dt^2} + 22\frac{dy}{dt} + 20y = 20u$$

When the initial conditions on $y(t)$ and its derivatives are zero, the output transform is

$$\mathscr{L}[y(t)] \equiv Y(s) = \frac{20}{s^3 + 12s^2 + 22s + 20}U(s)$$

where $U(s) = \mathscr{L}[u(t)]$. This can be rewritten as

$$Y(s) = \frac{10}{41}\left(\frac{1}{s+10} - \frac{s}{s^2 + 2s + 2}\right)U(s) + \frac{80}{41}\left(\frac{U(s)}{s^2 + 2s + 2}\right)$$

The constant factor $\frac{80}{41}$ of the second term is 8 times the constant factor $\frac{10}{41}$ of the first term. The output $y(t)$ will then be dominated by the time function

$$\frac{80}{41}\mathscr{L}^{-1}\left[\frac{U(s)}{s^2 + 2s + 2}\right]$$

The output transform $Y(s)$ can then be approximated by this second term; that is,

$$Y(s) \simeq \frac{80}{41}\left(\frac{U(s)}{s^2 + 2s + 2}\right) \simeq \left(\frac{2}{s^2 + 2s + 2}\right)U(s)$$

The second-order approximation is $d^2y/dt^2 + 2(dy/dt) + 2y = 2u$.

**4.37.** In Chapter 6 it will be shown that the output $y(t)$ of a time-invariant linear causal system with all initial conditions equal to zero is related to the input $u(t)$ in the Laplace transform domain by the equation $Y(s) = P(s)U(s)$, where $P(s)$ is called the *transfer function* of the system. Show that $p(t)$, the inverse Laplace transform of $P(s)$, is equal to the *weighting function* $w(t)$ of a system described by the constant-coefficient differential equation

$$\sum_{i=0}^{n} a_i \frac{d^i y}{dt^i} = u$$

The forced response for a system described by the above equation is given by Equation (*3.15*), with all $b_i = 0$ except $b_0 = 1$:

$$y(t) = \int_{0^+}^{t} w(t - \tau) u(\tau)\, d\tau$$

and $w(t - \tau)$ is the weighting function of the differential equation.

The inverse Laplace transform of $Y(s) = P(s)U(s)$ is easily determined from the convolution integral of Property 12 as

$$y(t) = \mathscr{L}^{-1}[Y(s)] = \mathscr{L}^{-1}[P(s)U(s)] = \int_{0^+}^{t} p(t - \tau) u(\tau)\, d\tau$$

Hence     $\displaystyle \int_{0^+}^{t} w(t - \tau) u(\tau)\, d\tau = \int_{0^+}^{t} p(t - \tau) u(\tau)\, d\tau$   or   $w(t) = p(t)$

## MISCELLANEOUS PROBLEMS

**4.38.** For the $R$-$C$ network in Fig. 4-7:

(*a*)  Find a differential equation which relates the output voltage $y$ and the input voltage $u$.

(*b*)  Let the initial voltage across the capacitor $C$ be $v_{c0} = 1$ volt with the polarity shown, and let $u = 2e^{-t}$. Using the Laplace transform technique, find $y$.

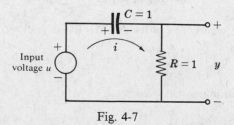

Fig. 4-7

(*a*)  From Kirchhoff's voltage law

$$u = v_{c0} + \frac{1}{C} \int_0^t i\, dt + Ri = v_{c0} + \int_0^t i\, dt + i$$

But $y = Ri = i$. Therefore $u = v_{c0} + \int_0^t y\, dt + y$. Differentiating both sides of this integral equation yields the differential equation $\dot{y} + y = \dot{u}$.

(*b*)  The Laplace transform of the differential equation found in part (*a*) is

$$sY(s) - y(0^+) + Y(s) = sU(s) - u(0^+)$$

where $U(s) = \mathscr{L}[2e^{-t}] = 2/(s+1)$ and $u(0^+) = \lim_{t \to 0} 2e^{-t} = 2$. To find $y(0^+)$, limits are taken on both sides of the original voltage equation:

$$u(0^+) = \lim_{t \to 0} u(t) = \lim_{t \to 0} \left[ v_{c0} + \int_0^t y\, dt + y(t) \right] = v_{c0} + y(0^+)$$

Hence $y(0^+) = u(0^+) - v_{c0} = 2 - 1 = 1$. The transform of $y(t)$ is then

$$Y(s) = \frac{2s}{(s+1)^2} - \frac{1}{s+1} = -\frac{2}{(s+1)^2} + \frac{2}{s+1} - \frac{1}{s+1} = -\frac{2}{(s+1)^2} + \frac{1}{s+1}$$

Finally,

$$y(t) = \mathscr{L}^{-1}\left[-\frac{2}{(s+1)^2}\right] + \mathscr{L}^{-1}\left[\frac{1}{s+1}\right] = -2te^{-t} + e^{-t}$$

**4.39.** Determine the Laplace transform of the output of the ideal sampler described in Problem 3.5.

From Definition 4.1 and Equation (*3.20*), the screening property of the unit impulse, we have

$$U^*(s) = \int_{0^+}^{\infty} e^{-st} U^*(t)\, dt = \int_{0^+}^{\infty} e^{-st} \sum_{k=0}^{\infty} u(t)\, \delta(t-kT)\, dt$$

$$= \sum_{k=0}^{\infty} \int_{0^+}^{\infty} e^{-st} u(t)\, \delta(t-kT)\, dt = \sum_{k=0}^{\infty} e^{-skT} u(kT)$$

**4.40.** Compare the result of Problem 4.39 with the $z$-transform of the sampled signal $u(kT)$, $k = 0, 1, 2, \ldots$.

By definition the $z$-transform of the sampled signal is

$$U(z) = \sum_{k=0}^{\infty} u(kT) z^{-k}$$

This result could have been obtained directly by substituting $z = e^{sT}$ in the result of Problem 4.39.

**4.41.** Prove the Shift Theorem (Property 6, Section 4.9).

By definition,

$$\mathscr{z}\{f(k)\} \equiv F(z) \equiv \sum_{k=0}^{\infty} f(k) z^{-k}$$

If we define a new, shifted sequence by $g(0) \equiv f(-1) = 0$ and $g(k) \equiv f(k-1)$, $k = 1, 2, \ldots$, then

$$\mathscr{z}\{g(k)\} \equiv \sum_{k=0}^{\infty} g(k) z^{-k} \equiv \sum_{j=0}^{\infty} g(j) z^{-j} = \sum_{j=0}^{\infty} f(j-1) z^{-j}$$

(see Remark 1 following Definition 4.4). Now let $k$ be redefined as $k \equiv j - 1$ in the last equation. Then

$$\mathscr{z}\{f(k-1)\} = \sum_{k=-1}^{\infty} f(k) z^{-k-1} = z^{-1} \sum_{k=-1}^{\infty} f(k) z^{-k}$$

$$= z^{-1} f(-1) z^{+1} + z^{-1} \sum_{k=0}^{\infty} f(k) z^{-k}$$

$$= z^0 \cdot 0 + z^{-1} \sum_{k=0}^{\infty} f(k) z^{-k} = z^{-1} F(z)$$

Note that repeated application of this result gives

$$Z[f(k-j)] = z^{-j} F(z)$$

# Supplementary Problems

**4.42.** Show that $\mathscr{L}[-tf(t)] = dF(s)/ds$, where $F(s) = \mathscr{L}[f(t)]$.

**4.43.** Using the convolution integral find the inverse transform of $1/s(s+2)$.

**4.44.** Determine the final value of the function $f(t)$ whose Laplace transform is

$$F(s) = \frac{2(s+1)}{s(s+3)(s+5)^2}$$

**4.45.** Determine the initial value of the function $f(t)$ whose Laplace transform is

$$F(s) = \frac{4s}{s^3 + 2s^2 + 9s + 6}$$

**4.46.** Find the partial fraction expansion of the function $F(s) = 10/(s+4)(s+2)^3$.

**4.47.** Find the inverse Laplace transform $f(t)$ of the function $F(s) = 10/(s+4)(s+2)^3$.

**4.48.** Solve Problem 3.24 using the Laplace transform technique.

**4.49.** Using the Laplace transform technique, find the forced response of the differential equation

$$\frac{d^2y}{dt^2} + 4\frac{dy}{dt} + 4y = 3\frac{du}{dt} + 2u$$

where $u(t) = e^{-3t}$, $t > 0$. Compare this solution with that obtained in Problem 3.26.

**4.50.** Using the Laplace transform technique, find the transient and steady state responses of the system described by the differential equation $d^2y/dt^2 + 3(dy/dt) + 2y = 1$ with initial conditions $y(0^+)$ and $(dy/dt)|_{t=0^+} = 1$.

**4.51.** Using the Laplace transform technique, find the unit impulse response of the system described by the differential equation $d^3y/dt^3 + dy/dt = u$.

# Answers to Some Supplementary Problems

**4.43.** $\frac{1}{2}[1 - e^{-2t}]$

**4.44.** $\frac{2}{75}$

**4.45.** 0

**4.46.** $F(s) = \dfrac{5}{(s+2)^3} - \dfrac{5}{2(s+2)^2} + \dfrac{5}{4(s+2)} - \dfrac{5}{4(s+4)}$

**4.47.** $f(t) = \dfrac{5t^2e^{-2t}}{2} - \dfrac{5te^{-2t}}{2} + \dfrac{5e^{-2t}}{4} - \dfrac{5e^{-4t}}{4}$

**4.49.** $y_b(t) = 7e^{-2t} - 7e^{-3t} - 7te^{-2t}$

**4.50.** Transient response $= 2e^{-t} - \frac{3}{2}e^{-2t}$. Steady state response $= \frac{1}{2}$.

**4.51.** $y_\delta(t) = 1 - \cos t$

# Chapter 5

# Stability

## 5.1 STABILITY DEFINITIONS

The stability of a continuous or discrete-time system is determined by its response to inputs or disturbances. Intuitively, a stable system is one that remains at rest unless excited by an external source and returns to rest if all excitations are removed. Stability can be precisely defined in terms of the impulse response $y_\delta(t)$ of a continuous system, or Kronecker delta response $y_\delta(k)$ of a discrete-time system (see Sections 3.13 and 3.16), as follows:

***Definition 5.1a:***   A continuous system (discrete-time system) is **stable** if its impulse response $y_\delta(t)$ (Kronecker delta response $y_\delta(k)$) approaches zero as time approaches infinity.

Alternatively, the definition of a stable system can be based upon the response of the system to **bounded inputs**, that is, inputs whose magnitudes are less than some finite value for all time.

***Definition 5.1b:***   A continuous or discrete-time system is **stable** if every bounded input produces a bounded output.

Consideration of the *degree* of stability of a system often provides valuable information about its behavior. That is, if it is stable, how close is it to being unstable? This is the concept of **relative stability**. Usually, relative stability is expressed in terms of some allowable variation of a particular system parameter, over which the system remains stable. More precise definitions of relative stability indicators are presented in later chapters. Stability of nonlinear systems is treated in Chapter 19.

## 5.2 CHARACTERISTIC ROOT LOCATIONS FOR CONTINUOUS SYSTEMS

A major result of Chapters 3 and 4 is that the impulse response of a linear time-invariant continuous system is a sum of exponential time functions whose exponents are the roots of the system characteristic equation (see Equation 4.12). *A necessary and sufficient condition for the system to be stable is that the real parts of the roots of the characteristic equation have negative real parts*. This ensures that the impulse response will decay exponentially with time.

If the system has some roots with real parts equal to zero, but none with positive real parts, the system is said to be **marginally stable**. In this instance, the impulse response does not decay to zero, although it is bounded, but certain other inputs will produce unbounded outputs. Therefore marginally stable systems are unstable.

**EXAMPLE 5.1.**   The system described by the Laplace transformed differential equation,

$$(s^2 + 1)Y(s) = U(s)$$

has the characteristic equation

$$s^2 + 1 = 0$$

This equation has the two roots $\pm j$. Since these roots have zero real parts, the system is not stable. It is, however, marginally stable since the equation has no roots with positive real parts. In response to most inputs or disturbances, the system oscillates with a bounded output. However, if the input is $u = \sin t$, the output will contain a term of the form: $y = t \cos t$, which is unbounded.

## 5.3  ROUTH STABILITY CRITERION

The Routh criterion is a method for determining continuous system stability, for systems with an $n$th-order characteristic equation of the form:

$$a_n s^n + a_{n-1} s^{n-1} + \cdots + a_1 s + a_0 = 0$$

The criterion is applied using a **Routh table** defined as follows:

$$
\begin{array}{c|cccc}
s^n & a_n & a_{n-2} & a_{n-4} & \cdots \\
s^{n-1} & a_{n-1} & a_{n-3} & a_{n-5} & \cdots \\
\cdot & b_1 & b_2 & b_3 & \cdots \\
\cdot & c_1 & c_2 & c_3 & \cdots \\
\cdot & \multicolumn{4}{c}{\cdots\cdots\cdots\cdots\cdots}
\end{array}
$$

where $a_n, a_{n-1}, \ldots, a_0$ are the coefficients of the characteristic equation and

$$b_1 \equiv \frac{a_{n-1}a_{n-2} - a_n a_{n-3}}{a_{n-1}} \qquad b_2 \equiv \frac{a_{n-1}a_{n-4} - a_n a_{n-5}}{a_{n-1}} \qquad \text{etc.}$$

$$c_1 \equiv \frac{b_1 a_{n-3} - a_{n-1}b_2}{b_1} \qquad c_2 \equiv \frac{b_1 a_{n-5} - a_{n-1}b_3}{b_1} \qquad \text{etc.}$$

The table is continued horizontally and vertically until only zeros are obtained. Any row can be multiplied by a positive constant before the next row is computed without disturbing the properties of the table.

**The Routh Criterion:** *All the roots of the characteristic equation have negative real parts if and only if the elements of the first column of the Routh table have the same sign. Otherwise, the number of roots with positive real parts is equal to the number of changes of sign.*

**EXAMPLE 5.2.**

$$s^3 + 6s^2 + 12s + 8 = 0$$

$$
\begin{array}{c|ccc}
s^3 & 1 & 12 & 0 \\
s^2 & 6 & 8 & 0 \\
s^1 & \frac{64}{6} & 0 & \\
s^0 & 8 & &
\end{array}
$$

Since there are no changes of sign in the first column of the table, all the roots of the equation have negative real parts.

Often it is desirable to determine a range of values of a particular system parameter for which the system is stable. This can be accomplished by writing the inequalities that ensure that there is no change of sign in the first column of the Routh table for the system. These inequalities then specify the range of allowable values of the parameter.

**EXAMPLE 5.3.**

$$s^3 + 3s^2 + 3s + 1 + K = 0$$

$$
\begin{array}{c|ccc}
s^3 & 1 & 3 & 0 \\
s^2 & 3 & 1+K & 0 \\
s^1 & \dfrac{8-K}{3} & 0 & \\
s^0 & 1+K & &
\end{array}
$$

For no sign changes in the first column, it is necessary that the conditions $8 - K > 0$, $1 + K > 0$ be satisfied. Thus the characteristic equation has roots with negative real parts if $-1 < K < 8$, the simultaneous solution of these two inequalities.

A row of zeros for the $s^1$ row of the Routh table indicates that the polynomial has a pair of roots which satisfy the **auxiliary equation** formed as follows:

$$As^2 + B = 0$$

where $A$ and $B$ are the first and second elements of the $s^2$ row.

To continue the table, the zeros in the $s^1$ row are replaced with the coefficients of the derivative of the auxiliary equation. The derivative of the auxiliary equation is

$$2As + 0 = 0$$

The coefficients $2A$ and $0$ are then entered into the $s^1$ row and the table is continued as described above.

**EXAMPLE 5.4.**   In the previous example, the $s^1$ row is zero if $K = 8$. In this case, the auxiliary equation is $3s^2 + 9 = 0$. Therefore two of the roots of the characteristic equation are $s = \pm j\sqrt{3}$.

## 5.4   HURWITZ STABILITY CRITERION

The Hurwitz criterion is another method for determining whether all the roots of the characteristic equation of a continuous system have negative real parts. This criterion is applied using determinants formed from the coefficients of the characteristic equation. It is assumed that the first coefficient, $a_n$, is positive. The determinants $\Delta_i$, $i = 1, 2, \ldots, n-1$, are formed as the principal minor determinants of the determinant

$$\Delta_n = \begin{vmatrix} a_{n-1} & a_{n-3} & \cdots & \begin{bmatrix} a_0 & \text{if } n \text{ odd} \\ a_1 & \text{if } n \text{ even} \end{bmatrix} & 0 & \cdots & 0 \\ a_n & a_{n-2} & \cdots & \begin{bmatrix} a_1 & \text{if } n \text{ odd} \\ a_0 & \text{if } n \text{ even} \end{bmatrix} & 0 & \cdots & 0 \\ 0 & a_{n-1} & a_{n-3} & \cdots\cdots\cdots\cdots\cdots\cdots\cdots\cdots\cdots\cdots\cdots & 0 \\ 0 & a_n & a_{n-2} & \cdots\cdots\cdots\cdots\cdots\cdots\cdots\cdots\cdots\cdots\cdots & 0 \\ \cdots\cdots\cdots\cdots\cdots\cdots\cdots\cdots\cdots\cdots\cdots\cdots\cdots\cdots\cdots\cdots\cdots\cdots \\ 0 & \cdots\cdots\cdots\cdots\cdots\cdots\cdots\cdots\cdots\cdots\cdots\cdots\cdots\cdots\cdots & a_0 \end{vmatrix}$$

The determinants are thus formed as follows:

$$\Delta_1 = a_{n-1}$$

$$\Delta_2 = \begin{vmatrix} a_{n-1} & a_{n-3} \\ a_n & a_{n-2} \end{vmatrix} = a_{n-1}a_{n-2} - a_n a_{n-3}$$

$$\Delta_3 = \begin{vmatrix} a_{n-1} & a_{n-3} & a_{n-5} \\ a_n & a_{n-2} & a_{n-4} \\ 0 & a_{n-1} & a_{n-3} \end{vmatrix} = a_{n-1}a_{n-2}a_{n-3} + a_n a_{n-1}a_{n-5} - a_n a_{n-3}^2 - a_{n-4}a_{n-1}^2$$

and so on up to $\Delta_{n-1}$.

**Hurwitz Criterion:** *All the roots of the characteristic equation have negative real parts if and only if $\Delta_i > 0$, $i = 1, 2, \ldots, n$.*

**EXAMPLE 5.5.**   For $n = 3$,

$$\Delta_3 = \begin{vmatrix} a_2 & a_0 & 0 \\ a_3 & a_1 & 0 \\ 0 & a_2 & a_0 \end{vmatrix} = a_2 a_1 a_0 - a_0^2 a_3, \qquad \Delta_2 = \begin{vmatrix} a_2 & a_0 \\ a_3 & a_1 \end{vmatrix} = a_2 a_1 - a_0 a_3, \qquad \Delta_1 = a_2$$

Thus all the roots of the characteristic equation have negative real parts if

$$a_2 > 0 \qquad\qquad a_2 a_1 - a_0 a_3 > 0 \qquad\qquad a_2 a_1 a_0 - a_0^2 a_3 > 0$$

## 5.5 CONTINUED FRACTION STABILITY CRITERION

This criterion is applied to the characteristic equation of a continuous system by forming a continued fraction from the odd and even portions of the equation, in the following manner. Let

$$Q(s) \equiv a_n s^n + a_{n-1} s^{n-1} + \cdots + a_1 s + a_0$$

$$Q_1(s) \equiv a_n s^n + a_{n-2} s^{n-2} + \cdots$$

$$Q_2(s) \equiv a_{n-1} s^{n-1} + a_{n-3} s^{n-3} + \cdots$$

Form the fraction $Q_1/Q_2$, and then divide the denominator into the numerator and invert the remainder, to form a continued fraction as follows:

$$\frac{Q_1(s)}{Q_2(s)} = \frac{a_n s}{a_{n-1}} + \frac{\left(a_{n-2} - \dfrac{a_n a_{n-3}}{a_{n-1}}\right) s^{n-2} + \left(a_{n-4} - \dfrac{a_n a_{n-5}}{a_{n-1}}\right) s^{n-4} + \cdots}{Q_2}$$

$$= h_1 s + \cfrac{1}{h_2 s + \cfrac{1}{h_3 s + \cfrac{1}{h_4 s + \cfrac{1}{\ddots \cfrac{1}{h_n s}}}}}$$

If $h_1, h_2, \ldots, h_n$ are all positive, then all the roots of $Q(s) = 0$ have negative real parts.

**EXAMPLE 5.6.**

$$Q(s) = s^3 + 6s^2 + 12s + 8$$

$$\frac{Q_1(s)}{Q_2(s)} = \frac{s^3 + 12s}{6s^2 + 8} = \frac{1}{6}s + \frac{\dfrac{32}{3}s}{6s^2 + 8}$$

$$= \frac{1}{6}s + \cfrac{1}{\dfrac{9}{16}s + \cfrac{1}{\dfrac{4}{3}s}}$$

Since all the coefficients of $s$ in the continued fraction are positive, that is, $h_1 = \frac{1}{6}$, $h_2 = \frac{9}{16}$, and $h_3 = \frac{4}{3}$, all the roots of the polynomial equation $Q(s) = 0$ have negative real parts.

## 5.6 STABILITY CRITERIA FOR DISCRETE-TIME SYSTEMS

The stability of discrete systems is determined by the roots of the discrete system characteristic equation

$$Q(z) = a_n z^n + a_{n-1} z^{n-1} + \cdots + a_1 z + a_0 = 0 \qquad (5.1)$$

However, in this case the stability region is defined by the *unit circle* $|z| = 1$ in the $z$-plane. **A necessary and sufficient condition for system stability is that all the roots of the characteristic equation have a magnitude less than one**, that is, be within the *unit circle*. This ensures that the Kronecker delta response decays with time.

A stability criterion for discrete systems similar to the Routh criterion is called the **Jury test**. For this test, the coefficients of the characteristic equation are first arranged in the *Jury array*:

| row | | | | | | |
|---|---|---|---|---|---|---|
| 1 | $a_0$ | $a_1$ | $a_2$ | $\cdots$ | $a_{n-1}$ | $a_n$ |
| 2 | $a_n$ | $a_{n-1}$ | $a_{n-2}$ | $\cdots$ | $a_1$ | $a_0$ |
| 3 | $b_0$ | $b_1$ | $b_2$ | $\cdots$ | $b_{n-1}$ | |
| 4 | $b_{n-1}$ | $b_{n-2}$ | $b_{n-3}$ | $\cdots$ | $b_0$ | |
| 5 | $c_0$ | $c_1$ | $c_2$ | $\cdot\cdot\ c_{n-2}$ | | |
| 6 | $c_{n-2}$ | $c_{n-3}$ | $c_{n-4}$ | $\cdot\cdot\ c_0$ | | |
| $\vdots$ | $\vdots$ | $\vdots$ | $\vdots$ | | | |
| $2n-5$ | $r_0$ | $r_1$ | $r_2$ | $r_3$ | | |
| $2n-4$ | $r_3$ | $r_2$ | $r_1$ | $r_0$ | | |
| $2n-3$ | $s_0$ | $s_1$ | $s_2$ | | | |

where

$$b_k = \begin{vmatrix} a_0 & a_{n-k} \\ a_n & a_k \end{vmatrix} \qquad c_k = \begin{vmatrix} b_0 & b_{n-1-k} \\ b_{n-1} & b_k \end{vmatrix}$$

$$s_0 = \begin{vmatrix} r_0 & r_3 \\ r_3 & r_0 \end{vmatrix} \qquad s_1 = \begin{vmatrix} r_0 & r_2 \\ r_3 & r_1 \end{vmatrix} \qquad s_2 = \begin{vmatrix} r_0 & r_1 \\ r_3 & r_2 \end{vmatrix}$$

The first two rows are written using the characteristic equation coefficients and the next two rows are computed using the determinant relationships shown above. The process is continued with each succeeding pair of rows having one less column than the previous pair until row $2n - 3$ is computed, which only has three entries. The array is then terminated.

**Jury Test:** *Necessary and sufficient conditions for the roots of $Q(z) = 0$ to have magnitudes less than one are:*

$$Q(1) > 0$$

$$Q(-1) \quad \begin{cases} > 0 & \text{for } n \text{ even} \\ < 0 & \text{for } n \text{ odd} \end{cases}$$

$$|a_0| < a_n$$

$$|b_0| > |b_{n-1}|$$

$$|c_0| > |c_{n-2}|$$

$$\vdots$$

$$|r_0| > |r_3|$$

$$|s_0| > |s_2|$$

Note that if the $Q(1)$ or $Q(-1)$ conditions above are not satisfied, the system is unstable and it is not necessary to construct the array.

**EXAMPLE 5.7.** For $Q(z) = 3z^4 + 2z^3 + z^2 + z + 1 = 0$ ($n$ even),

$$Q(1) = 3 + 2 + 1 + 1 + 1 = 8 > 0$$

$$Q(-1) = 3 - 2 + 1 - 1 + 1 = 2 > 0$$

Thus the Jury array must be completed as

| row | | | | |
|---|---|---|---|---|
| 1 | 1 | 1 | 1 | 2   3 |
| 2 | 3 | 2 | 1 | 1   1 |
| 3 | $-8$ | $-5$ | $-2$ | $-1$ |
| 4 | $-1$ | $-2$ | $-5$ | $-8$ |
| 5 | 63 | 38 | 11 | |

The remaining test condition constraints are therefore

$$|a_0| = 1 < 3 = a_n$$
$$|b_0| = |-8| > |-1| = |b_{n-1}|$$
$$|c_0| = 63 > 11 = |c_{n-2}|$$

Since all the constraints of the Jury test are satisfied, all the roots of the characteristic equation are within the unit circle and the system is stable.

### The $w$-Transform

The stability of a linear discrete-time system expressed in the $z$-domain also can be determined using the $s$-plane methods developed for continuous systems (e.g., Routh, Hurwitz). The following *bilinear transformation* of the complex variable $z$ into the new complex variable $w$ given by the equivalent expressions:

$$z = \frac{1+w}{1-w} \qquad (5.2)$$

$$w = \frac{z-1}{z+1} \qquad (5.3)$$

transforms the interior of the unit circle in the $z$-plane onto the left half of the $w$-plane. Therefore the stability of a discrete-time system with characteristic polynomial $Q(z)$ can be determined by examining the locations of the roots of

$$Q(w) = Q(z)\big|_{z=(1+w)/(1-w)} = 0$$

in the $w$-plane, treating $w$ like $s$ and using $s$-plane techniques to establish stability properties. This transformation is developed more extensively in Chapter 10 and is also used in subsequent frequency domain analysis and design chapters.

 **EXAMPLE 5.8.**   The polynomial equation

$$27z^3 + 27z^2 + 9z + 1 = 0$$

is the characteristic equation of a discrete-time system. To test for roots outside the unit circle $|z| = 1$, which would signify instability, we set

$$z = \frac{1+w}{1-w}$$

which, after some algebraic manipulation, leads to a new characteristic equation in $w$:

$$w^3 + 6w^2 + 12w + 8 = 0$$

This equation was found to have roots only in the left half of the complex plane in Example 5.2. Therefore the original discrete-time system is stable.

# Solved Problems

### STABILITY DEFINITIONS

**5.1.**   The impulse responses of several linear continuous systems are given below. For each case determine if the impulse response represents a stable or an unstable system.

(a) $h(t) = e^{-t}$, (b) $h(t) = te^{-t}$, (c) $h(t) = 1$, (d) $h(t) = e^{-t}\sin 3t$, (e) $h(t) = \sin \omega t$.

If the impulse response decays to zero as time approaches infinity, the system is stable. As can be seen in Fig. 5-1, the impulse responses (a), (b), and (d) decay to zero as time approaches infinity and therefore

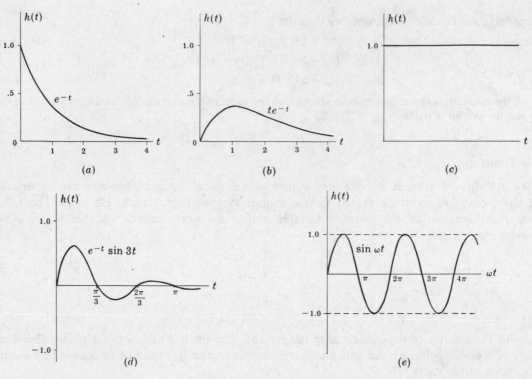

Fig. 5-1

represent *stable* systems. Since the impulse responses (*c*) and (*e*) do not approach zero, they represent *unstable* systems.

**5.2.** If a step function is applied at the input of a continuous system and the output remains below a certain level for all time, is the system stable?

The system is not necessarily stable since the output must be bounded for every bounded input. A bounded output to one specific bounded input does not ensure stability.

**5.3.** If a step function is applied at the input of a continuous system and the output is of the form $y = t$, is the system stable or unstable?

The system is unstable since a bounded input produced an unbounded output.

## CHARACTERISTIC ROOT LOCATIONS FOR CONTINUOUS SYSTEMS

**5.4.** The roots of the characteristic equations of several systems are given below. Determine in each case if the set of roots represents stable, marginally stable, or unstable systems.

| | | |
|---|---|---|
| (*a*)  $-1, -2$ | (*d*)  $-1+j, -1-j$ | (*g*)  $-6, -4, 7$ |
| (*b*)  $-1, +1$ | (*e*)  $-2+j, -2-j, 2j, -2j$ | (*h*)  $-2+3j, -2-3j, -2$ |
| (*c*)  $-3, -2, 0$ | (*f*)  $2, -1, -3$ | (*i*)  $-j, j, -1, 1$ |

The sets of roots (*a*), (*d*), and (*h*) represent stable systems since all the roots have negative real parts. The sets of roots (*c*) and (*e*) represent marginally stable systems since all the roots have nonpositive real parts, that is, zero or negative. The sets (*b*), (*f*), (*g*), and (*i*) represent unstable systems since each has at least one root with a positive real part.

**5.5.** A system has poles at $-1$ and $-5$ and zeros at 1 and $-2$. Is the system stable?

The system is stable since the poles are the roots of the system characteristic equation (Chapter 3) which have negative real parts. The fact that the system has a zero with a positive real part does not affect its stability.

**5.6.** Determine if the system with the following characteristic equation is stable:

$$(s+1)(s+2)(s-3) = 0.$$

This characteristic equation has the roots $-1$, $-2$, and 3 and therefore represents an unstable system since there is a positive real root.

**5.7.** The differential equation of an integrator may be written as follows: $dy/dt = u$. Determine if an integrator is stable.

The characteristic equation of this system is $s = 0$. Since the root does not have a negative real part, an integrator is not stable. Since it has no roots with positive real parts, an integrator is marginally stable.

**5.8.** Determine a bounded input which will produce an unbounded output from an integrator.

The input $u = 1$ will produce the output $y = t$, which is unbounded.

## ROUTH STABILITY CRITERION

**5.9.** Determine if the following characteristic equation represents a stable system:

$$s^3 + 4s^2 + 8s + 12 = 0$$

The Routh table for this system is

$$
\begin{array}{c|cc}
s^3 & 1 & 8 \\
s^2 & 4 & 12 \\
s^1 & 5 & 0 \\
s^0 & 12 &
\end{array}
$$

Since there are no changes of sign in the first column, all the roots of the characteristic equation have negative real parts and the system is stable.

**5.10.** Determine if the following characteristic equation has any roots with positive real parts:

$$s^4 + s^3 - s - 1 = 0$$

Note that the coefficient of the $s^2$ term is zero. The Routh table for this equation is

$$
\begin{array}{cc|ccc}
 & s^4 & 1 & 0 & -1 \\
 & s^3 & 1 & -1 & 0 \\
 & s^2 & 1 & -1 & \\
 & s^1 & 0 & 0 & \\
\text{new} & s^1 & 2 & 0 & \\
 & s^0 & -1 & &
\end{array}
$$

The presence of the zeros in the $s^1$ row indicates that the characteristic equation has two roots which satisfy the auxiliary equation formed from the $s^2$ row as follows: $s^2 - 1 = 0$. The roots of this equation are $+1$ and $-1$.

The new $s^1$ row was formed using the coefficients from the derivative of the auxiliary equation: $2s - 0 = 0$. Since there is one change of sign, the characteristic equation has one root with a positive real part, the one at $+1$ determined from the auxiliary equation.

**5.11.** The characteristic equation of a given system is

$$s^4 + 6s^3 + 11s^2 + 6s + K = 0$$

What restrictions must be placed upon the parameter $K$ in order to ensure that the system is stable?

The Routh table for this system is

$$
\begin{array}{c|ccc}
s^4 & 1 & 11 & K \\
s^3 & 6 & 6 & 0 \\
s^2 & 10 & K & 0 \\
s^1 & \dfrac{60 - 6K}{10} & 0 & \\
s^0 & K & &
\end{array}
$$

For the system to be stable, $60 - 6K > 0$, or $K < 10$, and $K > 0$. Thus $0 < K < 10$.

**5.12.** Construct a Routh table and determine the number of roots with positive real parts for the equation

$$2s^3 + 4s^2 + 4s + 12 = 0$$

The Routh table for this equation is given below. Here the $s^2$ row was divided by 4 before the $s^1$ row was computed. The $s^1$ row was then divided by 2 before the $s^0$ row was computed.

$$
\begin{array}{c|cc}
s^3 & 2 & 4 \\
s^2 & 1 & 3 \\
s^1 & -1 & 0 \\
s^0 & 3 &
\end{array}
$$

Since there are two changes of sign in the first column of the Routh table, the equation above has two roots with positive real parts.

## HURWITZ STABILITY CRITERION

**5.13.** Determine if the characteristic equation below represents a stable or an unstable system.

$$s^3 + 8s^2 + 14s + 24 = 0$$

The Hurwitz determinants for this system are

$$\Delta_3 = \begin{vmatrix} 8 & 24 & 0 \\ 1 & 14 & 0 \\ 0 & 8 & 24 \end{vmatrix} = 2112 \qquad \Delta_2 = \begin{vmatrix} 8 & 24 \\ 1 & 14 \end{vmatrix} = 88 \qquad \Delta_1 = 8$$

Since each determinant is positive, the system is stable. Note that the general formulation of Example 5.5 could have been used to check the stability in this case by substituting the appropriate values for the coefficients $a_0$, $a_1$, $a_2$, and $a_3$.

**5.14.** For what range of values of $K$ is the system with the following characteristic equation stable?

$$s^2 + Ks + 2K - 1 = 0$$

The Hurwitz determinants for this system are

$$\Delta_2 = \begin{vmatrix} K & 0 \\ 1 & 2K - 1 \end{vmatrix} = 2K^2 - K = K(2K - 1) \qquad \Delta_1 = K$$

In order for these determinants to be positive, it is necessary that $K > 0$ and $2K - 1 > 0$. Thus the system is stable if $K > \frac{1}{2}$.

**5.15.** A system is designed to give satisfactory performance when a particular amplifier gain $K = 2$. Determine how much $K$ can vary before the system becomes unstable if the characteristic equation is

$$s^3 + (4 + K)s^2 + 6s + 16 + 8K = 0$$

Substituting the coefficients of the given equation into the general Hurwitz conditions of Example 5.5 results in the following requirements for stability:

$$4 + K > 0 \qquad (4 + K)6 - (16 + 8K) > 0 \qquad (4 + K)(6)(16 + 8K) - (16 + 8K)^2 > 0$$

Assuming the amplifier gain $K$ cannot be negative, the first condition is satisfied. The second and third conditions are satisfied if $K$ is less than 4. Hence with an amplifier gain design value of 2, the system could tolerate an increase in gain of a factor of 2 before it would become unstable. The gain could also drop to zero without causing instability.

**5.16.** Determine the Hurwitz conditions for stability of the following general fourth-order characteristic equation, assuming $a_4$ is positive.

$$a_4 s^4 + a_3 s^3 + a_2 s^2 + a_1 s + a_0 = 0$$

The Hurwitz determinants are

$$\Delta_4 = \begin{vmatrix} a_3 & a_1 & 0 & 0 \\ a_4 & a_2 & a_0 & 0 \\ 0 & a_3 & a_1 & 0 \\ 0 & a_4 & a_2 & a_0 \end{vmatrix} = a_3(a_2 a_1 a_0 - a_3 a_0^2) - a_1^2 a_0 a_4$$

$$\Delta_3 = \begin{vmatrix} a_3 & a_1 & 0 \\ a_4 & a_2 & a_0 \\ 0 & a_3 & a_1 \end{vmatrix} = a_3 a_2 a_1 - a_0 a_3^2 - a_4 a_1^2$$

$$\Delta_2 = \begin{vmatrix} a_3 & a_1 \\ a_4 & a_2 \end{vmatrix} = a_3 a_2 - a_4 a_1$$

$$\Delta_1 = a_3$$

The conditions for stability are then

$$a_3 > 0 \qquad a_3 a_2 - a_4 a_1 > 0 \qquad a_3 a_2 a_1 - a_0 a_3^2 - a_4 a_1^2 > 0 \qquad a_3(a_2 a_1 a_0 - a_3 a_0^2) - a_1^2 a_0 a_4 > 0$$

**5.17.** Is the system with the following characteristic equation stable?

$$s^4 + 3s^3 + 6s^2 + 9s + 12 = 0$$

Substituting the appropriate values for the coefficients in the general conditions of Problem 5.16, we have

$$3 > 0 \qquad 18 - 9 > 0 \qquad 162 - 108 - 81 \not> 0 \qquad 3(648 - 432) - 972 \not> 0$$

Since the last two conditions are not satisfied, the system is unstable.

## CONTINUED FRACTION STABILITY CRITERION

**5.18.** Repeat Problem 5.9 using the continued fraction stability criterion.

The polynomial $Q(s) = s^3 + 4s^2 + 8s + 12$ is divided into the two parts:

$$Q_1(s) = s^3 + 8s \qquad Q_2(s) = 4s^2 + 12$$

The continued fraction for $Q_1(s)/Q_2(s)$ is

$$\frac{Q_1(s)}{Q_2(s)} = \frac{s^3 + 8s}{4s^2 + 12} = \frac{1}{4}s + \frac{5s}{4s^2 + 12} = \frac{1}{4}s + \cfrac{1}{\cfrac{4}{5}s + \cfrac{1}{\frac{5}{12}s}}$$

Since all the coefficients of $s$ are positive, the polynomial has all its roots in the left half-plane and the system with the characteristic equation $Q(s) = 0$ is stable.

**5.19.** Determine bounds upon the parameter $K$ for which a system with the following characteristic equation is stable:

$$s^3 + 14s^2 + 56s + K = 0$$

$$\frac{Q_1(s)}{Q_2(s)} = \frac{s^3 + 56s}{14s^2 + K} = \frac{1}{14}s + \frac{(56 - K/14)s}{14s^2 + K} = \frac{1}{14}s + \cfrac{1}{\left[\cfrac{14}{56 - K/14}\right]s + \cfrac{1}{\left[\cfrac{56 - K/14}{K}\right]s}}$$

For the system to be stable, the following conditions must be satisfied: $56 - K/14 > 0$ and $K > 0$, that is, $0 < K < 784$.

**5.20.** Derive conditions for all the roots of a general third-order polynomial to have negative real parts.

For $Q(s) = a_3 s^3 + a_2 s^2 + a_1 s + a_0$,

$$\frac{Q_1(s)}{Q_2(s)} = \frac{a_3 s^3 + a_1 s}{a_2 s^2 + a_0} = \frac{a_3}{a_2}s + \frac{[a_1 - a_3 a_0/a_2]s}{a_2 s^2 + a_0} = \frac{a_3}{a_2}s + \cfrac{1}{\left[\cfrac{a_2}{a_1 - a_3 a_0/a_2}\right]s + \cfrac{1}{\left[\cfrac{a_1 - a_3 a_0/a_2}{a_0}\right]s}}$$

The conditions for all the roots of $Q(s)$ to have negative real parts are then

$$\frac{a_3}{a_2} > 0 \qquad \frac{a_2}{a_1 - a_3 a_0/a_2} > 0 \qquad \frac{a_1 - a_3 a_0/a_2}{a_0} > 0$$

Thus if $a_3$ is positive, the required conditions are $a_2, a_1, a_0 > 0$ and $a_1 a_2 - a_3 a_0 > 0$. Note that if $a_3$ is not positive, $Q(s)$ should be multiplied by $-1$ before checking the above conditions.

**5.21.** Is the system with the following characteristic equation stable?

$$s^4 + 4s^3 + 8s^2 + 16s + 32 = 0$$

$$\frac{Q_1(s)}{Q_2(s)} = \frac{s^4 + 8s^2 + 32}{4s^3 + 16s} = \frac{1}{4}s + \frac{4s^2 + 32}{4s^3 + 16s}$$

$$= \frac{1}{4}s + \cfrac{1}{s + \cfrac{-16s}{4s^2 + 32}} = \frac{1}{4}s + \cfrac{1}{s + \cfrac{1}{-\frac{1}{4}s + \cfrac{1}{-\frac{1}{2}s}}}$$

Since the coefficients of $s$ are not all positive, the system is unstable.

## DISCRETE-TIME SYSTEMS

**5.22.** Is the system with the following characteristic equation stable?

$$Q(z) = z^4 + 2z^3 + 3z^2 + z + 1 = 0$$

Applying the Jury test, with $n = 4$ (even),

$$Q(1) = 1 + 2 + 3 + 1 + 1 = 8 > 0$$
$$Q(-1) = 1 - 2 + 3 - 1 + 1 = 2 > 0$$

The Jury array must be constructed, as follows:

| row | | | | | |
|---|---|---|---|---|---|
| 1 | 1 | 1 | 3 | 2 | 1 |
| 2 | 1 | 2 | 3 | 1 | 1 |
| 3 | 0 | -1 | 0 | 1 | |
| 4 | 1 | 0 | -1 | 0 | |
| 5 | -1 | 1 | 0 | | |

The Jury test constraints are

$$|a_0| = 1 \not< 1 = a_n$$
$$|b_0| = 0 \not> 1 = |b_{n-1}|$$
$$|c_0| = |-1| > 0 = |c_{n-2}|$$

Since all the constraints are not satisfied, the system is unstable.

**5.23.** Is the system with the following characteristic equation stable?

$$Q(z) = 2z^4 + 2z^3 + 3z^2 + z + 1 = 0$$

Applying the Jury test, with $n = 4$ (even),

$$Q(1) = 2 + 2 + 3 + 1 + 1 = 9 > 0$$
$$Q(-1) = 2 - 2 + 3 - 1 + 1 = 3 > 0$$

The Jury array must be constructed, as follows:

| row | | | | | |
|---|---|---|---|---|---|
| 1 | 1 | 1 | 3 | 2 | 2 |
| 2 | 2 | 2 | 3 | 1 | 1 |
| 3 | 3 | 3 | 2 | 0 | |
| 4 | 0 | 2 | 3 | 3 | |
| 5 | 9 | 7 | 0 | | |

The test constraints are

$$|a_0| = 1 < 2 = a_n$$
$$|b_0| = 3 > 0 = |b_{n-1}|$$
$$|c_0| = 9 > 0 = |c_{n-2}|$$

Since all the constraints are satisfied, the system is stable.

**5.24.** Is the system with the following characteristic equation stable?

$$Q(z) = z^5 + 3z^4 + 3z^3 + 3z^2 + 2z + 1 = 0$$

Applying the Jury test, with $n = 5$ (odd),

$$Q(1) = 1 + 3 + 3 + 3 + 2 + 1 = 13 > 0$$
$$Q(-1) = -1 + 3 - 3 + 3 - 2 + 1 = 1 > 0$$

Since $n$ is odd, $Q(-1)$ must be less than zero for the system to be stable. Therefore the system is unstable.

## MISCELLANEOUS PROBLEMS

**5.25.** If a zero appears in the first column of the Routh table, is the system necessarily unstable?

Strictly speaking, a zero in the first column must be interpreted as having no sign, that is, neither positive nor negative. Consequently, all the elements of the first column cannot have the same sign if one of them is zero, and the system is unstable. In some cases, a zero in the first column indicates the presence of two roots of equal magnitude but opposite sign (see Problem 5.10). In other cases, it indicates the presence of one or more roots with zero real parts. Thus a characteristic equation having one or more roots with zero real parts and no roots with positive real parts will produce a Routh table in which all the elements of the first column do not have the same sign and do not have any sign changes.

**5.26.** Prove that a continuous system is unstable if any coefficients of the characteristic equation are zero.

The characteristic equation may be written in the form

$$(s - s_1)(s - s_2)(s - s_3) \cdots (s - s_n) = 0$$

where $s_1, s_2, \ldots, s_n$ are the roots of the equation. If this equation is multiplied out, $n$ new equations can be obtained relating the roots and the coefficients of the characteristic equation in the usual form. Thus

$$a_n s^n + a_{n-1} s^{n-1} + \cdots + a_0 = 0 \qquad \text{or} \qquad s^n + \frac{a_{n-1}}{a_n} s^{n-1} + \cdots + \frac{a_0}{a_n} = 0$$

and the relations are

$$\frac{a_{n-1}}{a_n} = - \sum_{i=1}^{n} s_i, \quad \frac{a_{n-2}}{a_n} = \sum_{\substack{i=1 \\ i \neq j}}^{n} \sum_{j=1}^{n} s_i s_j, \quad \frac{a_{n-3}}{a_n} = - \sum_{\substack{i=1 \\ i \neq j \neq k}}^{n} \sum_{j=1}^{n} \sum_{k=1}^{n} s_i s_j s_k, \ldots, \frac{a_0}{a_n} = (-1)^n s_1 s_2 \cdots s_n$$

The coefficients $a_{n-1}, a_{n-2}, \ldots, a_0$ all have the same sign as $a_n$ and are nonzero if all the roots $s_1, s_2, \ldots, s_n$ have negative real parts. The only way any one of the coefficients can be zero is for one or more of the roots to have zero or positive real parts. In either case, the system would be unstable.

**5.27.** Prove that a continuous system is unstable if all the coefficients of the characteristic equation do not have the same sign.

From the relations presented in Problem 5.26, it can be seen that the coefficients $a_{n-1}, a_{n-2}, \ldots, a_0$ have the same sign as $a_n$ if all the roots $s_1, s_2, \ldots, s_n$ have negative real parts. The only way any of these coefficients may differ in sign from $a_n$ is for one or more of the roots to have a positive real part. Thus the system is necessarily unstable if all the coefficients do not have the same sign. Note that a system is *not* necessarily stable if all the coefficients do have the same sign.

**5.28.** Can the continuous system stability criteria presented in this chapter be applied to continuous systems which contain time delays?

No they cannot be directly applied because systems which contain time delays do not have characteristic equations of the required form, that is, finite polynomials in $s$. For example, the following characteristic equation represents a system which contains a time delay:

$$s^2 + s + e^{-sT} = 0$$

Strictly speaking, this equation has an infinite number of roots. However, in some cases an approximation may be employed for $e^{-sT}$ to give useful, although not entirely accurate, information concerning system stability. To illustrate, let $e^{-sT}$ in the equation above be replaced by the first two terms of its Taylor series. The equation then becomes

$$s^2 + s + 1 - sT = 0 \qquad \text{or} \qquad s^2 + (1 - T)s + 1 = 0$$

One of the stability criteria of this chapter may then be applied to this approximation of the characteristic equation.

**5.29.** Determine an approximate upper limit on the time delay in order that the system discussed in the solution of Problem 5.28 be stable.

Employing the approximate equation $s^2 + (1 - T)s + 1 = 0$, the Hurwitz determinants are $\Delta_1 = \Delta_2 = 1 - T$. Hence for the system to be stable, the time delay $T$ must be less than 1.

# Supplementary Problems

**5.30.** For each characteristic polynomial, determine if it represents a stable or an unstable system.

Mathcad

(a) $2s^4 + 8s^3 + 10s^2 + 10s + 20$  (c) $s^5 + 6s^4 + 10s^2 + 5s + 24$  (e) $s^4 + 8s^3 + 24s^2 + 32s + 16$

(b) $s^3 + 7s^2 + 7s + 46$  (d) $s^3 - 2s^2 + 4s + 6$  (f) $s^6 + 4s^4 + 8s^2 + 16$

**5.31.** For what values of $K$ does the polynomial $s^3 + (4 + K)s^2 + 6s + 12$ have roots with negative real parts?

**5.32.** How many roots with positive real parts does each polynomial have?

(a) $s^3 + s^2 - s + 1$   (b) $s^4 + 2s^3 + 2s^2 + 2s + 1$   (c) $s^3 + s^2 - 2$   (d) $s^4 - s^2 - 2s + 2$

(e) $s^3 + s^2 + s + 6$

**5.33.** For what positive value of $K$ does the polynomial $s^4 + 8s^3 + 24s^2 + 32s + K$ have roots with zero real parts? What are these roots?

# Answers to Supplementary Problems

**5.30.** (b) and (e) represent stable systems; (a), (c), (d), and (f) represent unstable systems.

**5.31.** $K > -2$

**5.32.** (a) 2, (b) 0, (c) 1, (d) 2, (e) 2

**5.33.** $K = 80$; $s = \pm j2$

# Chapter 6

# Transfer Functions

## 6.1 DEFINITION OF A CONTINUOUS SYSTEM TRANSFER FUNCTION

As shown in Chapters 3 and 4, the response of a time-invariant linear system can be separated into two parts: the forced response and the free response. This is true for both continuous and discrete systems. We consider continuous transfer functions first, and for single-input, single-output systems only. Equation ($4.8$) clearly illustrates this division for the most general constant-coefficient, linear, ordinary differential equation. The forced response includes terms due to initial values $u_0^k$ of the input, and the free response depends only on initial conditions $y_0^k$ on the output. If terms due to *all* initial values, that is, $u_0^k$ *and* $y_0^k$, are lumped together, Equation ($4.8$) can be written as

$$y(t) = \mathcal{L}^{-1}\left[ \left( \sum_{i=0}^{m} b_i s^i \Big/ \sum_{i=0}^{n} a_i s^i \right) U(s) + \left( \text{terms due to } all \text{ initial values } u_0^k,\ y_0^k \right) \right]$$

or, in transform notation, as

$$Y(s) = \left( \sum_{i=0}^{m} b_i s^i \Big/ \sum_{i=0}^{n} a_i s^i \right) U(s) + \left( \text{terms due to } all \text{ initial values } u_0^k,\ y_0^k \right)$$

The **transfer function** $P(s)$ of a continuous system is defined as that factor in the equation for $Y(s)$ multiplying the transform of the input $U(s)$. For the system described above, the transfer function is

$$P(s) = \sum_{i=0}^{m} b_i s^i \Big/ \sum_{i=0}^{n} a_i s^i = \frac{b_m s^m + b_{m-1} s^{m-1} + \cdots + b_0}{a_n s^n + a_{n-1} s^{n-1} + \cdots + a_0}$$

the denominator is the characteristic polynomial, and the transform of the response may be rewritten as

$$Y(s) = P(s)U(s) + \left( \text{terms due to } all \text{ initial values } u_0^k,\ y_0^k \right)$$

If the quantity (terms due to *all* initial values $u_0^k,\ y_0^k$) is zero, the Laplace transform of the output $Y(s)$ in response to an input $U(s)$ is given by

$$Y(s) = P(s)U(s)$$

If the system is at rest prior to application of the input, that is, $d^k y/dt^k = 0$, $k = 0, 1, \ldots, n-1$, for $t < 0$, then

$$\left( \text{terms due to } all \text{ initial values } u_0^k,\ y_0^k \right) = 0$$

and the output as a function of time $y(t)$ is simply the inverse transform of $P(s)U(s)$.

It is emphasized that not all transfer functions are rational algebraic expressions. For example, the transfer function of a continuous system including time delays contains terms of the form $e^{-sT}$ (e.g., Problem 5.28). The transfer function of an element representing a pure time delay is $P(s) = e^{-sT}$, where $T$ is the time delay in units of time.

Since the formation of the output transform $Y(s)$ is purely an algebraic multiplication of $P(s)$ and $U(s)$ when (terms due to *all* initial values $u_0^k,\ y_0^k) = 0$, the multiplication is commutative; that is,

$$Y(s) = U(s)P(s) = P(s)U(s) \tag{6.1}$$

128

## 6.2  PROPERTIES OF A CONTINUOUS SYSTEM TRANSFER FUNCTION

The transfer function of a continuous system has several useful properties:

1.  It is the Laplace transform of its impulse response $y_\delta(t)$, $t \geq 0$. That is, if the input to a system with transfer function $P(s)$ is an impulse and all initial values are zero the transform of the output is $P(s)$.

2.  The system transfer function can be determined from the system differential equation by taking the Laplace transform and ignoring all terms arising from initial values. The transfer function $P(s)$ is then given by

$$P(s) = \frac{Y(s)}{U(s)}$$

3.  The system differential equation can be obtained from the transfer function by replacing the $s$ variable with the differential operator $D$ defined by $D \equiv d/dt$.

4.  The stability of a time-invariant linear system can be determined from the characteristic equation (see Chapter 5). The denominator of the system transfer function is the *characteristic polynomial*. Consequently, for continuous systems, if all the roots of the denominator have negative real parts, the system is stable.

5.  The roots of the denominator are the system poles and the roots of the numerator are the system zeros (see Chapter 4). The system transfer function can then be specified to within a constant by specifying the system poles and zeros. This constant, usually denoted by $K$, is the system **gain factor**. As was described in Chapter 4, Section 4.11, the system poles and zeros can be represented schematically by a pole-zero map in the $s$-plane.

6.  If the system transfer function has no poles or zeros with positive real parts, the system is a **minimum phase** system.

**EXAMPLE 6.1.**   Consider the system with the differential equation $dy/dt + 2y = du/dt + u$.

The Laplace transform version of this equation with all initial values set equal to zero is $(s+2)Y(s) = (s+1)U(s)$.

The system transfer function is thus given by $P(s) = Y(s)/U(s) = (s+1)/(s+2)$.

**EXAMPLE 6.2.**   Given $P(s) = (2s+1)/(s^2 + s + 1)$, the system differential equation is

$$y = \left[ \frac{2D+1}{D^2 + D + 1} \right] u \quad \text{or} \quad D^2y + Dy + y = 2Du + u \quad \text{or} \quad \frac{d^2y}{dt^2} + \frac{dy}{dt} + y = 2\frac{du}{dt} + u$$

**EXAMPLE 6.3.**   The transfer function $P(s) = K(s+a)/(s+b)(s+c)$ can be specified by giving the zero location $-a$, the pole locations $-b$ and $-c$, and the gain factor $K$.

## 6.3  TRANSFER FUNCTIONS OF CONTINUOUS CONTROL SYSTEM COMPENSATORS AND CONTROLLERS

The transfer functions of four common control system components are presented below. Typical mechanizations of three of these transfer functions, using $R$-$C$ networks, are presented in the solved problems.

**EXAMPLE 6.4.**   The general transfer function of a **continuous system lead compensator** is

$$P_{\text{Lead}}(s) = \frac{s+a}{s+b} \qquad b > a \tag{6.2}$$

This compensator has a zero at $s = -a$ and a pole at $s = -b$.

**EXAMPLE 6.5.** The general transfer function of a **continuous system lag compensator** is

$$P_{\text{Lag}}(s) = \frac{a(s+b)}{b(s+a)} \qquad b > a \qquad\qquad (6.3)$$

However, in this case the zero is at $s = -b$ and the pole is at $s = -a$. The gain factor $a/b$ is included because of the way it is usually mechanized (Problem 6.13).

**EXAMPLE 6.6.** The general transfer function of a **continuous system lag-lead compensator** is

$$P_{\text{LL}}(s) = \frac{(s+a_1)(s+b_2)}{(s+b_1)(s+a_2)} \qquad b_1 > a_1, \, b_2 > a_2 \qquad\qquad (6.4)$$

This compensator has two zeros and two poles. For mechanization considerations, the restriction $a_1 b_2 = b_1 a_2$ is usually imposed (Problem 6.14).

**EXAMPLE 6.7.** The transfer function of the **PID controller** of Example 2.14 is

$$P_{\text{PID}}(s) \equiv \frac{U_{\text{PID}}(s)}{E(s)} = K_P + K_D s + \frac{K_I}{s} = \frac{K_D s^2 + K_P s + K_I}{s} \qquad\qquad (6.5)$$

This controller has two zeros and one pole. It is similar to the lag-lead compensator of the previous example except that the smallest pole is at the origin (an integrator) and it does not have the second pole. It is typically mechanized in an analog or digital computer.

## 6.4 CONTINUOUS SYSTEM TIME RESPONSE

The Laplace transform of the response of a continuous system to a specific input is given by

$$Y(s) = P(s)U(s)$$

*when all initial conditions are zero.* The inverse transform $y(t) = \mathscr{L}^{-1}[P(s)U(s)]$ is then the time response and $y(t)$ may be determined by finding the poles of $P(s)U(s)$ and evaluating the residues at these poles (when there are no multiple poles). Therefore $y(t)$ depends on both the poles and zeros of the transfer function and the poles and zeros of the input.

The residues can be determined graphically from a *pole-zero map* of $Y(s)$, constructed from the pole-zero map of $P(s)$ by simply adding the poles and zeros of $U(s)$. Graphical evaluation of the residues may then be performed as described in Chapter 4, Section 4.12.

## 6.5 CONTINUOUS SYSTEM FREQUENCY RESPONSE

The steady state response of a continuous system to sinusoidal inputs can be determined from the system transfer function. For the special case of a step function input of amplitude $A$, often called a **d.c. input**, the Laplace transform of the system output is given by

$$Y(s) = P(s)\frac{A}{s}$$

If the system is stable, the steady state response is a step function of amplitude $AP(0)$, since this is the residue at the input pole. The amplitude of the input signal is thus multiplied by $P(0)$ to determine the amplitude of the output. $P(0)$ is therefore the **d.c. gain** of the system.

Note that for an unstable system such as an integrator ($P(s) = 1/s$), a steady state response does not always exist. If the input to an integrator is a step function, the output is a ramp, which is unbounded (see Problems 5.7 and 5.8). For this reason, integrators are sometimes said to have infinite d.c. gain.

The steady state response of a stable system to an input $u = A \sin \omega t$ is given by

$$y_{\text{ss}} = A|P(j\omega)|\sin(\omega t + \phi)$$

where $|P(j\omega)|$ = magnitude of $P(j\omega)$, $\phi = \arg P(j\omega)$, and the complex number $P(j\omega)$ is determined from $P(s)$ by replacing $s$ by $j\omega$ (see Problem 6.20). The system output has the same frequency as the input and can be obtained by multiplying the magnitude of the input by $|P(j\omega)|$ and shifting the phase angle of the input by $\arg P(j\omega)$. The magnitude $|P(j\omega)|$ and angle $\arg P(j\omega)$ for all $\omega$ together define the **system frequency response**. The magnitude $|P(j\omega)|$ is the *gain* of the system for sinusoidal inputs with frequency $\omega$.

The system frequency response can be determined graphically in the $s$-plane from a pole-zero map of $P(s)$ in the same manner as the graphical calculation of residues. In this instance, however, the magnitude and phase angle of $P(s)$ are computed at a point on the $j\omega$ axis by measuring the magnitudes and angles of the vectors drawn from the poles and zeros of $P(s)$ to the point on the $j\omega$ axis.

**EXAMPLE 6.8.** Consider the system with the transfer function

$$P(s) = \frac{1}{(s+1)(s+2)}$$

Referring to Fig. 6-1, the magnitude and angle of $P(j\omega)$ for $\omega = 1$ are computed in the $s$-plane as follows. The magnitude of $P(j1)$ is

$$|P(j1)| = \frac{1}{\sqrt{5} \cdot \sqrt{2}} = 0.316$$

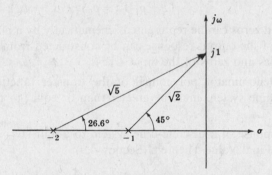

Fig. 6-1

and the angle is

$$\arg P(j1) = -26.6° - 45° = -71.6°$$

**EXAMPLE 6.9.** The system frequency response is usually represented by two graphs (see Fig. 6-2): one of $|P(j\omega)|$ as a function of $\omega$ and one of $\arg P(j\omega)$ as a function of $\omega$. For the transfer function of Example 6.8, $P(s) = 1/(s+1)(s+2)$, these graphs are easily determined by plotting the values of $|P(j\omega)|$ and $\arg P(j\omega)$ for several values of $\omega$ as shown below.

| $\omega$ | 0 | 0.5 | 1.0 | 2.0 | 4.0 | 8.0 |
|---|---|---|---|---|---|---|
| $|P(j\omega)|$ | 0.5 | 0.433 | 0.316 | 0.158 | 0.054 | 0.015 |
| $\arg P(j\omega)$ | 0 | $-40.6°$ | $-71.6°$ | $-108.5°$ | $-139.4°$ | $-158.9°$ |

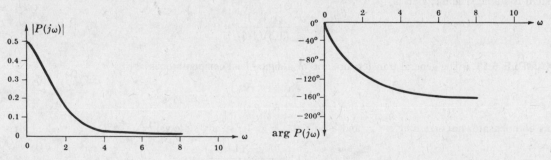

Fig. 6-2

## 6.6  DISCRETE-TIME SYSTEM TRANSFER FUNCTIONS, COMPENSATORS AND TIME RESPONSES

The transfer function $P(z)$ for a discrete-time system is defined as that factor in the equation for the transform of the output $Y(z)$ that multiplies the transform of the input $U(z)$. If all terms due to initial conditions are zero, then the system response to an input $U(z)$ is given by: $Y(z) = P(z)U(z)$ in the $z$-domain, and $\{y(k)\} = \mathcal{Z}^{-1}[P(z)U(z)]$ in the time-domain.

The transfer function of a discrete-time system has the following properties:

1.  $P(z)$ is the $z$-transform of its Kronecker delta response $y_\delta(k)$, $k = 0, 1, \ldots$.

2.  The system difference equation can be obtained from $P(z)$ by replacing the $z$ variable with the shift operator $Z$ defined for any integers $k$ and $n$ by

$$Z^n[y(k)] = y(k+n) \tag{6.6}$$

3.  The denominator of $P(z)$ is the system *characteristic polynomial*. Consequently, if all the roots of the denominator are within the unit circle of the $z$-plane, the system is stable.

4.  The roots of the denominator are system poles and the roots of the numerator are the system zeros. $P(z)$ can be specified by specifying the system poles and zeros and the gain factor $K$:

$$P(z) = \frac{K(z + z_1)(z + z_2) \cdots (z + z_m)}{(z + p_1)(z + p_2) \cdots (z + p_n)} \tag{6.7}$$

The system poles and zeros can be represented schematically by a pole-zero map in the $z$-plane. The pole-zero map of the output response can be constructed from the pole-zero map of $P(z)$ by including the poles and zeros of the input $U(z)$.

5.  The order of the denominator polynomial of the transfer function of a causal (physically realizable) discrete-time system must be greater than or equal to the order of the numerator polynomial.

6.  The steady state response of a discrete-time system to a unit step input is called the **d.c. gain** and is given by the Final Value Theorem (Section 4.9):

$$\lim_{k \to \infty} y(k) = \lim_{z \to 1} \left[ \frac{z-1}{z} P(z) \frac{z}{z-1} \right] = P(1) \tag{6.8}$$

**EXAMPLE 6.10.**  Consider a discrete-time system characterized by the difference equation

$$y(k+2) + 1.1y(k+1) + 0.3y(k) = u(k+2) + 0.2u(k+1)$$

The $z$-transform version of this equation with all initial conditions set equal to zero is

$$(z^2 + 1.1z + 0.3)Y(z) = (z^2 + 0.2z)U(z)$$

The system transfer function is given by

$$P(z) = \frac{z(z + 0.2)}{z^2 + 1.1z + 0.3} = \frac{z(z + 0.2)}{(z + 0.5)(z + 0.6)}$$

This system has a zero at $-0.2$ and two poles, at $-0.5$ and $-0.6$. Since the poles are inside the unit circle, the system is stable. The d.c. gain is

$$P(1) = \frac{1(1.2)}{(1.5)(1.6)} = 0.5$$

**EXAMPLE 6.11.**  The general transfer function of a **digital lead compensator** is

$$P_{\text{Lead}}(z) = \frac{K_{\text{Lead}}(z - z_c)}{z - p_c} \qquad z_c > p_c \tag{6.9}$$

This compensator has a zero at $z = z_c$ and a pole at $z = p_c$. Its steady state gain is

$$P_{\text{Lead}}(1) = \frac{K_{\text{Lead}}(1 - z_c)}{1 - p_c} \tag{6.10}$$

The gain factor $K_{\text{Lead}}$ is included in the transfer function to adjust its gain at a given $\omega$ to a desired value. In Problem 12.13, for example, $K_{\text{Lead}}$ is chosen to render the steady state gain of $P_{\text{Lead}}$ (at $\omega = 0$) equal to that of its analog counterpart.

**EXAMPLE 6.12.**   The general transfer function of a **digital lag compensator** is

$$P_{\text{Lag}}(z) = \frac{(1-p_c)(z-z_c)}{(1-z_c)(z-p_c)} \qquad z_c < p_c \qquad (6.11)$$

This compensator has a zero at $z = z_c$ and a pole at $z = p_c$. The gain factor $(1-p_c)/(1-z_c)$ is included so that the low frequency or steady state gain $P_{\text{Lag}}(1) = 1$, analogous to the continuous-time lag compensator.

**EXAMPLE 6.13.**   Digital lag and lead compensators can be designed directly from $s$-domain specifications by using the transform between the $s$- and $z$-domains defined by $z = e^{sT}$. That is, the poles and zeros of

$$P_{\text{Lead}}(s) = \frac{s+a}{s+b} \qquad \text{and} \qquad P_{\text{Lag}} = \frac{a(s+b)}{b(s+a)}$$

can be mapped according to $z = e^{sT}$. For the lead compensator, the zero at $s = -a$ maps into the zero at $z = z_c \equiv e^{-aT}$, and the pole at $s = -b$ maps into the pole at $z = p_c \equiv e^{-bT}$. This gives

$$P'_{\text{Lead}}(z) = \frac{z - e^{-aT}}{z - e^{-bT}} \qquad (6.12)$$

Similarly,

$$P'_{\text{Lag}}(z) = \left( \frac{1 - e^{-aT}}{1 - e^{-bT}} \right) \left( \frac{z - e^{-bT}}{z - e^{-aT}} \right) \qquad (6.13)$$

Note that $P'_{\text{Lag}}(1) = 1$.

This transformation is only one of many possible for digital lead and lag compensators, or any type of compensators for that matter. Another variant of the lead compensator is illustrated in Problems 12.13 through 12.15.

An example of how Equation (6.13) can be used in applications is given in Example 12.7.

## 6.7   DISCRETE-TIME SYSTEM FREQUENCY RESPONSE

The steady state response to an input sequence $\{u(k) = A \sin \omega kT\}$ of a stable discrete-time system with transfer function $P(z)$ is given by

$$y_{ss} = A \left| P(e^{j\omega T}) \right| \sin(\omega kT + \phi) \qquad k = 0, 1, 2, \ldots \qquad (6.14)$$

where $|P(e^{j\omega T})|$ is the magnitude of $P(e^{j\omega T})$, $\phi = \arg P(e^{j\omega T})$, and the complex function $P(e^{j\omega T})$ is determined from $P(z)$ by replacing $z$ by $e^{j\omega T}$ (see Problem 6.40). The system output is a sequence of samples of a sinusoid with the same frequency as the input sinusoid. The output sequence is obtained by multiplying the magnitude $A$ of the input by $|P(e^{j\omega T})|$ and shifting the phase angle of the input by $\arg P(e^{j\omega T})$. The magnitude $|P(e^{j\omega T})|$ and phase angle $\arg P(e^{j\omega T})$, for all $\omega$, together define the **discrete-time system frequency response function**. The magnitude $|P(e^{j\omega T})|$ is the **gain** of the system for sinusoidal inputs with angular frequency $\omega$.

A discrete-time system frequency response function can be determined in the $z$-plane from a pole-zero map of $P(z)$ in the same manner as the graphical calculation of residues (Section 4.12). In this instance, however, the magnitude and phase angle are computed on the $e^{j\omega T}$ circle (the unit circle), by measuring the magnitude and angle of the vectors drawn from the poles and zeros of $P$ to the point on the unit circle. Since $P(e^{j\omega T})$ is periodic in $\omega$, with period $2\pi/T$, the frequency response function need only be determined over the angular frequency range $-\pi/T \le \omega \le \pi/T$. Also, since the magnitude function is an even function of $\omega$, and the phase angle is an odd function of $\omega$, actual computations need only be performed over half this angular frequency range, that is, $0 \le \omega \le \pi/T$.

## 6.8  COMBINING CONTINUOUS-TIME AND DISCRETE-TIME ELEMENTS

Thus far the $z$-transform has been used mainly to describe systems and elements which operate on and produce only discrete-time signals, and the Laplace transform has been used only for continuous-time systems and elements, with continuous-time input and output signals. However, many control systems include both types of elements. Some of the important relationships between the $z$-transform and the Laplace transform are developed here, to facilitate analysis and design of mixed (continuous/discrete) systems.

Discrete-time signals arise either from the sampling of continuous-time signals, or as the output of inherently discrete-time system components, such as digital computers. If a continuous-time signal $y(t)$ with Laplace transform $Y(s)$ is sampled uniformly, with period $T$, the resulting sequence of samples $y(kT)$, $k = 0, 1, 2, \ldots$, can be written as

$$y(kT) = \frac{1}{2\pi j} \int_{c-j\infty}^{c+j\infty} Y(s) e^{skT} \, ds \qquad k = 0, 1, 2, \ldots$$

where $c > \sigma_0$ (see Definition 4.3). The $z$-transform of this sequence is $Y^*(z) = \sum_{k=0}^{\infty} y(kT) z^{-k}$ (Definition 4.4) which, as shown in Problem 6.41, can be written as

$$Y^*(z) = \frac{1}{2\pi j} \int_{c-j\infty}^{c+j\infty} Y(s) \left( \frac{1}{1 - e^{sT} z^{-1}} \right) ds \qquad (6.15)$$

for the region of convergence $|z| > e^{cT}$. This relationship between the Laplace transform and the $z$-transform can be evaluated by application of Cauchy's integral law [1]. However, in practice, it is usually not necessary to use this complex analysis approach.

The continuous-time function $y(t) = \mathcal{L}^{-1}[Y(s)]$ can be determined from $Y(s)$ and a table of Laplace transforms, and the time variable $t$ is then replaced by $kT$, providing the $k$th element of the desired sequence:

$$y(kT) = \mathcal{L}^{-1}[Y(s)]\big|_{t=kT}$$

Then the $z$-transform of the sequence $y(kT)$, $k = 0, 1, 2, \ldots$, is generated by referring to a table of $z$-transforms, which yields the desired result:

$$Y^*(z) = \mathcal{Z}\{y(kT)\} = \mathcal{Z}\left\{ \mathcal{L}^{-1}[Y(s)]\big|_{t=kT} \right\} \qquad (6.16)$$

Thus, in Equation (6.16), the symbolic operations $\mathcal{L}^{-1}$ and $\mathcal{Z}$ represent straightforward table lookups, and $|_{t=kT}$ generates the sequence to be $z$-transformed.

A common combination of discrete-time and continuous-time elements and signals is shown in Fig. 6-3.

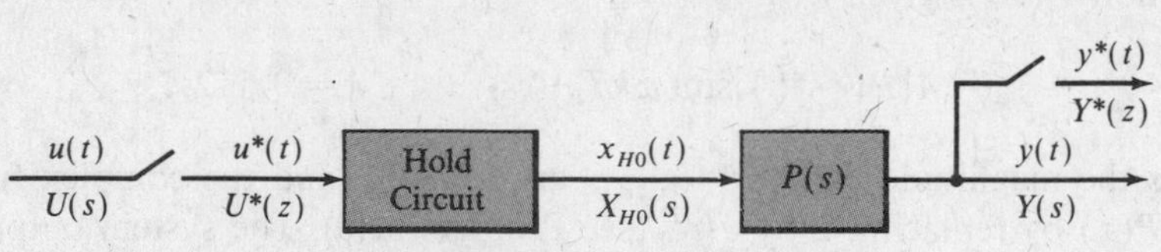

Fig. 6-3

If the hold circuit is a *zero-order hold*, then as shown in Problem 6.42, the discrete-time transfer function from $U^*(z)$ to $Y^*(z)$ is given by

$$\frac{Y^*(z)}{U^*(z)} = (1 - z^{-1}) \mathcal{Z}\left\{ \mathcal{L}^{-1} \frac{P(s)}{s} \bigg|_{t=kT} \right\} \qquad (6.17)$$

In practice, the sampler at the output, generating $y^*(t)$ in Fig. 6-3, may not exist. However, it is sometimes convenient to assume one exists at that point, for purposes of analysis (see, e.g., Problem 10.13). When this is done, the sampler is often called a **fictitious sampler**.

If both the input and output of a system like the one shown in Fig. 6-3 are continuous-time signals, and the input is subsequently sampled, then Equation (6.17) generates a discrete-time transfer function which relates the input at the sampling times $T, 2T, \ldots$ to the output at the same sampling times. However, this discrete-time system transfer function does *not* relate input and output signals at times $\tau$ *between* sampling times, that is, for $kT < \tau < (k+1)T$, $k = 0, 1, 2, \ldots$.

**EXAMPLE 6.14.** In Fig. 6-3, if the hold circuit is a zero-order hold and $P(s) = 1/(s+1)$, then from Equation (6.17), the discrete-time transfer function of the mixed-element subsystem is

$$
\begin{aligned}
\frac{Y^*(z)}{U^*(z)} &= (1 - z^{-1})\, \mathcal{Z}\left\{ \mathcal{L}^{-1}\left( \frac{1}{s(s+1)} \right)\Big|_{t=kT} \right\} \\[2mm]
&= (1 - z^{-1})\, \mathcal{Z}\left\{ \mathcal{L}^{-1}\left( \frac{1}{s} - \frac{1}{s+1} \right)\Big|_{t=kT} \right\} \\[2mm]
&= (1 - z^{-1})\, \mathcal{Z}\left\{ \left( \mathbf{1}(t) - e^{-t} \right)\big|_{t=kT} \right\} \\[2mm]
&= (1 - z^{-1})\, \mathcal{Z}\left\{ \mathbf{1}(kT) - e^{-kT} \right\} \\[2mm]
&= (1 - z^{-1})\left[ \mathcal{Z}\left\{ \mathbf{1}(kT) \right\} - \mathcal{Z}\left\{ e^{-kT} \right\} \right] \\[2mm]
&= (1 - z^{-1})\left[ \frac{1}{1 - z^{-1}} - \frac{1}{1 - e^{-T}z^{-1}} \right] \\[2mm]
&= \left( \frac{z-1}{z} \right)\left( \frac{z}{z-1} \right)\left[ \frac{1 - e^{-T}}{z - e^{-T}} \right] \\[2mm]
&= \frac{1 - e^{-T}}{z - e^{-T}}
\end{aligned}
$$

# Solved Problems

## TRANSFER FUNCTION DEFINITIONS

**6.1.** What is the transfer function of a system whose input and output are related by the following differential equation?

$$
\frac{d^2 y}{dt^2} + 3\frac{dy}{dt} + 2y = u + \frac{du}{dt}
$$

Taking the Laplace transform of this equation, ignoring terms due to initial conditions, we obtain

$$
s^2 Y(s) + 3sY(s) + 2Y(s) = U(s) + sU(s)
$$

This equation can be written as

$$
Y(s) = \left[ \frac{s+1}{s^2 + 3s + 2} \right] U(s)
$$

The transfer function of this system is therefore given by

$$
P(s) = \frac{s+1}{s^2 + 3s + 2}
$$

**6.2.** A particular system containing a time delay has the differential equation $(d/dt)y(t) + y(t) = u(t - T)$. Find the transfer function of this system.

The Laplace transform of the differential equation, ignoring terms due to initial conditions, is $sY(s) + Y(s) = e^{-sT}U(s)$. $Y(s)$ and $U(s)$ are related by the following function of $s$, which is the system transfer function

$$
P(s) = \frac{Y(s)}{U(s)} = \frac{e^{-sT}}{s+1}
$$

**6.3.** The position $y$ of a moving object of constant mass $M$ is related to the total force $f$ applied to the object by the differential equation $M(d^2y/dt^2) = f$. Determine the transfer function relating the position to the applied force.

Taking the Laplace transform of the differential equation, we obtain $Ms^2Y(s) = F(s)$. The transfer function relating $Y(s)$ to $F(s)$ is therefore $P(s) = Y(s)/F(s) = 1/Ms^2$.

**6.4.** A motor connected to a load with inertia $J$ and viscous friction $B$ produces a torque proportional to the input current $i$. If the differential equation for the motor and load is $J(d^2\theta/dt^2) + B(d\theta/dt) = Ki$, determine the transfer function between the input current $i$ and the shaft position $\theta$.

The Laplace transform version of the differential equation is $(Js^2 + Bs)\Theta(s) = KI(s)$, and the required transfer function is $P(s) = \Theta(s)/I(s) = K/s(Js + B)$.

## PROPERTIES OF TRANSFER FUNCTIONS

**6.5.** An impulse is applied at the input of a continuous system and the output is observed to be the time function $e^{-2t}$. Find the transfer function of this system.

The transfer function is $P(s) = Y(s)/U(s)$ and $U(s) = 1$ for $u(t) = \delta(t)$. Therefore

$$P(s) = Y(s) = \frac{1}{s+2}$$

**6.6.** The impulse response of a certain continuous system is the sinusoidal signal $\sin t$. Determine the system transfer function and differential equation.

The system transfer function is the Laplace transform of its impulse response, $P(s) = 1/(s^2 + 1)$. Then $P(D) = y/u = 1/(D^2 + 1)$, $D^2y + y = u$ or $d^2y/dt^2 + y = u$.

**6.7.** The step response of a given system is $y = 1 - \frac{7}{3}e^{-t} + \frac{3}{2}e^{-2t} - \frac{1}{6}e^{-4t}$. What is the transfer function of this system?

Since the derivative of a step is an impulse (see Definition 3.17), the impulse response for this system is $p(t) = dy/dt = \frac{7}{3}e^{-t} - 3e^{-2t} + \frac{2}{3}e^{-4t}$.
The Laplace transform of $p(t)$ is the desired transfer function. Thus

$$P(s) = \frac{\frac{7}{3}}{s+1} + \frac{-3}{s+2} + \frac{\frac{2}{3}}{s+4} = \frac{s+8}{(s+1)(s+2)(s+4)}$$

Note that an alternative solution would be to compute the Laplace transform of $y$ and then multiply by $s$ to determine $P(s)$, since a multiplication by $s$ in the $s$-domain is equivalent to differentiation in the time domain.

**6.8.** Determine if the transfer function $P(s) = (2s + 1)/(s^2 + s + 1)$ represents a stable or an unstable system.

The characteristic equation of the system is obtained by setting the denominator polynomial to zero, that is, $s^2 + s + 1 = 0$. The characteristic equation may then be tested using one of the stability criteria described in Chapter 5. The Routh table for this system is given by

$$
\begin{array}{c|cc}
s^2 & 1 & 1 \\
s^1 & 1 & \\
s^0 & 1 & \\
\end{array}
$$

Since there are no sign changes in the first column, the system is stable.

**6.9.** Does the transfer function $P(s) = (s + 4)/(s + 1)(s + 2)(s - 1)$ represent a stable or an unstable system?

The stability of the system is determined by the roots of the denominator polynomial, that is, the *poles* of the system. Here the denominator is in factored form and the poles are located at $s = -1, -2, +1$. Since there is one pole with a positive real part, the system is unstable.

**6.10.** What is the transfer function of a system with a gain factor of 2 and a pole-zero map in the $s$-plane as shown in Fig. 6-4?

The transfer function has a zero at $-1$ and poles at $-2$ and the origin. Hence the transfer function is $P(s) = 2(s + 1)/s(s + 2)$.

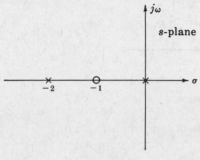
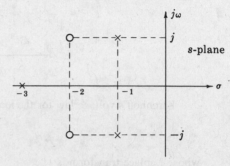

Fig. 6-4                                                    Fig. 6-5

**6.11.** Determine the transfer function of a system with a gain factor of 3 and the pole-zero map shown in Fig. 6-5

Mathcad

The transfer function has zeros at $-2 \pm j$ and poles at $-3$ and at $-1 \pm j$. The transfer function is therefore $P(s) = 3(s + 2 + j)(s + 2 - j)/(s + 3)(s + 1 + j)(s + 1 - j)$.

## TRANSFER FUNCTIONS OF CONTINUOUS CONTROL SYSTEM COMPONENTS

**6.12.** An $R$-$C$ network mechanization of a lead compensator is shown in Fig. 6-6. Find its transfer function.

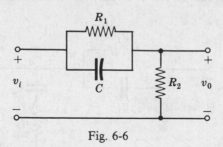

Fig. 6-6

Assuming the circuit is not loaded, that is, no current flows through the output terminals, Kirchhoff's current law for the output node yields

$$C \frac{d}{dt}(v_i - v_0) + \frac{1}{R_1}(v_i - v_0) = \frac{1}{R_2} v_0$$

The Laplace transform of this equation (with zero initial conditions) is

$$Cs[V_i(s) - V_0(s)] + \frac{1}{R_1}[V_i(s) - V_0(s)] = \frac{1}{R_2} V_0(s)$$

The transfer function is

$$P_{\text{Lead}} = \frac{V_0(s)}{V_i(s)} = \frac{Cs + 1/R_1}{Cs + 1/R_1 + 1/R_2} = \frac{s+a}{s+b}$$

where $a = 1/R_1C$ and $b = 1/R_1C + 1/R_2C$.

**6.13.** Determine the transfer function of the $R$-$C$ network mechanization of the lag compensator shown in Fig. 6-7.

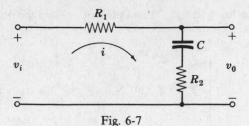

Fig. 6-7

Kirchhoff's voltage law for the loop yields the equation

$$iR_1 + \frac{1}{C}\int_0^t i\,dt + iR_2 = v_i$$

whose Laplace transform is

$$\left(R_1 + R_2 + \frac{1}{Cs}\right)I(s) = V_i(s)$$

The output voltage is given by

$$V_0(s) = \left(R_2 + \frac{1}{Cs}\right)I(s)$$

The transfer function of the lag network is therefore

$$P_{\text{Lag}} = \frac{V_0(s)}{V_i(s)} = \frac{R_2 + 1/Cs}{R_1 + R_2 + 1/Cs} = \frac{a(s+b)}{b(s+a)} \quad \text{where} \quad a = \frac{1}{(R_1+R_2)C} \qquad b = \frac{1}{R_2C}$$

**6.14.** Derive the transfer function of the $R$-$C$ network mechanization of the lag-lead compensator shown in Fig. 6-8.

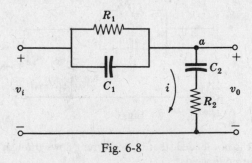

Fig. 6-8

Equating currents at the output node $a$ yields

$$\frac{1}{R_1}(v_i - v_0) + C_1\frac{d}{dt}(v_i - v_0) = i$$

The voltage $v_0$ and the current $i$ are related by

$$\frac{1}{C_2}\int_0^t i\,dt + iR_2 = v_0$$

Taking the Laplace transform of these two equations (with zero initial conditions) and eliminating $I(s)$ results in the equation

$$\left(\frac{1}{R_1} + C_1 s\right)[V_i(s) - V_0(s)] = \frac{V_0(s)}{1/sC_2 + R_2}$$

The transfer function of the network is therefore

$$P_{LL} = \frac{V_0(s)}{V_i(s)} = \frac{\left(s + \dfrac{1}{R_1 C_1}\right)\left(s + \dfrac{1}{R_2 C_2}\right)}{s^2 + \left(\dfrac{1}{R_2 C_2} + \dfrac{1}{R_2 C_1} + \dfrac{1}{R_1 C_1}\right)s + \dfrac{1}{R_1 C_1 R_2 C_2}} = \frac{(s + a_1)(s + b_2)}{(s + b_1)(s + a_2)}$$

where

$$a_1 = \frac{1}{R_1 C_1} \qquad b_1 a_2 = a_1 b_2 \qquad b_1 + a_2 = a_1 + b_2 + \frac{1}{R_2 C_1} \qquad b_2 = \frac{1}{R_2 C_2}$$

**6.15.** Find the transfer function of the *simple* lag network shown in Fig. 6-9.

This network is a special case of the lag compensation network of Problem 6.13 with $R_2$ set equal to zero. Hence the transfer function is given by

$$P(s) = \frac{V_0(s)}{V_i(s)} = \frac{1/Cs}{R + 1/Cs} = \frac{1/RC}{s + 1/RC}$$

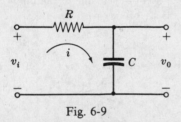

Fig. 6-9

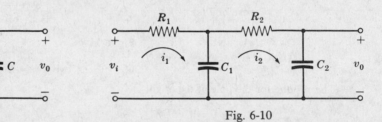

Fig. 6-10

Mathcad

**6.16.** Determine the transfer function of two simple lag networks connected in series as shown in Fig. 6-10.

The two loop equations are

$$R_1 i_1 + \frac{1}{C_1} \int_0^t (i_1 - i_2)\, dt = v_i$$

$$R_2 i_2 + \frac{1}{C_2} \int_0^t i_2\, dt + \frac{1}{C_1} \int_0^t (i_2 - i_1)\, dt = 0$$

Using the Laplace transformation and solving the two loop equations for $I_2(s)$, we obtain

$$I_2(s) = \frac{C_2 s V_i(s)}{R_1 R_2 C_1 C_2 s^2 + (R_1 C_1 + R_1 C_2 + R_2 C_2)s + 1}$$

The output voltage is given by $v_0 = (1/C_2)\int_0^t i_2\, dt$. Thus

$$\frac{V_0(s)}{V_i(s)} = \frac{1}{R_1 R_2 C_1 C_2 s^2 + (R_1 C_1 + R_1 C_2 + R_2 C_2)s + 1}$$

## CONTINUOUS SYSTEM TIME RESPONSE

**6.17.** What is the unit step response of a continuous system whose transfer function has a zero at $-1$, a pole at $-2$, and a gain factor of 2?

The Laplace transform of the output is given by $Y(s) = P(s)U(s)$. Here

$$P(s) = \frac{2(s+1)}{s+2} \qquad U(s) = \frac{1}{s} \qquad Y(s) = \frac{2(s+1)}{s(s+2)} = \frac{1}{s} + \frac{1}{s+2}$$

Evaluating the inverse transform of the partial fraction expansion of $Y(s)$ gives $y(t) = 1 + e^{-2t}$.

**6.18.** Graphically evaluate the unit step response of a continuous system whose transfer function is given by

$$P(s) = \frac{(s+2)}{(s+0.5)(s+4)}$$

The pole-zero map of the output is obtained by adding the poles and zeros of the input to the pole-zero map of the transfer function. The output pole-zero map therefore has poles at 0, $-0.5$, and $-4$ and a zero at $-2$ as shown in Fig. 6-11.

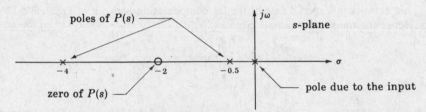

Fig. 6-11

The residue for the pole at the origin is

$$|R_1| = \frac{2}{0.5(4)} = 1 \qquad \arg R_1 = 0°$$

For the pole at $-0.5$,

$$|R_2| = \frac{1.5}{0.5(3.5)} = 0.857 \qquad \arg R_2 = -180°$$

For the pole at $-4$,

$$|R_3| = \frac{2}{4(3.5)} = 0.143 \qquad \arg R_3 = -180°$$

The time response is therefore $y(t) = R_1 + R_2 e^{-0.5t} + R_3 e^{-4t} = 1 - 0.857 e^{-0.5t} - 0.143 e^{-4t}$.

**6.19.** Evaluate the unit step response of the system of Problem 6.11.

Mathcad

The Laplace transform of the system output is

$$Y(s) = P(s)U(s) = \frac{3(s+2+j)(s+2-j)}{s(s+3)(s+1+j)(s+1-j)}$$

Expanding $Y(s)$ into partial fractions yields

$$Y(s) = \frac{R_1}{s} + \frac{R_2}{s+3} + \frac{R_3}{s+1+j} + \frac{R_4}{s+1-j}$$

where

$$R_1 = \frac{3(2+j)(2-j)}{3(1+j)(1-j)} = \frac{5}{2} \qquad R_3 = \frac{3(1)(1-2j)}{(-1-j)(2-j)(-2j)} = \frac{-3}{20}(7+j)$$

$$R_2 = \frac{3(-1+j)(-1-j)}{-3(-2+j)(-2-j)} = \frac{-2}{5} \qquad R_4 = \frac{3(1+2j)(1)}{(2+j)(-1+j)(2j)} = \frac{-3}{20}(7-j)$$

Evaluating the inverse Laplace transform,

$$y = \frac{5}{2} - \frac{2}{5}e^{-3t} - \frac{3\sqrt{2}}{4}e^{-t}\left[e^{-j(t+\theta)} + e^{j(t+\theta)}\right] = \frac{5}{2} - \frac{2}{5}e^{-3t} - \frac{3\sqrt{2}}{2}e^{-t}\cos(t+\theta)$$

where $\theta = -\tan^{-1}[\frac{1}{7}] = -8.13°$.

## CONTINUOUS SYSTEM FREQUENCY RESPONSE

**6.20.** Prove that the steady state output of a stable system with transfer function $P(s)$ and input $u = A\sin\omega t$ is given by

$$y_{ss} = A|P(j\omega)|\sin(\omega t + \phi) \qquad \text{where} \quad \phi = \arg P(j\omega)$$

The Laplace transform of the output is $Y(s) = P(s)U(s) = P(s)[A\omega/(s^2 + \omega^2)]$.

When this transform is expanded into partial fractions, there will be terms due to the poles of $P(s)$ and two terms due to the poles of the input ($s = \pm j\omega$). Since the system is stable, all time functions resulting from the poles of $P(s)$ decay to zero as time approaches infinity. Thus the steady state output contains only the time functions resulting from the terms in the partial fraction expansion due to the poles of the input. The Laplace transform of the steady state output is therefore

$$Y_{ss}(s) = \frac{AP(j\omega)}{2j(s-j\omega)} + \frac{AP(-j\omega)}{-2j(s+j\omega)}$$

The inverse transform of this equation is

$$y_{ss} = A|P(j\omega)|\left[\frac{e^{j\phi}e^{j\omega t} - e^{-j\phi}e^{-j\omega t}}{2j}\right] = A|P(j\omega)|\sin(\omega t + \phi) \qquad \text{where} \quad \phi = \arg P(j\omega)$$

**6.21.** Find the *d.c. gain* of each of the systems represented by the following transfer functions:

$(a) \quad P(s) = \frac{1}{s+1} \qquad (b) \quad P(s) = \frac{10}{(s+1)(s+2)} \qquad (c) \quad P(s) = \frac{(s+8)}{(s+2)(s+4)}$

The d.c. gain is given by $P(0)$. Then $(a)$ $P(0) = 1$, $(b)$ $P(0) = 5$, $(c)$ $P(0) = 1$.

**6.22.** Evaluate the gain and phase shift of $P(s) = 2/(s+2)$ for $\omega = 1, 2,$ and $10$.

The gain of $P(s)$ is given by $|P(j\omega)| = 2/\sqrt{\omega^2+4}$. For $\omega = 1$, $|P(j1)| = 2/\sqrt{5} = 0.894$; for $\omega = 2$, $|P(j2)| = 2/\sqrt{8} = 0.707$; for $\omega = 10$, $|P(j10)| = 2/\sqrt{104} = 0.196$.

The phase shift of the transfer function is the phase angle of $P(j\omega)$, $\arg P(j\omega) = -\tan^{-1}\omega/2$. For $\omega = 1$, $\arg P(j1) = -\tan^{-1}\frac{1}{2} = -26.6°$; for $\omega = 2$, $\arg P(j2) = -\tan^{-1}1 = -45°$; for $\omega = 10$, $\arg P(j10) = -\tan^{-1}5 = -78.7°$.

**6.23.** Sketch the graphs of $|P(j\omega)|$ and $\arg P(j\omega)$ as a function of frequency for the transfer function of Problem 6.22.

In addition to the values calculated in Problem 6.22 for $|P(j\omega)|$ and $\arg P(j\omega)$, the values for $\omega = 0$ will also be useful: $|P(j0)| = 2/2 = 1$, $\arg P(j0) = -\tan^{-1}0 = 0$.

As $\omega$ becomes large, $|P(j\omega)|$ asymptotically approaches zero while $\arg P(j\omega)$ asymptotically approaches $-90°$. The graphs representing the frequency response of $P(s)$ are shown in Fig. 6-12.

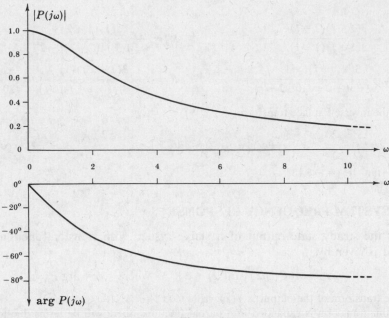

Fig. 6-12

## DISCRETE-TIME SYSTEM TRANSFER FUNCTIONS AND TIME RESPONSES

**6.24.** The Kronecker delta response of a discrete-time system is given by $y_\delta(k) = 1$ for all $k \geq 0$. What is its transfer function?

The transfer function is the $z$-transform of the Kronecker delta response, as given in Example 4.26:

$$P(z) = 1 + z^{-1} + z^{-2} + z^{-3} + \cdots$$

To determine a pole-zero representation of $P(z)$, note that

$$zP(z) - z = P(z)$$

or

$$(z - 1)P(z) = z$$

so that

$$P(z) = \frac{z}{z - 1}$$

Alternatively, note that the Kronecker delta response is the unit step sequence, which has the $z$-transform

$$P(z) = \frac{z}{z - 1}$$

(see Table 4.2).

**6.25.** The Kronecker delta response of a particular discrete system is given by $y_\delta(k) = (0.5)^k$ for $k \geq 0$. What is its transfer function?

The form of the Kronecker delta response indicates the presence of a single pole at 0.5. The Kronecker delta response of a system with a single pole and no zero has no output at $k = 0$. That is,

$$\frac{1}{z - 0.5} = z^{-1} + 0.5z^{-2} + 0.25z^{-3} + \cdots + (0.5)^{n-1}z^{-n} + \cdots$$

Consequently, the transfer function must have a zero in the numerator to advance the output sequence one sample interval. That is,

$$P(z) = \frac{z}{z - 0.5}$$

**6.26.** What is the difference equation for a system whose transfer function is

$$P(z) = \frac{z - 0.1}{z^2 + 0.3z + 0.2}$$

Replacing $z^n$ with $Z^n$, we get

$$P(Z) = \frac{Z - 0.1}{Z^2 + 0.3Z + 0.2}$$

Then

$$y(k) = P(Z)u(k) = \frac{(Z - 0.1)u(k)}{Z^2 + 0.3Z + 0.2} = \frac{u(k+1) - 0.1u(k)}{Z^2 + 0.3Z + 0.2}$$

and, by cross multiplying,

$$y(k + 2) + 0.3y(k + 1) + 0.2y(k) = u(k + 1) - 0.1u(k)$$

**6.27.** What is the transfer function of a discrete system with a gain factor of 2, zeros at 0.2 and $-0.5$, and poles at 0.5, 0.6, and $-0.4$? Is it stable?

The transfer function is

$$P(z) = \frac{2(z - 0.2)(z + 0.5)}{(z - 0.5)(z - 0.6)(z + 0.4)}$$

Since all the system poles are inside the unit circle, the system is stable.

## MISCELLANEOUS PROBLEMS

**6.28.** A *d.c.* (*direct current*) *motor* is shown schematically in Fig. 6-13. $L$ and $R$ represent the inductance and resistance of the motor armature circuit, and the voltage $v_b$ represents the generated back e.m.f. (electromotive force) which is proportional to the shaft velocity $d\theta/dt$. The torque $T$ generated by the motor is proportional to the armature current $i$. The inertia $J$ represents the combined inertia of the motor armature and the load, and $B$ is the total viscous friction acting on the output shaft. Determine the transfer function between the input voltage $V$ and the angular position $\Theta$ of the output shaft.

**Motor Armature Circuit**                    **Inertial Load**

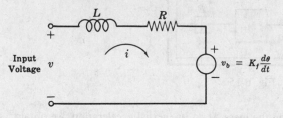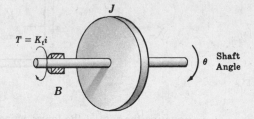

Fig. 6-13

The differential equations of the motor armature circuit and the inertial load are

$$Ri + L\frac{di}{dt} = v - K_f\frac{d\theta}{dt} \quad \text{and} \quad K_t i = J\frac{d^2\theta}{dt^2} + B\frac{d\theta}{dt}$$

Taking the Laplace transform of each equation, ignoring initial conditions,

$$(R + sL)I = V - K_f s\Theta \quad \text{and} \quad K_t I = (Js^2 + Bs)\Theta$$

Solving these equations simultaneously for the transfer function between $V$ and $\Theta$, we have

$$\frac{\Theta}{V} = \frac{K_t}{(Js^2 + Bs)(Ls + R) + K_t K_f s} = \frac{K_t/JL}{s\left[s^2 + (B/J + R/L)s + BR/JL + K_t K_f/JL\right]}$$

**6.29.** The back e.m.f generated by the armature circuit of a d.c. machine is proportional to the angular velocity of its shaft, as noted in the problem above. This principle is utilized in the *d.c. tachometer* shown schematically in Fig. 6-14, where $v_b$ is the voltage generated by the armature, $L$ is the armature inductance, $R_a$ is the armature resistance, and $v_0$ is the output voltage. If $K_f$ is the proportionality constant between $v_b$ and shaft velocity $d\theta/dt$, that is, $v_b = K_f(d\theta/dt)$, determine the transfer function between the shaft position $\Theta$ and the output voltage $V_0$. The output load is represented by a resistance $R_L$ and $R_L + R_a \equiv R$.

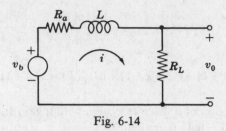

Fig. 6-14

The Laplace transformed equation representing the tachometer is $I(R + sL) = K_f s\Theta$. The output voltage is given by

$$V_0 = IR_L = \frac{R_L K_f s\Theta}{R + sL}$$

The transfer function of the d.c. tachometer is then

$$\frac{V_0}{\Theta} = \frac{R_L K_f}{L}\left(\frac{s}{s + R/L}\right)$$

**6.30.** A simple mechanical *accelerometer* is shown in Fig. 6-15. The position $y$ of the mass $M$ with respect to the accelerometer case is proportional to the acceleration of the case. What is the transfer function between the input acceleration $A$ $(a = d^2x/dt^2)$ and the output $Y$?

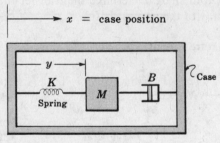

Fig. 6-15

Equating the sum of the forces acting on the mass $M$ to its inertial acceleration, we obtain

$$-B\frac{dy}{dt} - Ky = M\frac{d^2}{dt^2}(y - x)$$

or

$$M\frac{d^2y}{dt^2} + B\frac{dy}{dt} + Ky = M\frac{d^2x}{dt^2} = Ma$$

where $a$ is the input acceleration. The zero initial condition transformed equation is

$$(Ms^2 + Bs + K)Y = MA$$

The transfer function of the accelerometer is therefore

$$\frac{Y}{A} = \frac{1}{s^2 + (B/M)s + K/M}$$

**6.31.** A differential equation describing the dynamic operation of the *one-degree-of-freedom gyroscope* shown in Fig. 6-16 is

$$J\frac{d^2\theta}{dt^2} + B\frac{d\theta}{dt} + K\theta = H\omega$$

where $\omega$ is the angular velocity of the gyroscope about the input axis, $\theta$ is the angular position of the spin axis—the measured output of the gyroscope, $H$ is angular momentum stored in the spinning wheel, $J$ is the inertia of the wheel about the output axis, $B$ is the viscous friction coefficient about the output axis, and $K$ is the spring constant of the restraining spring attached to the spin axis.

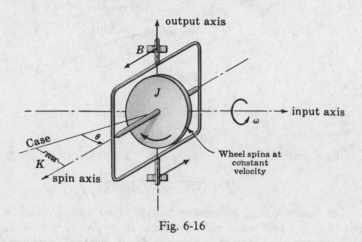

Fig. 6-16

(a) Determine the transfer function relating the Laplace transforms of $\omega$ and $\theta$, and show that the steady state output is proportional to the magnitude of a constant rate input. This type of gyroscope is called a *rate gyro*.

(b) Determine the transfer function between $\omega$ and $\theta$ with the restraining spring removed ($K = 0$). Since here the output is proportional to the integral of the input rate, this type of gyroscope is called an *integrating gyro*.

(a) The zero initial condition transform of the gyroscope differential equation is

$$(Js^2 + Bs + K)\Theta = H\Omega$$

where $\Theta$ and $\Omega$ are the Laplace transforms of $\theta$ and $\omega$, respectively. The transfer function relating $\Theta$ and $\Omega$ is therefore

$$\frac{\Theta}{\Omega} = \frac{H}{(Js^2 + Bs + K)}$$

For a constant or d.c. rate input $\omega_K$, the magnitude of the steady state output $\theta_{ss}$ can be obtained by multiplying the input by the d.c. gain of the transfer function, which in this case is $H/K$. Thus the steady state output is proportional to the magnitude of the rate input, that is, $\theta_{ss} = (H/K)\omega_K$.

(b) Setting $K$ equal to zero in the transfer function of (a) yields $\Theta/\Omega = H/s(Js + B)$. This transfer function now has a pole at the origin, so that an integration is obtained between the input $\Omega$ and the output $\Theta$. The output is thus proportional to the integral of the input rate or, equivalently, the input angle.

**6.32.** A differential equation approximating the rotational dynamics of a rigid vehicle moving in the atmosphere is

$$J\frac{d^2\theta}{dt^2} - NL\theta = T$$

where $\theta$ is the vehicle attitude angle, $J$ is its inertia, $N$ is the normal-force coefficient, $L$ is the distance from the center of gravity to the center of pressure, and $T$ is any applied torque (see Fig. 6-17). Determine the transfer function between an applied torque and the vehicle attitude angle.

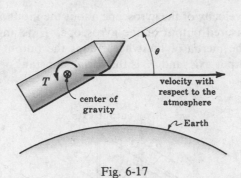

center of gravity

velocity with respect to the atmosphere

Earth

Fig. 6-17

The zero initial condition, transformed system differential equation is

$$(Js^2 - NL)\Theta = T$$

The desired transfer function is

$$\frac{\Theta}{T} = \frac{1}{Js^2 - NL} = \frac{1/J}{s^2 - NL/J}$$

Note that if $NL$ is positive (center of pressure forward of the vehicle center of gravity), the system is *unstable* because there is a pole in the right half-plane at $s = \sqrt{NL/J}$. If $NL$ is negative, the poles are imaginary and the system is *oscillatory* (marginally stable). However, aerodynamic damping terms not included in the differential equation are actually present and perform the function of damping out any oscillations.

**6.33.** Pressure receptors called *baroreceptors* measure changes in arterial blood pressure, as outlined in Problem 2.14. They are shown as a block in the feedback path of the block diagram determined in the solution of that problem. The frequency $b(t)$ at which signals (action potentials) move along the vagus and glossopharyngeal nerves from the baroreceptors to the vasomotor center (VMC) in the brain is proportional to arterial blood pressure $p$ plus the time rate of change of blood pressure. Determine the form of the transfer function for the baroreceptors.

From the description given above, the equation for $b$ is

$$b = k_1 p + k_2 \frac{dp}{dt}$$

where $k_1$ and $k_2$ are constants, and $p$ is blood pressure. [$p$ should not be confused here with the notation $p(t)$, the inverse Laplace transform of $P(s)$ introduced in this chapter as a general representation for a transfer function.] The Laplace transform of the above equation, with zero initial conditions, is

$$B = k_1 P + k_2 sP = P(k_1 + k_2 s)$$

The transfer function of the baroreceptors is therefore $B/P = k_1 + k_2 s$. We again remind the reader that $P$ represents the transform of arterial blood pressure in this problem.

**6.34.** Consider the transfer function $C_k/R_k$ for the biological system described in Problem 3.4($a$) by the equations

$$c_k(t) = r_k(t) - \sum_{i=1}^{n} a_{k-i} c_i(t - \Delta t)$$

for $k = 1, 2, \ldots, n$. Explain how $C_k/R_k$ may be computed.

Taking the Laplace transform of the above equations, ignoring initial conditions, yields the following set of equations:

$$C_k = R_k - \sum_{i=1}^{n} a_{k-i} C_i e^{-s \Delta t}$$

for $k = 1, 2, \ldots, n$. If all $n$ equations were written down, we would have $n$ equations in $n$ unknowns ($C_k$ for $k = 1, 2, \ldots, n$). The general solution for any $C_k$ in terms of the inputs $R_k$ can then be determined using the standard techniques for solving simultaneous equations. Let $D$ represent the determinant of the coefficient matrix:

$$D \equiv \begin{vmatrix} 1 + a_0 e^{-s \Delta t} & a_{-1} e^{-s \Delta t} & \cdots & a_{1-n} e^{-s \Delta t} \\ a_1 e^{-s \Delta t} & 1 + a_0 e^{-s \Delta t} & \cdots & a_{2-n} e^{-s \Delta t} \\ \cdots\cdots\cdots\cdots\cdots\cdots\cdots\cdots\cdots\cdots\cdots\cdots\cdots \\ a_{n-1} e^{-s \Delta t} & \cdots & a_1 e^{-s \Delta t} & 1 + a_0 e^{-s \Delta t} \end{vmatrix}$$

Then in general,

$$C_k = \frac{D_k}{D}$$

where $D_k$ is the determinant of the coefficient matrix with the $k$th column replaced by

$$\begin{matrix} R_1 \\ R_2 \\ \vdots \\ R_n \end{matrix}$$

The transfer function $C_k / R_k$ is then determined by setting all the inputs except $R_k$ equal to zero, computing $C_k$ from the formula above, and dividing $C_k$ by $R_k$.

**6.35.** Can you determine the $s$-domain transfer function of the ideal sampler described in Problems 3.5 and 4.39? Why?

No. From the results of Problem 4.39, the output transform $U(s)$ of the ideal sampler is

$$U^*(s) = \sum_{k=0}^{\infty} e^{-skT} u(kT)$$

It is not possible to factor out the transform $U(s)$ of the input signal $u(t)$ applied to the sampler, because the sampler is not a time-invariant system element. Therefore it cannot be described by an ordinary transfer function.

**6.36.** Based on the developments of the sampler and zero-order hold function given in Problems 3.5, 3.6, 3.7, and 4.39, design an idealization of the zero-order hold transfer function.

In Problem 3.7, impulses in $m_{\text{IT}}(t)$ replaced the current pulses modulated by $m_s(t)$ in Problem 3.6. Then, by the screening property of the unit impulse, Equation (3.20), the integral of each impulse is the value of $u(t)$ at the sampling instant $kT$, $k = 0, 1, \ldots$, etc. Therefore it is logical to replace the capacitor (and resistor) in the approximate hold circuit of Problem 3.6 by an integrator, which has the Laplace transform $1/s$. To complete the design, the output of the hold must be equal to $u$ at each sampling time, not $u - y_{H0}$; therefore we need a function that automatically resets the integrator to zero after each sampling period. The transfer function of such a device is given by the "pulse" transfer function:

$$P_{H0}(s) = \frac{1}{s}(1 - e^{-sT})$$

Then we can write the transform of the output of the ideal hold device as

$$Y_{H0}(s) = P_{H0}(s) U^*(s) = \frac{1}{s}(1 - e^{-sT}) \sum_{k=0}^{\infty} e^{-sT} u(kT)$$

**6.37.** Can you determine the $s$-domain transfer function of the ideal sampler and ideal zero-order hold combination of the previous problem? Why?

No. It is not possible to factor out the transform $U(s)$ of $u(t)$ applied to the sampler. Again, the sampler is not a time-invariant device.

**6.38.** The simple lag circuit of Fig. 6-3, with a switch $S$ in the input line, was described in Problem 3.6 as an approximate sample and zero-order hold device, and idealized in Problem 6.36. Why is this the case, and under what circumstances?

The transfer function of the simple lag was shown in Problem 6-15 to be

$$P(s) = \frac{1/RC}{s + 1/RC}$$

If $RC \ll 1$, $P(s)$ can be approximated as $P(s) \approx 1$, and the capacitor ideally holds the output constant until the next sample time.

**6.39.** Show that for a rational function $P(z)$ to be the transfer function of a *causal* discrete-time system, the order of its denominator polynomial must be equal to or greater than the order of its numerator polynomial (Property 6, Section 6.6).

In Section 3.16 we saw that a discrete-time system is causal if its weighting sequence $w(k) = 0$ for $k < 0$. Let $P(z)$, the system transfer function, have the form:

$$P(z) = \frac{b_m z^m + b_{m-1} z^{m-1} + \cdots + b_1 z + b_0}{a_n z^n + a_{n-1} z^{n-1} + \cdots + a_1 z + a_0}$$

where $a_n \neq 0$ and $b_m \neq 0$. The weighting sequence $w(k)$ can be generated by inverting $P(z)$, using the long division technique of Section 4.9.

We first divide the numerator and denominator of $P(z)$ by $z^m$, thus forming:

$$P(z) = \frac{b_m + b_{m-1} z^{-1} + \cdots + b_0 z^{-m}}{a_n z^{n-m} + a_{n-1} z^{n-m-1} + \cdots + a_0 z^{-m}}$$

Dividing the denominator of $P(z)$ into its numerator then gives

$$P(z) = \left(\frac{b_m}{a_n}\right) z^{m-n} + \left(b_{m-1} - \frac{b_m a_{n-1}}{a_n}\right) z^{m-n-1} + \cdots$$

The coefficient of $z^{-k}$ in this expansion of $P(z)$ is $w(k)$, and we see that $w(k) = 0$ for $k < n - m$ and

$$w(n-m) = \frac{b_m}{a_n} \neq 0$$

For causality, $w(k) = 0$ for $k < 0$, therefore $n - m \geq 0$ and $n \geq m$.

**6.40.** Show that the steady state response of a stable discrete-time system to an input sequence $u(k) = A \sin \omega kT$, $k = 0, 1, 2, \ldots$, is given by

$$y_{ss} = A \left| P\left(e^{j\omega T}\right) \right| \sin(\omega kT + \phi) \qquad k = 0, 1, 2, \ldots \qquad (6.14)$$

where $P(z)$ is the system transfer function.

Since the system is linear, if this result is true for $A = 1$, then it is true for arbitrary values of $A$. To simplify the arguments, an input $u'(k) = e^{j\omega kT}$, $k = 0, 1, 2, \ldots$, is used. By noting that

$$u'(k) = e^{j\omega kT} = \cos \omega kT + j \sin \omega kT$$

the response of the system to $\{u'(k)\}$ is a complex combination of the responses to $\{\cos \omega kT\}$ and

$\{\sin \omega kT\}$, where the imaginary part is the response to $\{\sin \omega kT\}$. From Table 4.2 the $z$-transform of $\{e^{j\omega kT}\}$ is

$$\frac{z}{z - e^{j\omega T}}$$

Thus the $z$-transform of the system output $Y'(z)$ is

$$Y'(z) = P(z)\frac{z}{z - e^{j\omega T}}$$

To invert $Y'(z)$, we form the partial fraction expansion of

$$\frac{Y'(z)}{z} = P(z)\frac{1}{z - e^{j\omega T}}$$

This expansion consists of terms due to the poles of $P(z)$ and a term due to the pole at $z = e^{j\omega T}$. Therefore

$$Y'(z) = z\left[\sum \text{terms due to poles of } P(z) + \frac{P(e^{j\omega T})}{z - e^{j\omega T}}\right]$$

and

$$\{y'(k)\} = \mathcal{Z}^{-1}\left[z\sum \text{terms due to poles of } P(z)\right] + \left\{P(e^{j\omega T})e^{j\omega kT}\right\}$$

Since the system is stable, the first term vanishes as $k$ becomes large and

$$y_{ss} = P(e^{j\omega T})e^{j\omega kT} = \left|P(e^{j\omega T})\right|e^{j(\omega kT + \phi)}$$

$$= \left|P(e^{j\omega T})\right|\left[\cos(\omega kT + \phi) + j\sin(\omega kT + \phi)\right] \qquad k = 0,1,2,\dots$$

where $\phi = \arg P(e^{j\omega T})$. The steady state response to the input $\sin \omega kT$ is the imaginary part of $y_{ss}$, or

$$y_{ss} = \left|P(e^{j\omega T})\right|\sin(\omega kT + \phi) \qquad k = 0,1,2,\dots$$

**6.41.** Show that, if a continuous-time function $y(t)$ with Laplace transform $Y(s)$ is sampled uniformly with period $T$, the $z$-transform of the resulting sequence of samples $Y^*(z)$ is related to $Y(s)$ by Equation (6.15).

From Definition 4.3:

$$y(t) = \frac{1}{2\pi j}\int_{c-j\infty}^{c+j\infty} Y(s)e^{st}\,ds$$

where $c > \sigma_0$. Uniformly sampling $y(t)$ generates the samples $y(kT)$, $k = 0,1,2,\dots$. Therefore

$$y(kT) = \frac{1}{2\pi j}\int_{c-j\infty}^{c+j\infty} Y(s)e^{skT}\,ds \qquad k = 0,1,2,\dots$$

The $z$-transform of this sequence is

$$Y^*(z) = \sum_{k=0}^{\infty} y(kT)z^{-k} = \sum_{k=0}^{\infty} \frac{z^{-k}}{2\pi j}\int_{c-j\infty}^{c+j\infty} Y(s)e^{skT}\,ds$$

and after interchanging summation and integration,

$$Y^*(z) = \frac{1}{2\pi j}\int_{c-j\infty}^{c+j\infty} Y(s)\sum_{k=0}^{\infty} e^{skT}z^{-k}\,ds$$

Now

$$\sum_{k=0}^{\infty} e^{skT}z^{-k} = \sum_{k=0}^{\infty} \left(e^{sT}z^{-1}\right)^k$$

is a geometric series, which converges if $|e^{sT}z^{-1}| < 1$. In this case,

$$\sum_{k=0}^{\infty} \left(e^{sT}z^{-1}\right)^k = \frac{1}{1 - e^{sT}z^{-1}}$$

The inequality $|e^{sT}z^{-1}| < 1$ implies that $|z| > |e^{sT}|$. On the integration contour, $|e^{sT}| = |e^{(c+j\omega)T}| = e^{cT}$

Thus the series converges for $|z| > e^{cT}$. Therefore

$$Y^*(z) = \frac{1}{2\pi j} \int_{c-j\infty}^{c+j\infty} Y(s) \frac{1}{1 - e^{sT}z^{-1}} \, ds$$

for $|z| > e^{cT}$, which is Equation (6.15).

**6.42.** Show that if the hold circuit in Fig. 6-3 is a zero-order hold, the discrete-time transfer function is given by Equation (6.17).

Let $p(t) = \mathcal{L}^{-1}[P(s)]$. Then, using the convolution integral (Definition 3.23), the output of $P(s)$ can be written as

$$y(t) = \int_0^t p(t - \tau) x_{H0}(\tau) \, d\tau$$

Since $x_{H0}(t)$ is the output of a zero-order hold, it is constant over each sampling interval. Thus $y(t)$ can be written as

$$y(t) = \int_0^T p(t - \tau) x(0) \, d\tau + \int_T^{2T} p(t - \tau) x(1) \, d\tau + \cdots$$

$$+ \int_{(j-2)T}^{(j-1)T} p(t - \tau) x[(j-2)T] \, d\tau + \int_{(j-1)T}^t p(t - \tau) x[(j-1)T] \, d\tau$$

where $(j-1)T \le t \le jT$. Now

$$y(jT) = \sum_{i=0}^{j-1} \left( \int_{iT}^{(i+1)T} p(jT - \tau) \, d\tau \right) x(iT)$$

By letting $\theta \equiv jT - \tau$, the integral can be rewritten as

$$\int_{iT}^{(i+1)T} p(jT - \tau) \, d\tau = \int_{(j-i-1)T}^{(j-i)T} p(\theta) \, d\theta$$

where $i = 0, 1, 2, 3, \ldots, j-1$. Now, defining $h(t) \equiv \int_0^t p(\theta) \, d\theta$ and $k = j - 1$ or $j = k + 1$ yields

$$\int_{(j-i-1)T}^{(j-i)T} p(\theta) \, d\theta = \int_0^{(j-i)T} p(\theta) \, d\theta - \int_0^{(j-i-1)T} p(\theta) \, d\theta = \int_0^{(k-i+1)T} p(\theta) \, d\theta - \int_0^{(k-i)T} p(\theta) \, d\theta$$

$$= h[(k-i+1)T] - h[(k-i)T]$$

Therefore we can write

$$y[(k+1)T] = \sum_{i=0}^k h[(k-i+1)T] x(iT) - \sum_{i=0}^k h[(k-i)T] x(iT)$$

Using the relationship between the convolution sum and the product of $z$-transforms in Section 4.9, the Shift Theorem (Property 6, Section 4.9), and the definition of the $z$-transform, the $z$-transform of the last equation is

$$zY^*(z) = zH^*(z) X^*(z) - H^*(z) X^*(z)$$

where $Y^*(z)$ is the $z$-transform of the sequence $y(kT)$, $k = 0, 1, 2, \ldots$, $H^*(z)$ is the $z$-transform of $\int_0^{kT} p(\theta) \, d\theta$, $k = 0, 1, 2, \ldots$, and $X^*(z)$ is the $z$-transform of $x(kT)$, $k = 0, 1, 2, \ldots$. Rearranging terms yields

$$\frac{Y^*(z)}{X^*(z)} = (1 - z^{-1}) H^*(z)$$

Then, since $h(t) = \int_0^t p(\theta)\, d\theta$, $\mathscr{L}[h(t)] = P(s)/s$ and

$$\frac{Y^*(z)}{X^*(z)} = (1 - z^{-1})\, \mathcal{Z}\left\{\mathscr{L}^{-1}\left(\frac{P(s)}{s}\right)\bigg|_{t=kT}\right\}$$

**6.43.** Compare the solution in Problem 6.42 with that in Problem 6.37. What is fundamentally different about Problem 6.42, thereby permitting the use of linear frequency domain methods on this problem?

      The presence of a sampler at the output of $P(s)$ permits the use of $z$-domain transfer functions for the combination of the sampler, zero-order hold, and $P(s)$.

# Supplementary Problems

**6.44.** Determine the transfer function of the $R$-$C$ network shown in Fig. 6-18

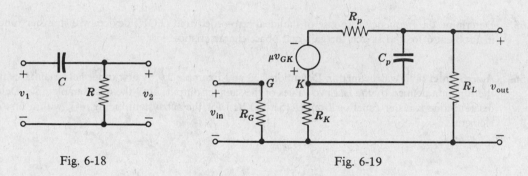

Fig. 6-18                                     Fig. 6-19

**6.45.** An equivalent circuit of an electronic amplifier is shown in Fig. 6-19. What is its transfer function?

**6.46.** Find the transfer function of a system having the impulse response $p(t) = e^{-t}(1 - \sin t)$.

**6.47.** A sinusoidal input $x = 2\sin 2t$ is applied to a system with the transfer function $P(s) = 2/s(s + 2)$. Determine the steady state output $y_{ss}$.

**6.48.** Find the step response of a system having the transfer function $P(s) = 4/(s^2 - 1)(s^2 + 1)$.

**6.49.** Determine which of the following transfer functions represent stable systems and which represent unstable systems:

(a)  $P(s) = \dfrac{(s - 1)}{(s + 2)(s^2 + 4)}$        (c)  $P(s) = \dfrac{(s + 2)(s - 2)}{(s + 1)(s - 1)(s + 4)}$

(b)  $P(s) = \dfrac{(s - 1)}{(s + 2)(s + 4)}$        (d)  $P(s) = \dfrac{6}{(s^2 + s + 1)(s + 1)^2}$

(e)  $P(s) = \dfrac{5(s + 10)}{(s + 5)(s^2 - s + 10)}$

**6.50.** Use the Final Value Theorem (Chapter 4) to show that the steady state value of the output of a stable system in response to a unit step input is equal to the d.c. gain of the system.

**6.51.** Determine the transfer function of two of the networks shown in Problem 6.44 connected in cascade (series).

**6.52.** Examine the literature for the transfer functions of two- and three-degree-of-freedom gyros and compare them with the one-degree-of-freedom gyro of Problem 6.31.

**6.53.** Determine the ramp response of a system having the transfer function $P(s) = (s+1)/(s+2)$.

**6.54.** Show that if a system described by

$$\sum_{i=0}^{n} a_i \frac{d^i y}{dt^i} = \sum_{i=0}^{m} b_i \frac{d^i u}{dt^i}$$

for $m \le n$ is at rest prior to application of the input, that is, $d^k y/dt^k = 0$, $k = 0, 1, \ldots, n-1$, for $t < 0$, then (terms due to *all* initial values $u_0^k$, $y_0^k$) = 0.
(*Hint:* Integrate the differential equation $n$ times from $0^- \equiv \lim_{\epsilon \to 0, \, \epsilon < 0} \epsilon$ to $t$, and then let $t \to 0^+$.)

**6.55.** Determine the frequency response of the ideal zero-order hold (ZOH) device, with transfer function given in Problem 6.36, and sketch the gain and phase characteristics.

**6.56.** A zero-order hold was defined in Definition 2.13 and Example 2.9. A **first-order hold** maintains the **slope** of the function defined by the last two values of the sampler output, until the next sample time. Determine the discrete-time transfer function from $U^*(z)$ to $Y^*(z)$ for the subsystem in Fig. 6-3, with a first-order hold element.

# Answers to Supplementary Problems

**6.44.** $\dfrac{V_2}{V_1} = \dfrac{s}{s + 1/RC}$

**6.45.** $\dfrac{V_{out}}{V_{in}} = \dfrac{-\mu R_L}{(R_k + R_L) R_p C_p s + (\mu + 1) R_k + R_p + R_L}$

**6.46.** $P(s) = \dfrac{s^2 + s + 1}{(s+1)(s^2 + 2s + 2)}$

**6.47.** $y_{ss} = 0.707 \sin(2t - 135°)$

**6.48.** $y = -4 + e^{-t} + e^t + 2\cos t$

**6.49.** $(b)$ and $(d)$ represent stable systems; $(a)$, $(c)$, and $(e)$ represent unstable systems.

**6.51.** $\dfrac{V_2}{V_1} = \dfrac{s^2}{s^2 + (3/RC)s + 1/R^2C^2}$

**6.53.** $y = \frac{1}{4} - \frac{1}{4}e^{-2t} + \frac{1}{2}t$

**6.55.**
$$P(j\omega) = \left[\frac{T\sin(\omega T/2)}{\omega T/2}\right]e^{-j\omega T/2}$$

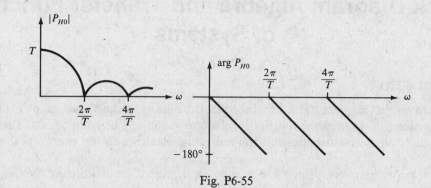

Fig. P6-55

**6.56.**
$$\frac{Y^*(z)}{U^*(z)} = (1 - z^{-1})^2 \mathcal{Z}\left\{\mathcal{L}^{-1}\left(\frac{G(s)}{s} + \frac{1}{T}\frac{G(s)}{s^2}\right)\bigg|_{t=kT}\right\}$$

# Block Diagram Algebra and Transfer Functions of Systems

## 7.1 INTRODUCTION

It is pointed out in Chapters 1 and 2 that the block diagram is a shorthand, graphical representation of a physical system, illustrating the functional relationships among its components. This latter feature permits evaluation of the contributions of the individual elements to the overall performance of the system.

In this chapter we first investigate these relationships in more detail, utilizing the frequency domain and transfer function concepts developed in preceding chapters. Then we develop methods for reducing complicated block diagrams to manageable forms so that they may be used to predict the overall performance of a system.

## 7.2 REVIEW OF FUNDAMENTALS

In general, a block diagram consists of a specific configuration of four types of elements: blocks, summing points, takeoff points, and arrows representing unidirectional signal flow:

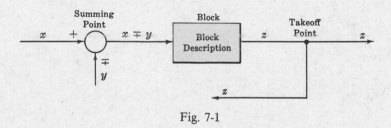

Fig. 7-1

The meaning of each element should be clear from Fig. 7-1.

Time-domain quantities are represented by lowercase letters.

**EXAMPLE 7.1.** $r = r(t)$ for continuous signals, and $r(t_k)$ or $r(k)$, $k = 1, 2, \ldots$, for discrete-time signals.

Capital letters in this chapter are used for Laplace transforms, or $z$-transforms. The argument $s$ or $z$ is often suppressed, to simplify the notation, if the context is clear, or if the results presented are the same for both Laplace (continuous-time system) and $z$-(discrete-time system) transfer function domains.

**EXAMPLE 7.2.** $R = R(s)$ or $R = R(z)$.

The basic feedback control system configuration presented in Chapter 2 is reproduced in Fig. 7-2, with all quantities in abbreviated transform notation.

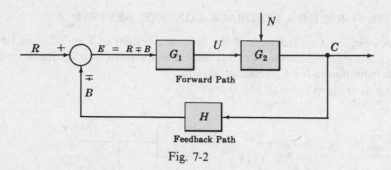

Fig. 7-2

The quantities $G_1$, $G_2$, and $H$ are the transfer functions of the components in the blocks. They may be either Laplace or $z$-transform transfer functions.

**EXAMPLE 7.3.**    $G_1 = U/E$ or $U = G_1 E$.

It is important to note that these results apply *either* to Laplace transform *or* to $z$-transform transfer functions, but not necessarily to *mixed* continuous/discrete block diagrams that include *samplers*. Samplers are linear devices, but they are not time-invariant. Therefore they cannot be characterized by an ordinary $s$-domain transfer function, as defined in Chapter 6. See Problem 7.38 for some exceptions, and Section 6.8 for a more extensive discussion of mixed continuous/discrete systems.

## 7.3  BLOCKS IN CASCADE

Any finite number of blocks in series may be algebraically combined by multiplication of transfer functions. That is, $n$ components or blocks with transfer functions $G_1, G_2, \ldots, G_n$ connected in cascade are equivalent to a single element $G$ with a transfer function given by

$$G = G_1 \cdot G_2 \cdot G_3 \cdots G_n = \prod_{i=1}^{n} G_i \qquad (7.1)$$

The symbol for multiplication "$\cdot$" is omitted when no confusion results.

**EXAMPLE 7.4.**

Fig. 7-3

Multiplication of transfer functions is *commutative*; that is,

$$G_i G_j = G_j G_i \qquad (7.2)$$

for any $i$ or $j$.

**EXAMPLE 7.5.**

Fig. 7-4

Loading effects (interaction of one transfer function upon its neighbor) must be accounted for in the derivation of the individual transfer functions before blocks can be cascaded. (See Problem 7.4.)

## 7.4   CANONICAL FORM OF A FEEDBACK CONTROL SYSTEM

The two blocks in the forward path of the feedback system of Fig. 7-2 may be combined. Letting $G \equiv G_1 G_2$, the resulting configuration is called the **canonical form** of a feedback control system. $G$ and $H$ are not necessarily unique for a particular system.

The following definitions refer to Fig. 7-5.

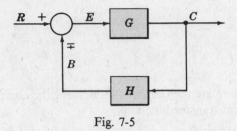

Fig. 7-5

**Definition 7.1:**    $G \equiv$ direct transfer function $\equiv$ forward transfer function

**Definition 7.2:**    $H \equiv$ feedback transfer function

**Definition 7.3:**    $GH \equiv$ loop transfer function $\equiv$ open-loop transfer function

**Definition 7.4:**    $C/R \equiv$ closed-loop transfer function $\equiv$ control ratio

**Definition 7.5:**    $E/R \equiv$ actuating signal ratio $\equiv$ error ratio

**Definition 7.6:**    $B/R \equiv$ primary feedback ratio

In the following equations, the $-$ sign refers to a *positive* feedback system, and the $+$ sign refers to a *negative* feedback system:

$$\frac{C}{R} = \frac{G}{1 \pm GH} \qquad (7.3)$$

$$\frac{E}{R} = \frac{1}{1 \pm GH} \qquad (7.4)$$

$$\frac{B}{R} = \frac{GH}{1 \pm GH} \qquad (7.5)$$

The denominator of $C/R$ determines the *characteristic equation* of the system, which is usually determined from $1 \pm GH = 0$ or, equivalently,

$$D_{GH} \pm N_{GH} = 0 \qquad (7.6)$$

where $D_{GH}$ is the denominator and $N_{GH}$ is the numerator of $GH$, unless a pole of $G$ cancels a zero of $H$ (see Problem 7.9). Relations (7.1) through (7.6) are valid for both continuous (*s*-domain) and discrete (*z*-domain) systems.

## 7.5   BLOCK DIAGRAM TRANSFORMATION THEOREMS

Block diagrams of complicated control systems may be simplified using easily derivable transformations. The first important transformation, combining blocks in cascade, has already been presented in Section 7.3. It is repeated for completeness in the chart illustrating the transformation theorems (Fig. 7-6). The letter $P$ is used to represent any transfer function, and $W, X, Y, Z$ denote any transformed signals.

| | Transformation | Equation | Block Diagram | Equivalent Block Diagram |
|---|---|---|---|---|
| 1 | Combining Blocks in Cascade | $Y = (P_1 P_2)X$ | 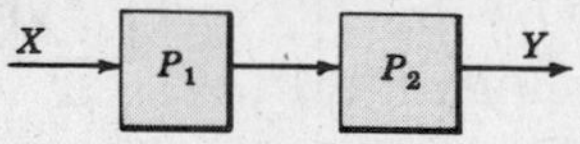 | 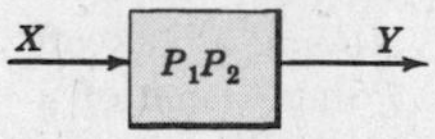 |
| 2 | Combining Blocks in Parallel; or Eliminating a Forward Loop | $Y = P_1 X \pm P_2 X$ | 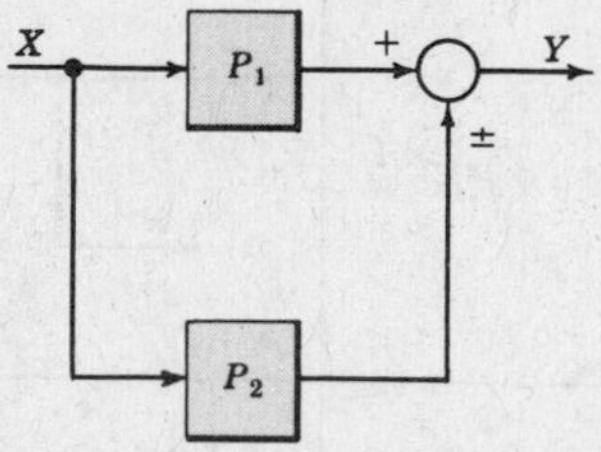 | 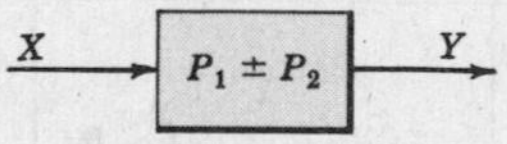 |
| 3 | Removing a Block from a Forward Path | $Y = P_1 X \pm P_2 X$ | | 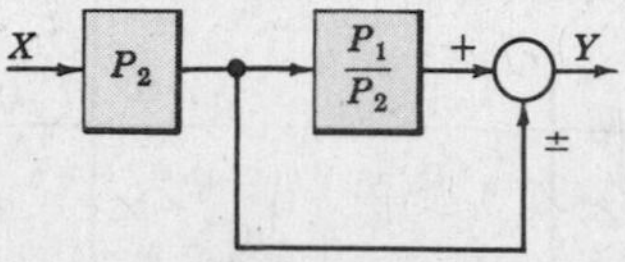 |
| 4 | Eliminating a Feedback Loop | $Y = P_1(X \mp P_2 Y)$ | 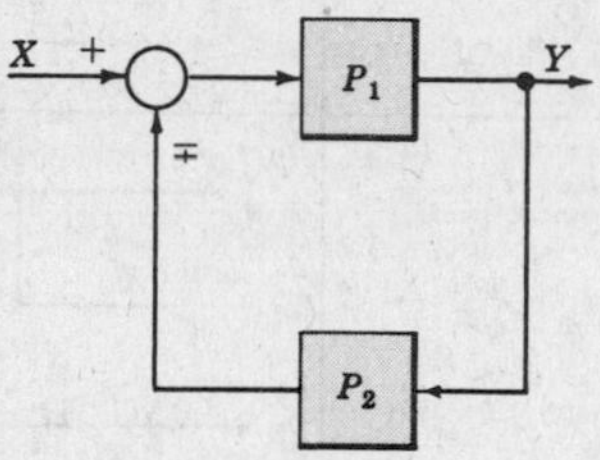 | 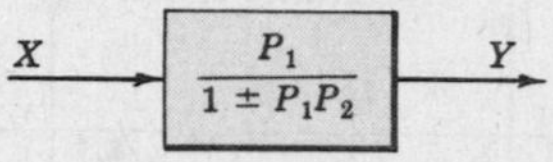 |
| 5 | Removing a Block from a Feedback Loop | $Y = P_1(X \mp P_2 Y)$ | | 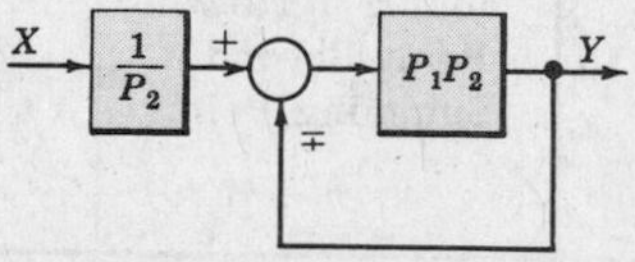 |
| 6a | Rearranging Summing Points | $Z = W \pm X \pm Y$ | 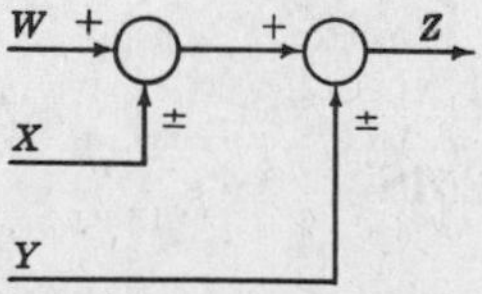 | 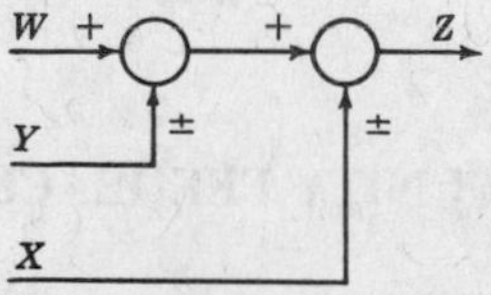 |
| 6b | Rearranging Summing Points | $Z = W \pm X \pm Y$ | 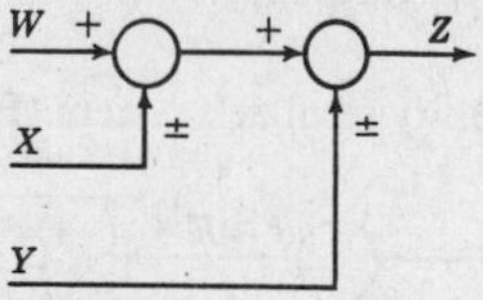 | 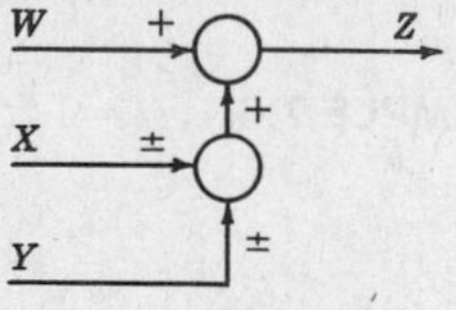 |
| 7 | Moving a Summing Point Ahead of a Block | $Z = PX \pm Y$ | 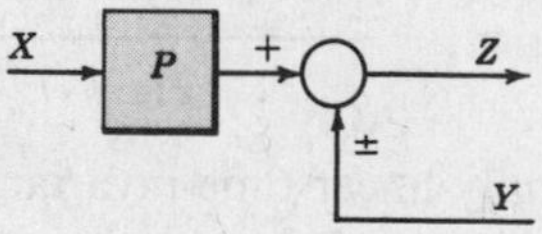 | 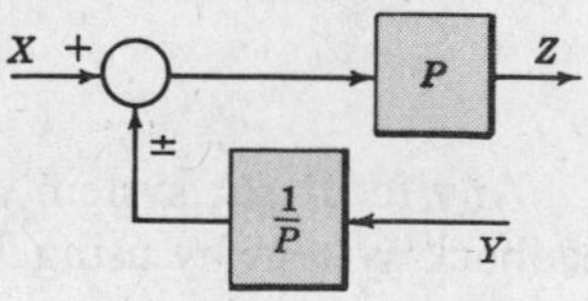 |
| 8 | Moving a Summing Point Beyond a Block | $Z = P[X \pm Y]$ | 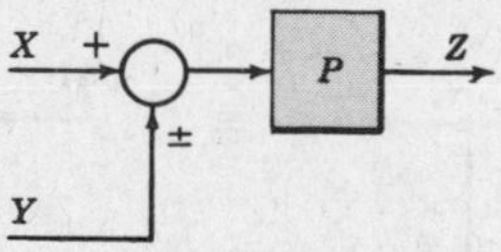 | 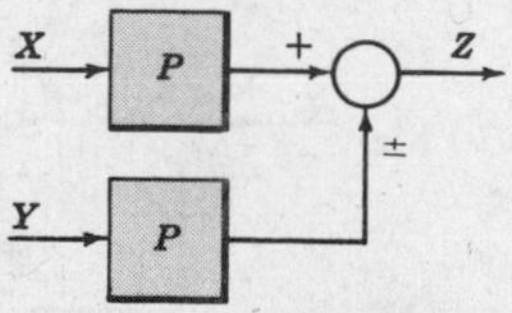 |

Fig. 7-6

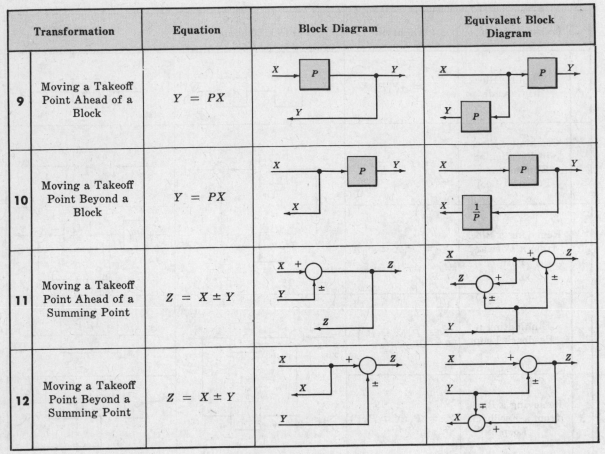

| | Transformation | Equation | Block Diagram | Equivalent Block Diagram |
|---|---|---|---|---|
| 9 | Moving a Takeoff Point Ahead of a Block | $Y = PX$ | | |
| 10 | Moving a Takeoff Point Beyond a Block | $Y = PX$ | | |
| 11 | Moving a Takeoff Point Ahead of a Summing Point | $Z = X \pm Y$ | | |
| 12 | Moving a Takeoff Point Beyond a Summing Point | $Z = X \pm Y$ | | |

Fig. 7-6   *Continued*

## 7.6  UNITY FEEDBACK SYSTEMS

***Definition 7.7***:     A **unity feedback system** is one in which the primary feedback $b$ is identically equal to the controlled output $c$.

**EXAMPLE 7.6.**   $H = 1$ for a linear, unity feedback system (Fig. 7-7).

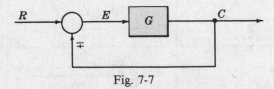

Fig. 7-7

Any feedback system with only linear time-invariant elements can be put into the form of a unity feedback system by using Transformation 5.

**EXAMPLE 7.7.**

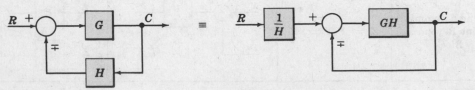

Fig. 7-8

The characteristic equation for the unity feedback system, determined from $1 \pm G = 0$, is

$$D_G \pm N_G = 0 \qquad\qquad (7.7)$$

where $D_G$ is the denominator and $N_G$ the numerator of $G$.

## 7.7 SUPERPOSITION OF MULTIPLE INPUTS

Sometimes it is necessary to evaluate system performance when several inputs are simultaneously applied at different points of the system.

When multiple inputs are present in a *linear* system, each is treated independently of the others. The output due to all stimuli acting together is found in the following manner. We assume zero initial conditions, as we seek the system response only to inputs.

**Step 1:** Set all inputs except one equal to zero.

**Step 2:** Transform the block diagram to canonical form, using the transformations of Section 7.5.

**Step 3:** Calculate the response due to the chosen input acting alone.

**Step 4:** Repeat Steps 1 to 3 for each of the remaining inputs.

**Step 5:** Algebraically add all of the responses (outputs) determined in Steps 1 to 4. This sum is the total output of the system with all inputs acting simultaneously.

We reemphasize here that the above superposition process is dependent on the system being linear.

**EXAMPLE 7.8.** We determine the output $C$ due to inputs $U$ and $R$ for Fig. 7-9.

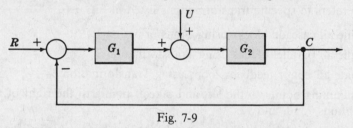

Fig. 7-9

**Step 1:** Put $U \equiv 0$.
**Step 2:** The system reduces to

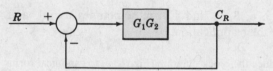

**Step 3:** By Equation (7.3), the output $C_R$ due to input $R$ is $C_R = [G_1G_2/(1 + G_1G_2)]R$.
**Step 4a:** Put $R \equiv 0$.
**Step 4b:** Put $-1$ into a block, representing the negative feedback effect:

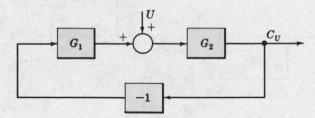

Rearrange the block diagram:

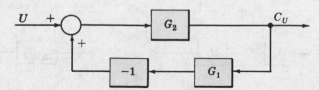

Let the $-1$ block be absorbed into the summing point:

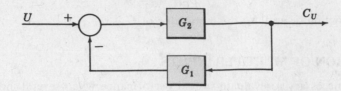

**Step 4c:**   By Equation (*7.3*), the output $C_U$ due to input $U$ is $C_U = [G_2/(1 + G_1G_2)]U$.
**Step 5:**    The total output is

$$C = C_R + C_U = \left[\frac{G_1G_2}{1 + G_1G_2}\right]R + \left[\frac{G_2}{1 + G_1G_2}\right]U = \left[\frac{G_2}{1 + G_1G_2}\right][G_1R + U]$$

## 7.8   REDUCTION OF COMPLICATED BLOCK DIAGRAMS

The block diagram of a practical feedback control system is often quite complicated. It may include several feedback or feedforward loops, and multiple inputs. By means of systematic block diagram reduction, every multiple loop linear feedback system may be reduced to canonical form. The techniques developed in the preceding paragraphs provide the necessary tools.

The following general steps may be used as a basic approach in the reduction of complicated block diagrams. Each step refers to specific transformations listed in Fig. 7-6.

**Step 1:**   Combine all cascade blocks using Transformation 1.

**Step 2:**   Combine all parallel blocks using Transformation 2.

**Step 3:**   Eliminate all minor feedback loops using Transformation 4.

**Step 4:**   Shift summing points to the left and takeoff points to the right of the major loop, using Transformations 7, 10, and 12.

**Step 5:**   Repeat Steps 1 to 4 until the canonical form has been achieved for a particular input.

**Step 6:**   Repeat Steps 1 to 5 for each input, as required.

Transformations 3, 5, 6, 8, 9, and 11 are sometimes useful, and experience with the reduction technique will determine their application.

**EXAMPLE 7.9.**   Let us reduce the block diagram (Fig. 7-10) to canonical form.

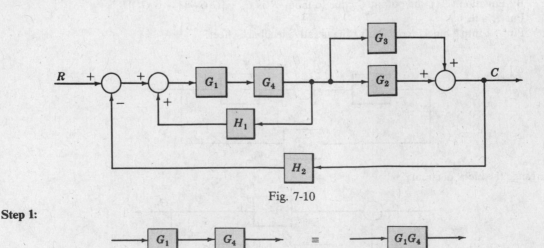

Fig. 7-10

**Step 1:**

**Step 2:**

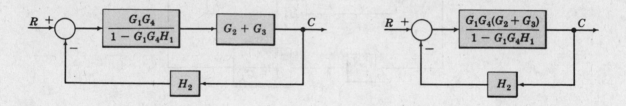

**Step 3:**

$$\text{[diagram]} \quad \equiv \quad \boxed{\dfrac{G_1 G_4}{1 - G_1 G_4 H_1}}$$

**Step 4:**   Does not apply.
**Step 5:**

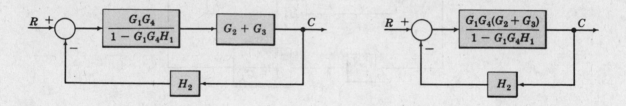

**Step 6:**   Does not apply.


An occasional requirement of block diagram reduction is the isolation of a particular block in a feedback or feedforward loop. This may be desirable to more easily examine the effect of a particular block on the overall system.

Isolation of a block generally may be accomplished by applying the same reduction steps to the system, but usually in a different order. Also, the block to be isolated cannot be combined with any others.

Rearranging Summing Points (Transformation 6) and Transformations 8, 9, and 11 are especially useful for isolating blocks.


**EXAMPLE 7.10.**   Let us reduce the block diagram of Example 7.9, isolating block $H_1$.
**Steps 1 and 2:**

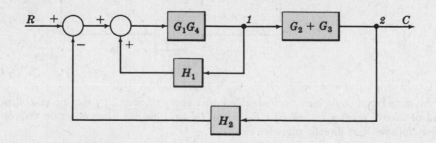

We do not apply Step 3 at this time, but go directly to Step 4, moving takeoff point *1* beyond block $G_2 + G_3$:

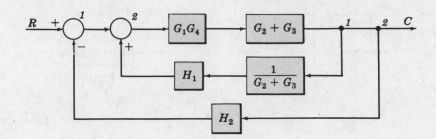

We may now rearrange summing points *1* and *2* and combine the cascade blocks in the forward loop using Transformation 6, then Transformation 1:

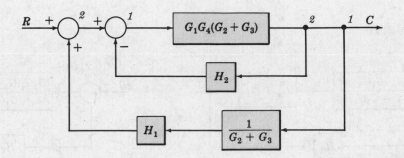

**Step 3:**

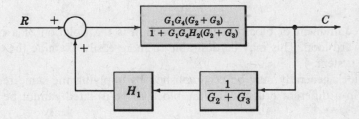

Finally, we apply Transformation 5 to remove $1/(G_2 + G_3)$ from the feedback loop:

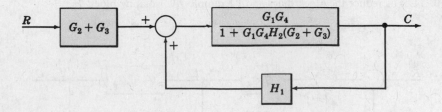

Note that the same result could have been obtained after applying Step 2 by moving takeoff point *2 ahead* of $G_2 + G_3$, instead of takeoff point *1 beyond* $G_2 + G_3$. Block $G_2 + G_3$ has the same effect on the control ratio $C/R$ whether it directly follows $R$ or directly precedes $C$.

# Solved Problems

## BLOCKS IN CASCADE

**7.1.** Prove Equation (*7.1*) for blocks in cascade.

The block diagram for $n$ transfer functions $G_1, G_2, \ldots, G_n$ in cascade is given in Fig. 7-11.

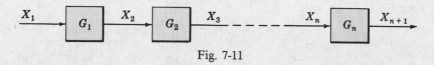

Fig. 7-11

The output transform for any block is equal to the input transform multiplied by the transfer function (see Section 6.1). Therefore $X_2 = X_1 G_1$, $X_3 = X_2 G_2, \ldots, X_n = X_{n-1} G_{n-1}$, $X_{n+1} = X_n G_n$. Combining these equations, we have

$$X_{n+1} = X_n G_n = X_{n-1} G_{n-1} G_n = \cdots = X_1 G_1 G_2 \cdots G_{n-1} G_n$$

Dividing both sides by $X_1$, we obtain $X_{n+1}/X_1 = G_1 G_2 \cdots G_{n-1} G_n$.

**7.2.** Prove the commutativity of blocks in cascade, Equation (*7.2*).

Consider two blocks in cascade (Fig. 7-12):

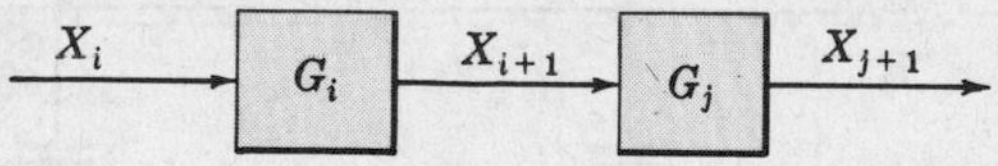

Fig. 7-12

From Equation (*6.1*) we have $X_{i+1} = X_i G_i = G_i X_i$ and $X_{j+1} = X_{i+1} G_j = G_j X_{i+1}$. Therefore $X_{j+1} = (X_i G_i) G_j = X_i G_i G_j$. Dividing both sides by $X_i$, $X_{j+1}/X_i = G_i G_j$.

Also, $X_{j+1} = G_j (G_i X_i) = G_j G_i X_i$. Dividing again by $X_i$, $X_{j+1}/X_i = G_j G_i$. Thus $G_i G_j = G_j G_i$.

This result is extended by mathematical induction to any finite number of transfer functions (blocks) in cascade.

**7.3.** Find $X_n/X_1$ for each of the systems in Fig. 7-13.

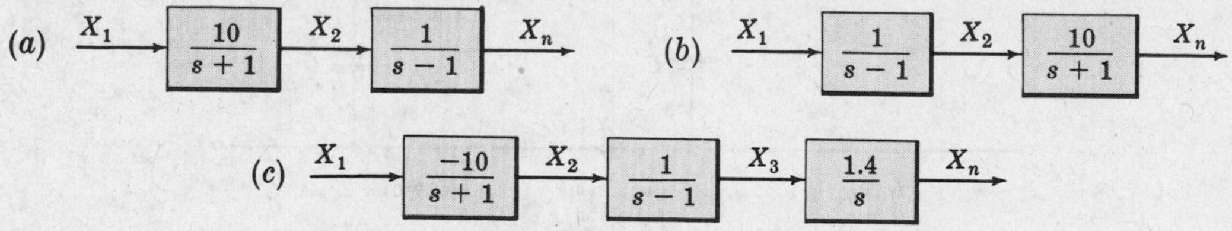

(*a*)

(*b*)

(*c*)

Fig. 7-13

(*a*)  One way to work this problem is to first write $X_2$ in terms of $X_1$:

$$X_2 = \left( \frac{10}{s+1} \right) X_1$$

Then write $X_n$ in terms of $X_2$:

$$X_n = \left( \frac{1}{s-1} \right) X_2 = \left( \frac{1}{s-1} \right) \left( \frac{10}{s+1} \right) X_1$$

Multiplying out and dividing both sides by $X_1$, we have $X_n/X_1 = 10/(s^2 - 1)$.

A shorter method is as follows. We know from Equation (7.1) that two blocks can be reduced to one by simply multiplying their transfer functions. Also, the transfer function of a single block is its output-to-input transform. Hence

$$\frac{X_n}{X_1} = \left(\frac{1}{s-1}\right)\left(\frac{10}{s+1}\right) = \frac{10}{s^2-1}$$

(b) This system has the same transfer function determined in part (a) because multiplication of transfer functions is commutative.

(c) By Equation (7.1), we have

$$\frac{X_n}{X_1} = \left(\frac{-10}{s+1}\right)\left(\frac{1}{s-1}\right)\left(\frac{1.4}{s}\right) = \frac{-14}{s(s^2-1)}$$

7.4. The transfer function of Fig. 7-14a is $\omega_0/(s+\omega_0)$, where $\omega_0 = 1/RC$. Is the transfer function of Fig. 7-14b equal to $\omega_0^2/(s+\omega_0)^2$? Why?

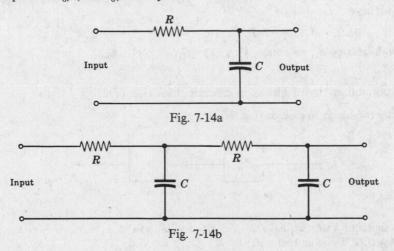

Fig. 7-14a

Fig. 7-14b

No. If two networks are connected in series (Fig. 7-15) the second loads the first by drawing current from it. Therefore Equation (7.1) cannot be directly applied to the combined system. The correct transfer function for the connected networks is $\omega_0^2/(s^2 + 3\omega_0 s + \omega_0^2)$ (see Problem 6.16), and this is *not* equal to $(\omega_0/(s+\omega_0))^2$.

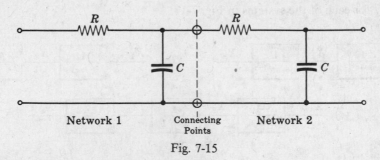

Fig. 7-15

## CANONICAL FEEDBACK CONTROL SYSTEMS

7.5. Prove Equation (7.3), $C/R = G/(1 \pm GH)$.

The equations describing the canonical feedback system are taken directly from Fig. 7-16. They are given by $E = R \mp B$, $B = HC$, and $C = GE$. Substituting one into the other, we have

$$C = G(R \mp B) = G(R \mp HC)$$
$$= GR \mp GHC = GR + (\mp GHC)$$

Subtracting $(\mp GHC)$ from both sides, we obtain $C \pm GHC = GR$ or $C/R = G/(1 \pm GH)$.

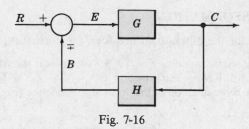

Fig. 7-16

**7.6.** Prove Equation (7.4), $E/R = 1/(1 \pm GH)$.

From the preceding problem, we have $E = R \mp B$, $B = HC$, and $C = GE$.
Then $E = R \mp HC = R \mp HGE$, $E \pm GHE = R$, and $E/R = 1/(1 \pm GH)$.

**7.7.** Prove Equation (7.5), $B/R = GH/(1 \pm GH)$.

From $E = R \mp B$, $B = HC$, and $C = GE$, we obtain $B = HGE = HG(R \mp B) = GHR \mp GHB$.
Then $B \pm GHB = GHR$, $B = GHR/(1 \pm GH)$, and $B/R = GH/(1 \pm GH)$.

**7.8.** Prove Equation (7.6), $D_{GH} \pm N_{GH} = 0$.

The characteristic equation is usually obtained by setting $1 \pm GH = 0$. (See Problem 7.9 for an exception.) Putting $GH \equiv N_{GH}/D_{GH}$, we obtain $D_{GH} \pm N_{GH} = 0$.

**7.9.** Determine (a) the loop transfer function, (b) the control ratio, (c) the error ratio, (d) the primary feedback ratio, (e) the characteristic equation, for the feedback control system in which $K_1$ and $K_2$ are constants (Fig. 7-17).

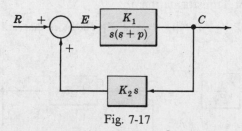

Fig. 7-17

(a) The loop transfer function is equal to $GH$.

Hence
$$GH = \left[ \frac{K_1}{s(s+p)} \right] K_2 s = \frac{K_1 K_2}{s+p}$$

(b) The control ratio, or closed-loop transfer function, is given by Equation (7.3) (with a minus sign for positive feedback):
$$\frac{C}{R} = \frac{G}{1 - GH} = \frac{K_1}{s(s+p-K_1K_2)}$$

(c) The error ratio, or actuating signal ratio, is given by Equation (7.4):
$$\frac{E}{R} = \frac{1}{1 - GH} = \frac{1}{1 - K_1K_2/(s+p)} = \frac{s+p}{s+p-K_1K_2}$$

(d) The primary feedback ratio is given by Equation (7.5):
$$\frac{B}{R} = \frac{GH}{1 - GH} = \frac{K_1K_2}{s+p-K_1K_2}$$

(e) The characteristic equation is given by the denominator of $C/R$ above, $s(s+p-K_1K_2) = 0$. In this case, $1 - GH = s+p-K_1K_2 = 0$, which is *not* the characteristic equation, because the pole $s$ of $G$ cancels the zero $s$ of $H$.

## BLOCK DIAGRAM TRANSFORMATIONS

**7.10.** Prove the equivalence of the block diagrams for Transformation 2 (Section 7.5).

The equation in the second column, $Y = P_1 X \pm P_2 X$, governs the construction of the block diagram in the third column, as shown. Rewrite this equation as $Y = (P_1 \pm P_2) X$. The equivalent block diagram in the last column is clearly the representation of this form of the equation (Fig. 7-18)

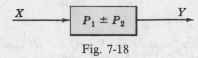

Fig. 7-18

**7.11.** Repeat Problem 7.10 for Transformation 3.

Rewrite $Y = P_1 X \pm P_2 X$ as $Y = (P_1/P_2) P_2 X \pm P_2 X$. The block diagram for this form of the equation is clearly given in Fig. 7-19.

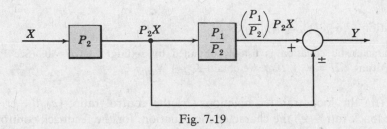

Fig. 7-19

**7.12.** Repeat Problem 7.10 for Transformation 5.

We have $Y = P_1[X \mp P_2 Y] = P_1 P_2[(1/P_2) X \mp Y]$. The block diagram for the latter form is given in Fig. 7-20.

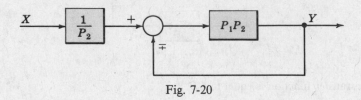

Fig. 7-20

**7.13.** Repeat Problem 7.10 for Transformation 7.

We have $Z = PX \pm Y = P[X \pm (1/P)Y]$, which yields the block diagram given in Fig. 7-21.

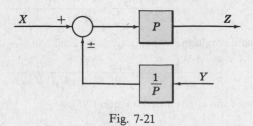

Fig. 7-21

**7.14.** Repeat Problem 7.10 for Transformation 8.

We have $Z = P(X \pm Y) = PX \pm PY$, whose block diagram is clearly given in Fig. 7-22.

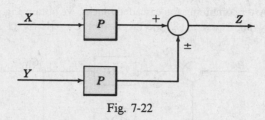

Fig. 7-22

## UNITY FEEDBACK SYSTEMS

**7.15.** Reduce the block diagram given in Fig. 7-23 to unity feedback form and find the system characteristic equation.

Mathcad

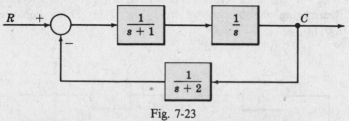

Fig. 7-23

Combining the blocks in the forward path, we obtain Fig. 7-24.

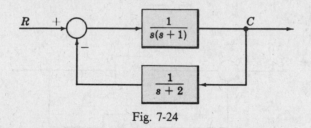

Fig. 7-24

Applying Transformation 5, we have Fig. 7-25.

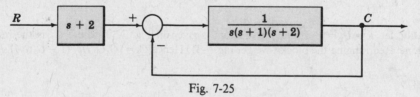

Fig. 7-25

By Equation (7.7), the characteristic equation for this system is $s(s + 1)(s + 2) + 1 = 0$ or $s^3 + 3s^2 + 2s + 1 = 0$.

## MULTIPLE INPUTS AND OUTPUTS

**7.16.** Determine the output $C$ due to $U_1$, $U_2$, and $R$ for Fig. 7-26.

Mathcad

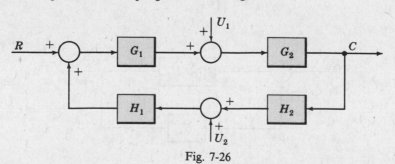

Fig. 7-26

Let $U_1 = U_2 = 0$. After combining the cascaded blocks, we obtain Fig. 7-27, where $C_R$ is the output due to $R$ acting alone. Applying Equation $(7.3)$ to this system, $C_R = [G_1G_2/(1 - G_1G_2H_1H_2)]R$.

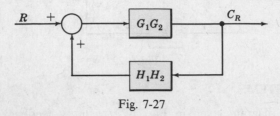

Fig. 7-27

Now let $R = U_2 = 0$. The block diagram is now given in Fig. 7-28, where $C_1$ is the response due to $U_1$ acting alone. Rearranging the blocks, we have Fig. 7-29. From Equation $(7.3)$, we get $C_1 = [G_2/(1 - G_1G_2H_1H_2)]U_1$.

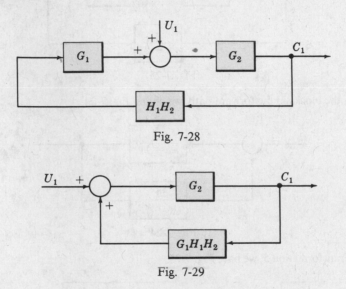

Fig. 7-28

Fig. 7-29

Finally, let $R = U_1 = 0$. The block diagram is given in Fig. 7-30, where $C_2$ is the response due to $U_2$ acting alone. Rearranging the blocks, we get Fig. 7-31. Hence $C_2 = [G_1G_2H_1/(1 - G_1G_2H_1H_2)]U_2$.

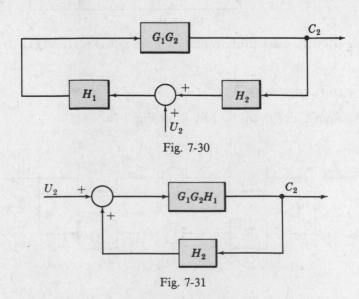

Fig. 7-30

Fig. 7-31

By superposition, the total output is

$$C = C_R + C_1 + C_2 = \frac{G_1 G_2 R + G_2 U_1 + G_1 G_2 H_1 U_2}{1 - G_1 G_2 H_1 H_2}$$

**7.17.** Figure 7-32 is an example of a multiinput-multioutput system. Determine $C_1$ and $C_2$ due to $R_1$ and $R_2$.

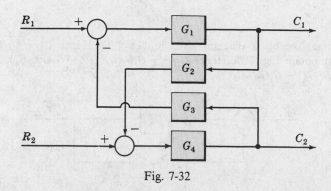

Fig. 7-32

First put the block diagram in the form of Fig. 7-33, ignoring the output $C_2$.

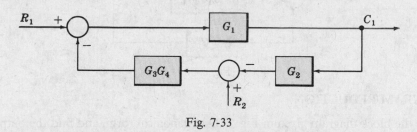

Fig. 7-33

Letting $R_2 = 0$ and combining the summing points, we get Fig. 7-34.

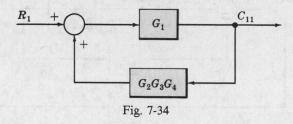

Fig. 7-34

Hence $C_{11}$, the output at $C_1$ due to $R_1$ alone, is $C_{11} = G_1 R_1 / (1 - G_1 G_2 G_3 G_4)$. For $R_1 = 0$, we have Fig. 7-35.

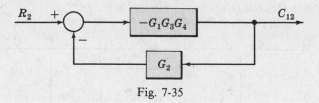

Fig. 7-35

Hence $C_{12} = -G_1 G_3 G_4 R_2 / (1 - G_1 G_2 G_3 G_4)$ is the output at $C_1$ due to $R_2$ alone. Thus $C_1 = C_{11} + C_{12} = (G_1 R_1 - G_1 G_3 G_4 R_2)/(1 - G_1 G_2 G_3 G_4)$.

Now we reduce the original block diagram, ignoring output $C_1$. First we obtain Fig. 7-36.

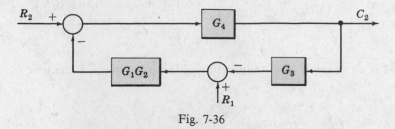

Fig. 7-36

Then we obtain the block diagram given in Fig. 7-37. Hence $C_{22} = G_4 R_2/(1 - G_1 G_2 G_3 G_4)$. Next, letting $R_2 = 0$, we obtain Fig. 7-38. Hence $C_{21} = -G_1 G_2 G_4 R_1/(1 - G_1 G_2 G_3 G_4)$. Finally, $C_2 = C_{22} + C_{21} = (G_4 R_2 - G_1 G_2 G_4 R_1)/(1 - G_1 G_2 G_3 G_4)$.

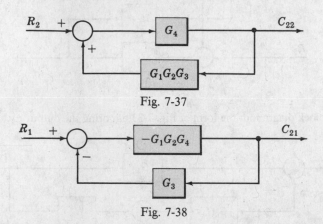

Fig. 7-37

Fig. 7-38

## BLOCK DIAGRAM REDUCTION

**7.18.** Reduce the block diagram given in Fig. 7-39 to canonical form, and find the output transform $C$. $K$ is a constant.

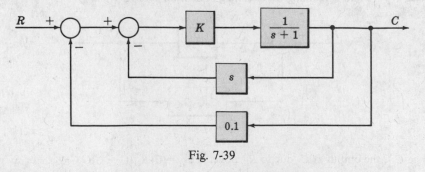

Fig. 7-39

First we combine the cascade blocks of the forward path and apply Transformation 4 to the innermost feedback loop to obtain Fig. 7-40.

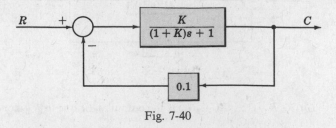

Fig. 7-40

Equation (7.3) or the reapplication of Transformation 4 yields $C = KR/[(1 + K)s + (1 + 0.1K)]$.

**7.19.** Reduce the block diagram of Fig. 7-39 to canonical form, isolating block $K$ in the forward loop.

By Transformation 9 we can move the takeoff point ahead of the $1/(s+1)$ block (Fig. 7-41):

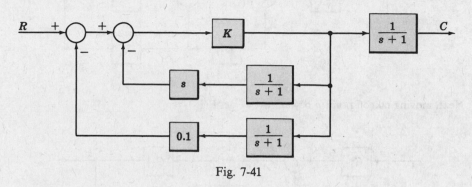

Fig. 7-41

Applying Transformations 1 and 6*b*, we get Fig. 7-42.

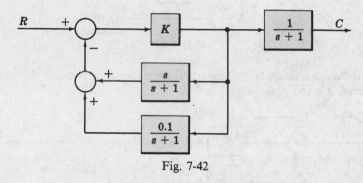

Fig. 7-42

Now we can apply Transformation 2 to the feedback loops, resulting in the final form given in Fig. 7-43.

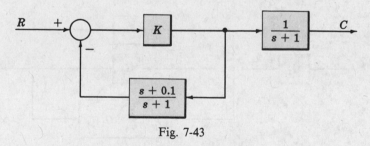

Fig. 7-43

**7.20.** Reduce the block diagram given in Fig. 7-44 to open-loop form.

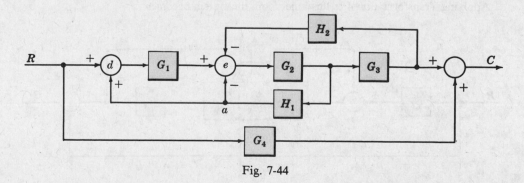

Fig. 7-44

First, moving the leftmost summing point beyond $G_1$ (Transformation 8), we obtain Fig. 7-45.

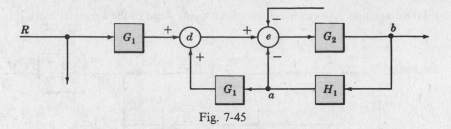

Fig. 7-45

Next, moving takeoff point $a$ beyond $G_1$, we get Fig. 7-46.

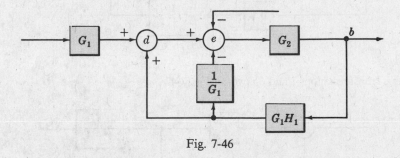

Fig. 7-46

Now, using Transformation 6$b$, and then Transformation 2, to combine the two lower feedback loops (from $G_1 H_1$) entering $d$ and $e$, we obtain Fig. 7-47.

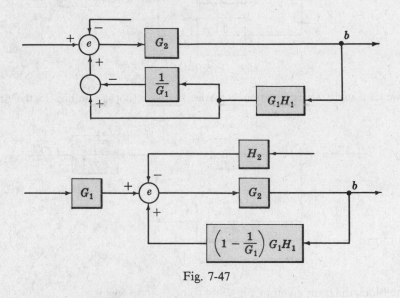

Fig. 7-47

Applying Transformation 4 to this inner loop, the system becomes

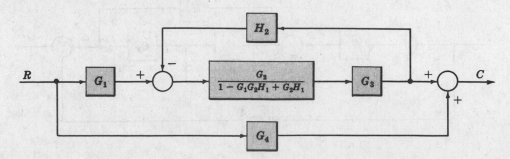

Again, applying Transformation 4 to the remaining feedback loop yields

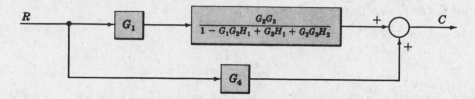

Finally, Transformation 1 and 2 give the open-loop block diagram:

$$R \quad \boxed{\frac{G_1 G_2 G_3 + G_4 - G_1 G_2 G_4 H_1 + G_2 G_4 H_1 + G_2 G_3 G_4 H_2}{1 - G_1 G_2 H_1 + G_2 H_1 + G_2 G_3 H_2}} \quad C$$

## MISCELLANEOUS PROBLEMS

**7.21.** Show that simple block diagram Transformation 1 of Section 7.5 (combining blocks in cascade) is not valid if the first block is (or includes) a *sampler*.

The output transform $U^*(s)$ of an ideal sampler was determined in Problem 4.39 as

$$U^*(s) = \sum_{k=0}^{\infty} e^{-skT} u(kT)$$

Taking $U^*(s)$ as the input of block $P_2$ of Transformation 1 of the table, the output transform $Y(s)$ of block $P_2$ is

$$Y(s) = P_2(s) U^*(s) = P_2(s) \sum_{k=0}^{\infty} e^{-skT} u(kT)$$

Clearly, the input transform $X(s) = U(s)$ cannot be factored from the right-hand side of $Y(s)$, that is, $Y(s) \neq F(s)U(s)$. The same problem occurs if $P_1$ includes other elements, as well as a sampler.

**7.22.** Why is the characteristic equation invariant under block diagram transformation?

Block diagram transformations are determined by *rearranging* the input-output equations of one or more of the subsystems that make up the total system. Therefore the final transformed system is governed by the same equations, probably arranged in a different manner than those for the original system.

Now, the characteristic equation is determined from the denominator of the overall system transfer function set equal to zero. Factoring or other rearrangement of the numerator and denominator of the system transfer function clearly does not change it, nor does it alter the denominator set equal to zero.

**7.23.** Prove that the transfer function represented by $C/R$ in Equation (7.3) can be approximated by $\pm 1/H$ when $|G|$ or $|GH|$ are very large.

Dividing the numerator and denominator of $G/(1 \pm GH)$ by $G$, we get $1 \big/ \left( \dfrac{1}{G} \pm H \right)$. Then

$$\lim_{|G| \to \infty} \left[ \frac{C}{R} \right] = \lim_{|G| \to \infty} \left[ \frac{1}{\dfrac{1}{G} \pm H} \right] = \pm \frac{1}{H}$$

Dividing by $GH$ and taking the limit, we obtain

$$\lim_{|GH| \to \infty} \left[ \frac{C}{R} \right] = \lim_{|GH| \to \infty} \left[ \frac{\dfrac{1}{H}}{\dfrac{1}{GH} \pm 1} \right] = \pm \frac{1}{H}$$

**7.24.** Assume that the characteristics of $G$ change radically or unpredictably during system operation. Using the results of the previous problem, show how the system should be designed so that the output $C$ can always be predicted reasonably well.

In problem 7.23 we found that

$$\lim_{|GH| \to \infty} \left[ \frac{C}{R} \right] = \pm \frac{1}{H}$$

Thus $C \to \pm R/H$ as $|GH| \to \infty$, or $C$ is independent of $G$ for large $|GH|$. Hence the system should be designed so that $|GH| \gg 1$.

**7.25.** Determine the transfer function of the system in Fig. 7-48. Then let $H_1 = 1/G_1$ and $H_2 = 1/G_2$.

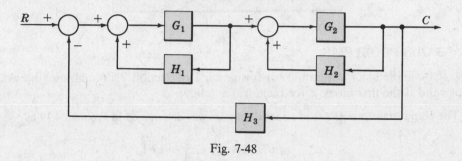

Fig. 7-48

Reducing the inner loops, we have Fig. 7-49.

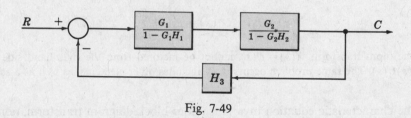

Fig. 7-49

Applying Transformation 4 again, we obtain Fig. 7-50.

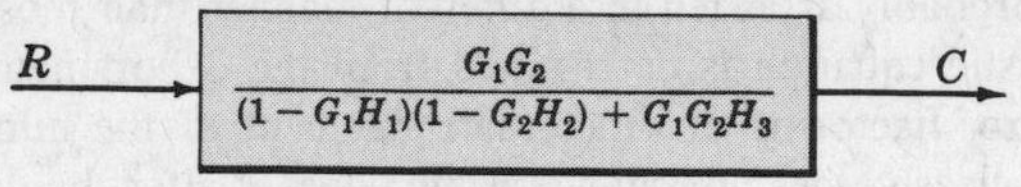

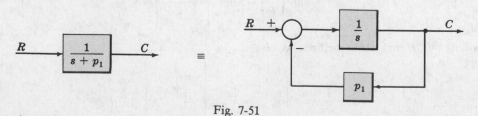

Fig. 7-50

Now put $H_1 = 1/G_1$ and $H_2 = 1/G_2$. This yields

$$\frac{C}{R} = \frac{G_1 G_2}{(1-1)(1-1) + G_1 G_2 H_3} = \frac{1}{H_3}$$

**7.26.** Show that Fig. 7-51 is valid.

Fig. 7-51

From the open-loop diagram, we have $C = R/(s + p_1)$. Rearranging, $(s + p_1)C = R$ and $C = (1/s)(R - p_1C)$. The closed-loop diagram follows from this equation.

**7.27.** Prove Fig. 7-52.

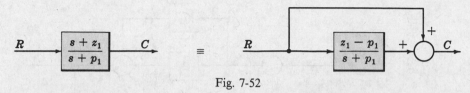

Fig. 7-52

This problem illustrates how a finite zero may be removed from a block.

From the forward-loop diagram, $C = R + (z_1 - p_1)R/(s + p_1)$. Rearranging,

$$C = \left(1 + \frac{z_1 - p_1}{s + p_1}\right)R = \left(\frac{s + p_1 + z_1 - p_1}{s + p_1}\right)R = \left(\frac{s + z_1}{s + p_1}\right)R$$

This mathematical equivalence clearly proves the equivalence of the block diagrams.

**7.28.** Assume that linear approximations in the form of transfer functions are available for each block of the Supply and Demand System of Problem 2.13, and that the system can be represented by Fig. 7-53.

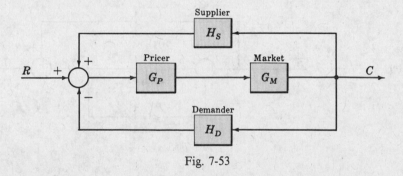

Fig. 7-53

Determine the overall transfer function of the system.

Block diagram Transformation 4, applied twice to this system, gives Fig. 7-54.

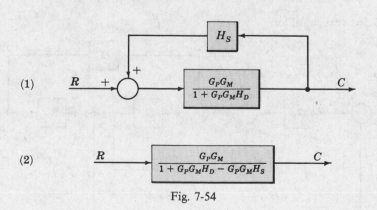

Fig. 7-54

Hence the transfer function for the linearized Supply and Demand model is: $\dfrac{G_P G_M}{1 + G_P G_M(H_D - H_S)}$.

# Supplementary Problems

**7.29.** Determine $C/R$ for each system in Fig. 7-55.

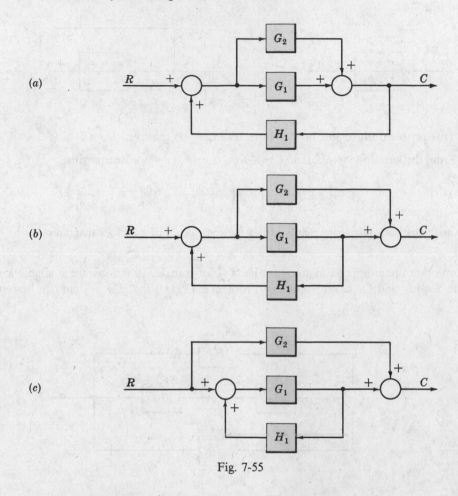

Fig. 7-55

**7.30.** Consider the blood pressure regulator described in Problem 2.14. Assume the vasomotor center (VMC) can be described by a linear transfer function $G_{11}(s)$, and the baroreceptors by the transfer function $k_1 s + k_2$ (see Problem 6.33). Transform the block diagram into its simplest, unity feedback form.

**7.31.** Reduce Fig. 7-56 to canonical form.

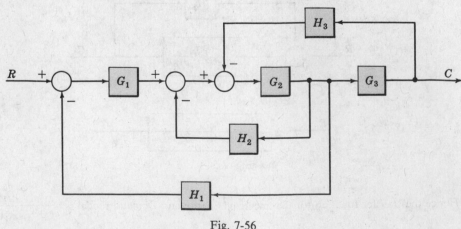

Fig. 7-56

**7.32.**    Determine $C$ for the system represented by Fig. 7-57.

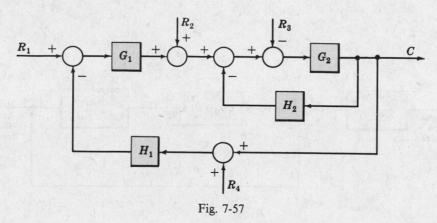

Fig. 7-57

**7.33.**    Give an example of two feedback systems in canonical form having identical control ratios $C/R$ but different $G$ and $H$ components.

**7.34.**    Determine $C/R_2$ for the system given in Fig. 7-58.

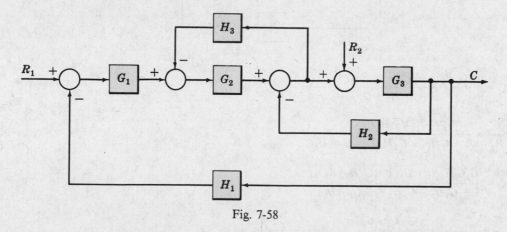

Fig. 7-58

**7.35.**    Determine the complete output $C$, with both inputs $R_1$ and $R_2$ acting simultaneously, for the system given in the preceding problem.

**7.36.**    Determine $C/R$ for the system represented by Fig. 7-59.

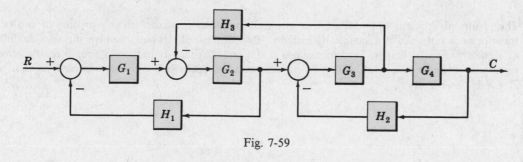

Fig. 7-59

**7.37.**    Determine the characteristic equation for each of the systems of Problems (a) 7.32, (b) 7.35, (c) 7.36.

**7.38.**    What block diagram transformation rules in the table of Section 7.5 permit the inclusion of a sampler?

# Answers to Supplementary Problems

**7.29.**   See Problem 8.15.

**7.30.**

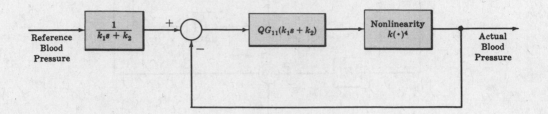

**7.31.**

**7.32.**   $C = \dfrac{G_1 G_2 R_1 + G_2 R_2 - G_2 R_3 - G_1 G_2 H_1 R_4}{1 + G_2 H_2 + G_1 G_2 H_1}$

**7.34.**   $\dfrac{C}{R_2} = \dfrac{G_3 (1 + G_2 H_3)}{1 + G_3 H_2 + G_2 H_3 + G_1 G_2 G_3 H_1}$

**7.35.**   $C = \dfrac{G_1 G_2 G_3 R_1 + G_3 (1 + G_2 H_3) R_2}{1 + G_3 H_2 + G_2 H_3 + G_1 G_2 G_3 H_1}$

**7.36.**   $\dfrac{C}{R} = \dfrac{G_1 G_2 G_3 G_4}{(1 + G_1 G_2 H_1)(1 + G_3 G_4 H_2) + G_2 G_3 H_3}$

**7.37.**   (*a*)   $1 + G_2 H_2 + G_1 G_2 H_1 = 0$

      (*b*)   $1 + G_3 H_2 + G_2 H_3 + G_1 G_2 G_3 H_1 = 0$

      (*c*)   $(1 + G_1 G_2 H_1)(1 + G_3 G_4 H_2) + G_2 G_3 H_3 = 0.$

**7.38.**   The results of Problem 7.21 indicate that any transformation that involves any *product* of two or more transforms is not valid if a sampler is included. But all those that simply involve the sum or difference of signals are valid, that is, Transformations 6, 11, and 12. Each represents a simple rearrangement of signals as a linear-sum, and addition is a commutative operation, even for sampled signals, that is $Z = X \pm Y = Y \pm X$.

# Chapter 8

## Signal Flow Graphs

### 8.1 INTRODUCTION

The most extensively used graphical representation of a feedback control system is the block diagram, presented in Chapters 2 and 7. In this chapter we consider another model, the signal flow graph.

A **signal flow graph** is a pictorial representation of the simultaneous equations describing a system. It graphically displays the transmission of signals through the system, as does the block diagram. But it is easier to draw and therefore easier to manipulate than the block diagram.

The properties of signal flow graphs are presented in the next few sections. The remainder of the chapter treats applications.

### 8.2 FUNDAMENTALS OF SIGNAL FLOW GRAPHS

Let us first consider the simple equation

$$X_i = A_{ij} X_j \tag{8.1}$$

The variables $X_i$ and $X_j$ can be functions of time, complex frequency, or any other quantity. They may even be constants, which are "variables" in the mathematical sense.

For signal flow graphs, $A_{ij}$ is a mathematical operator mapping $X_j$ into $X_i$, and is called the **transmission function**. For example, $A_{ij}$ may be a constant, in which case $X_i$ is a constant times $X_j$ in Equation ($8.1$); if $X_i$ and $X_j$ are functions of $s$ or $z$, $A_{ij}$ may be a transfer function $A_{ij}(s)$ or $A_{ij}(z)$.

The signal flow graph for Equation ($8.1$) is given in Fig. 8-1. This is the simplest form of a signal flow graph. Note that the variables $X_i$ and $X_j$ are represented by a small dot called a **node**, and the transmission function $A_{ij}$ is represented by a line with an arrow, called a **branch**.

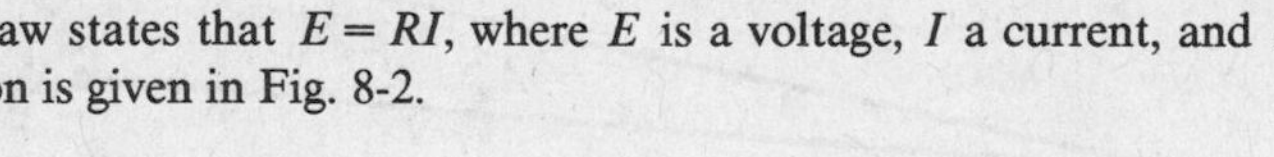

Node      $A_{ij}$      Node

$X_j$     Branch     $X_i$

Fig. 8-1

Every variable in a signal flow graph is designated by a node, and every transmission function by a branch. Branches are always unidirectional. The arrow denotes the direction of signal flow.

**EXAMPLE 8.1.** Ohm's law states that $E = RI$, where $E$ is a voltage, $I$ a current, and $R$ a resistance. The signal flow graph for this equation is given in Fig. 8-2.

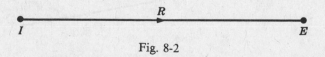

$I$            $R$            $E$

Fig. 8-2

## 8.3   SIGNAL FLOW GRAPH ALGEBRA

### 1.   The Addition Rule

The value of the variable designated by a node is equal to the sum of all signals entering the node. In other words, the equation

$$X_i = \sum_{j=1}^{n} A_{ij} X_j$$

is represented by Fig. 8-3.

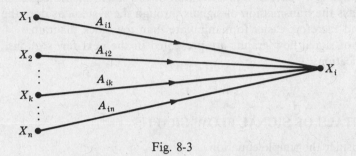

Fig. 8-3

**EXAMPLE 8.2.**   The signal flow graph for the equation of a line in rectangular coordinates, $Y = mX + b$, is given in Fig. 8-4. Since $b$, the $Y$-axis intercept, is a constant it may represent a node (variable) or a transmission function.

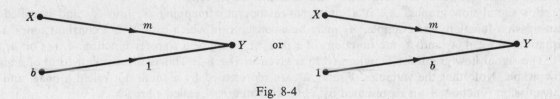

Fig. 8-4

### 2.   The Transmission Rule

The value of the variable designated by a node is transmitted on every branch leaving that node. In other words, the equation

$$X_i = A_{ik} X_k \qquad i = 1, 2, \ldots, n, \ k \text{ fixed}$$

is represented by Fig. 8-5.

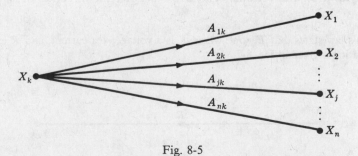

Fig. 8-5

**EXAMPLE 8.3.**   The signal flow graph of the simultaneous equations $Y = 3X$, $Z = -4X$ is given in Fig. 8-6.

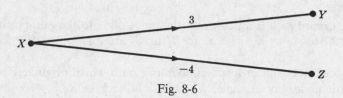

Fig. 8-6

### 3.  The Multiplication Rule

A cascaded (series) connection of $n-1$ branches with transmission functions $A_{21}, A_{32}, A_{43}, \ldots, A_{n(n-1)}$ can be replaced by a single branch with a new transmission function equal to the product of the old ones. That is,

$$X_n = A_{21} \cdot A_{32} \cdot A_{43} \cdots A_{n(n-1)} \cdot X_1$$

The signal flow graph equivalence is represented by Fig. 8-7.

$$\begin{array}{cc} \underset{X_1}{\bullet} \xrightarrow{A_{21}} \underset{X_2}{\bullet} \cdots \to \cdots \underset{X_{n-1}}{\bullet} \xrightarrow{A_{n(n-1)}} \underset{X_n}{\bullet} & \equiv & \underset{X_1}{\bullet} \xrightarrow{A_{21}A_{32} \; \cdots \; A_{n(n-1)}} \underset{X_n}{\bullet} \end{array}$$

Fig. 8-7

**EXAMPLE 8.4.**   The signal flow graph of the simultaneous equations $Y = 10X$, $Z = -20Y$ is given in Fig. 8-8.

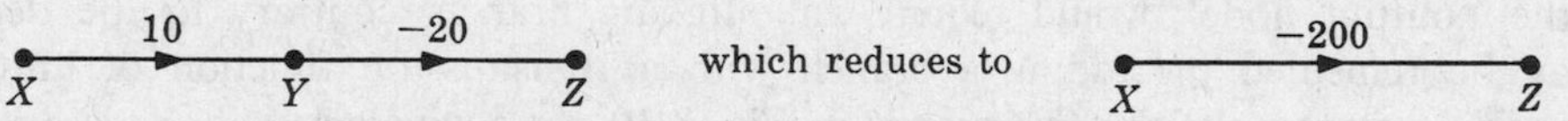

$$\underset{X}{\bullet} \xrightarrow{10} \underset{Y}{\bullet} \xrightarrow{-20} \underset{Z}{\bullet} \qquad \text{which reduces to} \qquad \underset{X}{\bullet} \xrightarrow{-200} \underset{Z}{\bullet}$$

Fig. 8-8

## 8.4  DEFINITIONS

The following terminology is frequently used in signal flow graph theory. The examples associated with each definition refer to Fig. 8-9.

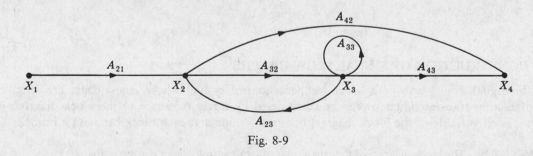

Fig. 8-9

*Definition 8.1*:     A **path** is a continuous, unidirectional succession of branches along which no node is passed more than once. For example, $X_1$ to $X_2$ to $X_3$ to $X_4$, $X_2$ to $X_3$ and back to $X_2$, and $X_1$ to $X_2$ to $X_4$ are paths.

*Definition 8.2*:     An **input node** or **source** is a node with only outgoing branches. For example, $X_1$ is an input node.

**Definition 8.3:**  An **output node** or **sink** is a node with only incoming branches. For example, $X_4$ is an output node.

**Definition 8.4:**  A **forward path** is a path from the input node to the output node. For example, $X_1$ to $X_2$ to $X_3$ to $X_4$, and $X_1$ to $X_2$ to $X_4$ are forward paths.

**Definition 8.5:**  A **feedback path** or **feedback loop** is a path which originates and terminates on the same node. For example, $X_2$ to $X_3$ and back to $X_2$ is a feedback path.

**Definition 8.6:**  A **self-loop** is a feedback loop consisting of a single branch. For example, $A_{33}$ is a self-loop.

**Definition 8.7:**  The **gain** of a branch is the transmission function of that branch when the transmission function is a multiplicative operator. For example, $A_{33}$ is the gain of the self-loop if $A_{33}$ is a constant or transfer function.

**Definition 8.8:**  The **path gain** is the product of the branch gains encountered in traversing a path. For example, the path gain of the forward path from $X_1$ to $X_2$ to $X_3$ to $X_4$ is $A_{21}A_{32}A_{43}$.

**Definition 8.9:**  The **loop gain** is the product of the branch gains of the loop. For example, the loop gain of the feedback loop from $X_2$ to $X_3$ and back to $X_2$ is $A_{32}A_{23}$.

Very often, a variable in a system is a function of the output variable. The canonical feedback system is an obvious example. In this case, if the signal flow graph were to be drawn directly from the equations, the "output node" would require an outgoing branch, contrary to the definition. This problem may be remedied by adding a branch with a transmission function of unity entering a "dummy" node. For example, the two graphs in Fig. 8-10 are equivalent, and $Y_4$ is an output node. Note that $Y_4 = Y_3$.

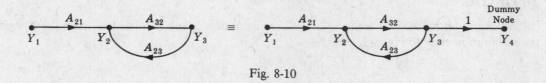

Fig. 8-10

## 8.5  CONSTRUCTION OF SIGNAL FLOW GRAPHS

The signal flow graph of a linear feedback control system whose components are specified by noninteracting transfer functions can be constructed by direct reference to the block diagram of the system. Each variable of the block diagram becomes a node and each block becomes a branch.

**EXAMPLE 8.5.**  The block diagram of the canonical feedback control system is given in Fig. 8-11.

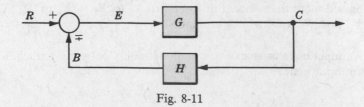

Fig. 8-11

The signal flow graph is easily constructed from Fig. 8-12. Note that the − or + sign of the summing point is associated with $H$.

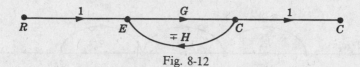

Fig. 8-12

The signal flow graph of a system described by a set of simultaneous equations can be constructed in the following general manner.

1.  Write the system equations in the form

$$X_1 = A_{11}X_1 + A_{12}X_2 + \cdots + A_{1n}X_n$$

$$X_2 = A_{21}X_1 + A_{22}X_2 + \cdots + A_{2n}X_n$$

$$\cdots\cdots\cdots\cdots\cdots\cdots\cdots\cdots\cdots\cdots\cdots$$

$$X_m = A_{m1}X_1 + A_{m2}X_2 + \cdots + A_{mn}X_n$$

An equation for $X_1$ is not required if $X_1$ is an input node.

2.  Arrange the $m$ or $n$ (whichever is larger) nodes from left to right. The nodes may be rearranged if the required loops later appear too cumbersome.

3.  Connect the nodes by the appropriate branches $A_{11}$, $A_{12}$, etc.

4.  If the desired output node has outgoing branches, add a dummy node and a unity gain branch.

5.  Rearrange the nodes and/or loops in the graph to achieve maximum pictorial clarity.

**EXAMPLE 8.6.**  Let us construct a signal flow graph for the simple resistance network given in Fig. 8-13. There are five variables, $v_1$, $v_2$, $v_3$, $i_1$, and $i_2$. $v_1$ is known. We can write four independent equations from Kirchhoff's voltage and current laws. Proceeding from left to right in the schematic, we have

$$i_1 = \left(\frac{1}{R_1}\right)v_1 - \left(\frac{1}{R_1}\right)v_2 \qquad v_2 = R_3i_1 - R_3i_2 \qquad i_2 = \left(\frac{1}{R_2}\right)v_2 - \left(\frac{1}{R_2}\right)v_3 \qquad v_3 = R_4i_2$$

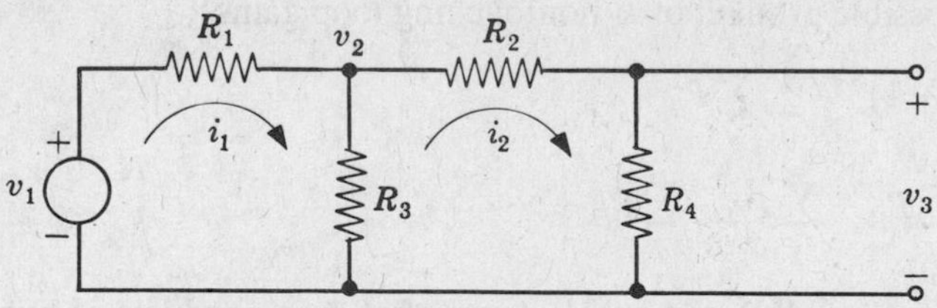

Fig. 8-13

Laying out the five nodes in the same order with $v_1$ as an input node, and connecting the nodes with the appropriate branches, we get Fig. 8-14. If we wish to consider $v_3$ as an output node, we must add a unity gain

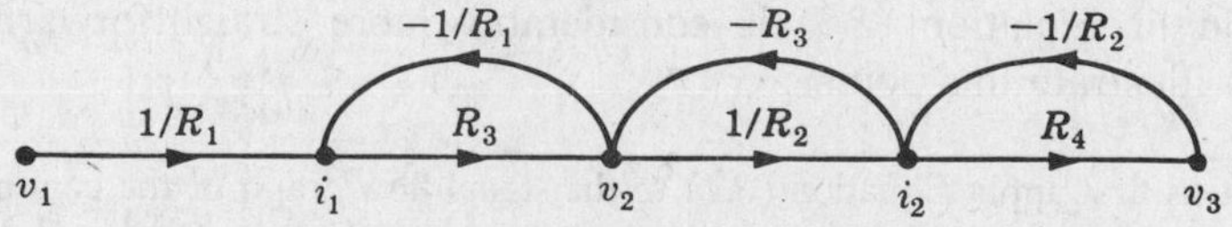

Fig. 8-14

branch and another node, yielding Fig. 8-15. No rearrangement of the nodes is necessary. We have one forward path and three feedback loops clearly in evidence.

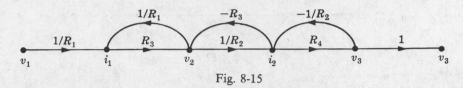

Fig. 8-15

Note that signal flow graph representations of equations are not unique. For example, the addition of a unity gain branch followed by a dummy node changes the graph, but not the equations it represents.

## 8.6   THE GENERAL INPUT-OUTPUT GAIN FORMULA

We found in Chapter 7 that we can reduce complicated block diagrams to canonical form, from which the control ratio is easily written as

$$\frac{C}{R} = \frac{G}{1 \pm GH}$$

It is possible to simplify signal flow graphs in a manner similar to that of block diagram reduction. But it is also possible, and much less time-consuming, to write down the input-output relationship *by inspection* from the original signal flow graph. This can be accomplished using the formula presented below. This formula can also be applied directly to block diagrams, but the signal flow graph representation is easier to read—especially when the block diagram is very complicated.

Let us denote the ratio of the input variable to the output variable by $T$. For linear feedback control systems, $T = C/R$. For the general signal flow graph presented in preceding paragraphs $T = X_n/X_1$, where $X_n$ is the output and $X_1$ is the input.

The general formula for any signal flow graph is

$$T = \frac{\sum_i P_i \Delta_i}{\Delta} \qquad\qquad (8.2)$$

where   $P_i$ = the $i$th forward path gain

$P_{jk}$ = $j$th possible product of $k$ nontouching loop gains

$$\Delta = 1 - (-1)^{k+1} \sum_k \sum_j P_{jk}$$

$$= 1 - \sum_j P_{j1} + \sum_j P_{j2} - \sum_j P_{j3} + \cdots$$

= 1 − (sum of all loop gains) + (sum of all gain products of two nontouching loops)
    − (sum of all gain products of three nontouching loops) + · · ·

$\Delta_i$ = $\Delta$ evaluated with all loops touching $P_i$ eliminated

Two loops, paths, or a loop and a path are said to be **nontouching** if they have no nodes in common.

$\Delta$ is called the **signal flow graph determinant** or **characteristic function**, since $\Delta = 0$ is the system characteristic equation.

The application of Equation (*8.2*) is considerably more straightforward than it appears. The following examples illustrate this point.

**EXAMPLE 8.7.**   Let us first apply Equation (*8.2*) to the signal flow graph of the canonical feedback system (Fig. 8-16).

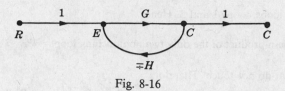

Fig. 8-16

There is only one forward path; hence

$$P_1 = G$$
$$P_2 = P_3 = \cdots = 0$$

There is only one (feedback) loop. Hence

$$P_{11} = \mp GH$$
$$P_{jk} = 0 \qquad j \neq 1 \qquad k \neq 1$$

Then

$$\Delta = 1 - P_{11} = 1 \pm GH \quad \text{and} \quad \Delta_1 = 1 - 0 = 1$$

Finally,

$$T = \frac{C}{R} = \frac{P_1 \Delta_1}{\Delta} = \frac{G}{1 \pm GH}$$

**EXAMPLE 8.8.** The signal flow graph of the resistance network of Example 8.6 is shown in Fig. 8-17. Let us apply Equation (8.2) to this graph and determine the voltage gain $T = v_3/v_1$ of the resistance network.

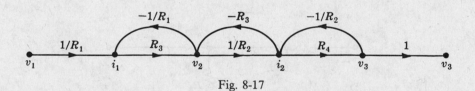

Fig. 8-17

There is one forward path (Fig. 8-18). Hence the forward path gain is

$$P_1 = \frac{R_3 R_4}{R_1 R_2}$$

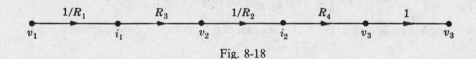

Fig. 8-18

There are three feedback loops (Fig. 8-19). Hence the loop gains are

$$P_{11} = -\frac{R_3}{R_1} \qquad P_{21} = -\frac{R_3}{R_2} \qquad P_{31} = -\frac{R_4}{R_2}$$

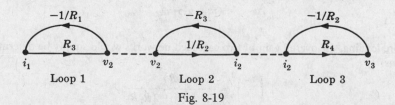

Loop 1          Loop 2          Loop 3

Fig. 8-19

There are two nontouching loops, loops 1 and 3. Hence

$$P_{12} = \text{gain product of the only two nontouching loops} = P_{11} \cdot P_{31} = \frac{R_3 R_4}{R_1 R_2}$$

There are no three loops that do not touch. Therefore

$$\Delta = 1 - (P_{11} + P_{21} + P_{31}) + P_{12} = 1 + \frac{R_3}{R_1} + \frac{R_3}{R_2} + \frac{R_4}{R_2} + \frac{R_3 R_4}{R_1 R_2}$$

$$= \frac{R_1 R_2 + R_1 R_3 + R_1 R_4 + R_2 R_3 + R_3 R_4}{R_1 R_2}$$

Since all loops touch the forward path, $\Delta_1 = 1$. Finally,

$$\frac{v_3}{v_1} = \frac{P_1 \Delta_1}{\Delta} = \frac{R_3 R_4}{R_1 R_2 + R_1 R_3 + R_1 R_4 + R_2 R_3 + R_3 R_4}$$

## 8.7    TRANSFER FUNCTION COMPUTATION OF CASCADED COMPONENTS

Loading effects of interacting components require little special attention using signal flow graphs. Simply combine the graphs of the components at their normal joining points (output node of one to the input node of another), account for loading by adding new loops at the joined nodes, and compute the overall gain using Equation (8.2). This procedure is best illustrated by example.

**EXAMPLE 8.9.**    Assume that two identical resistance networks are to be cascaded and used as the control elements in the forward loop of a control system. The networks are simple voltage dividers of the form given in Fig. 8-20.

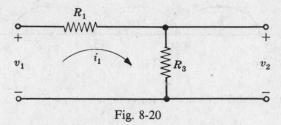

Fig. 8-20

Two independent equations for this network are

$$i_1 = \left(\frac{1}{R_1}\right) v_1 - \left(\frac{1}{R_1}\right) v_2 \quad \text{and} \quad v_2 = R_3 i_1$$

The signal flow graph is easily drawn (Fig. 8-21). The gain of this network is, by inspection, equal to

$$\frac{v_2}{v_1} = \frac{R_3}{R_1 + R_3}$$

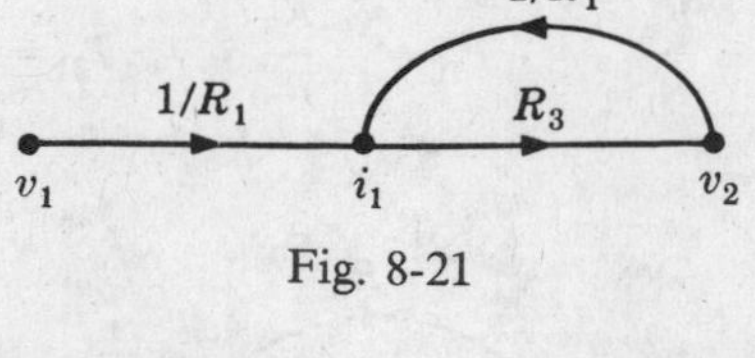

Fig. 8-21

If we were to ignore loading, the overall gain of two cascaded networks would simply be determined by multiplying the individual gains:

$$\left(\frac{v_2}{v_1}\right)^2 = \frac{R_3^2}{R_1^2 + R_3^2 + 2R_1 R_3}$$

*This answer is incorrect.* We prove this in the following manner. When the two identical networks are cascaded, we note that the result is equivalent to the network of Example 8.6, with $R_2 = R_1$ and $R_4 = R_3$ (Fig. 8-22).

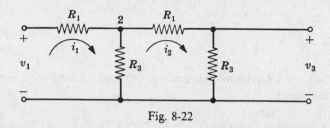

Fig. 8-22

The signal flow graph of this network was also determined in Example 8.6 (Fig. 8-23).

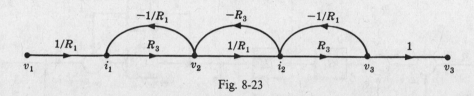

Fig. 8-23

We observe that the feedback branch $-R_3$ in Fig. 8-23 does not appear in the signal flow graph of the cascaded signal flow graphs of the individual networks connected from node $v_2$ to $v_1'$ (Fig. 8-24). This means that, as a result of connecting the two networks, the second one loads the first, changing the equation for $v_2$ from

$$v_2 = R_3 i_1 \qquad \text{to} \qquad v_2 = R_3 i_1 - R_3 i_2$$

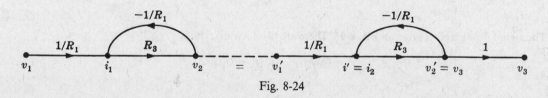

Fig. 8-24

This result could also have been obtained by directly writing the equations for the combined networks. In this case, only the equation for $v_2$ would have changed form.

The gain of the combined networks was determined in Example 8.8 as

$$\frac{v_3}{v_1} = \frac{R_3^2}{R_1^2 + R_3^2 + 3R_1 R_3}$$

when $R_2$ is set equal to $R_1$ and $R_4$ is set equal to $R_3$. We observe that

$$\left(\frac{v_2}{v_1}\right)^2 = \frac{R_3^2}{R_1^2 + R_3^2 + 2R_1 R_3} \neq \frac{v_3}{v_1}$$

It is good general practice to calculate the gain of cascaded networks directly from the *combined* signal flow graph. Most practical control system components load each other when connected in series.

## 8.8   BLOCK DIAGRAM REDUCTION USING SIGNAL FLOW GRAPHS
## AND THE GENERAL INPUT-OUTPUT GAIN FORMULA

Often, the easiest way to determine the control ratio of a complicated block diagram is to translate the block diagram into a signal flow graph and apply Equation (*8.2*). Takeoff points and summing points must be separated by a unity gain branch in the signal flow graph when using Equation (*8.2*).

If the elements $G$ and $H$ of a canonical feedback representation are desired, Equation $(8.2)$ also provides this information. The direct transfer function is

$$G = \sum_i P_i \Delta_i \qquad (8.3)$$

The loop transfer function is

$$GH = \Delta - 1 \qquad (8.4)$$

Equations $(8.3)$ and $(8.4)$ are solved simultaneously for $G$ and $H$, and the canonical feedback control system is drawn from the result.

**EXAMPLE 8.10.** Let us determine the control ratio $C/R$ and the canonical block diagram of the feedback control system of Example 7.9 (Fig. 8-25).

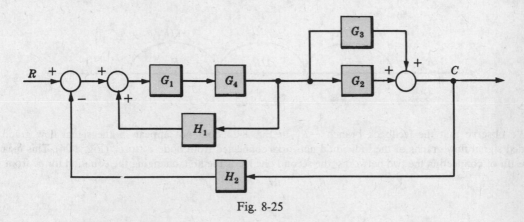

Fig. 8-25

The signal flow graph is given in Fig. 8-26. There are two forward paths:

$$P_1 = G_1 G_2 G_4 \qquad P_2 = G_1 G_3 G_4$$

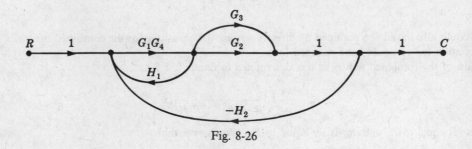

Fig. 8-26

There are three feedback loops:

$$P_{11} = G_1 G_4 H_1 \qquad P_{21} = -G_1 G_2 G_4 H_2 \qquad P_{31} = -G_1 G_3 G_4 H_2$$

There are no nontouching loops, and all loops touch both forward paths; then

$$\Delta_1 = 1 \qquad \Delta_2 = 1$$

Therefore the control ratio is

$$T = \frac{C}{R} = \frac{P_1 \Delta_1 + P_2 \Delta_2}{\Delta} = \frac{G_1 G_2 G_4 + G_1 G_3 G_4}{1 - G_1 G_4 H_1 + G_1 G_2 G_4 H_2 + G_1 G_3 G_4 H_2}$$

$$= \frac{G_1 G_4 (G_2 + G_3)}{1 - G_1 G_4 H_1 + G_1 G_2 G_4 H_2 + G_1 G_3 G_4 H_2}$$

From Equations ($8.3$) and ($8.4$), we have

$$G = G_1G_4(G_2 + G_3) \quad \text{and} \quad GH = G_1G_4(G_3H_2 + G_2H_2 - H_1)$$

Therefore

$$H = \frac{GH}{G} = \frac{(G_2 + G_3)H_2 - H_1}{G_2 + G_3}$$

The canonical block diagram is therefore given in Fig. 8-27.

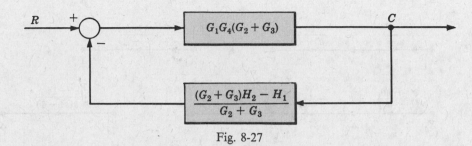

Fig. 8-27

The negative summing point sign for the feedback loop is a result of using a positive sign in the $GH$ formula above. If this is not obvious, refer to Equation ($7.3$) and its explanation in Section 7.4.

The block diagram above may be put into the final form of Examples 7.9 or 7.10 by using the transformation theorems of Section 7.5.

# Solved Problems

## SIGNAL FLOW GRAPH ALGEBRA AND DEFINITIONS

**8.1.** Simplify the signal flow graphs given in Fig. 8-28.

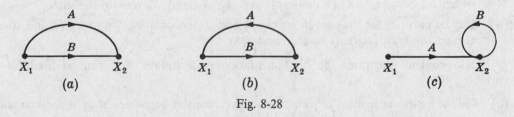

Fig. 8-28

(a)  Clearly, $X_2 = AX_1 + BX_1 = (A + B)X_1$. Therefore we have

$$\underset{X_1}{\bullet} \xrightarrow{\;A + B\;} \underset{X_2}{\bullet}$$

(b)  We have $X_2 = BX_1$ and $X_1 = AX_2$. Hence $X_2 = BAX_2$, or $X_1 = ABX_1$, yielding

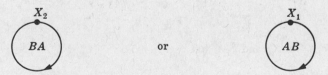

(c)  If $A$ and $B$ are multiplicative operators (e.g., constants or transfer functions), we have $X_2 = AX_1 + BX_2 = (A/(1 - B))X_1$. Hence the signal flow graph becomes

$$\underset{X_1}{\bullet} \xrightarrow{\;\frac{A}{1 - B}\;} \underset{X_2}{\bullet}$$

**8.2.** Draw signal flow graphs for the block diagrams in Problem 7.3 and reduce them by the multiplication rule (Fig. 8-29).

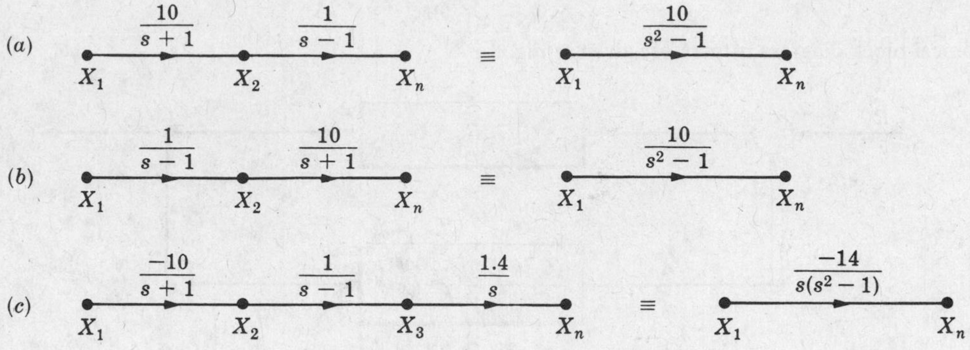

(a)

$X_1 \xrightarrow{\frac{10}{s+1}} X_2 \xrightarrow{\frac{1}{s-1}} X_n \equiv X_1 \xrightarrow{\frac{10}{s^2-1}} X_n$

(b)

$X_1 \xrightarrow{\frac{1}{s-1}} X_2 \xrightarrow{\frac{10}{s+1}} X_n \equiv X_1 \xrightarrow{\frac{10}{s^2-1}} X_n$

(c)

$X_1 \xrightarrow{\frac{-10}{s+1}} X_2 \xrightarrow{\frac{1}{s-1}} X_3 \xrightarrow{\frac{1.4}{s}} X_n \equiv X_1 \xrightarrow{\frac{-14}{s(s^2-1)}} X_n$

Fig. 8-29

**8.3.** Consider the signal flow graph in Fig. 8-30.

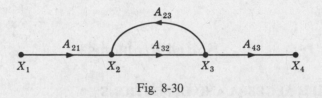

Fig. 8-30

(a) Draw the signal flow graph for the system equivalent to that graphed in Fig. 8-30, but in which $X_3$ becomes $kX_3$ ($k$ constant) and $X_1$, $X_2$, and $X_4$ remain the same.

(b) Repeat part (a) for the case in which $X_2$ and $X_3$ become $k_2X_2$ and $k_3X_3$, and $X_1$ and $X_4$ remain the same ($k_2$ and $k_3$ are constants).

This problem illustrates the fundamentals of a technique that can be used for *scaling* variables.

(a) For the system to remain the same when a node variable is multiplied by a constant, all signals entering the node must be multiplied by the same constant, and all signals leaving the node divided by that constant. Since $X_1$, $X_2$, and $X_4$ must remain the same, the *branches* are modified (Fig. 8-31).

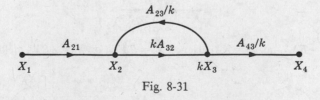

Fig. 8-31

(b) Substitute $k_2X_2$ for $X_2$, and $k_3X_3$ for $X_3$ (Fig. 8-32)

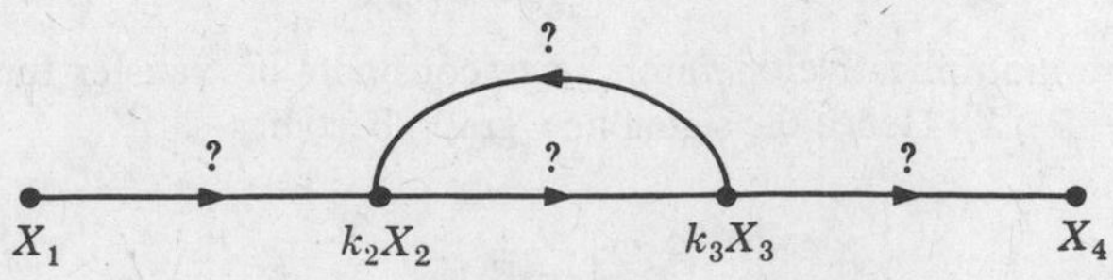

Fig. 8-32

It is clear from the graph that $A_{21}$ becomes $k_2 A_{21}$, $A_{32}$ becomes $(k_3/k_2)A_{32}$, $A_{23}$ becomes $(k_2/k_3)A_{23}$, and $A_{43}$ becomes $(1/k_3)A_{43}$ (Fig. 8-33).

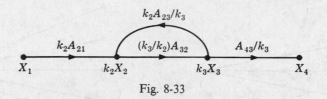

Fig. 8-33

**8.4.** Consider the signal flow graph given in Fig. 8-34.

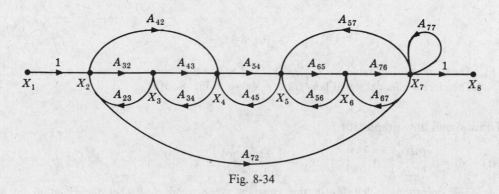

Fig. 8-34

Identify the ($a$) input node, ($b$) output node, ($c$) forward paths, ($d$) feedback paths, ($e$) self-loop. Determine the ($f$) loop gains of the feedback loops, ($g$) path gains of the forward paths.

($a$)   $X_1$

($b$)   $X_8$

($c$)   $X_1$ to $X_2$ to $X_3$ to $X_4$ to $X_5$ to $X_6$ to $X_7$ to $X_8$
        $X_1$ to $X_2$ to $X_7$ to $X_8$
        $X_1$ to $X_2$ to $X_4$ to $X_5$ to $X_6$ to $X_7$ to $X_8$

($d$)   $X_2$ to $X_3$ to $X_2$;  $X_3$ to $X_4$ to $X_3$;  $X_4$ to $X_5$ to $X_4$;  $X_2$ to $X_4$ to $X_3$ to $X_2$;
        $X_2$ to $X_7$ to $X_5$ to $X_4$ to $X_3$ to $X_2$;  $X_5$ to $X_6$ to $X_5$;  $X_6$ to $X_7$ to $X_6$;
        $X_5$ to $X_6$ to $X_7$ to $X_5$;  $X_7$ to $X_7$;  $X_2$ to $X_7$ to $X_6$ to $X_5$ to $X_4$ to $X_3$ to $X_2$

($e$)   $X_7$ to $X_7$

($f$)   $A_{32}A_{23}$;  $A_{43}A_{34}$;  $A_{54}A_{45}$;  $A_{65}A_{56}$;  $A_{76}A_{67}$;  $A_{65}A_{76}A_{57}$;  $A_{77}$;  $A_{42}A_{34}A_{23}$;
        $A_{72}A_{57}A_{45}A_{34}A_{23}$;  $A_{72}A_{67}A_{56}A_{45}A_{34}A_{23}$

($g$)   $A_{32}A_{43}A_{54}A_{65}A_{76}$;  $A_{72}$;  $A_{42}A_{54}A_{65}A_{76}$

## SIGNAL FLOW GRAPH CONSTRUCTION

**8.5.** Consider the following equations in which $x_1, x_2, \ldots, x_n$ are variables and $a_1, a_2, \ldots, a_n$ are coefficients or mathematical operators:

($a$)   $x_3 = a_1 x_1 + a_2 x_2 \mp 5$      ($b$)   $x_n = \sum_{k=1}^{n-1} a_k x_k + 5$

What are the minimum number of nodes and the minimum number of branches required to construct the signal flow graphs of these equations? Draw the graphs.

($a$)   There are four variables in this equation: $x_1$ $x_2$, $x_3$, and $\pm 5$. Therefore a minimum of four nodes are required. There are three coefficients or transmission functions on the right-hand side of the equation:

$a_1$, $a_2$, and $\mp 1$. Hence a minimum of three branches are required. A minimal signal flow graph is shown in Fig. 8-35($a$).

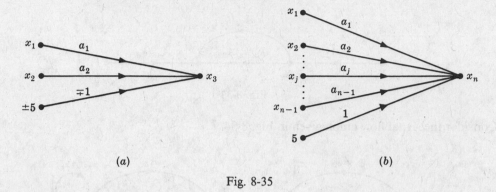

$(a)$          $(b)$

Fig. 8-35

($b$)  There are $n + 1$ variables: $x_1, x_2, \ldots, x_n$, and 5; and there are $n$ coefficients: $a_1, a_2, \ldots, a_{n-1}$, and 1. Therefore a minimal signal flow graph is shown in Fig. 8-35($b$).

**8.6.**  Draw signal flow graphs for

$(a)$   $x_2 = a_1\left(\dfrac{dx_1}{dt}\right)$      $(b)$   $x_3 = \dfrac{d^2 x_2}{dt^2} + \dfrac{dx_1}{dt} - x_1$      $(c)$   $x_4 = \int x_3\, dt$

($a$)  The operations called for in this equation are $a_1$ and $d/dt$. Let the equation be written as $x_2 = a_1 \cdot (d/dt)(x_1)$. Since there are two operations, we may define a new variable $dx_1/dt$ and use it as an intermediate node. The signal flow graph is given in Fig. 8-36.

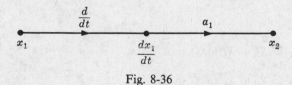

Fig. 8-36

($b$)  Similarly, $x_3 = (d^2/dt^2)(x_2) + (d/dt)(x_1) - x_1$. Therefore we obtain Fig. 8-37

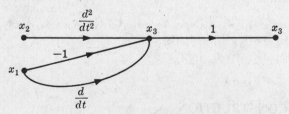

Fig. 8-37

($c$)  The operation is integration. Let the operator be denoted by $\int dt$. The signal flow graph is given in Fig. 8-38.

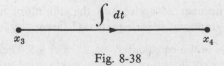

Fig. 8-38

**8.7.**   Construct the signal flow graph for the following set of simultaneous equations:

$$x_2 = A_{21}x_1 + A_{23}x_3 \qquad x_3 = A_{31}x_1 + A_{32}x_2 + A_{33}x_3 \qquad x_4 = A_{42}x_2 + A_{43}x_3$$

There are four variables: $x_1, \ldots, x_4$. Hence four nodes are required. Arranging them from left to right and connecting them with the appropriate branches, we obtain Fig. 8-39.

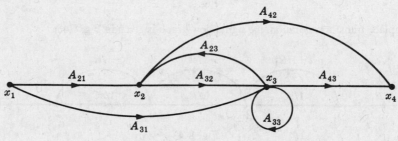

Fig. 8-39

A neater way to arrange this graph is shown in Fig. 8-40.

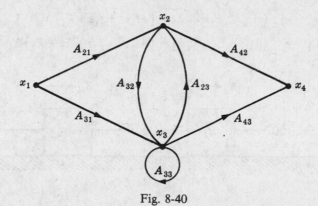

Fig. 8-40

**8.8.**   Draw a signal flow graph for the resistance network shown in Fig. 8-41 in which $v_2(0) = v_3(0) = 0$. $v_2$ is the voltage across $C_1$.

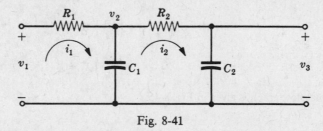

Fig. 8-41

The five variables are $v_1, v_2, v_3, i_1$, and $i_2$; and $v_1$ is the input. The four independent equations derived from Kirchhoff's voltage and current laws are

$$i_1 = \left(\frac{1}{R_1}\right)v_1 - \left(\frac{1}{R_1}\right)v_2 \qquad v_2 = \frac{1}{C_1}\int_0^t i_1\, dt - \frac{1}{C_1}\int_0^t i_2\, dt$$

$$i_2 = \left(\frac{1}{R_2}\right)v_2 - \left(\frac{1}{R_2}\right)v_3 \qquad v_3 = \frac{1}{C_2}\int_0^t i_2\, dt$$

The signal flow graph can be drawn directly from these equations (Fig. 8-42).

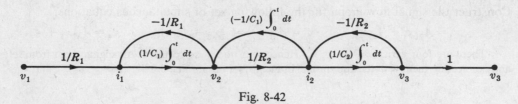

Fig. 8-42

In Laplace transform notation, the signal flow graph is given in Fig. 8-43.

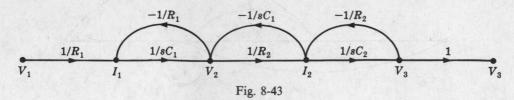

Fig. 8-43

## THE GENERAL INPUT-OUTPUT GAIN FORMULA

**8.9.** The transformed equations for the mechanical system given in Fig. 8-44 are

$$\text{(i)} \qquad F + k_1 X_2 = \left( M_1 s^2 + f_1 s + k_1 \right) X_1$$

$$\text{(ii)} \qquad k_1 X_1 = \left( M_2 s^2 + f_2 s + k_1 + k_2 \right) X_2$$

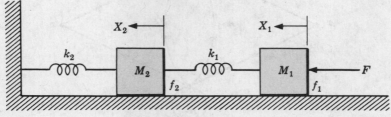

Fig. 8-44

where $F$ is force, $M$ is mass, $k$ is spring constant, $f$ is friction, and $X$ is displacement. Determine $X_2/F$ using Equation (8.2).

There are three variables: $X_1$, $X_2$, and $F$. Therefore we need three nodes. In order to draw the signal flow graph, divide Equation (i) by $A$ and Equation (ii) by $B$, where $A \equiv M_1 s^2 + f_1 s + k_1$, and $B \equiv M_2 s^2 + f_2 s + k_1 + k_2$:

$$\text{(iii)} \qquad \left( \frac{1}{A} \right) F + \left( \frac{k_1}{A} \right) X_2 = X_1$$

$$\text{(iv)} \qquad \left( \frac{k_1}{B} \right) X_1 = X_2$$

Therefore the signal flow graph is given in Fig. 8-45.

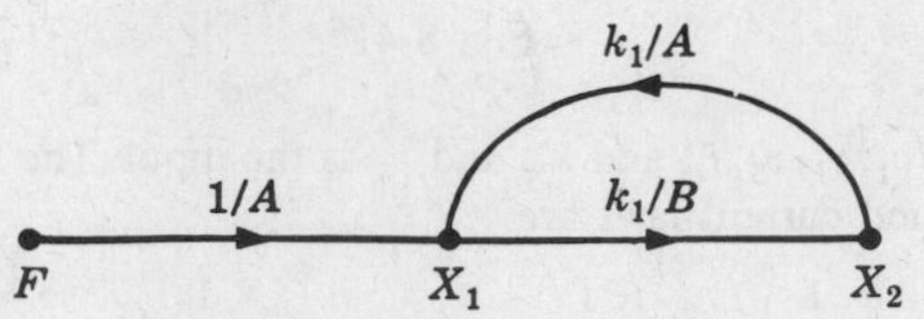

Fig. 8-45

The forward path gain is $P_1 = k_1/AB$. The feedback loop gain is $P_{11} = k_1^2/AB$. then $\Delta = 1 - P_{11} = (AB - k_1^2)/AB$ and $\Delta_1 = 1$. Finally,

$$\frac{X_2}{F} = \frac{P_1 \Delta_1}{\Delta} = \frac{k_1}{AB - k_1^2} = \frac{k_1}{\left( M_1 s^2 + f_1 s + k_1 \right)\left( M_2 s^2 + f_2 s + k_1 + k_2 \right) - k_1^2}$$

**8.10.** Determine the transfer function for the block diagram in Problem 7.20 by signal flow graph techniques.

The signal flow graph, Fig. 8-46, is drawn directly from Fig. 7-44. There are two forward paths. The path gains are $P_1 = G_1G_2G_3$ and $P_2 = G_4$. The three feedback loop gains are $P_{11} = -G_2H_1$, $P_{21} = G_1G_2H_1$, and $P_{31} = -G_2G_3H_2$. No loops are nontouching. Hence $\Delta = 1 - (P_{11} + P_{21} + P_{31})$. Also, $\Delta_1 = 1$; and since no loops touch the nodes of $P_2$, $\Delta_2 = \Delta$. Thus

$$T = \frac{P_1\Delta_1 + P_2\Delta_2}{\Delta} = \frac{G_1G_2G_3 + G_4 + G_2G_4H_1 - G_1G_2G_4H_1 + G_2G_3G_4H_2}{1 + G_2H_1 - G_1G_2H_1 + G_2G_3H_2}$$

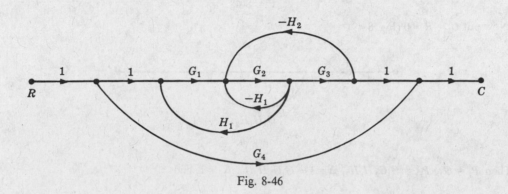

Fig. 8-46

**8.11.** Determine the transfer function $V_3/V_1$ from the signal flow graph of Problem 8.8.

Mathcad

The single forward path gain is $1/(s^2R_1R_2C_1C_2)$. The loop gains of the three feedback loops are $P_{11} = -1/(sR_1C_1)$, $P_{21} = -1/(sR_2C_1)$, and $P_{31} = -1/(sR_2C_2)$. The gain product of the only two nontouching loops is $P_{12} = P_{11} \cdot P_{31} = 1/(s^2R_1R_2C_1C_2)$. Hence

$$\Delta = 1 - (P_{11} + P_{21} + P_{31}) + P_{12} = \frac{s^2R_1R_2^2C_1^2C_2 + s(R_2^2C_1C_2 + R_1R_2C_1C_2 + R_1R_2C_1^2) + R_2C_1}{s^2R_1R_2^2C_1^2C_2}$$

Since all loops touch the forward path, $\Delta_1 = 1$. Finally,

$$\frac{V_3}{V_1} = \frac{P_1\Delta_1}{\Delta} = \frac{1}{s^2R_1R_2C_1C_2 + s(R_2C_2 + R_1C_2 + R_1C_1) + 1}$$

**8.12.** Solve Problem 7.16 with signal flow graph techniques.

The signal flow graph is drawn directly from Fig. 7-26, as shown in Fig. 8-47:

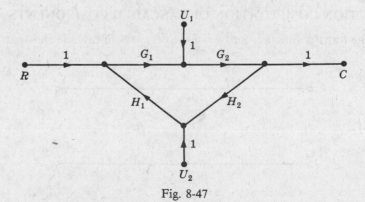

Fig. 8-47

With $U_1 = U_2 = 0$, we have Fig. 8-48. Then $P_1 = G_1G_2$ and $P_{11} = G_1G_2H_1H_2$. Hence $\Delta = 1 - P_{11} = 1 - G_1G_2H_1H_2$, $\Delta_1 = 1$, and

$$C_R = TR = \frac{P_1\Delta_1 R}{\Delta} = \frac{G_1G_2 R}{1 - G_1G_2H_1H_2}$$

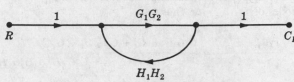

Fig. 8-48

Now put $U_2 = R = 0$ (Fig. 8-49).

Fig. 8-49

Then $P_1 = G_2$, $P_{11} = G_1G_2H_1H_2$, $\Delta = 1 - G_1G_2H_1H_2$, $\Delta_1 = 1$, and

$$C_1 = TU_1 = \frac{G_2U_1}{1 - G_1G_2H_1H_2}$$

Now put $R = U_1 = 0$ (Fig. 8-50).

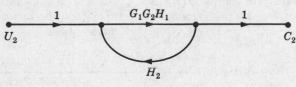

Fig. 8-50

Then $P_1 = G_1G_2H_1$, $P_{11} = G_1G_2H_1H_2$, $\Delta = 1 - G_1G_2H_1H_2$, $\Delta_1 = 1$, and

$$C_2 = TU_2 = \frac{P_1\Delta_1 U_2}{\Delta} = \frac{G_1G_2H_1U_2}{1 - G_1G_2H_1H_2}$$

Finally, we have

$$C = C_R + C_1 + C_2 = \frac{G_1G_2 R + G_2U_1 + G_1G_2H_1U_2}{1 - G_1G_2H_1H_2}$$

## TRANSFER FUNCTION COMPUTATION OF CASCADED COMPONENTS

**8.13.** Determine the transfer function for two of the networks in cascade shown in Fig. 8-51.

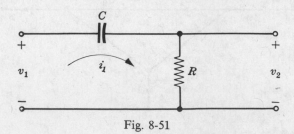

Fig. 8-51

In Laplace transform notation the network becomes Fig. 8-52.

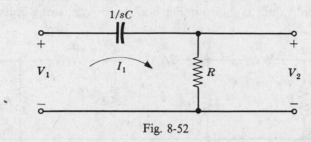

Fig. 8-52

By Kirchhoff's laws, we have $I_1 = sCV_1 - sCV_2$ and $V_2 = RI_1$. The signal flow graph is given in Fig. 8.53.

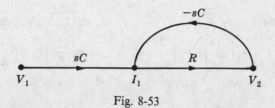

Fig. 8-53

For two networks in cascade (Fig. 8-54) the $V_2$ equation is also dependent on $I_2$: $V_2 = RI_1 - RI_2$. Hence two networks are joined at node 2 (Fig. 8-55) and a feedback loop ($-RI_2$) is added between $I_2$ and $V_2$ (Fig. 8-56).

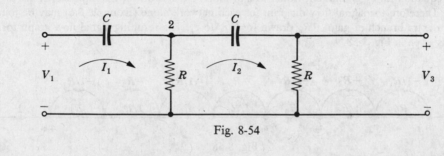

Fig. 8-54

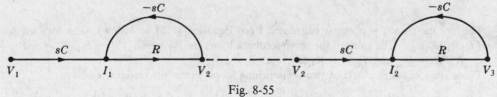

Fig. 8-55

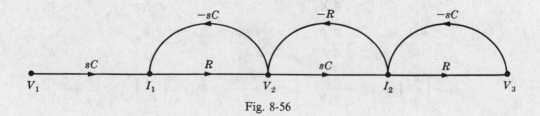

Fig. 8-56

Then $P_1 = s^2R^2C^2$, $P_{11} = P_{31} = -sRC$, $P_{12} = P_{11} \cdot P_{31} = s^2R^2C^2$, $\Delta = 1 - (P_{11} + P_{21} + P_{31}) + P_{12} = 1 + 3sRC + s^2R^2C^2$, $\Delta_1 = 1$, and

$$T = \frac{P_1\Delta_1}{\Delta} = \frac{s^2}{s^2 + (3/RC)s + 1/(RC)^2}$$

**8.14.** Two resistance networks in the form of that in Example 8.6 are to be used for control elements in the forward path of a control system. They are to be cascaded and shall have identical respective component values as shown in Fig. 8-57. Find $v_5/v_1$ using Equation (8.2).

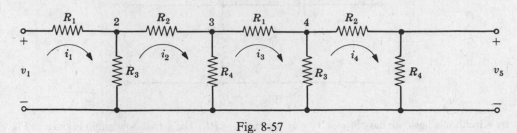

Fig. 8-57

There are nine variables: $v_1$, $v_2$, $v_3$, $v_4$, $v_5$, $i_1$, $i_2$, $i_3$, and $i_4$. Eight independent equations are

$$i_1 = \left(\frac{1}{R_1}\right)v_1 - \left(\frac{1}{R_1}\right)v_2 \qquad i_3 = \left(\frac{1}{R_1}\right)v_3 - \left(\frac{1}{R_1}\right)v_4$$

$$v_2 = R_3 i_1 - R_3 i_2 \qquad v_4 = R_3 i_3 - R_3 i_4$$

$$i_2 = \left(\frac{1}{R_2}\right)v_2 - \left(\frac{1}{R_2}\right)v_3 \qquad i_4 = \left(\frac{1}{R_2}\right)v_4 - \left(\frac{1}{R_2}\right)v_5$$

$$v_3 = R_4 i_2 - R_4 i_3 \qquad v_5 = R_4 i_4$$

Only the equation for $v_3$ is different from those of the single network of Example 8.6; it has an extra term, $(-R_4 i_3)$. Therefore the signal flow diagram for each network alone (Example 8.6) may be joined at node $v_3$, and an extra branch of gain $-R_4$ drawn from $i_3$ to $v_3$. The resulting signal flow graph for the double network is given in Fig. 8-58.

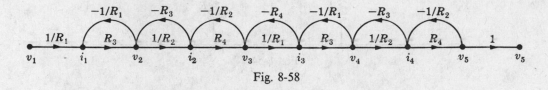

Fig. 8-58

The voltage gain $T = v_5/v_1$ is calculated from Equation (8.2) as follows. One forward path yields $P_1 = (R_3 R_4 / R_1 R_2)^2$. The gains of the seven feedback loops are $P_{11} = -R_3/R_1 = P_{51}$, $P_{21} = -R_3/R_2 = P_{61}$, $P_{31} = -R_4/R_2 = P_{71}$, and $P_{41} = -R_4/R_1$.

There are 15 gain products of two nontouching loops. From left to right, we have

$$P_{12} = \frac{R_3 R_4}{R_1 R_2} \qquad P_{42} = \frac{R_3^2}{R_1 R_2} \qquad P_{72} = \frac{R_3^2}{R_1 R_2} \qquad P_{10,2} = \frac{R_3 R_4}{R_1 R_2} \qquad P_{13,2} = \frac{R_3 R_4}{R_1 R_2}$$

$$P_{22} = \frac{R_3 R_4}{R_1^2} \qquad P_{52} = \frac{R_3 R_4}{R_1 R_2} \qquad P_{82} = \left(\frac{R_3}{R_2}\right)^2 \qquad P_{11,2} = \frac{R_3 R_4}{R_2^2} \qquad P_{14,2} = \frac{R_4^2}{R_1 R_2}$$

$$P_{32} = \left(\frac{R_3}{R_1}\right)^2 \qquad P_{62} = \frac{R_3 R_4}{R_1 R_2} \qquad P_{92} = \frac{R_3 R_4}{R_2^2} \qquad P_{12,2} = \left(\frac{R_4}{R_2}\right)^2 \qquad P_{15,2} = \frac{R_3 R_4}{R_1 R_2}$$

There are 10 gain products of three nontouching loops. From left to right, we have

$$P_{13} = \frac{R_3^2 R_4}{R_1^2 R_2} \qquad P_{33} = -\frac{R_3 R_4^2}{R_1 R_2^2} \qquad P_{63} = -\frac{R_3^2 R_4}{R_1^2 R_2} \qquad P_{83} = -\frac{R_3 R_4^2}{R_1 R_2^2} \qquad P_{53} = -\frac{R_3 R_4^2}{R_1^2 R_2}$$

$$P_{23} = -\frac{R_3^2 R_4}{R_1 R_2^2} \qquad P_{43} = -\frac{R_3^2 R_4}{R_1^2 R_2} \qquad P_{73} = -\frac{R_3^2 R_4}{R_1 R_2^2} \qquad P_{93} = -\frac{R_3^2 R_4}{R_1 R_2^2} \qquad P_{10,3} = -\frac{R_3 R_4^2}{R_1 R_2^2}$$

There is one gain product of four nontouching loops: $P_{14} = P_{11}P_{31}P_{51}P_{71} = (R_3R_4/R_1R_2)^2$. Therefore the determinant is

$$\Delta = 1 - \sum_{j=1}^{7} P_{j1} + \sum_{j=1}^{15} P_{j2} - \sum_{j=1}^{10} P_{j3} + P_{14}$$

$$= 1 + \frac{R_1R_3 + R_1R_4 + R_2R_3 + R_2R_4 + 6R_3R_4 + 2R_3^2 + R_4^2}{R_1R_2} + \frac{R_3R_4 + R_3^2}{R_1^2} + \frac{R_3^2 + R_4^2 + R_3R_4}{R_2^2}$$

Since all loops touch the forward path, $\Delta_1 = 1$ and

$$T = \frac{P_1\Delta_1}{\Delta} = \frac{(R_3R_4)^2}{(R_1R_2)^2 + R_1^2(R_2R_3 + R_2R_4 + R_3R_4 + R_3^2 + R_4^2) + R_2^2(R_3^2 + R_1R_3 + R_1R_4 + R_3R_4)}$$
$$+ 2R_1R_2R_3^2 + R_1R_2R_4^2 + 6R_1R_2R_3R_4$$

## BLOCK DIAGRAM REDUCTION

**8.15.** Determine $C/R$ for each system shown in Fig. 8-59 using Equation (8.2).

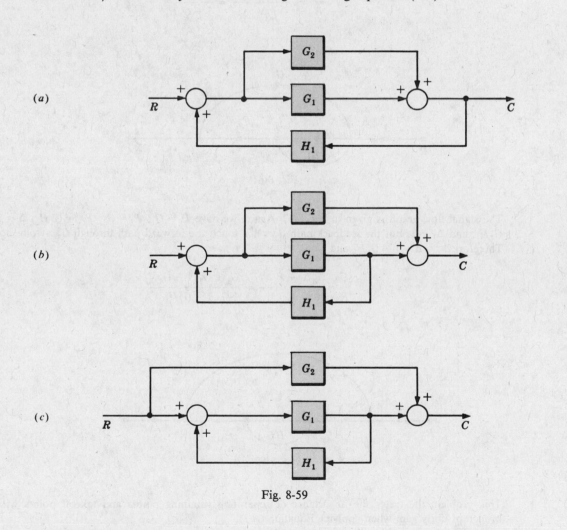

Fig. 8-59

(a) The signal flow graph is given in Fig. 8-60. The two forward path gains are $P_1 = G_1$, $P_2 = G_2$. The two feedback loop gains are $P_{11} = G_1H_1$, $P_{21} = G_2H_1$. Then

$$\Delta = 1 - (P_{11} + P_{21}) = 1 - G_1H_1 - G_2H_1$$

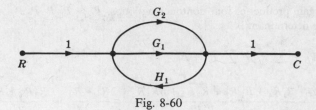

Fig. 8-60

Now, $\Delta_1 = 1$ and $\Delta_2 = 1$ because both paths touch the feedback loops at both interior nodes. Hence

$$\frac{C}{R} = \frac{P_1 \Delta_1 + P_2 \Delta_2}{\Delta} = \frac{G_1 + G_2}{1 - G_1 H_1 - G_2 H_1}$$

(b)  The signal flow graph is given in Fig. 8-61. Again, we have $P_1 = G_1$ and $P_2 = G_2$. But now there is only one feedback loop, and $P_{11} = G_1 H_1$; then $\Delta = 1 - G_1 H_1$. The forward path through $G_1$ clearly touches the feedback loop at nodes $a$ and $b$; thus $\Delta_1 = 1$. The forward path through $G_2$ touches the feedback loop at node $a$; then $\Delta_2 = 1$. Hence

$$\frac{C}{R} = \frac{P_1 \Delta_1 + P_2 \Delta_2}{\Delta} = \frac{G_1 + G_2}{1 - G_1 H_1}$$

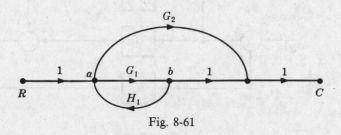

Fig. 8-61

(c)  The signal flow graph is given in Fig. 8-62. Again, we have $P_1 = G_1$, $P_2 = G_2$, $P_{11} = G_1 H_1$, $\Delta = 1 - G_1 H_1$, and $\Delta_1 = 1$. But the feedback path *does not* touch the forward path through $G_2$ at *any* node. Therefore $\Delta_2 = \Delta = 1 - G_1 H_1$ and

$$\frac{C}{R} = \frac{P_1 \Delta_1 + P_2 \Delta_2}{\Delta} = \frac{G_1 + G_2(1 - G_1 H_1)}{1 - G_1 H_1}$$

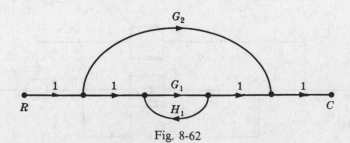

Fig. 8-62

This problem illustrates the importance of separating summing points and takeoff points with a branch of unity gain when applying Equation (8.2).

**8.16.**  Find the transfer function $C/R$ for the system shown in Fig. 8-63 in which $K$ is a constant.

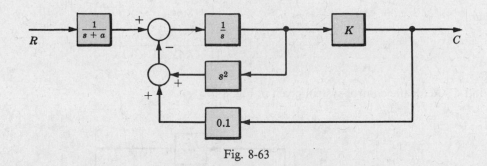

Fig. 8-63

The signal flow graph is given in Fig. 8-64. The only forward path gain is

$$P_1 = \left(\frac{1}{s+a}\right) \cdot \left(\frac{1}{s}\right) K = \frac{K}{s(s+a)}$$

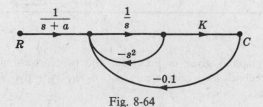

Fig. 8-64

The two feedback loop gains are $P_{11} = (1/s) \cdot (-s^2) = -s$ and $P_{21} = -0.1K/s$. There are no nontouching loops. Hence

$$\Delta = 1 - (P_{11} + P_{21}) = \frac{s^2 + s - 0.1K}{s} \qquad \Delta_1 = 1 \qquad \frac{C}{R} = \frac{P_1 \Delta_1}{\Delta} = \frac{K}{(s+a)(s^2 + s + 0.1K)}$$

**8.17.** Solve Problem 7.18 using signal flow graph techniques.

The signal flow graph is given in Fig. 8-65.

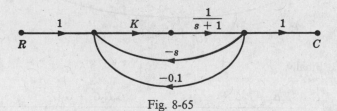

Fig. 8-65

Applying the multiplication and addition rules, we obtain Fig. 8-66. Now

$$P_1 = \frac{K}{s+1} \qquad P_{11} = -\frac{K(s+0.1)}{s+1} \qquad \Delta = 1 + \frac{K(s+0.1)}{s+1} \qquad \Delta_1 = 1,$$

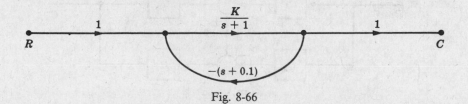

Fig. 8-66

and

$$C = TR = \frac{P_1 \Delta_1 R}{\Delta} = \frac{KR}{(1+K)s + 1 + 0.1K}$$

**8.18.** Find $C/R$ for the control system given in Fig. 8-67.

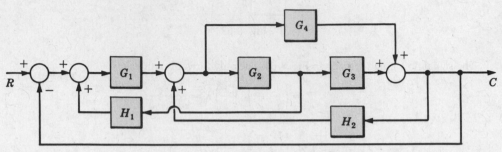

Fig. 8-67

The signal flow graph is given in Fig. 8.68. The two forward path gains are $P_1 = G_1 G_2 G_3$ and $P_2 = G_1 G_4$. The five feedback loop gains are $P_{11} = G_1 G_2 H_1$, $P_{21} = G_2 G_3 H_2$, $P_{31} = -G_1 G_2 G_3$, $P_{41} = G_4 H_2$, and $P_{51} = -G_1 G_4$. Hence

$$\Delta = 1 - (P_{11} + P_{21} + P_{31} + P_{41} + P_{51}) = 1 + G_1 G_2 G_3 - G_1 G_2 H_1 - G_2 G_3 H_2 - G_4 H_2 + G_1 G_4$$

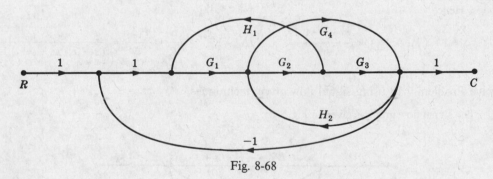

Fig. 8-68

and $\Delta_1 = \Delta_2 = 1$. Finally,

$$\frac{C}{R} = \frac{P_1 \Delta_1 + P_2 \Delta_2}{\Delta} = \frac{G_1 G_2 G_3 + G_1 G_4}{1 + G_1 G_2 G_3 - G_1 G_2 H_1 - G_2 G_3 H_2 - G_4 H_2 + G_1 G_4}$$

**8.19.** Determine $C/R$ for the system given in Fig. 8-69. *Then* put $G_3 = G_1 G_2 H_2$.

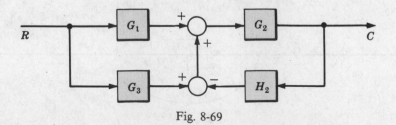

Fig. 8-69

The signal flow graph is given in Fig. 8-70. We have $P_1 = G_1G_2$, $P_2 = G_2G_3$, $P_{11} = -G_2H_2$, $\Delta = 1 + G_2H_2$, $\Delta_1 = \Delta_2 = 1$, and

$$\frac{C}{R} = \frac{P_1\Delta_1 + P_2\Delta_2}{\Delta} = \frac{G_2(G_1 + G_3)}{1 + G_2H_2}$$

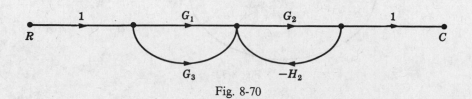

Fig. 8-70

Putting $G_3 = G_1G_2H_2$, we obtain $C/R = G_1G_2$ and the system transfer function becomes open-loop.

**8.20.** Determine the elements for a canonical feedback system for the system of Problem 8.10.

From Problem 8.10, $P_1 = G_1G_2G_3$, $P_2 = G_4$, $\Delta = 1 + G_2H_1 - G_1G_2H_1 + G_2G_3H_2$, $\Delta_1 = 1$, and $\Delta_2 = \Delta$. From Equation (8.3) we have

$$G = \sum_{i=1}^{2} P_i\Delta_i = G_1G_2G_s + G_4 + G_2G_4H_1 - G_1G_2G_4H_1 + G_2G_3G_4H_2$$

and from Equation (8.4) we obtain

$$H = \frac{\Delta - 1}{G} = \frac{G_2H_1 - G_1G_2H_1 + G_2G_3H_2}{G_1G_2G_3 + G_4 + G_2G_4H_1 - G_1G_2G_4H_1 + G_2G_3G_4H_2}$$

# Supplementary Problems

**8.21.** Find $C/R$ for Fig. 8-71, using Equation (8.2).

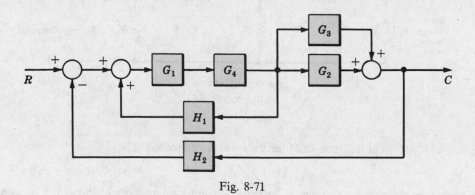

Fig. 8-71

**8.22.** Determine a set of canonical feedback system transfer functions for the preceding problem, using Equations (8.3) and (8.4).

**8.23.** Scale the signal flow graph in Fig. 8-72 so that $X_3$ becomes $X_3/2$ (see Problem 8.3).

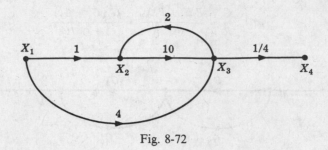

Fig. 8-72

**8.24.** Draw a signal flow graph for several nodes of the lateral inhibition system described in Problem 3.4 by the equation

$$c_k = r_k - \sum_{i=1}^{n} a_{k-i} c_i$$

**8.25.** Draw a signal flow graph for the system presented in Problem 7.31.

**8.26.** Draw a signal flow graph for the system presented in Problem 7.32.

**8.27.** Determine $C/R_4$ from Equation (8.2) for the signal flow graph drawn in Problem 8.26.

**8.28.** Draw a signal flow graph for the electrical network in Fig. 8-73.

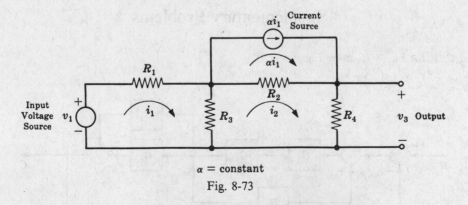

Fig. 8-73

**8.29.** Determine $V_3/V_1$ from Equation (8.2) for the network of Problem 8.28.

**8.30.** Determine the elements for a canonical feedback system for the network of Problem 8.28, using Equations (8.3) and (8.4).

**8.31.** Draw the signal flow graph for the analog computer circuit in Fig 8-74.

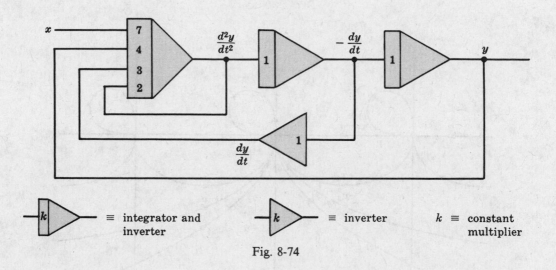

Fig. 8-74

**8.32.** Scale the analog computer circuit of Problem 8.31 so that $y$ becomes $10y$, $dy/dt$ becomes $20(dy/dt)$, and $d^2y/dt^2$ becomes $5(d^2y/dt^2)$.

# Answers to Supplementary Problems

**8.21.** $P_1 = G_1G_2G_4$; $P_2 = G_1G_3G_4$, $P_{11} = G_1G_4H_1$, $P_{21} = -G_1G_2G_4H_2$, $P_{31} = -G_1G_3G_4H_2$, $\Delta = 1 - G_1G_4H_1 + G_1G_2G_4H_2 + G_1G_3G_4H_2$, and $\Delta_1 = \Delta_2 = 1$. Therefore

$$\frac{C}{R} = \frac{P_1\Delta_1 + P_2\Delta_2}{\Delta} = \frac{G_1G_4(G_2 + G_3)}{1 - G_1G_4[H_1 - H_2(G_2 + G_3)]}$$

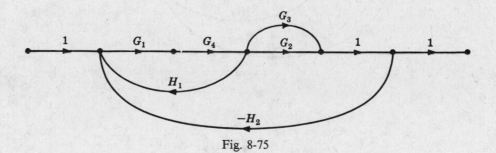

Fig. 8-75

**8.22.** $G = P_1\Delta_1 + P_2\Delta_2 = G_1G_4(G_2 + G_3)$      $H = \dfrac{\Delta - 1}{G} = H_2 - \dfrac{H_1}{G_2 + G_3}$

**8.23.**

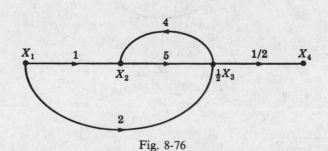

Fig. 8-76

**8.24.**

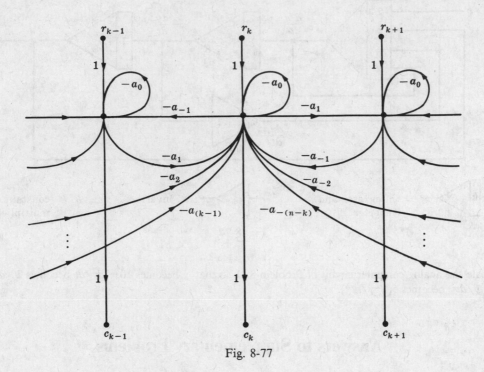

Fig. 8-77

**8.25.**

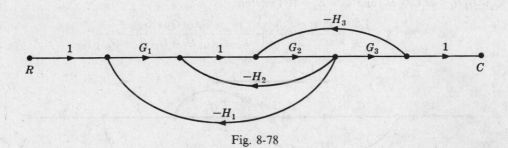

Fig. 8-78

**8.26.**

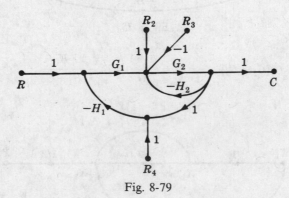

Fig. 8-79

**8.27.** $\dfrac{C}{R_4} = \dfrac{-G_1 G_2 H_1}{1 + G_2 H_2 + G_1 G_2 H_1}$

**8.28.**

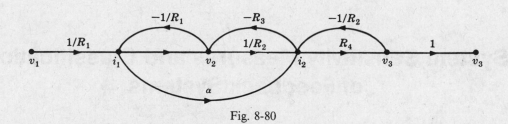

Fig. 8-80

**8.29.** $\dfrac{V_3}{V_1} = \dfrac{R_3 R_4 + \alpha R_2 R_4}{R_1 R_2 + R_1 R_3 + R_1 R_4 + R_2 R_3 + R_3 R_4 - \alpha R_2 R_3}$

**8.30.** $G = R_4(R_3 + \alpha R_2)$

$H = \dfrac{R_1(R_2 + R_3 + R_4) + R_3 R_4 + R_2 R_3(1 - \alpha)}{R_4(R_3 + \alpha R_2)}$

**8.31.**

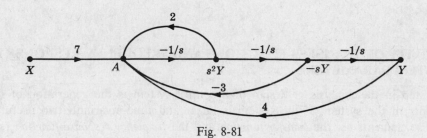

Fig. 8-81

**8.32.**

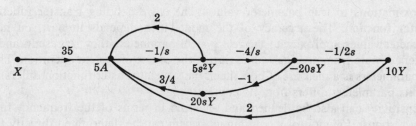

Fig. 8-82

# Chapter 9

# System Sensitivity Measures and Classification of Feedback Systems

## 9.1 INTRODUCTION

In earlier chapters the concepts of feedback and feedback systems have been emphasized. Since a system with a given transfer function can be synthesized in either an open-loop or a closed-loop configuration, a closed-loop (feedback) configuration must have some desirable properties which an open-loop configuration does not have.

In this chapter some of the properties of feedback and feedback systems are further discussed, and quantitative measures of the effectiveness of feedback are developed in terms of the concepts of *sensitivity* and *error constants*.

## 9.2 SENSITIVITY OF TRANSFER FUNCTIONS AND FREQUENCY RESPONSE FUNCTIONS TO SYSTEM PARAMETERS

An early step in the analysis or design of a control system is the generation of models for the various elements in the system. If the system is linear and time-invariant, two useful mathematical models for these elements are the *transfer function* and the *frequency response function* (see Chapter 6).

The transfer function is fixed when its parameters are specified, and the values given to these parameters are called **nominal values**. They are rarely, if ever, known exactly, so nominal values are actually approximations to true parameter values. The corresponding transfer function is called the **nominal transfer function**. The accuracy of the model then depends in part, on how closely these nominal parameter values approximate the real system parameters they represent, and also how much these parameters deviate from nominal values during the course of system operation. The **sensitivity** of a system to its parameters is a measure of how much the system transfer function differs from its nominal when each of its parameters differs from its nominal value.

System sensitivity can also be defined and analyzed in terms of the frequency response function. The frequency response function of a continuous system can be determined directly from the transfer function of the system, if it is known, by replacing the complex variable $s$ in the transfer function by $j\omega$. For discrete-time systems, the frequency response function is obtained by replacing $z$ by $e^{j\omega T}$. Thus the frequency response function is defined by the same parameters as those of the transfer function, and its accuracy is determined by the accuracy of these parameters. The frequency response function can alternatively be defined by graphs of its magnitude and phase- angle, both plotted as a function of the real frequency $\omega$. These graphs are often determined experimentally, and in many cases cannot be defined by a finite number of parameters. Hence an infinite number of values of amplitude and phase angle (values for all frequencies) define the frequency response function. The **sensitivity** of the system is in this case a measure of the amount by which its frequency response function differs from its nominal when the frequency response function of an element of the system differs from its nominal value.

Consider the mathematical model $T(k)$ (transfer function or frequency response function) of a linear time-invariant system, written in polar form as

$$T(k) = |T(k)|e^{j\phi_T} \qquad (9.1)$$

where $k$ is a parameter upon which $T(k)$ depends. Usually both $|T(k)|$ and $\phi_T$ depend on $k$, and $k$ is a real or complex parameter of the system.

208

***Definition 9.1:***     For the mathematical model $T(k)$, with $k$ regarded as the only parameter, the **sensitivity of $T(k)$ with respect to the parameter $k$** is defined by

$$S_k^{T(k)} \equiv \frac{d \ln T(k)}{d \ln k} = \frac{dT(k)/T(k)}{dk/k} = \frac{dT(k)}{dk} \frac{k}{T(k)} \qquad (9.2)$$

In some treatments of this subject, $S_k^{T(k)}$ is called the **relative sensitivity**, or **normalized** sensitivity, because it represents the variation $dT$ relative to the nominal $T$, for a variation $dk$ relative to the nominal $k$. $S_k^{T(k)}$ is also sometimes called the *Bode sensitivity*.

***Definition 9.2:***     The **sensitivity of the magnitude of $T(k)$ with respect to the parameter $k$** is defined by

$$S_k^{|T(k)|} \equiv \frac{d \ln |T(k)|}{d \ln k} = \frac{d|T(k)|/|T(k)|}{dk/k} = \frac{d|T(k)|}{dk} \frac{k}{|T(k)|} \qquad (9.3)$$

***Definition 9.3:***     The **sensitivity of the phase angle $\phi_T$ of $T(k)$ with respect to the parameter $k$** is defined by

$$S_k^{\phi_T} \equiv \frac{d \ln \phi_T}{d \ln k} = \frac{d\phi_T/\phi_T}{dk/k} = \frac{d\phi_T}{dk} \frac{k}{\phi_T} \qquad (9.4)$$

The sensitivities of $T(k) = |T(k)|e^{j\phi_T}$, the magnitude $|T(k)|$, and the phase angle $\phi_T$ with respect to the parameter $k$ are related by the expression

$$S_k^{T(k)} = S_k^{|T(k)|} + j\phi_T S_k^{\phi_T} \qquad (9.5)$$

Note that, in general, $S_k^{|T(k)|}$ and $S_k^{\phi_T}$ are complex numbers. In the special but very important case where $k$ is real, then both $S_k^{|T(k)|}$ and $S_k^{\phi_T}$ are real. When $S_k^{T(k)} = 0$, $T(k)$ is **insensitive** to $k$.

**EXAMPLE 9.1.**     Consider the frequency response function

$$T(\mu) = e^{-j\omega\mu}$$

where $\mu \equiv k$. The magnitude of $T(\mu)$ is $|T(\mu)| = 1$, and the phase angle of $T(\mu)$ is $\phi_T = -\omega\mu$.
  *The sensitivity of $T(\mu)$ with respect to the parameter $\mu$ is*

$$S_\mu^{T(\mu)} = \frac{d(e^{-j\omega\mu})}{d_\mu} \frac{\mu}{e^{-j\omega\mu}} = -j\omega\mu$$

*The sensitivity of the magnitude of $T(\mu)$ with respect to the parameter $\mu$ is*

$$S_\mu^{|T(\mu)|} = \frac{d|T(\mu)|}{\mu} \frac{\mu}{|T(\mu)|} = 0$$

*The sensitivity of the phase angle of $T(\mu)$ with respect to the parameter $\mu$ is*

$$S_\mu^{\phi_T} = \frac{d\phi_T}{d\mu} \frac{\mu}{\phi_T} = -\omega \cdot \frac{\mu}{-\omega\mu} = 1$$

Note that

$$S_\mu^{|T(\mu)|} + j\phi_T S_\mu^{\phi_T} = -j\omega\mu = S_\mu^{T(\mu)}$$

The following development is in terms of transfer functions. However, everything is applicable to frequency response functions (for continuous systems) by simply replacing $s$ in all equations by $j\omega$, or $z = e^{j\omega T}$ for discrete systems.
  A special but very important class of system transfer functions has the form:

$$T = \frac{A_1 + kA_2}{A_3 + kA_4}. \qquad (9.6)$$

where $k$ is a parameter and $A_1$, $A_2$, $A_3$, and $A_4$ are polynomials in $s$ (or $z$). This type of dependence between a parameter $k$ and a transfer function $T$ is general enough to include many of the systems considered in this book.

For a transfer function with the form of Equation (9.6), the *sensitivity of $T$ with respect to the parameter $k$* is given by

$$S_k^T \equiv \frac{dT}{dk} \cdot \frac{k}{T} = \frac{k(A_2 A_3 - A_1 A_4)}{(A_3 + k A_4)(A_1 + k A_2)} \tag{9.7}$$

In general, $S_k^T$ is a function of the complex variable $s$ (or $z$).

Mathcad

**EXAMPLE 9.2.**    The transfer function of the discrete-time system given in Fig. 9-1 is

$$T \equiv \frac{C}{R} = \frac{K}{z^3 + (a+b)z^2 + abz + K}$$

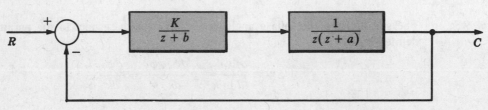

Fig. 9-1

If $K$ is the parameter of interest ($k \equiv K$), we group terms in $T$ as follows:

$$T = \frac{K}{\left[ z^3 + (a+b)z^2 + abz \right] + K}$$

Comparing $T$ with Equation (9.6), we see that

$$A_1 = 0 \qquad A_2 = 1 \qquad A_3 = z^3 + (a+b)z^2 + abz \qquad A_4 = 1$$

If $a$ is the parameter of interest ($k \equiv a$), $T$ can be rewritten as

$$T = \frac{K}{\left[ z^3 + bz^2 + K \right] + a\left[ z^2 + bz \right]}$$

Comparing this expression with Equation (9.6) we see that

$$A_1 = K \qquad A_2 = 0 \qquad A_3 = z^3 + bz^2 + K \qquad A_4 = z^2 + bz$$

If $b$ is the parameter of interest ($k \equiv b$), $T$ can be rewritten as

$$T = \frac{K}{\left[ z^3 + az^2 + K \right] + b\left[ z^2 + az \right]}$$

Again comparing this expression with Equation (9.6), we see that

$$A_1 = K \qquad A_2 = 0 \qquad A_3 = z^3 + az^2 + K \qquad A_4 = z^2 + az$$

**EXAMPLE 9.3.**    For the lead network shown in Fig. 9-2 the transfer function is

$$T \equiv \frac{E_0}{E_i} = \frac{1 + RCs}{2 + RCs}$$

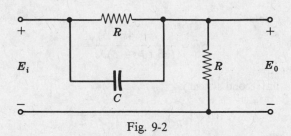

Fig. 9-2

If $C$ (capacitance) is the parameter of interest, we write $T = (1 + C(Rs)]/[2 + C(Rs)]$. Comparing this expression with Equation (9.6), we see that $A_1 = 1$, $A_2 = Rs$, $A_3 = 2$, $A_4 = Rs$.

**EXAMPLE 9.4.**  For the system of Example 9.2 *the sensitivity of $T$ with respect to $K$ is*

$$S_K^T = \frac{K[z^3 + (a+b)z^2 + abz]}{K[z^3 + (a+b)z^2 + abz + K]} = \frac{1}{1 + \dfrac{K}{z^3 + (a+b)z^2 + abz}}$$

*The sensitivity of $T$ with respect to the parameter $a$ is*

$$S_a^T = \frac{-aK(z^2 + bz)}{K[z^3 + bz^2 + K + a(z^2 + bz)]} = \frac{-1}{1 + \dfrac{z^3 + bz^2 + K}{a(z^2 + bz)}}$$

*The sensitivity of $T$ with respect to the parameter $b$ is*

$$S_b^T = \frac{-bK(z^2 + az)}{K[z^3 + az^2 + K + b(z^2 + az)]} = \frac{-1}{1 + \dfrac{z^3 + az^2 + K}{b(z^2 + az)}}$$

**EXAMPLE 9.5.**  For the lead network of Fig. 9-2 the sensitivity of $T$ with respect to the capacitance $C$ is

$$S_C^T = \frac{C(2Rs - Rs)}{(2 + RCs)(1 + RCs)} = \frac{RCs}{(2 + RCs)(1 + RCs)} = \frac{1}{(1 + 2/RCs)(1 + 1/RCs)}$$

**EXAMPLE 9.6.**  The open-loop and closed-loop systems given in Fig. 9-3 have the same plant and the same overall system transfer function for $K = 2$.

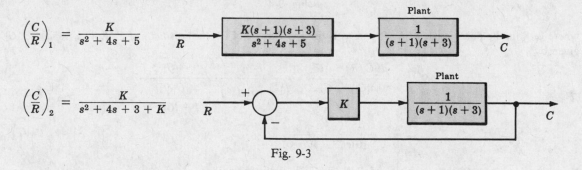

$$\left(\frac{C}{R}\right)_1 = \frac{K}{s^2 + 4s + 5}$$

$$\left(\frac{C}{R}\right)_2 = \frac{K}{s^2 + 4s + 3 + K}$$

Fig. 9-3

Although these systems are precisely equivalent for $K = 2$, their properties differ significantly for small (and large) deviations of $K$ from $K = 2$. The transfer function of the first system is

$$T_1 \equiv \left(\frac{C}{R}\right) = \frac{K}{s^2 + 4s + 5}$$

Comparing this expression with Equation (9.6) gives $A_1 = 0$, $A_2 = 1$, $A_3 = s^2 + 4s + 5$, $A_4 = 0$. Substituting these

values into Equation (9.7), we obtain

$$S_K^{T_1} = \frac{K(s^2 + 4s + 5)}{(s^2 + 4s + 5)K} = 1$$

for all $K$.

The transfer function of the second system is

$$T_2 \equiv \left(\frac{C}{R}\right)_2 = \frac{K}{s^2 + 4s + 3 + K}$$

Comparing this expression with Equation (9.6) yields $A_1 = 0$, $A_2 = 1$, $A_3 = s^2 + 4s + 3$, $A_4 = 1$. Substituting these values into Equation (9.7), we obtain

$$S_K^{T_2} = \frac{K(s^2 + 4s + 3)}{(s^2 + 4s + 3 + K)(K)} = \frac{1}{1 + K/(s^2 + 4s + 3)}$$

For $K = 2$, $S_{K^2}^{T_2} = 1/[1 + 2/(s^2 + s + 3)]$.

Note that the sensitivity of the open-loop system $T_1$ is fixed at 1 for all values of gain $K$. On the other hand, the closed-loop sensitivity is a function of $K$ and the complex variable $s$. Thus $S_K^{T_2}$ may be adjusted in a design problem by varying $K$ or maintaining the frequencies of the input function within an appropriate range.

For $\omega < \sqrt{3}$ rad/sec, the sensitivity of the closed-loop system is

$$S_K^{T_2} \simeq \frac{1}{1 + \frac{2}{3}} = \frac{3}{5} = 0.6$$

Thus the feedback system is 40% less sensitive than the open-loop system for low frequencies. For high frequencies, the sensitivity of the closed-loop system approaches 1, the same as that of the open-loop system.

Mathcad
**EXAMPLE 9.7.** Suppose $G$ is a frequency response function, either $G(j\omega)$ for a continuous system, or $G(e^{j\omega T})$ for a discrete-time system. The frequency response function for the unity feedback system (continuous or discrete-time) given in Fig. 9-4 is related to the forward-loop frequency response function $G$ by

$$\frac{C}{R} = \left|\frac{C}{R}\right| e^{j\phi_{C/R}} = \frac{G}{1 + G} = \frac{|G| e^{j\phi_G}}{1 + |G| e^{j\phi_G}}$$

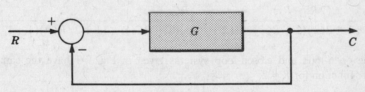

Fig. 9-4

where $\phi_{C/R}$ is the phase angle of $C/R$ and $\phi_G$ is the phase angle of $G$. *The sensitivity of $C/R$ with respect to $|G|$ is given by*

$$S_{|G|}^{C/R} = \frac{d(C/R)}{d|G|} \cdot \frac{|G|}{C/R} = \frac{e^{j\phi_G}}{\left(1 + |G| e^{j\phi_G}\right)^2} \cdot \frac{|G|}{\dfrac{|G| e^{j\phi_G}}{1 + |G| e^{j\phi_G}}}$$

$$= \frac{1}{1 + |G| e^{j\phi_G}} = \frac{1}{1 + G} \qquad (9.8)$$

Note that for large $|G|$ the sensitivity of $C/R$ to $|G|$ is relatively small.

Mathcad
**EXAMPLE 9.8.** Suppose the system of Example 9.7 is continuous, that $\omega = 1$, and for some given $G(j\omega)$, $G(j1) = 1 + j$. Then $|G(j\omega)| = \sqrt{2}$, $\phi_G = \pi/4$ rad, $(C/R)(j\omega) = \frac{3}{5} + j\frac{1}{5}$, $|(C/R)(j\omega)| = \sqrt{10}/5$, and $\phi_{C/R} = 0.3215$ rad.

Using the result of the previous example, the *sensitivity* of $(C/R)(j\omega)$ *with respect to* $|G(j\omega)|$ is

$$S_{|G(j\omega)|}^{(C/R)(j\omega)} = \frac{1}{2 + j} = \frac{2}{5} - j\frac{1}{5}$$

Then from Equation (9.5) we have

$$S_{|G(j\omega)|}^{|(C/R)(j\omega)|} = \frac{2}{5} = 0.4 \qquad \phi_{C/R} S_{|G(j\omega)|}^{\phi_{C/R}} = -\frac{1}{5} \qquad S_{|G(j\omega)|}^{\phi_{C/R}} = -\frac{1}{5(0.3215)} = -0.622$$

These real values of sensitivity mean that a 10% change in $|G(j\omega)|$ will produce a 4% change in $|(C/R)(j\omega)|$ and a $-6.22\%$ change in $\phi_{C/R}$.

A qualitative attribute of a system related to its sensitivity is its *robustness*. A system is said to be **robust** when its operation is insensitive to parameter variations. Robustness may be characterized in terms of the sensitivity of its transfer or frequency response function, or of a set of performance indices to system parameters.

## 9.3  OUTPUT SENSITIVITY TO PARAMETERS FOR DIFFERENTIAL AND DIFFERENCE EQUATION MODELS

The concept of sensitivity is also applicable to system models expressed in the time domain. **The sensitivity of the model output $y$ to any parameter $p$** is given by

$$S_p^{y(t)} \equiv S_p^y = \frac{d(\ln y)}{d(\ln p)} = \frac{dy/y}{dp/p} = \frac{dy}{dp}\frac{p}{y}$$

Since the model is defined in the time domain, the sensitivity is usually found by solving for the output $y(t)$ in the time domain. The derivative $dy/dp$ is sometimes called the **output sensitivity coefficient**, which is generally a function of time, as is the sensitivity $S_p^y$.

**EXAMPLE 9.9.**  We determine the sensitivity of the output $y(t) = x(t)$ to the parameter $a$ for the differential system $\dot{x} = ax + u$. The sensitivity is

$$S_a^y = \frac{dy}{da}\frac{a}{y} = \frac{dx}{da}\frac{a}{x}$$

To determine $S_a^y$, consider the *time derivative* of $dx/da$, and interchange the order of differentiation, that is,

$$\frac{d}{dt}\left(\frac{dx}{da}\right) = \frac{d}{da}\left(\frac{dx}{dt}\right) = \frac{d}{da}(ax + u)$$

Now define a new variable $v \equiv dx/da$. Then

$$\dot{v} = \frac{d}{da}(ax + u) = a\frac{dx}{da} + 1 \cdot x = av + x$$

The sensitivity function $S_a^y$ can then be found by first solving the system differential equation for $x(t)$, because $x(t)$ is the forcing function in the differential equation for $v(t)$ above. The required solutions were developed in Section 3.15 as

$$x(t) = e^{at}x(0) + \int_0^t e^{a(t-\tau)}u(\tau)\,d\tau$$

and

$$v(t) = \int_0^t e^{a(t-\tau)}x(\tau)\,d\tau$$

because $v(0) = 0$. The time-varying output sensitivity is computed from these two functions as

$$S_a^y = \frac{dx}{da}\frac{a}{x} = \frac{av(t)}{x(t)}$$

**EXAMPLE 9.10.**  For the discrete system defined by

$$x(k+1) = ax(k) + u(k)$$
$$y(k) = cx(k)$$

we determine the sensitivity of the output $y$ to the parameter $a$ as follows. Let

$$v(k) \equiv \frac{\partial x(k)}{\partial a}$$

Then

$$v(k+1) = \frac{\partial x(k+1)}{\partial a} = \frac{\partial}{\partial a}[ax(k) + u(k)]$$

$$= x(k) + a\frac{\partial x(k)}{\partial a} = av(k) + x(k)$$

and

$$\frac{\partial y(k)}{\partial a} = \frac{\partial cx(k)}{\partial a} = c\frac{\partial x(k)}{\partial a} = cv(k)$$

Thus, to determine $S_a^y$, we first solve the two discrete equations:

$$x(k+1) = ax(k) + u(k)$$
$$v(k+1) = av(k) + x(k)$$

(e.g., see Section 3.17). Then

$$S_a^y = \frac{\partial y(k)}{\partial a} \cdot \frac{a}{y(k)} = \frac{av(k)}{x(k)}$$

## 9.4  CLASSIFICATION OF CONTINUOUS FEEDBACK SYSTEMS BY TYPE

Consider the class of canonical feedback systems defined by Fig. 9-5. For continuous systems, the open-loop transfer function may be written as

$$GH = \frac{K\prod_{i=1}^{m}(s+z_i)}{\prod_{i=1}^{n}(s+p_i)}$$

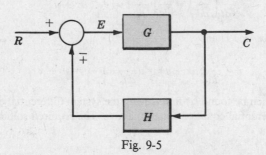

Fig. 9-5

where $K$ is a constant, $m \leq n$, and $-z_i$ and $-p_i$ are the finite zeros and poles, respectively, of $GH$. If there are $a$ zeros and $b$ poles at the origin, then

$$GH = \frac{Ks^a\prod_{i=1}^{m-a}(s+z_i)}{s^b\prod_{i=1}^{n-b}(s+p_i)}$$

In the remainder of this chapter, only systems for which $b \geq a$ are considered, and $l \equiv b - a$.

***Definition 9.4:***     A canonical feedback system whose open-loop transfer function can be written in the form:

$$GH = \frac{K \prod_{i=1}^{m-a} (s + z_i)}{s^l \prod_{i=1}^{n-a-l} (s + p_i)} \equiv \frac{KB_1(s)}{s^l B_2(s)} \qquad (9.9)$$

where $l \geq 0$ and $-z_i$ and $-p_i$ are the nonzero finite zeros and poles of $GH$, respectively, is called a **type $l$ system**.

**EXAMPLE 9.11.**     The system defined by Fig. 9-6 is a *type 2 system*.

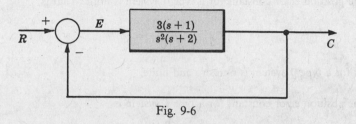

Fig. 9-6

**EXAMPLE 9.12.**     The system defined by Fig. 9-7 is a *type 1 system*.

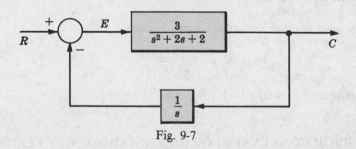

Fig. 9-7

**EXAMPLE 9.13.**     The system defined by Fig. 9-8 is a *type 0 system*.

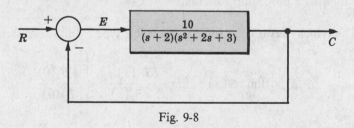

Fig. 9-8

## 9.5  POSITION ERROR CONSTANTS FOR CONTINUOUS UNITY FEEDBACK SYSTEMS

One criterion of the effectiveness of feedback in a *stable type $l$ unity feedback system* is the *position* (*step*) *error constant*. It is a measure of the steady state error between the input and output when the input is a unit step function, that is, the difference between the input and output when the system is in steady state and the input is a step.

*Definition 9.5:*        The **position error constant** $K_p$ of a type $l$ unity feedback system is defined as

$$K_p \equiv \lim_{s \to 0} G(s) = \lim_{s \to 0} \frac{KB_1(s)}{s^l B_2(s)} = \begin{cases} \dfrac{KB_1(0)}{B_2(0)} & \text{for } l = 0 \\[2mm] \infty & \text{for } l > 0 \end{cases} \qquad (9.10)$$

The steady state error of a stable type $l$ unity feedback system when the input is a unit step function $[e(\infty) = 1 - c(\infty)]$ is related to the position error constant by

$$e(\infty) = \lim_{t \to \infty} e(t) = \frac{1}{1 + K_p} \qquad (9.11)$$

**EXAMPLE 9.14.**   The position error constant for a type 0 system is finite. That is,

$$|K_p| = \left| \frac{KB_1(0)}{B_2(0)} \right| < \infty$$

The steady state error for a type 0 system is nonzero and finite.

**EXAMPLE 9.15.**   The position error constant for a type 1 system is

$$K_p = \lim_{s \to 0} \frac{KB_1(0)}{sB_2(0)} = \infty$$

Therefore the steady state error is $e(\infty) = 1/(1 + K_p) = 0$.

**EXAMPLE 9.16.**   The position error constant for a type 2 system is

$$K_p = \lim_{s \to 0} \frac{KB_1(s)}{s^2 B_2(s)} = \infty$$

Therefore the steady state error is $e(\infty) = 1/(1 + K_p) = 0$.

## 9.6   VELOCITY ERROR CONSTANTS FOR CONTINUOUS UNITY FEEDBACK SYSTEMS

Another criterion of the effectiveness of feedback in a *stable type l unity feedback system* is the *velocity (ramp) error constant*. It is a measure of the steady state error between the input and output of the system when the input is a unit ramp function.

*Definition 9.6:*        The **velocity error constant** $K_v$ of a stable type $l$ unity feedback system is defined as

$$K_v \equiv \lim_{s \to 0} sG(s) = \lim_{s \to 0} \frac{KB_1(s)}{s^{l-1} B_2(s)} = \begin{cases} 0 & \text{for } l = 0 \\[2mm] \dfrac{KB_1(0)}{B_2(0)} & \text{for } l = 1 \\[2mm] \infty & \text{for } l > 1 \end{cases} \qquad (9.12)$$

The steady state error of a stable type $l$ unity feedback system when the input is a unit ramp function is related to the velocity error constant by

$$e(\infty) = \lim_{t \to \infty} e(t) = \frac{1}{K_v} \qquad (9.13)$$

**EXAMPLE 9.17.**   The velocity error constant for a type 0 system is $K_v = 0$. Hence the steady state error is infinite.

**EXAMPLE 9.18.**   The velocity error constant for a type 1 system, $K_v = KB_1(0)/B_2(0)$, is finite. Therefore the steady state error is nonzero and finite.

**EXAMPLE 9.19.** The velocity error constant for a type 2 system is infinite. Therefore the steady state error is zero.

## 9.7  ACCELERATION ERROR CONSTANTS FOR CONTINUOUS UNITY FEEDBACK SYSTEMS

A third criterion of the effectiveness of feedback in a *stable type l unity feedback* system is the *acceleration* (*parabolic*) *error constant*. It is a measure of the steady state error of the system when the input is a unit parabolic function; that is, $r = t^2/2$ and $R = 1/s^3$.

***Definition 9.7:***      The **acceleration error constant** $K_a$ of a stable type $l$ unity feedback system is defined as

$$K_a \equiv \lim_{s \to 0} s^2 G(s) = \lim_{s \to 0} \frac{KB_1(s)}{s^{l-2}B_2(s)} = \begin{cases} 0 & \text{for} \quad l = 0,1 \\ \dfrac{KB_1(0)}{B_2(0)} & \text{for} \quad l = 2 \\ \infty & \text{for} \quad l > 2 \end{cases} \qquad (9.14)$$

The steady state error of a stable type $l$ unity feedback system when the input is a unit parabolic function is related to the acceleration error constant by

$$e(\infty) = \lim_{t \to \infty} e(t) = \frac{1}{K_a} \qquad (9.15)$$

**EXAMPLE 9.20.** The acceleration error constant for a type 0 system is $K_a = 0$. Hence the steady state error is infinite.

**EXAMPLE 9.21.** The acceleration error constant for a type 1 system is $K_a = 0$. Hence the steady state error is infinite.

**EXAMPLE 9.22.** The acceleration error constant for a type 2 system, $K_a = KB_1(0)/B_2(0)$, is finite. Hence the steady state error is nonzero and finite.

## 9.8  ERROR CONSTANTS FOR DISCRETE UNITY FEEDBACK SYSTEMS

The open-loop transfer function for a type $l$ discrete system can be written as

$$GH = \frac{K(z + z_1) \cdots (z + z_m)}{(z - 1)^l (z + p_1) \cdots (z + p_n)} = \frac{KB_1(z)}{(z - 1)^l B_2(z)}$$

where $l \geq 0$ and $-z_i$ and $-p_i$ are the nonunity zeros and poles of $GH$ in the $z$-plane.

All the results developed for continuous unity feedback systems in Sections 9.5 through 9.7 are the same for discrete systems with this open-loop transfer function.

## 9.9  SUMMARY TABLE FOR CONTINUOUS AND DISCRETE-TIME UNITY FEEDBACK SYSTEMS

In Table 9.1 the error constants are given in terms of $\alpha$, where $\alpha = 0$ for continuous systems, and $\alpha = 1$ for discrete-time systems. For continuous systems $T = 1$ in the steady state error.

**TABLE 9.1**

| Input | Unit Step | | Unit Ramp | | Unit Parabola | |
|---|---|---|---|---|---|---|
| System Type | $K_p$ | Steady State Error | $K_v$ | Steady State Error | $K_a$ | Steady State Error |
| Type 0 | $\dfrac{KB_1(\alpha)}{B_2(\alpha)}$ | $\dfrac{1}{1+K_p}$ | 0 | $\infty$ | 0 | $\infty$ |
| Type 1 | $\infty$ | 0 | $\dfrac{KB_1(\alpha)}{B_2(\alpha)}$ | $\dfrac{T}{K_v}$ | 0 | $\infty$ |
| Type 2 | $\infty$ | 0 | $\infty$ | 0 | $\dfrac{KB_1(\alpha)}{B_2(\alpha)}$ | $\dfrac{T^2}{K_a}$ |

## 9.10 ERROR CONSTANTS FOR MORE GENERAL SYSTEMS

The results of Sections 9.5 through 9.9 are only applicable to stable unity feedback linear systems. They can be readily extended, however, to more general stable linear systems. In Fig. 9-9, $T_d$ represents the transfer function of a desired (ideal) system, and $C/R$ represents the transfer function of the actual system (an approximation of $T_d$). $R$ is the input to both systems, and $E$ is the difference (the error) between the desired output and the actual output. For this more general system, three error constants are defined below and are related to the steady state error.

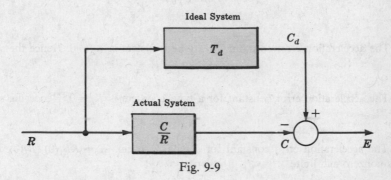

Fig. 9-9

***Definition 9.8:*** The **step error constant** $K_s$ is defined for continuous systems as

$$K_s \equiv \frac{1}{\lim\limits_{s \to 0} \left[ T_d - \dfrac{C}{R} \right]} \qquad (9.16)$$

The steady state error for the general system when the input is a unit step function is related to $K_s$ by

$$e(\infty) = \lim_{t \to \infty} e(t) = \frac{1}{K_s} \qquad (9.17)$$

***Definition 9.9:*** The **ramp error constant** $K_r$ is defined for continuous systems as

$$K_r \equiv \frac{1}{\lim\limits_{s \to 0} \dfrac{1}{s} \left[ T_d - \dfrac{C}{R} \right]} \qquad (9.18)$$

The steady state error for the general system when the input is a unit ramp function is related to $K_r$ by

$$e(\infty) = \lim_{t \to \infty} e(t) = \frac{1}{K_r} \qquad (9.19)$$

***Definition 9.10:***	The **parabolic error constant** $K_{pa}$ is defined for continuous systems as

$$K_{pa} \equiv \frac{1}{\lim\limits_{s \to 0} \dfrac{1}{s^2}\left[T_d - \dfrac{C}{R}\right]} \tag{9.20}$$

The steady state error for the general system when the input is a unit parabolic function is related to $K_{pa}$ by

$$e(\infty) = \lim_{t \to \infty} e(t) = \frac{1}{K_{pa}} \tag{9.21}$$

**EXAMPLE 9.23.**  The nonunity feedback system given in Fig. 9-10 has the transfer function $C/R = 2/(s^2 + 2s + 4)$. If the desired transfer function which $C/R$ approximates is $T_d = \frac{1}{2}$, then

$$T_d - \frac{C}{R} = \frac{s(s+2)}{2(s^2 + 2s + 4)}$$

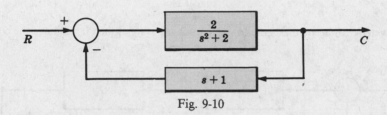

Fig. 9-10

Therefore

$$K_s = \frac{1}{\lim\limits_{s \to 0}\left[\dfrac{s(s+2)}{2(s^2+2s+4)}\right]} = \infty \qquad K_r = \frac{1}{\lim\limits_{s \to 0}\dfrac{1}{s}\left[\dfrac{s(s+2)}{2(s^2+2s+4)}\right]} = 4$$

$$K_{pa} = \frac{1}{\lim\limits_{s \to 0}\dfrac{1}{s^2}\left[\dfrac{s(s+2)}{2(s^2+2s+4)}\right]} = 0$$

**EXAMPLE 9.24.**  For the system of Example 9.23 the steady state errors due to a unit step input, a unit ramp input, and a unit parabolic input can be found using the results of that example. For a unit step input, $e(\infty) = 1/K_s = 0$. For a unit ramp input, $e(\infty) = 1/K_r = \frac{1}{4}$. For a unit parabolic input, $e(\infty) = 1/K_{pa} = \infty$.

To establish relationships between the general error constants $K_s$, $K_r$, and $K_{pa}$ and the error constants $K_p$, $K_v$, and $K_a$ for unity feedback systems, we let the actual system be a continuous unity feedback system and let the desired system have a unity transfer function. That is, we let

$$T_d = 1 \qquad \text{and} \qquad \frac{C}{R} = \frac{G}{1+G}$$

Therefore

$$K_s = \frac{1}{\lim\limits_{s \to 0}\left[\dfrac{1}{1+G}\right]} = 1 + \lim_{s \to 0} G(s) = 1 + K_p \tag{9.22}$$

$$K_r = \frac{1}{\lim\limits_{s \to 0}\left[\dfrac{1}{s}\left(\dfrac{1}{1+G}\right)\right]} = \lim_{s \to 0} sG(s) = K_v \tag{9.23}$$

$$K_{pa} = \frac{1}{\lim\limits_{s \to 0}\left[\dfrac{1}{s^2}\left(\dfrac{1}{1+G}\right)\right]} = \lim_{s \to 0} s^2 G(s) = K_a \tag{9.24}$$

# Solved Problems

## SYSTEM CONFIGURATIONS

**9.1.** A given plant has the transfer function $G_2$. A system is desired which includes $G_2$ as the output element and has a transfer function $C/R$. Show that, if no constraints (such as stability) are placed on the compensating elements, then such a system can be synthesized as either an open-loop or a unity feedback system.

     If the system can be synthesized as an open-loop system, then it will have the configuration given in Fig. 9-11, where $G_1$ is an unknown compensating element. The system transfer function is $C/R = G_1'G_2$, from which $G_1' = (C/R)/G_2$. This value for $G_1'$ permits synthesis of $C/R$ as an open-loop system.

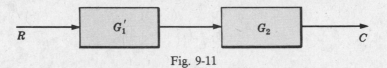

Fig. 9-11

     If the system can be synthesized as a unity feedback system, then it will have the configuration given in Fig. 9-12.

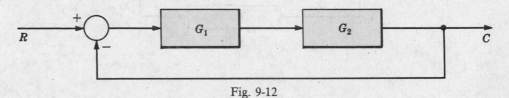

Fig. 9-12

The system transfer function is $C/R = G_1G_2/(1 + G_1G_2)$ from which

$$G_1 = \frac{1}{G_2}\left(\frac{C/R}{1 - C/R}\right)$$

This value for $G_1$ permits synthesis of $C/R$ as a unity feedback system.

**9.2.** Using the results of Problem 9.1, show how the system transfer function $C/R = 2/(s^2 + s + 2)$ which includes as its output element the plant $G_2 = 1/s(s + 1)$ can be synthesized as ($a$) an open-loop system, ($b$) a unity feedback system.

($a$)   For the open-loop system,

$$G_1' = \frac{C/R}{G_2} = \frac{2s(s + 1)}{s^2 + s + 2}$$

and the system block diagram is given in Fig. 9-13.

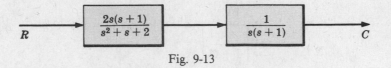

Fig. 9-13

($b$)   For the unity feedback system,

$$G_1 = \frac{1}{G_2}\left(\frac{C/R}{1 - C/R}\right) = s(s + 1)\left[\frac{2/(s^2 + s + 2)}{(s^2 + s + 2 - 2)/(s^2 + s + 2)}\right] = 2$$

and the system block diagram is given in Fig. 9-14.

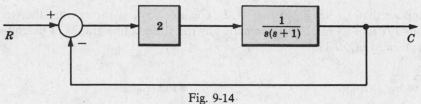

Fig. 9-14

## TRANSFER FUNCTION SENSITIVITY

**9.3.** The two systems given in Fig. 9-15 have the same transfer function when $K_1 = K_2 = 100$.

$$T_1 = \left( \frac{C}{R} \right)_1 \Bigg|_{\substack{K_1 = 100 \\ K_2 = 100}} = \frac{K_1 K_2}{1 + 0.0099 K_1 K_2} = 100$$

$$T_2 = \left( \frac{C}{R} \right)_2 \Bigg|_{\substack{K_1 = 100 \\ K_2 = 100}} = \left( \frac{K_1}{1 + 0.09 K_1} \right)\left( \frac{K_2}{1 + 0.09 K_2} \right) = 100$$

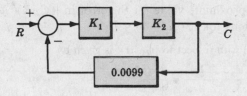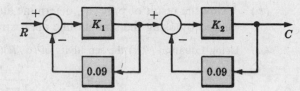

Fig. 9-15

Compare the sensitivities of these two systems with respect to parameter $K_1$ for nominal values $K_1 = K_2 = 100$.

For the first system, $T_1 = K_1 K_2 / [1 + K_1(0.0099 K_2)]$. Comparing this expression with Equation (9.6) yields $A_1 = 0$, $A_2 = K_2$, $A_3 = 1$, $A_4 = 0.0099 K_2$. Substituting these values into Equation (9.7), we obtain

$$S_{K_1}^{T_1} = \frac{K_1 K_2}{(1 + 0.0099 K_1 K_2)(K_1 K_2)} = \frac{1}{1 + 0.0099 K_1 K_2} = 0.01 \quad \text{for} \quad K_1 = K_2 = 100$$

For the second system,

$$T_2 = \left( \frac{K_1}{1 + 0.09 K_1} \right)\left( \frac{K_2}{1 + 0.09 K_2} \right) = \frac{K_1 K_2}{1 + 0.09 K_1 + 0.09 K_2 + 0.0081 K_1 K_2}$$

Comparing this expression with Equation (9.6) yields $A_1 = 0$, $A_2 = K_2$, $A_3 = 1 + 0.09 K_2$, $A_4 = 0.09 + 0.0081 K_2$. Substituting these values into Equation (9.7), we have

$$S_{K_1}^{T_2} = \frac{K_1 K_2 (1 + 0.09 K_2)}{(1 + 0.09 K_1)(1 + 0.09 K_2)(K_1 K_2)} = \frac{1}{1 + 0.09 K_1} = 0.1 \quad \text{for} \quad K_1 = K_2 = 100$$

A 10% variation in $K_1$ will approximately produce a 0.1% variation in $T_1$ and a 1% variation in $T_2$. Thus the second system $T_2$ is 10 times more sensitive to variations in $K_1$ than is the first system $T_1$.

**9.4.** The closed-loop system given in Fig. 9-16 is defined in terms of the frequency response function of the feedforward element $G(j\omega)$.

$$\frac{C}{R}(j\omega) = \frac{G(j\omega)}{1 + G(j\omega)}$$

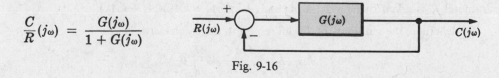

Fig. 9-16

Suppose that $G(j\omega) = 1/(j\omega + 1)$. In Chapter 15 it is shown that the frequency response functions $1/(j\omega + 1)$ can be approximated by the straight line graphs of magnitude and phase of $G(j\omega)$ given in Fig. 9-17.

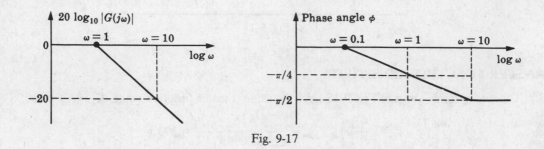

Fig. 9-17

At $\omega = 1$ the true values of $20\log_{10}|G(j\omega)|$ and $\phi$ are $-3$ and $-\pi/4$, respectively. For $\omega = 1$, find:

(a)   The sensitivity of $|(C/R)(j\omega)|$ with respect to $|G(j\omega)|$.

(b)   Using the result of part (a), determine an approximate value for the error in $|(C/R)(j\omega)|$ caused by using the straight-line approximations for $1/(j\omega + 1)$.

(a)   Using Equation (9.8) the sensitivity of $(C/R)(j\omega)$ with respect to $|G(j\omega)|$ is given by

$$S_{|G(j\omega)|}^{(C/R)(j\omega)} = \frac{1}{1 + G(j\omega)} = \frac{1}{2 + j\omega} = \frac{2 - j\omega}{4 + \omega^2}$$

Since $|G(j\omega)|$ is real,

$$S_{|G(j\omega)|}^{|(C/R)(j\omega)|} = \text{Re}\, S_{|G(j\omega)|}^{(C/R)(j\omega)} = \frac{2}{4 + \omega^2}$$

For $\omega = 1$, $S_{|G(j\omega)|}^{|(C/R)(j\omega)|} = 0.4$.

(b)   For $\omega = 1$, the exact value of $|G(j\omega)|$ is $|G(j\omega)| = 1/\sqrt{2} = 0.707$. The approximate value taken from the graph is $|G(j\omega)| = 1$. Then the percentage error in the approximation is $100(1 - 0.707)/0.707 = 41.4\%$. The approximate percentage error in $|(C/R)(j\omega)|$ is $41.4\, S_{|G(j\omega)|}^{|(C/R)(j\omega)|} = 16.6\%$.

**9.5.**   Show that the sensitivities of $T(k) = |T(k)|e^{j\phi_T}$, the magnitude $|T(k)|$, and the phase angle $\phi_T$ with respect to parameter $k$ are related by

$$S_k^{T(k)} = S_k^{|T(k)|} + j\phi_T \cdot S_k^{\phi_T} \qquad [\text{Equation } (9.5)]$$

Using Equation (9.2),

$$S_k^{T(k)} = \frac{d\ln T(k)}{d\ln k} = \frac{d\ln\left[|T(k)|e^{j\phi_T}\right]}{d\ln k} = \frac{d\left[\ln|T(k)| + j\phi_T\right]}{d\ln k}$$

$$= \frac{d\ln|T(k)|}{d\ln k} + j\frac{d\phi_T}{d\ln k} = \frac{d\ln|T(k)|}{d\ln k} + j\phi_T\frac{d\ln\phi_T}{d\ln k} = S_k^{|T(k)|} + j\phi_T S_k^{\phi_T}$$

Note that if $k$ is real, then $S_k^{|T(k)|}$ and $S_k^{\phi_T}$ are both real, and

$$S_k^{|T(k)|} = \text{Re}\, S_k^{T(k)} \qquad \phi_T S_k^{\phi_T} = \text{Im}\, S_k^{T(k)}$$

**9.6.**   Show that the sensitivity of the transfer function $T = (A_1 + kA_2)/(A_3 + kA_4)$ with respect to the parameter $k$ is given by $S_k^T = k(A_2A_3 - A_1A_4)/(A_3 + kA_4)(A_1 + kA_2)$.

By definition, the sensitivity of $T$ with respect to the parameter $k$ is

$$S_k^T = \frac{d\ln T}{d\ln k} = \frac{dT}{dk} \cdot \frac{k}{T}$$

Now

$$\frac{dT}{dk} = \frac{A_2(A_3 + kA_4) - A_4(A_1 + kA_2)}{(A_3 + kA_4)^2} = \frac{A_2 A_3 - A_1 A_4}{(A_3 + kA_4)^2}$$

Thus

$$S_k^T = \frac{A_2 A_3 - A_1 A_4}{(A_3 + kA_4)^2} \cdot \frac{k(A_3 + kA_4)}{A_1 + kA_2} = \frac{k(A_2 A_3 - A_1 A_4)}{(A_3 + kA_4)(A_1 + kA_2)}$$

**9.7.** Consider the system of Example 9.6 with the addition of a load disturbance and a noise input as shown in Fig. 9-18. Show that the feedback controller improves the output sensitivity to the noise input and the load disturbance.

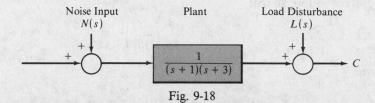

Fig. 9-18

For the open-loop system, the output due to the noise input and load disturbance is

$$C(s) = L(s) + \frac{1}{(s + 1)(s + 3)} N(s)$$

independent of the action of open-loop controller. For the closed-loop system,

$$C(s) = \frac{(s + 1)(s + 3)}{s^2 + 4s + 5} L(s) + \frac{1}{s^2 + 4s + 5} N(s)$$

For low frequencies the closed-loop system attenuates both the load disturbance and the noise input, compared to the open-loop system. In particular, the closed-loop system has steady state or d.c. gain:

$$C(0) = \frac{3}{5} L(0) + \frac{1}{5} N(0)$$

while the open-loop system has

$$C(0) = L(0) + \frac{1}{3} N(0)$$

At high frequencies these gains are approximately equal.

## SYSTEM OUTPUT SENSITIVITY IN THE TIME DOMAIN

**9.8.** For the system defined by

$$\dot{x} = A(\mathbf{p})x + B(\mathbf{p})u$$
$$y = C(\mathbf{p})x$$

show that the matrix of output sensitivities

$$\left[ \frac{\partial y_i}{\partial p_j} \right]$$

is determined by solution of the differential equations

$$\dot{x} = Ax + u \qquad\qquad (9.25)$$

$$\dot{V} = AV + \frac{\partial A}{\partial \mathbf{p}} x + \frac{\partial B}{\partial \mathbf{p}} u \qquad\qquad (9.26)$$

with

$$\left[\frac{\partial y_i}{\partial p_j}\right] = CV + \frac{\partial C}{\partial \mathbf{p}}\mathbf{x} \qquad (9.27)$$

where

$$V \equiv [v_{ij}] \equiv \frac{\partial \mathbf{x}}{\partial \mathbf{p}} \equiv \left[\frac{\partial x_i}{\partial p_j}\right]$$

that is, $V$ is the matrix of sensitivity functions. The derivative of the sensitivity function $v_{ij}$ is given by

$$\dot{v}_{ij} = \frac{d}{dt}\left(\frac{\partial x_i}{\partial p_j}\right)$$

Assuming the state variables have continuous derivatives, we can interchange the order of total and partial differentiation, so that

$$\dot{v}_{ij} = \frac{\partial}{\partial p_j}\left(\frac{dx_i}{dt}\right)$$

In matrix form,

$$\dot{V} = \frac{\partial \dot{\mathbf{x}}}{\partial \mathbf{p}} = \frac{\partial}{\partial \mathbf{p}}[A\mathbf{x} + B\mathbf{u}] = \frac{\partial A}{\partial \mathbf{p}}\mathbf{x} + A\frac{\partial \mathbf{x}}{\partial \mathbf{p}} + \frac{\partial B}{\partial \mathbf{p}}\mathbf{u}$$

Since $V = \partial \mathbf{x}/\partial \mathbf{p}$, we have

$$\dot{V} = AV + \frac{\partial A}{\partial \mathbf{p}}\mathbf{x} + \frac{\partial B}{\partial \mathbf{p}}\mathbf{u}$$

Then

$$\frac{\partial \mathbf{y}}{\partial \mathbf{p}} = \frac{\partial C\mathbf{x}}{\partial \mathbf{p}} = \frac{\partial C}{\partial \mathbf{p}}\mathbf{x} + C\frac{\partial \mathbf{x}}{\partial \mathbf{p}} = CV + \frac{\partial C}{\partial \mathbf{p}}\mathbf{x}$$

Note that, in the above equations, the partial derivative of a matrix with respect to the vector $\mathbf{p}$ is understood to generate a series of matrices, each one of which, when multiplied by $\mathbf{x}$, generates a column in the resulting matrix. That is, $(\partial A/\partial \mathbf{p})\mathbf{x}$ is a matrix with $j$th column $(\partial A/\partial p_j)\mathbf{x}$. This is easily verified by writing out all the scalar equations explicitly and differentiating term by term.

## SYSTEMS CLASSIFICATION BY TYPE

**9.9.** The canonical feedback system is represented by Fig. 9-19.

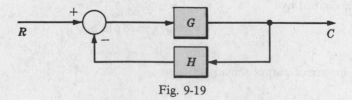

Fig. 9-19

Classify this system according to type if

(a)    $G = \dfrac{1}{s}$      $H = 1$

(b)    $G = \dfrac{5}{s(s+3)}$      $H = \dfrac{s+1}{s+2}$

(c)   $G = \dfrac{2}{s^2 + 2s + 5}$    $H = s + 5$

(d)   $G = \dfrac{24}{(2s + 1)(4s + 1)}$    $H = \dfrac{4}{4s(3s + 1)}$

(e)   $G = \dfrac{4}{s(s + 3)}$    $H = \dfrac{1}{s}$

(a)   $GH = \dfrac{1}{s}$; *type 1*

(b)   $GH = \dfrac{5(s + 1)}{s(s + 2)(s + 3)}$; *type 1*

(c)   $GH = \dfrac{2(s + 5)}{s^2 + 2s + 5}$; *type 0*

(d)   $GH = \dfrac{96}{4s(2s + 1)(3s + 1)(4s + 1)} = \dfrac{1}{s\left(s + \frac{1}{2}\right)\left(s + \frac{1}{3}\right)\left(s + \frac{1}{4}\right)}$; *type 1*

(e)   $GH = \dfrac{4}{s^2(s + 3)}$; *type 2*

**9.10.**   Classify the system given in Fig. 9-20 by type.

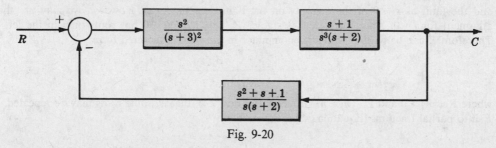

Fig. 9-20

The open-loop transfer function of this system is

$$GH = \frac{s^2(s + 1)(s^2 + s + 1)}{s^4(s + 2)^2(s + 3)^2} = \frac{(s + 1)(s^2 + s + 1)}{s^2(s + 2)^2(s + 3)^2}$$

Therefore it is a *type 2* system.

## ERROR CONSTANTS AND STEADY STATE ERRORS

**9.11.**   Show that the steady state error $e(\infty)$ of a stable type *1* unity feedback system when the input is a unit step function is related to the position error constant by

$$e(\infty) = \lim_{t \to \infty} e(t) = \frac{1}{1 + K_p}$$

The error ratio (Definition 7.5) for a unity negative feedback system is given by Equation (7.4) with $H = 1$, that is, $E/R = 1/(1 + G)$. For $R = 1/s$, $E = (1/s)(1/(1 + G))$. From the Final Value Theorem, we obtain

$$e(\infty) = \lim_{s \to 0} sE(s) = \lim_{s \to 0} \left( \frac{s}{s[1 + G(s)]} \right) = \frac{1}{1 + \lim_{s \to 0} G(s)} = \frac{1}{1 + K_p}$$

where we have used the definition $K_p \equiv \lim_{s \to 0} G(s)$.

**9.12.** Show that the steady state error $e(\infty)$ of a stable type $l$ unity feedback system with a unit ramp function input is related to the velocity error constant by $e(\infty) = \lim_{t \to \infty} e(t) = 1/K_v$.

We have $E/R = 1/(1 + G)$, and $E = (1/s^2)(1/(1 + G))$ for $R = 1/s^2$. Since $G = KB_1(s)/s^l B_2(s)$ by Definition 9.4,

$$E = \frac{1}{s^2}\left[\frac{s^l B_2(s)}{s^l B_2(s) + KB_1(s)}\right]$$

For $l > 0$, we have

$$sE(s) = \frac{B_2(s)}{sB_2(s) + KB_1(s)/s^{l-1}}$$

where $l - 1 \geq 0$. Now we can use the Final Value Theorem, as was done in the previous problem, because the condition for the application of this theorem is satisfied. That is, for $l > 0$ we have

$$e(\infty) = \lim_{s \to 0} sE(s) = \begin{cases} 0 & \text{for } l > 1 \\ \dfrac{B_2(0)}{KB_i(0)} & \text{for } l = 1 \end{cases}$$

$B_1(0)$ and $B_2(0)$ are nonzero and finite by Definition 9.4; hence the limit exists (i.e., it is finite).

We cannot evoke the Final Value Theorem for the case $l = 0$ because

$$sE(s)|_{l=0} = \frac{1}{s}\left[\frac{B_2(s)}{B_2(s) + KB_1(s)}\right]$$

and the limit as $s \to 0$ of the quantity on the right does not exist. However, we may use the following argument for $l = 0$. Since the system is stable, $B_2(s) + KB_1(s) = 0$ has roots only in the left half-plane. Therefore $E$ can be written with its denominator in the general factored form:

$$E = \frac{B_2(s)}{s^2\prod_{i=1}^{r}(s + p_i)^{n_i}}$$

where $\text{Re}(p_i) > 0$ and $\sum_{i=1}^{r} n_i = n - a$ (see Definition 9.4), that is, some roots may be repeated. Expanding $E$ into partial fractions [Equation (4.10a)], we obtain

$$E = \frac{c_{20}}{s^2} + \frac{c_{10}}{s} + \sum_{i=1}^{r}\sum_{k=1}^{n_i}\frac{c_{ik}}{(s + p_i)^k}$$

where $b_n$ in Equation (4.10a) is zero because the degree of the denominator is greater than that of the numerator ($m < n$). Inverting $E(s)$ (Section 4.8), we get

$$e(t) = c_{20}t + c_{10} + \sum_{i=1}^{r}\sum_{k=1}^{n_i}\frac{c_{ik}}{(k-1)!}t^{k-1}e^{-p_i t}$$

Since $\text{Re}(p_i) > 0$ and $c_{20}$ and $c_{10}$ are finite nonzero constants ($E$ is a rational algebraic expression), then

$$e(\infty) = \lim_{t \to \infty} e(t) = \lim_{t \to \infty}(c_{20}t) + c_{10} = \infty$$

Collecting results, we have

$$e(\infty) = \begin{cases} \infty & \text{for } l = 0 \\ \dfrac{B_2(0)}{KB_1(0)} & \text{for } l = 1 \\ 0 & \text{for } l > 1 \end{cases}$$

Equivalently,

$$\frac{1}{e(\infty)} = \begin{cases} 0 & \text{for } l = 0 \\ \dfrac{KB_1(0)}{B_2(0)} & \text{for } l = 1 \\ \infty & \text{for } l > 1 \end{cases}$$

These three values for $1/e(\infty)$ define $K_v$; thus

$$e(\infty) = \frac{1}{K_v}$$

**9.13.** For Fig. 9-21 find the position, velocity, and acceleration error constants.

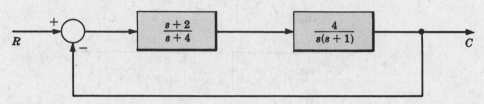

Fig. 9-21

Position error constant:

$$K_p = \lim_{s \to 0} G(s) = \lim_{s \to 0} \frac{4(s+2)}{s(s+1)(s+4)} = \infty$$

Velocity error constant:

$$K_v = \lim_{s \to 0} sG(s) = \lim_{s \to 0} \frac{4(s+2)}{(s+1)(s+4)} = 2$$

Acceleration error constant:

$$K_a = \lim_{s \to 0} s^2 G(s) = \lim_{s \to 0} \frac{4s(s+2)}{(s+1)(s+4)} = 0$$

**9.14.** For the system in Problem 9.13, find the steady state error for ($a$) a unit step input, ($b$) a unit ramp input, ($c$) a unit parabolic input.

($a$)   The steady state error for a unit step input is given by $e(\infty) = 1/(1 + K_p)$. Using the result of Problem 9.13 yields $e(\infty) = 1/(1 + \infty) = 0$.

($b$)   The steady state error for a unit ramp input is given by $e(\infty) = 1/K_v$. Again using the result of Problem 9.13, we get $e(\infty) = \frac{1}{2}$.

($c$)   The steady state error for a unit parabolic input is given by $e(\infty) = 1/K_a$. Then $e(\infty) = 1/0 = \infty$.

**9.15.** Figure 9-22 approximately represents a differentiator. Its transfer function is $C/R = Ks/[s(\tau s + 1) + K]$. Note that $\lim_{\tau \to 0 \ K \to \infty} C/R = s$, that is, $C/R$ is a pure differentiator in the limit. Find the step, ramp, and parabolic error constants for this system, where the ideal system $T_d$ is assumed to be a differentiator.

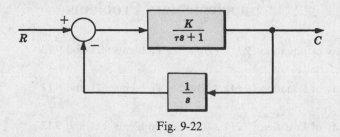

Fig. 9-22

Using the notation of Section 9.10, $T_d = s$ and $T_d - C/R = s^2(\tau s + 1)/[s(\tau s + 1) + K]$. Applying Definitions 9.8, 9.9, and 9.10 yields

$$K_s = \frac{1}{\lim\limits_{s \to 0}\left[T_d - \dfrac{C}{R}\right]} = \frac{1}{\lim\limits_{s \to 0}\left[\dfrac{s^2(\tau s + 1)}{s(\tau s + 1) + K}\right]} = \infty$$

$$K_r = \frac{1}{\lim\limits_{s \to 0}\dfrac{1}{s}\left[T_d - \dfrac{C}{R}\right]} = \frac{1}{\lim\limits_{s \to 0}\left[\dfrac{s(\tau s + 1)}{s(\tau s + 1) + K}\right]} = \infty$$

$$K_{pa} = \frac{1}{\lim\limits_{s \to 0}\dfrac{1}{s^2}\left[T_d - \dfrac{C}{R}\right]} = \frac{1}{\lim\limits_{s \to 0}\left[\dfrac{\tau s + 1}{s(\tau s + 1) + K}\right]} = K$$

**9.16.** Find the steady state value of the difference (error) between the outputs of a pure differentiator and the approximate differentiator of the previous problem for (a) a unit step input, (b) a unit ramp input, (c) a unit parabolic input.

   From Problem 9.15, $K_s = \infty$, $K_r = \infty$, and $K_{pa} = K$.

   (a)   The steady state error for a unit step input is $e(\infty) = 1/K_s = 0$.
   (b)   The steady state error for a unit ramp input is $e(\infty) = 1/K_r = 0$.
   (c)   The steady state error for a unit parabolic input is $e(\infty) = 1/K_{pa} = 1/K$.

**9.17.** Given the stable type 2 unity feedback system shown in Fig. 9-23, find (a) the position, velocity, and acceleration error constants, (b) the steady state error when the input is $R = \dfrac{3}{s} - \dfrac{1}{s^2} + \dfrac{1}{2s^3}$.

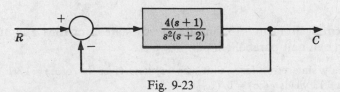

Fig. 9-23

   (a)   Using the last row of Table 9.1 (*type 2* systems), the error constants are $K_p = \infty$, $K_v = \infty$, $K_a = (4)(1)/2 = 2$.

   (b)   The steady state errors for unit step, unit ramp, and unit parabolic inputs are obtained from the same row of the table and are given by: $e_1(\infty) = 0$ for a unit step; $e_2(\infty) = 0$ for a unit ramp; $e_3(\infty) = \frac{1}{2}$ for a unit parabola.

   Since the system is linear, the errors can be superimposed. Thus the steady state error when the input is $R = \dfrac{3}{s} - \dfrac{1}{s^2} + \dfrac{1}{2s^3}$ is given by $e(\infty) = 3e_1(\infty) - e_2(\infty) + \frac{1}{2}e_3(\infty) = \frac{1}{4}$.

# Supplementary Problems

**9.18.** Prove the validity of Equation (*9.17*). (*Hint:* See Problems 9.11 and 9.12.)

**9.19.** Prove the validity of Equation (*9.19*). (*Hint:* See Problems 9.11 and 9.12.)

**9.20.** Prove the validity of Equation (*9.21*). (*Hint:* See Problems 9.11 and 9.12.)

**9.21.**  Determine the sensitivity of the system in Problem 7.9, to variations in each of the parameters $K_1$, $K_2$ and $p$ individually.

**9.22.**  Generate an expression, in terms of the sensitivities determined in Problem 9.21, which relates the total variation in the transfer function of the system in Problem 7.9 to variations in $K_1$, $K_2$, and $p$.

**9.23.**  Show that the steady state error $e(\infty)$ of a stable type $l$ unity feedback system with a unit parabolic input is related to the acceleration error constant by $e(\infty) = \lim_{t \to \infty} e(t) = 1/K_a$. (*Hint:* See Problem 9.12.)

**9.24.**  Verify Equations (*9.26*) and (*9.27*) by performing all differentiations on the full set of scalar simultaneous differential equations making up Equation (*9.25*).

# Answers to Some Supplementary Problems

**9.21.**  $S_{K_1}^{C/R} = \dfrac{s+p}{s+p-K_1K_2}$     $S_{K_2}^{C/R} = \dfrac{K_1K_2}{s+p-K_1K_2}$     $S_p^{C/R} = \dfrac{-p}{s+p-K_1K_2}$

**9.22.**  $\Delta\dfrac{C}{R} = \dfrac{(s+p)\Delta K_1 + (K_1K_2)\Delta K_2 - p\Delta p}{s+p-K_1K_2}$

# Chapter 10

## Analysis and Design of Feedback Control Systems: Objectives and Methods

### 10.1 INTRODUCTION

The basic concepts, mathematical tools, and properties of feedback control systems have been presented in the first nine chapters. Attention is now focused on our major goal: *analysis and design* of feedback control systems.

The methods presented in the next eight chapters are linear techniques, applicable to linear models. However, under appropriate circumstances, one or more can also be used for some nonlinear control system problems, thereby generating approximate designs when the particular method is sufficiently robust. Techniques for solving control system problems represented by nonlinear models are introduced in Chapter 19.

This chapter is mainly devoted to making explicit the objectives and to describing briefly the methodology of analysis and design. It also includes one digital system design approach, in Section 10.8, that can be considered independently of the several approaches developed in subsequent chapters.

### 10.2 OBJECTIVES OF ANALYSIS

The three predominant objectives of feedback control systems analysis are the determination of the following system characteristics:

1. The degree or extent of system stability
2. The steady state performance
3. The transient performance

Knowing whether a system is absolutely stable or not is insufficient information for most purposes. If a system is stable, we usually want to know how close it is to being unstable. We need to determine its *relative stability*.

In Chapter 3 we learned that the complete solution of the equations describing a system may be split into two parts. The first, the steady state response, is that part of the complete solution which does not approach zero as time approaches infinity. The second, the transient response, is that part of the complete solution which approaches zero (or decays) as time approaches infinity. We shall soon see that there is a strong correlation between relative stability and transient response of feedback control systems.

### 10.3 METHODS OF ANALYSIS

The general procedure for analyzing a linear control system is the following:

1. Determine the equations or transfer function for each system component.
2. Choose a scheme for representing the system (block diagram or signal flow graph).
3. Formulate the system model by appropriately connecting the components (blocks, or nodes and branches).
4. Determine the system response characteristics.

Several methods are available for determining the response characteristics of linear systems. Direct solution of the system equations may be employed to find the steady state and transient solutions

(Chapters 3 and 4). This technique can be cumbersome for higher than second-order systems, and relative stability is difficult to study in the time-domain.

Four primarily graphical methods are available to the control system analyst which are simpler and more direct than time-domain methods for practical linear models of feedback control systems. They are:

1.   The Root-Locus Method
2.   Bode-Plot Representations
3.   Nyquist Diagrams
4.   Nichols Charts

The latter three are frequency-domain techniques. All four are considered in detail in Chapters 13, 15, 11, and 17, respectively.

## 10.4   DESIGN OBJECTIVES

The basic goal of control system design is meeting *performance specifications*. Performance specifications are the constraints put on system response characteristics. They may be stated in any number of ways. Generally they take two forms:

1.   Frequency-domain specifications (pertinent quantities expressed as functions of frequency)
2.   Time-domain specifications (in terms of time response)

The desired system characteristics may be prescribed in either or both of the above forms. In general, they specify three important properties of dynamic systems:

1.   Speed of response
2.   Relative stability
3.   System accuracy or allowable error

**Frequency-domain specifications** for both continuous and discrete-time systems are often stated in one or more of the following seven ways. To maintain generality, we define a **unified open-loop frequency response function** $GH(\omega)$:

$$GH(\omega) \equiv \begin{cases} GH(j\omega) & \text{for continuous systems} \\ GH(e^{j\omega T}) & \text{for discrete-time systems} \end{cases} \qquad (10.1)$$

**1.   Gain Margin**

Gain margin, a measure of relative stability, is defined as the magnitude of the reciprocal of the open-loop transfer function, evaluated at the frequency $\omega_\pi$ at which the phase angle (see chapter 6) is $-180°$. That is,

$$\text{gain margin} \equiv \frac{1}{|GH(\omega_\pi)|} \qquad (10.2)$$

where $\arg GH(\omega_\pi) = -180° = -\pi$ radians and $\omega_\pi$ is called the **phase crossover** frequency.

**2.   Phase Margin** $\phi_{PM}$

Phase margin $\phi_{PM}$, a measure of relative stability, is defined as 180° plus the phase angle $\phi_1$ of the open-loop transfer function at unity gain. That is,

$$\phi_{PM} \equiv [180 + \arg GH(\omega_1)] \text{ degrees} \qquad (10.3)$$

where $|GH(\omega_1)| = 1$ and $\omega_1$ is called the **gain crossover** frequency.

**EXAMPLE 10.1.** The gain and phase margins of a typical continuous-time feedback control system are illustrated in Fig. 10-1.

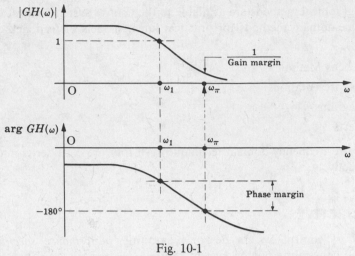

Fig. 10-1

## 3. Delay Time $T_d$

Delay time $T_d$, interpreted as a frequency-domain specification, is a measure of the speed of response, and is given by

$$T_d(\omega) = -\frac{d\gamma}{d\omega} \qquad (10.4)$$

where $\gamma = \arg(C/R)$. The average value of $T_d(\omega)$ over the frequencies of interest is usually specified.

## 4. Bandwidth (BW)

Roughly speaking, the bandwidth of a system was defined in Chapter 1 as that range of frequencies over which the system responds satisfactorily.

Satisfactory performance is determined by the application and the characteristics of the particular system. For example, audio amplifiers are often compared on the basis of their bandwidth. An ideal high-fidelity audio amplifier has a *flat frequency response* from 20 to 20,000 Hz. That is, it has a passband or bandwidth of 19,980 Hz (usually rounded off to 20,000 Hz). Flat frequency response means that the *magnitude ratio* of output to input is essentially constant over the bandwidth. Hence signals in the audio spectrum are faithfully reproduced by a 20,000-Hz bandwidth amplifier. The magnitude ratio is the absolute value of the system frequency response function.

The frequency response of a high-fidelity audio amplifier is shown in Fig. 10-2. The magnitude ratio is 0.707 of, or approximately 3 db below, its maximum at the **cutoff frequencies**

$$f_{c1} = 20 \text{ Hz} \qquad f_{c2} = 20,000 \text{ Hz}$$

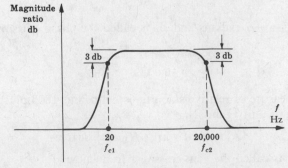

Fig. 10-2

"db" is the abbreviation for **decibel**, defined by the following equation:

$$db \equiv 20 \log_{10}(\text{magnitude ratio}) \qquad (10.5)$$

Often the bandwidth of a system is defined as that range of frequencies over which the magnitude ratio does not differ by more than $-3$ db from is value at a specified frequency. But not always. In general, the precise meaning of bandwidth is made clear by the problem description. In any case, bandwidth is generally a measure of the speed of response of a system.

The gain crossover frequency $\omega_1$ defined in Equation (10.3) is often a good approximation for the bandwidth of a closed-loop system.

The notion of signal sampling, and of *uniform sampling time T*, were introduced in Chapters 1 and 2 (especially in Section 2.4), for systems containing both discrete-time and continuous-time signals, and both types of elements, including samplers, hold devices and computers. The value of $T$ is a design parameter for such systems and its choice is governed by both accuracy and cost considerations. The *sampling theorem* [9,10] provides an upper bound on $T$, by requiring the sampling rate to be at least twice that of the highest frequency component $f_{\max}$ of the sampled signal, that is, $T \leq \dfrac{1}{2f_{\max}}$. In practice, we might use the cutoff frequency $f_{c2}$ (as in Fig. 10-2) for $f_{\max}$, and a practical rule-of-thumb might be to choose $T$ in the range $\dfrac{1}{10f_{c2}} \leq T \leq \dfrac{1}{6f_{c2}}$. Other design requirements, however, may require even smaller $T$ values. On the other hand, the largest value of $T$ consistent with the specifications usually yields the lowest cost for system components.

## 5.  Cutoff Rate

The cutoff rate is the frequency rate at which the magnitude ratio decreases beyond the cutoff frequency $\omega_c$. For example, the cutoff rate may be specified as 6 db/octave. An octave is a factor-of-two change in frequency.

## 6.  Resonance Peak $M_p$

The resonance peak $M_p$, a measure of relative stability, is the maximum value of the magnitude of the closed-loop frequency response. That is,

$$M_p \equiv \max_{\omega} \left| \frac{C}{R} \right| \qquad (10.6)$$

## 7.  Resonant Frequency $\omega_p$

The resonant frequency $\omega_p$ is the frequency at which $M_p$ occurs.

**Mathcad EXAMPLE 10.2.**  The bandwidth BW, cutoff frequency $\omega_c$, resonance peak $M_p$, and resonant frequency $\omega_p$ for an underdamped second-order continuous system are illustrated in Fig. 10-3.

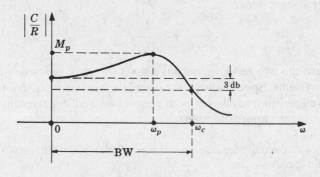

Fig. 10-3

**Time-domain specifications** are customarily defined in terms of unit step, ramp, and parabolic responses. Each response has a steady state and a transient component.

*Steady state performance*, in terms of steady state error, is a measure of system accuracy when a specific input is applied. Figures of merit for steady state performance are, for example, the error constants $K_p$, $K_v$, and $K_a$ defined in Chapter 9.

*Transient performance* is often described in terms of the unit step function response. Typical specifications are:

## 1. Overshoot

The overshoot is the maximum difference between the transient and steady state solutions for a unit step input. It is a measure of relative stability and is often represented as a percentage of the final value of the output (steady state solution).

The following four specifications are measures of the speed of response.

## 2. Delay Time $T_d$

The delay time $T_d$, interpreted as a time-domain specification, is often defined as the time required for the response to a unit step input to reach 50% of its final value.

## 3. Rise Time $T_r$

The rise time $T_r$ is customarily defined as the time required for the response to a unit step input to rise from 10 to 90 percent of its final value.

## 4. Settling Time $T_s$

The settling time $T_s$ is most often defined as the time required for the response to a unit step input to reach and remain within a specified percentage (frequently 2 or 5%) of its final value.

## 5. Dominant Time Constant

The dominant time constant $\tau$, an alternative measure for settling time, is often defined as the time constant associated with the term that dominates the transient response.

The dominant time constant is defined in terms of the exponentially decaying character of the transient response. For example, for first and second-order underdamped continuous systems, the transient terms have the form $Ae^{-\alpha t}$ and $Ae^{-\alpha t}\cos(\omega_d t + \phi)$, respectively ($\alpha > 0$). In each case, the decay is governed by $e^{-\alpha t}$. The time constant $\tau$ is defined as the time at which the exponent $-\alpha t = -1$, that is, when the exponential reaches 37% of its initial value. Hence $\tau = 1/\alpha$.

For continuous feedback control systems of order higher than two, the dominant time constant can sometimes be estimated from the time constant of an underdamped second-order system which approximates the higher system. Since

$$\tau \le \frac{1}{\zeta \omega_n} \qquad (10.7)$$

$\zeta$ and $\omega_n$ (Chapter 3) are the two most significant figures of merit, defined for second-order but often useful for higher-order systems. Specifications are often given in terms of $\zeta$ and $\omega_n$.

This concept is developed more fully for both continuous and discrete-time systems in Chapter 14, in terms of dominant pole-zero approximations.

Mathcad **EXAMPLE 10.3.** The plot of the unit step response of an underdamped continuous second-order system in Fig. 10-4 illustrates time-domain specifications.

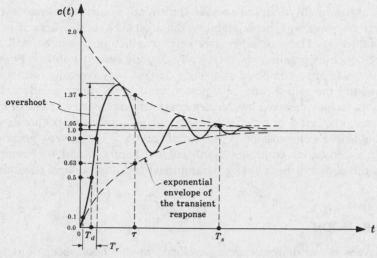

Fig. 10-4

## 10.5  SYSTEM COMPENSATION

We assume first that $G$ and $H$ are fixed configurations of components over which the designer has no control. To meet performance specifications for feedback control systems, appropriate *compensation* components (sometimes called *equalizers*) are normally introduced into the system. Compensation components may consist of either passive or active elements, several of which were discussed in Chapters 2 and 6. They may be introduced into the forward path (*cascade compensation*), or the feedback path (*feedback compensation*), as shown in Fig. 10-5:

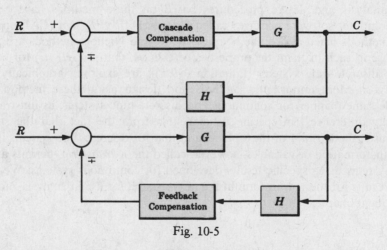

Fig. 10-5

Feedback compensation may also occur in minor feedback loops (Fig. 10-6).

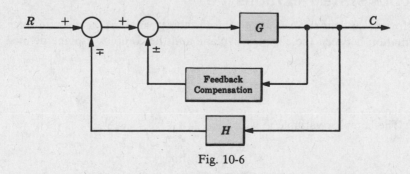

Fig. 10-6

Compensators are normally designed so that the overall system (continuous or discrete) has an acceptable transient response, and hence stability characteristics, and a desired or acceptable steady state accuracy (Chapter 9). These objectives are often conflicting, because small steady state errors usually require large open-loop gains, which typically degrade system stability. For this reason, simple compensator elements are often combined in a single design. They typically consist of combinations of components that modify the gain $K$ and/or time constants $\tau$, or otherwise add zeros or poles to $GH$. *Passive* compensators include passive physical elements such as resistive-capacitive networks, to modify $K$ ($K < 1$), time constants, zeros, or poles; *lag*, *lead*, and *lag-lead* networks are examples (Chapter 6). The most common active compensator is the amplifier ($K > 1$). A very general one is the PID (proportional-integral-derivative) controller discussed in Chapter 2 and 6 (Examples 2.14 and 6.7), commonly used in the design of both analog (continuous) and discrete-time (digital) systems.

## 10.6   DESIGN METHODS

Design by analysis is the design scheme developed in this book, because it is generally a more practical approach, with the exception that direct design of digital systems, discussed in Section 10.8, is a true synthesis technique. The previously mentioned analysis methods, reiterated below, are applied to design in Chapters 12, 14, 16, and 18.

1.   Nyquist Plot (Chapter 12)

2.   Root-Locus (Chapter 14)

3.   Bode Plot (Chapters 16)

4.   Nichols Chart (Chapter 18)

Control system analysis and design procedures based on these methods have been automated in special-purpose computer software packages called *Computer-Aided Design* (CAD) packages.

Of the four methods listed above, the Nyquist, Bode, and Nichols methods are *frequency response* techniques, because in each of them the properties of $GH(\omega)$, that is, $GH(j\omega)$ for continuous systems or $GH(e^{j\omega T})$ for discrete-time systems [Equation (*10.1*)], are explored graphically as a function of angular frequency $\omega$. More importantly, analysis and design using these methods is performed in fundamentally the same manner for continuous and discrete-time systems, as illustrated in subsequent chapters. The only differences (in specific details) stem from the fact that the stability region for continuous systems is the left half of the $s$-plane, and that for discrete-time systems is the unit circle in the $z$-plane. A transformation of variables, however, called the *w-transform*, permits analysis and design of discrete-time systems using specific results developed for continuous systems. We present the major features and the results for the $w$-transform in the next section, for use in analysis and design of control systems in subsequent chapters.

## 10.7   THE *w*-TRANSFORM FOR DISCRETE-TIME SYSTEMS ANALYSIS AND DESIGN USING CONTINUOUS SYSTEM METHODS

The $w$-transform was defined in Chapter 5 for stability analysis of discrete-time systems. It is a bilinear transformation between the complex $w$-plane and the complex $z$-plane defined by the pair:

$$w = \frac{z-1}{z+1} \qquad z = \frac{1+w}{1-w} \qquad (10.8)$$

where $z = \mu + j\nu$. The complex variable $w$ is defined as

$$w = \mathrm{Re}\, w + j\, \mathrm{Im}\, w \qquad (10.9)$$

The following relations among these variables are useful in the analysis and design of discrete-time control systems:

1. $$\operatorname{Re} w = \frac{\mu^2 + \nu^2 - 1}{\mu^2 + \nu^2 + 2\mu + 1} \tag{10.10}$$

2. $$\operatorname{Im} w = \frac{2\nu}{\mu^2 + \nu^2 + 2\mu + 1} \tag{10.11}$$

3. $\qquad\qquad$ If $|z| < 1$, then $\operatorname{Re} w < 0$ $\qquad\qquad\qquad$ (10.12)

4. $\qquad\qquad$ If $|z| = 1$, then $\operatorname{Re} w = 0$ $\qquad\qquad\qquad$ (10.13)

5. $\qquad\qquad$ If $|z| > 1$, then $\operatorname{Re} w > 0$ $\qquad\qquad\qquad$ (10.14)

6. $\qquad$ On the unit circle of the $z$-plane:

$$z = e^{j\omega T} = \cos \omega T + j \sin \omega T \tag{10.15}$$

$$\mu^2 + \nu^2 = \cos^2 \omega T + \sin^2 \omega T = 1 \tag{10.16}$$

$$w = j\frac{\nu}{\mu + 1} \tag{10.17}$$

Thus the region inside the unit circle in the $z$-plane maps into the left half of the $w$-plane (LHP); the region outside the unit circle maps into the right half of the $w$-plane (RHP); and the unit circle maps onto the imaginary axis of the $w$-plane. Also, rational functions of $z$ map into rational functions of $w$.

For these reasons, absolute and relative stability properties of discrete systems can be determined using methods developed for continuous systems in the $s$-plane. Specifically, for frequency response analysis and design of discrete-time systems in the $w$-plane, we generally treat the $w$-plane as if it were the $s$-plane. However, we must account for distortions in certain mappings, particularly angular frequency, when interpreting the results.

From Equation (10.17), we define an angular frequency $\omega_w$ on the imaginary axis in the $w$-plane by

$$\omega_w \equiv \frac{\nu}{\mu + 1} \tag{10.18}$$

This new angular frequency $\omega_w$ in the $w$-plane is related to the true angular frequency $\omega$ in the $z$-plane by

$$\omega_w = \tan \frac{\omega T}{2} \qquad \text{or} \qquad \omega = \frac{2}{T}\tan^{-1}\omega_w \tag{10.19}$$

The following properties of $\omega_w$ are useful in plotting functions for frequency response analysis in the $w$-plane:

1. $\qquad\qquad$ If $\omega = 0$, then $\omega_w = 0$ $\qquad\qquad\qquad$ (10.20)

2. $\qquad\qquad$ If $\omega \to \dfrac{\pi}{T}$, then $\omega_w \to +\infty$ $\qquad\qquad$ (10.21)

3. $\qquad\qquad$ If $\omega \to -\dfrac{\pi}{T}$, then $\omega_w \to -\infty$ $\qquad\qquad$ (10.22)

4. $\qquad$ The range $-\dfrac{\pi}{T} < \omega < \dfrac{\pi}{T}$ is mapped into the range $-\infty < \omega_w < +\infty$ $\qquad$ (10.23)

### Algorithm for Frequency Response Analysis and Design Using the $w$-Transform

The procedure is summarized as follows:

1. Substitute $(1 + w)/(1 - w)$ for $z$ in the open-loop transfer function $GH(z)$:

$$GH(z)\big|_{z=(1+w)/(1-w)} \equiv GH'(w) \tag{10.24}$$

2.   Generate frequency response curves, that is, Nyquist Plots, Bode Plots, etc., for

$$GH'(w)|_{w=j\omega_w} \equiv GH'(j\omega_w) \qquad (10.25)$$

3.   Analyze relative stability properties of the system in the $w$-plane (as if it were the $s$-plane). For example, determine gain and phase margins, crossover frequencies, the closed-loop frequency response, the bandwidth, or any other desired frequency-response-related characteristics.

4.   Transform $w$-plane critical frequencies (values of $\omega_w$) determined in Step 3 into angular frequencies (values of $\omega$) in the true frequency domain ($z$-plane), using Equation (10.19).

5.   If this is a design problem, design appropriate compensators to modify $GH'(j\omega_w)$ to satisfy performance specifications.

This algorithm is developed further and applied in Chapters 15 through 18.

**EXAMPLE 10.4.**   The open-loop transfer function

$$GH(z) = \frac{(z+1)^2/100}{(z-1)(z+\frac{1}{3})(z+\frac{1}{2})} \qquad (10.26)$$

is transformed into the $w$-domain by substituting $z = (1+w)/(1-w)$ in the expression for $GH(z)$, which yields

$$GH'(w) = \frac{-6(w-1)/100}{w(w+2)(w+3)} \qquad (10.27)$$

Relative stability analysis of $GH'(w)$ is postponed until Chapter 15.

## 10.8   ALGEBRAIC DESIGN OF DIGITAL SYSTEMS, INCLUDING DEADBEAT SYSTEMS

When digital computers or microprocessors are components of a discrete-time system, compensators can be readily implemented in software or firmware, thereby facilitating direct design of the system by algebraic solution for the transfer function of the compensator that satisfies given design objectives. For example, suppose we wish to construct a system having a given closed-loop transfer function $C/R$, which might be defined by requisite closed-loop characteristics such as bandwidth, steady state gain, response time, etc. Then, given the plant transfer function $G_2(z)$, the required forward loop compensator $G_1(z)$ can be determined from the relation for the closed-loop transfer function of the canonical system given in Section 7.5:

$$\frac{C}{R} = \frac{G_1 G_2}{1 + G_1 G_2 H} \qquad (10.28)$$

Then the required compensator is determined by solving for $G_1(z)$:

$$G_1 = \frac{C/R}{G_2(1 - HC/R)} \qquad (10.29)$$

**EXAMPLE 10.5.**   The unity feedback ($H \equiv 1$) system in Fig. 10-7, with $T = 0.1$-sec uniform and synchronous sampling, is required to have a steady state gain $(C/R)(1) = 1$ and a rise time $T_r$ of 2 sec or less.

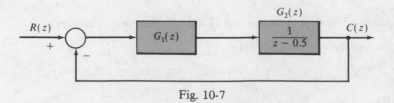

Fig. 10-7

The simplest $C/R$ that satisfies the requirements is $(C/R) = 1$. However, the required compensator would be

$$G_1 = \frac{\dfrac{C}{R}}{G_2\left(1 - \dfrac{C}{R}\right)} = \frac{1}{\dfrac{1}{z - 0.5}(1 - 1)} = \frac{z - 0.5}{0}$$

which has infinite gain, a zero at $z = 0.5$, and no poles, which is unrealizable. For realizability (Section 6.6), $G_1$ must have at least as many poles as zeros. Consequently, even with cancellation of the poles and zeros of $G_2$ by zeros and poles of $G_1$, $C/R$ must contain at least $n - m$ poles, where $n$ is the number of poles and $m$ is the number of zeros of $G_2$.

The simplest realizable $C/R$ has the form:

$$\frac{C}{R} = \frac{K}{z - a}$$

As shown in Problem 10.10, the rise time for a first-order discrete-time system, like the one given by $C/R$ above, is

$$T_r \leq \frac{T \ln\frac{1}{9}}{\ln a}$$

Solving for $a$, we get

$$a = \left[\frac{1}{9}\right]^{T_r/T} = \left[\frac{1}{9}\right]^{20} = 0.8959$$

Then

$$\frac{C}{R} = \frac{K}{z - a} = \frac{K}{z - 0.8959}$$

and, for the steady state gain $(C/R)(1)$ to be 1, $K = 1 - 0.8959 = 0.1041$. Therefore the required compensator is

$$G_1 = \frac{\dfrac{C}{R}}{G_2\left(1 - \dfrac{C}{R}\right)} = \frac{\dfrac{0.1041}{z - 0.8959}}{\dfrac{1}{z - 0.5}\left(1 - \dfrac{0.1041}{z - 0.8959}\right)} = \frac{0.1041(z - 0.5)}{z - 1}$$

We see that $G_1$ has added a pole to $G_1G_2$ at $z = 1$, making the system type 1. This is due to the requirement that the steady state gain equal 1.

*Deadbeat systems* are a class of discrete-time systems that can be readily designed using the direct approach described above. By definition, the closed-loop *transient* response of a **deadbeat system** has finite length, that is, it becomes zero, and remains zero, after a finite number of sample times. In response to a step input, the output of such a system is constant at each sample time after a finite period. This is termed a **deadbeat response**.

**EXAMPLE 10.6.**   For a unity feedback system with forward transfer function

$$G_2(z) = \frac{K_1(z + z_1)}{(z + p_1)(z + p_2)}$$

introduction of a feedforward compensator with

$$G_1(z) = \frac{(z + p_1)(z + p_2)}{(z - K_1)(z + z_1)}$$

results in the closed-loop transfer function:

$$\frac{C}{R} = \frac{G_1G_2}{1 + G_1G_2} = \frac{K_1}{z}$$

The impulse response of this system is $c(0) = K_1$ and $c(k) = 0$ for $k > 0$. The step response is $c(0) = 0$ and $c(k) = K_1$ for $k > 0$.

In general, systems can be designed to exhibit a deadbeat response with a transient response $n - m$ samples long, where $m$ is the number of zeros and $n$ is the number of poles of the plant. However, to avoid *intersample ripple* (periodic or aperiodic variations) in mixed continuous/discrete-time systems, where $G_2(z)$ has a continuous input and/or output, the zeros of $G_2(z)$ should not be cancelled by the compensator as in Example 10.5. The transient response in these cases is a minimum of $n$ samples in length and the closed loop transfer function has $n$ poles at $z = 0$.

**EXAMPLE 10.7.**   For a system with

$$G_2(z) = \frac{K(z + 0.5)}{(z - 0.2)(z - 0.4)}$$

let

$$G_1(z) = \frac{(z - 0.2)(z - 0.4)}{(z + a)(z + b)}$$

Then

$$\frac{C}{R} = \frac{G_1 G_2}{1 + G_1 G_2} = \frac{K(z + 0.5)}{(z + a)(z + b) + K(z + 0.5)}$$

$$= \frac{K(z + 0.5)}{z^2 + (a + b + K)z + ab + 0.5K}$$

For a deadbeat response, we choose

$$\frac{C}{R} \equiv \frac{K(z + 0.5)}{z^2}$$

and therefore

$$a + b + K \equiv 0$$
$$ab + 0.5K \equiv 0$$

There are many possible solutions for $a$, $b$, and $K$ and one is $a = 0.3$, $b = -0.75$, and $K = 0.45$.

If it is required that the closed-loop system be type $l$, it is necessary that $G_1(z)G_2(z)$ contain $l$ poles at $z = 1$. If $G_2(z)$ has the required number of poles, they should be retained, that is, not cancelled by zeros of $G_1(z)$. If $G_2(z)$ does not have all the required poles at $z = 1$, they can be added in $G_1(z)$.

**EXAMPLE 10.8.**   For the system with

$$G_2(z) = \frac{K}{z - 1}$$

suppose a type 2 closed-loop system with deadbeat response is desired. This can be achieved with a compensator of the form:

$$G_1(z) = \frac{z + a}{z - 1}$$

which adds a pole at $z = 1$. Then

$$\frac{C}{R} = \frac{G_1 G_2}{1 + G_1 G_2} = \frac{K(z + a)}{(z - 1)^2 + K(z + a)} = \frac{K(z + a)}{z^2 + (K - 2)z + 1 + Ka}$$

If a deadbeat response is desired, we must have

$$\frac{C}{R} = \frac{K(z + a)}{z^2}$$

and therefore $K - 2 = 0$ and $1 + Ka = 0$, giving $K = 2$ and $a = -0.5$.

# Solved Problems

**10.1.** The graph of Fig. 10-8 represents the input-output characteristic of a controller-amplifier for a feedback control system whose other components are linear. What is the linear range of $e(t)$ for this system?

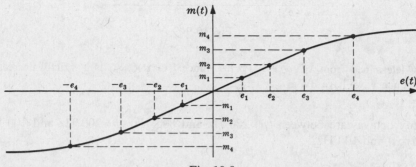

Fig. 10-8

The amplifier-controller operates linearly over the approximate range $-e_3 \leq e \leq e_3$.

**10.2.** Determine the gain margin for the system in which $GH(j\omega) = 1/(j\omega + 1)^3$.

Writing $GH(j\omega)$ in polar form, we have

$$GH(j\omega) = \frac{1}{(\omega^2 + 1)^{3/2}} \underline{/-3\tan^{-1}\omega} \qquad \arg GH(j\omega) = -3\tan^{-1}\omega$$

Then $-3\tan^{-1}\omega_\pi = -\pi$, $\omega_\pi = \tan(\pi/3) = 1.732$. Hence, by Equation (10.2), gain margin $= 1/|GH(j\omega_\pi)| = 8$.

**10.3.** Determine the phase margin for the system of Problem 10.2.

We have

$$|GH(j\omega)| = \frac{1}{(\omega^2 + 1)^{3/2}} = 1$$

only when $\omega = \omega_1 = 0$. Therefore

$$\phi_{\text{PM}} = 180° + (-3\tan^{-1}0) = 180° = \pi \text{ radians}$$

**10.4.** Determine the average value of $T_d(\omega)$ over the frequency range $0 \leq \omega \leq 10$ for $C/R = j\omega/(j\omega + 1)$. $T_d(\omega)$ is given by Equation (10.4).

$$\gamma = \arg\frac{C}{R}(j\omega) = \frac{\pi}{2} - \tan^{-1}\omega \qquad \text{and} \qquad T_d(\omega) = \frac{-d\gamma}{d\omega} = \frac{d}{d\omega}[\tan^{-1}\omega] = \frac{1}{1 + \omega^2}$$

Therefore

$$\text{Avg } T_d(\omega) = \frac{1}{10}\int_0^{10} \frac{d\omega}{1 + \omega^2} = 0.147 \text{ sec}$$

**10.5.** Determine the bandwidth for the system with transfer function $(C/R)(s) = 1/(s + 1)$.

We have

$$\left|\frac{C}{R}(j\omega)\right| = \frac{1}{\sqrt{\omega^2 + 1}}$$

A sketch of $|(C/R)(j\omega)|$ versus $\omega$ is given in Fig. 10-9.

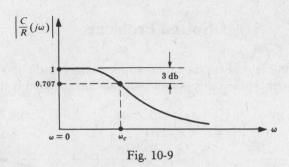

Fig. 10-9

$\omega_c$ is determined from $1/\sqrt{\omega_c^2 + 1} = 0.707$. Since $|(C/R)(j\omega)|$ is a strictly decreasing function of positive frequency, we have BW $= \omega_c = 1$ rad.

**10.6.** How many octaves are between (a) 200 Hz and 800 Hz, (b) 200 Hz and 100 Hz, (c) 10,048 rad/sec (rps) and 100 Hz?

(a)  Two octaves.

(b)  One octave.

(c)  $f = \omega/2\pi = 10{,}048/2\pi = 1600$ Hz. Hence there are four octaves between 10,048 rps and 100 Hz.

**10.7.** Determine the resonance peak $M_p$ and the resonant frequency $\omega_p$ for the system whose transfer function is $(C/R)(s) = 5/(s^2 + 2s + 5)$.

$$\left| \frac{C}{R}(j\omega) \right| = \frac{5}{|-\omega^2 + 2j\omega + 5|} = \frac{5}{\sqrt{\omega^4 - 6\omega^2 + 25}}$$

Setting the derivative of $|(C/R)(j\omega|$ equal to zero, we get $\omega_p = \pm\sqrt{3}$. Therefore

$$M_p = \max_\omega \left| \frac{C}{R}(j\omega) \right| = \left| \frac{C}{R}(j\sqrt{3}) \right| = \frac{5}{4}$$

**10.8.** The output in response to a unit step function input for a particular continuous control system is $c(t) = 1 - e^{-t}$. What is the delay time $T_d$?

The output is given as a function of time. Therefore, the time-domain definition of $T_d$ presented in Section 10.4 is applicable. The final value of the output is $\lim_{t \to \infty} c(t) = 1$. Hence $T_d$ (at 50% of the final value) is the solution of $0.5 = 1 - e^{-T_d}$, and is equal to $\log_e(2)$, or 0.693.

**10.9.** Find the rise time $T_r$ for $c(t) = 1 - e^{-t}$.

At 10% of the final value, $0.1 = 1 - e^{-t_1}$; hence $t_1 = 0.104$ sec. At 90% of the final value, $0.9 = 1 - e^{-t_2}$; thus $t_2 = 2.302$ sec. Then $T_r = 2.302 - 0.104 = 2.198$ sec.

**10.10.** Determine the rise time of the first-order discrete system

$$P(z) = (1 - a)/(z - a) \text{ with } |a| < 1.$$

For a step input, the output transform is

$$Y(z) = P(z)U(z) = \frac{(1 - a)z}{(z - 1)(z - a)}$$

and the time response is $y(k) = 1 - a^k$ for $k = 0, 1, \ldots$. Since $y(\infty) = 1$, the rise time $T_r$ is the time required for this unit step response to go from 0.1 to 0.9. Since the sampled response may not have the exact values 0.1 and 0.9, we must find the sampled values that bound these values. Thus, for the lower value, $y(k) \leq 0.1$, or $1 - a^k \leq 0.1$ and therefore $a^k \geq 0.9$. Similarly for $y(k + T_r/T) = 1 - a^{k + T_r/T} \geq 0.9$, $a^{k + T_r/T} \leq 0.1$.

Dividing the two expressions, we get

$$\frac{a^{k+T_r/T}}{a^k} \leq \frac{1}{9}$$

or

$$a^{T_r/T} \leq \frac{1}{9}$$

Then, by taking logarithms of both sides, we get

$$T_r \leq \frac{T \ln\frac{1}{9}}{\ln a}$$

**10.11.** Verify the six properties of the $w$-transform in Section 10.7, Equations $(10.10)$ through $(10.17)$.

From $w = (z-1)/(z+1)$ and $z = \mu + j\nu$,

$$w = \frac{\mu + j\nu - 1}{\mu + j\nu + 1} = \frac{(\mu - 1 + j\nu)(\mu + 1 - j\nu)}{(\mu + 1 + j\nu)(\mu + 1 - j\nu)} = \left(\frac{\mu^2 + \nu^2 - 1}{\mu^2 + \nu^2 + 2\mu + 1}\right) + j\left(\frac{2\nu}{\mu^2 + \nu^2 + 2\mu + 1}\right)$$

Thus

1. $$\text{Re } w = \frac{\mu^2 + \nu^2 - 1}{\mu^2 + \nu^2 + 2\mu + 1} \equiv \sigma_w$$

2. $$\text{Im } w = \frac{2\nu}{\mu^2 + \nu^2 + 2\mu + 1} \equiv \omega_w$$

3. $\quad |z| < 1$ means $\mu^2 + \nu^2 < 1$, which implies $\sigma_w < 0$

4. $\quad |z| = 1$ means $\mu^2 + \nu^2 = 1$, which implies $\sigma_w = 0$

5. $\quad |z| > 1$ means $\mu^2 + \nu^2 > 1$, which implies $\sigma_w > 0$

The sixth property follows from elementary trigonometric identities.

**10.12.** Show that the transformed angular frequency $\omega_w$ is related to the real frequency $\omega$ by Equation $(10.19)$.

From Problem 10.11, $|z| = 1$ also implies that $w = j[\nu/(\mu+1)] \equiv j\omega_w$ [Equation $(10.17)$]. But $|z| = 1$ implies that $z = e^{j\omega T} \equiv \cos \omega T + j \sin \omega T = \mu + j\nu$ [Equation $(10.15)$]. Therefore

$$\omega_w = \frac{\sin \omega T}{\cos \omega T + 1}$$

Finally, substituting the following half-angle identities of trigonometry into the last expression:

$$2 \sin\left(\frac{\omega T}{2}\right) \cos\left(\frac{\omega T}{2}\right) = \sin \omega T$$

$$\cos^2\left(\frac{\omega T}{2}\right) - \sin^2\left(\frac{\omega T}{2}\right) = \cos \omega T$$

$$\cos^2\left(\frac{\omega T}{2}\right) + \sin^2\left(\frac{\omega T}{2}\right) = 1$$

we have

$$\omega_w = \frac{2 \sin\left(\frac{\omega T}{2}\right) \cos\left(\frac{\omega T}{2}\right)}{2 \cos^2\left(\frac{\omega T}{2}\right)} = \frac{\sin\left(\frac{\omega T}{2}\right)}{\cos\left(\frac{\omega T}{2}\right)} = \tan\left(\frac{\omega T}{2}\right)$$

**10.13.** For the uniformly and synchronously sampled system given in Fig. 10-10, determine $G_1(z)$ so that the system is type 1 with a deadbeat response.

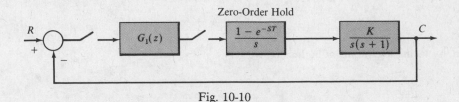

Fig. 10-10

The forward loop $z$-transform, assuming *fictitious* sampling of the output $c(t)$ (see Section 6.8), is determined from Equation (6.9):

$$G_2(z) = \frac{z-1}{z} \mathcal{Z} \left\{ \mathcal{L}^{-1} \left( \frac{G(s)}{s} \right) \Big|_{t=kT} \right\} = \frac{K_1(z+z_1)}{(z-1)(z-e^{-T})}$$

where

$$K_1 \equiv K(T + e^{-T} - 1) \quad \text{and} \quad z_1 \equiv \frac{1 - e^{-T} - Te^{-T}}{T + e^{-T} - 1}$$

Let $G_1(z)$ have the form $G_1(z) = (z - e^{-T})/(z+b)$. Then, if we also assume a fictitious sampler at the input $r(t)$, we can determine the closed-loop $z$-domain transfer function:

$$\frac{C}{R} = \frac{G_1 G_2}{1 + G_1 G_2} = \frac{K_1(z+z_1)}{(z-1)(z+b) + K_1(z+z_1)}$$

$$= \frac{K_1(z+z_1)}{z^2 + (b-1+K_1)z - b + K_1 z_1}$$

For a deadbeat response, $b - 1 + K_1 = 0$ ($b = 1 - K_1$) and $-b + K_1 z_1 = 0$ ($-1 + K_1 + K_1 z_1 = 0$). Then

$$K_1 = \frac{1}{1 + z_1}$$

and

$$b = 1 - K_1 = \frac{z_1}{1 + z_1}$$

Since $K_1 = K(T + e^{-T} - 1)$,

$$K = \frac{K_1}{T + e^{-T} - 1} = \frac{1}{(1 + z_1)(T + e^{-T} - 1)} = \frac{1}{T(1 - e^{-T})}$$

For this system, with continuous input and output signals, $(C/R)(z)$ determined above gives the closed-loop input-output relationship at the sampling times only.

# Supplementary Problems

**10.14.** Determine the phase margin for $GH = 2(s+1)/s^2$.

**10.15.** Find the bandwidth for $GH = 60/s(s+2)(s+6)$ for the closed-loop system.

**10.16.** Calculate the gain and phase margin for $GH = 432/s(s^2 + 13s + 115)$.

**10.17.** Calculate the phase margin and bandwidth for $GH = 640/s(s+4)(s+16)$ for the closed-loop system.

# Answers to Supplementary Problems

**10.14.**  $\phi_{PM} = 65.5°$

**10.15.**  BW = 3 rad/sec

**10.16.**  Gain margin = 3.4, phase margin = 65°

**10.17.**  $\phi_{PM} = 17°$, BW = 8.8 rad/sec

# Chapter 11

# Nyquist Analysis

## 11.1 INTRODUCTION

Nyquist analysis, a frequency response method, is essentially a graphical procedure for determining absolute and relative stability of closed-loop control systems. Information about stability is available directly from a graph of the open-loop frequency response function $GH(\omega)$, once the feedback system has been put into canonical form.

Nyquist methods are applicable to both continuous and discrete-time control systems, and the methodological development for Nyquist analysis is presented here for both types of systems, with some emphasis given to continuous systems, for pedagogical purposes.

There are several reasons why the Nyquist method may be chosen to determine information about system stability. The methods of Chapter 5 (Routh, Hurwitz, etc.) are often inadequate because, with few exceptions, they can only be used for determining *absolute* stability, and are only applicable to systems whose characteristic equation is a *finite polynomial* in $s$ or $z$. For example, when a signal is delayed by $T$ seconds somewhere in the loop of a continuous system, exponential terms of the form $e^{-Ts}$ appear in the characteristic equation. The methods of Chapter 5 can be applied to such systems if $e^{-Ts}$ is approximated by a few terms of the power series

$$e^{-Ts} = 1 - Ts + \frac{T^2 s^2}{2!} - \frac{T^3 s^3}{3!} + \cdots$$

but this technique yields only *approximate* stability information. The Nyquist method handles systems with time delays without the necessity of approximations, and hence yields *exact* results about both absolute and relative stability of the system.

Nyquist techniques are also useful for obtaining information about transfer functions of components or systems from experimental frequency response data. The Polar Plot (Section 11.5) may be directly graphed from sinusoidal steady state measurements on the components making up the open-loop transfer function. This feature is very useful in the determination of system stability characteristics when transfer functions of loop components are not available in analytic form, or when physical systems are to be tested and evaluated experimentally.

In the next several sections we present the mathematical preliminaries and techniques necessary for generating Polar Plots and Nyquist Stability Plots of feedback control systems, and the mathematical basis and properties of the Nyquist Stability Criterion. The remaining sections of this chapter deal with the interpretation and uses of Nyquist analysis for the determination of *relative* stability and evaluation of the closed-loop frequency response.

## 11.2 PLOTTING COMPLEX FUNCTIONS OF A COMPLEX VARIABLE

A real function of a real variable is easily graphed on a single set of coordinate axes. For example, the real function $f(x)$, $x$ real, is easily plotted in rectangular coordinates with $x$ as the abscissa and $f(x)$ as the ordinate. A complex function of a complex variable, such as the transfer function $P(s)$ with $s = \sigma + j\omega$, cannot be plotted on a single set of coordinates.

The complex variable $s = \sigma + j\omega$ depends on two independent quantities, the real and imaginary parts of $s$. Hence $s$ cannot be represented by a line. The complex function $P(s)$ also has real and imaginary parts. It too cannot be graphed in a single dimension. Similarly, the complex variable $z = \mu + j\nu$ and discrete-time system complex transfer functions $P(z)$ cannot be graphed in one dimension.

In general, in order to plot $P(s)$ with $s = \sigma + j\omega$, two two-dimensional graphs are required. The first is a graph of $j\omega$ versus $\sigma$ called the **s-plane**, the same set of coordinates as those used for plotting pole-zero maps in Chapter 4. The second is the imaginary part of $P(s)$ (Im $P$) versus the real part of $P(s)$ (Re $P$) called the **$P(s)$-plane**. The corresponding coordinate planes for discrete-time systems are the **z-plane** and the **$P(z)$-plane**.

The correspondence between points in the two planes is called a **mapping** or **transformation**. For example, points in the s-plane are *mapped* into points of the $P(s)$-plane by the function $P$ (Fig. 11-1).

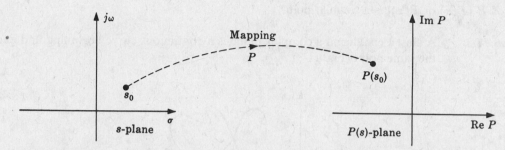

Fig. 11-1

In general, only a very specific locus of points in the s-plane (or the z-plane) is mapped into the $P(s)$-plane [or the $P(z)$-plane]. For Nyquist Stability Plots this locus is called the *Nyquist Path*, the subject of Section 11.7.

For the special case $\sigma = 0$, $s = j\omega$, the s-plane degenerates into a line, and $P(j\omega)$ may be represented in a $P(j\omega)$-plane with $\omega$ as a parameter. *Polar Plots* are constructed in the $P(j\omega)$-plane from this line ($s = j\omega$) in the s-plane.

**EXAMPLE 11.1.** Consider the complex function $P(s) = s^2 + 1$. The point $s_0 = 2 + j4$ is mapped into the point $P(s_0) = P(2 + j4) = (2 + j4)^2 + 1 = -11 + j16$ (Fig. 11-2).

Fig. 11-2

## 11.3  DEFINITIONS

The following definitions are required in subsequent sections.

***Definition 11.1:***     If the *derivative* of $P$ at $s_0$ defined by

$$\frac{dP}{ds}\bigg|_{s=s_0} \equiv \lim_{s \to s_0} \left[ \frac{P(s) - P(s_0)}{s - s_0} \right]$$

exists at all points in a region of the s-plane, that is, if the limit is finite and unique, then $P$ is **analytic** in that region [same definition for $P(z)$ in the z-plane, with $z$ replacing $s$ and $z_0$ replacing $s_0$].

Transfer functions of practical physical systems (those considered in this book) are analytic in the finite $s$-plane (or finite $z$-plane) except at the poles of $P(s)$ [or poles of $P(z)$]. In subsequent developments, when there is no danger of ambiguity, and when a given statement applies to both $P(s)$ and $P(z)$, then $P(s)$ or $P(z)$ may be abbreviated as $P$ with no argument.

**Definition 11.2:**    A point at which $P$ [$P(s)$ or $P(z)$] is not analytic is a **singular point** or **singularity** of $P$ [$P(s)$ or $P(z)$].

A *pole* of $P$ [$P(s)$ or $P(z)$] is a singular point.

**Definition 11.3:**    A **closed contour** in a complex plane is a continuous curve beginning and ending at the same point (Fig. 11-3).

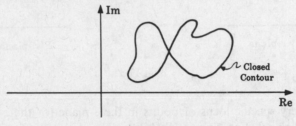

Fig. 11-3

**Definition 11.4:**    All points to the right of a contour as it is traversed in a prescribed direction are said to be **enclosed** by it (Fig. 11-4).

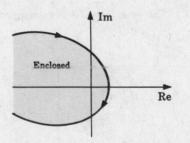

Fig. 11-4

**Definition 11.5:**    A *clockwise* (CW) traverse around a contour is defined as the **positive direction** (Fig. 11-5).

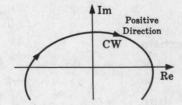 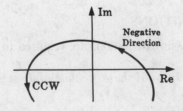

Fig. 11-5

**Definition 11.6:**    A closed contour in the $P$-plane is said to make $n$ **positive encirclements** of the origin if a radial line drawn from the origin to a point on the $P$ curve rotates in a clockwise (CW) direction through $360n$ degrees in completely traversing the closed path. If the path is traversed in a counterclockwise (CCW) direction, a **negative encirclement** is obtained. The **total number of encirclements** $N_0$ is equal to the CW minus the CCW encirclements.

**EXAMPLE 11.2.** The $P$-plane contour in Fig. 11-6 encircles the origin once. That is, $N_0 = 1$. Beginning at point $a$, we rotate a radial line from the origin to the contour in a CW direction to point $c$. The angle subtended is $+270°$. From $c$ to $d$ the angle increases, then decreases, and the sum total is $0°$. From $d$ to $e$ and back to $d$ again, the angle swept out by the radial line is again $0°$. $d$ to $c$ is $0°$ and $c$ to $a$ is clearly $+90°$. Hence the total angle is $270° + 90° = 360°$. Therefore $N_0 = 1$.

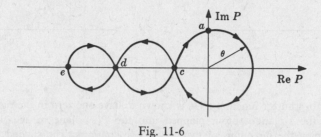

Fig. 11-6

## 11.4   PROPERTIES OF THE MAPPING $P(s)$ or $P(z)$

All mappings $P$ [$P(s)$ or $P(z)$] considered in the remainder of this chapter have the following properties.

1.  $P$ is a *single-valued function*. That is, every point in the $s$-plane (or the $z$-plane) maps into one and only one point in the $P$-plane.

2.  $s$-plane ($z$-plane) contours avoid singular points of $P$.

3.  $P$ is *analytic* except possibly at a finite number of points (singularities) in the $s$-plane (or the $z$-plane).

4.  Every closed contour in the $s$-plane (or the $z$-plane) maps into a closed contour in the $P$-plane.

5.  $P$ is a *conformal mapping*. This means that the direction of and the angle between any two intersecting curves at their point of intersection in the $s$-plane (or the $z$-plane) are preserved by the mapping of these curves into the $P$-plane.

6.  The mapping $P$ obeys the *principle of arguments*. That is, the *total number of encirclements $N_0$ of the origin* made by a closed $P$ contour in the $P$-plane, mapped from a closed $s$-plane (or $z$-plane) contour, is equal to the number of zeros $Z_0$ minus the number of poles $P_0$ of $P$ enclosed by the the $s$-plane (or $z$-plane) contour. That is,

$$N_0 = Z_0 - P_0 \qquad\qquad (11.1)$$

7.  If the origin is *enclosed* by the $P$ contour, then $N_0 > 0$. If the origin is *not enclosed* by the $P$ contour, then $N_0 \le 0$. That is,

$$\text{enclosed} \Rightarrow N_0 > 0$$
$$\text{not enclosed} \Rightarrow N_0 \le 0$$

The *sign of $N_0$* is easily determined by shading the region to the right of the contour in the prescribed direction. If the origin falls in a shaded region, $N_0 > 0$; if not, $N_0 \le 0$.

**EXAMPLE 11.3.** The principle of conformal mapping is illustrated in Fig. 11-7. Curves $C_1$ and $C_2$ are mapped into $C_1'$ and $C_2'$. The angle between the tangents to these curves at $s_0$ and $P(s_0)$ is equal to $\alpha$, and the curves turn right at $s_0$ and at $P(s_0)$, as indicated by the arrows in both graphs.

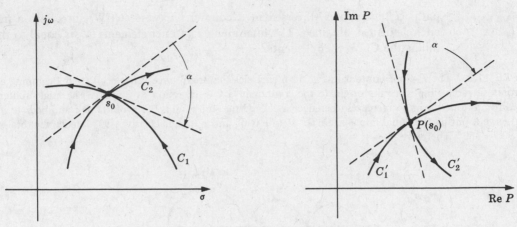

Fig. 11-7

**EXAMPLE 11.4.** A certain transfer function $P(s)$ is known to have one zero in the right half of the $s$-plane, and this zero is enclosed by the $s$-plane contour mapped into the $P(s)$-plane in Fig. 11-8. Points $s_1, s_2, s_3$ and $P(s_1), P(s_2), P(s_3)$ determine the directions of their respective contours. The shaded region to the right of the $P(s)$-plane contour indicates that $N_0 \leq 0$, since the origin does not lie in the shaded region. But, clearly, the $P(s)$ contour encircles the origin once in a CCW direction. Hence $N_0 = -1$. Thus the number of poles of $P(s)$ enclosed by the $s$-plane contour is $P_0 = Z_0 - N_0 = 1 - (-1) = 2$.

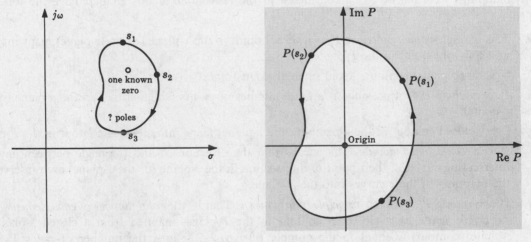

Fig. 11-8

## 11.5  POLAR PLOTS

A continuous system transfer function $P(s)$ may be represented in the frequency domain as a sinusoidal transfer function by substituting $j\omega$ for $s$ in the expression for $P(s)$. The resulting form $P(j\omega)$ is a complex function of the single variable $\omega$. Therefore it may be plotted in two dimensions, with $\omega$ as a parameter, and written in the following equivalent forms:

**Polar Form:**   $P(j\omega) = |P(j\omega)| \underline{/\phi(\omega)}$                    (11.2)

**Euler Form:**   $P(j\omega) = |P(j\omega)|(\cos\phi(\omega) + j\sin\phi(\omega))$          (11.3)

$|P(j\omega)|$ is the **magnitude** of the complex function $P(j\omega)$, and $\phi(j\omega)$ is its **phase angle**, $\arg P(j\omega)$.

$|P(j\omega)|\cos\phi(\omega)$ is the *real part*, and $|P(j\omega)|\sin\phi(\omega)$ is the *imaginary part* of $P(j\omega)$. Therefore $P(j\omega)$ may also be written as

**Rectangular or Complex Form**: $\qquad P(j\omega) = \operatorname{Re} P(j\omega) + j \operatorname{Im} P(j\omega) \qquad\qquad (11.4)$

A **Polar Plot** of $P(j\omega)$ is a graph of $\operatorname{Im} P(j\omega)$ versus $\operatorname{Re} P(j\omega)$ in the finite portion of the $P(j\omega)$-plane for $-\infty < \omega < \infty$. At singular points of $P(j\omega)$ (poles on the $j\omega$-axis), $|P(j\omega)| \to \infty$. A Polar Plot may also be generated on polar coordinate paper. The magnitude and phase angle of $P(j\omega)$ are plotted with $\omega$ varying from $-\infty$ to $+\infty$.

The locus of $P(j\omega)$ is identical on either rectangular or polar coordinates. The choice of coordinate system may depend on whether $P(j\omega)$ is available in analytic form or as experimental data. If $P(j\omega)$ is expressed analytically, the choice of coordinates depends on whether it is easier to write $P(j\omega)$ in the form of Equation (11.2), in which case polar coordinates are used, or in the form of Equation (11.4) for rectangular coordinates. Experimental data on $P(j\omega)$ are usually expressed in terms of magnitude and phase angle. In this case, polar coordinates are the natural choice.

**EXAMPLE 11.5.** The Polar Plots in Fig. 11-9 are identical; only the coordinate systems are different.

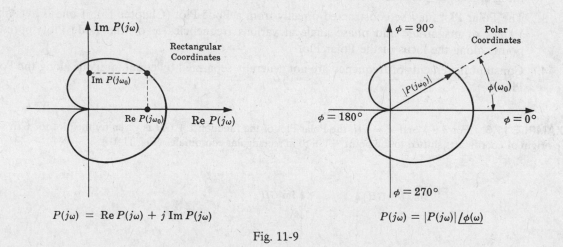

$$P(j\omega) = \operatorname{Re} P(j\omega) + j \operatorname{Im} P(j\omega) \qquad\qquad P(j\omega) = |P(j\omega)|\underline{/\phi(\omega)}$$

Fig. 11-9

For *discrete-time systems*, Polar Plots are defined in the frequency domain in the same manner. Recall that we can write $z \equiv e^{sT}$ (see Section 4.9). Therefore a discrete transfer function $P(z) \equiv P(e^{sT})$ and, if we set $s = j\omega$, $P(z)$ becomes $P(e^{j\omega T})$. The **Polar Plot of** $P(e^{j\omega T})$ is a graph of $\operatorname{Im} P(e^{j\omega T})$ versus $\operatorname{Re} P(e^{j\omega T})$ in the finite portion of the $P(e^{j\omega T})$-plane, for $-\infty < \omega < \infty$.

We often discuss Polar Plots, their properties, and many results dependent on these in subsequent sections in a unified manner for both continuous and discrete-time systems. To do this, we adopt for our general transfer function $P$ the unified representation for frequency response functions given in Equation (10.1) for $GH$, that is, we use the generic representation $P(\omega)$ defined by

$$P(\omega) = \begin{cases} P(j\omega) & \text{for continuous systems} \\ P(e^{j\omega T}) & \text{for discrete-time systems} \end{cases}$$

In these terms, Equations (11.2) through (11.4) become

$$P(\omega) = |P(\omega)|\underline{/\phi(\omega)} = |P(\omega)|(\cos\phi(\omega) + j\sin\phi(\omega)) = \operatorname{Re} P(\omega) + j \operatorname{Im} P(\omega)$$

We use this unified notation in much of the remainder of this chapter, and in subsequent chapters, particularly where the results are applicable to both continuous and discrete-time systems.

## 11.6  PROPERTIES OF POLAR PLOTS

The following are several useful properties of Polar Plots of $P(\omega)$ [$P(j\omega)$ or $P(e^{j\omega T})$].

1.  The Polar Plot for

$$P(\omega) + a$$

    where $a$ is any complex constant, is identical to the plot for $P(\omega)$ with the origin of coordinates shifted to the point $-a = -(\text{Re }a + j\text{ Im }a)$.

2.  The Polar Plot of the transfer function of a time-invariant, linear system exhibits *conjugate symmetry*. That is, the graph for $-\infty < \omega < 0$ is the mirror image about the horizontal axis of the graph for $0 \le \omega < \infty$.

3.  The Polar Plot may be constructed directly from a Bode Plot (Chapter 15), if one is available. Values of magnitude and phase angle at various frequencies $\omega$ on the Bode Plot represent points along the locus of the Polar Plot.

4.  Constant increments of frequency are not generally separated by equal intervals along the Polar Plot.

**EXAMPLE 11.6.**   For $a = 1$ and $P = GH$, the Polar Plot of the function $1 + GH$ is given by the plot for $GH$, with the origin of coordinates shifted to the point $-1 + j0$ in rectangular coordinates (Fig. 11-10).

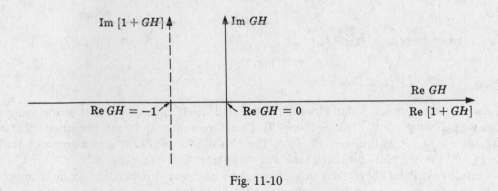

Fig. 11-10

**EXAMPLE 11.7.**   To illustrate plotting of a transfer function, consider the open-loop continuous system transfer function

$$GH(s) = \frac{1}{s + 1}$$

Letting $s = j\omega$ and rewriting $GH(j\omega)$ in the form of Equation (*11.2*) (polar form), we have

$$GH(j\omega) = \frac{1}{j\omega + 1} = \frac{1}{\sqrt{\omega^2 + 1}} \underline{/-\tan^{-1}\omega}$$

For $\omega = 0$, $\omega = 1$, and $\omega \to \infty$:

$$GH(j0) = 1\underline{/0°}$$

$$GH(j1) = \left(1/\sqrt{2}\right)\underline{/-45°}$$

$$\lim_{\omega \to \infty} GH(j\omega) = 0\underline{/-90°}$$

Substitution of several other positive values of $\omega$ yields a semicircular locus for $0 \le \omega < \infty$. The graph for $-\infty < \omega < 0$ is the mirror image about the diameter of this semicircle. It is shown in Fig. 11-11 by a dashed line. Note the strikingly unequal increments of frequency between the arcs $\overline{ab}$ and $\overline{bc}$.

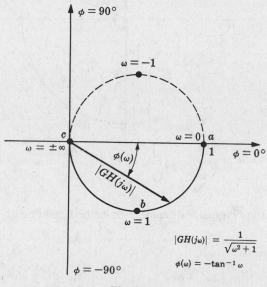

$$|GH(j\omega)| = \frac{1}{\sqrt{\omega^2 + 1}}$$

$$\phi(\omega) = -\tan^{-1}\omega$$

Fig. 11-11

Polar Plots are not very difficult to sketch for very simple transfer functions, although they are usually a little more difficult to determine for discrete-time systems, as illustrated in Example 11.11. But the computations can be very laborious for complicated $P(s)$ or $P(z)$. On the other hand, widely available computer programs for frequency response analysis, or more generally for plotting complex functions of a complex variable, typically generate accurate Polar Plots quite conveniently.

## 11.7   THE NYQUIST PATH

For continuous systems, the **Nyquist Path** is a closed contour in the $s$-plane, enclosing the entire right half of the $s$-plane (RHP). For discrete-time systems, the corresponding **Nyquist Path** encloses the entire $z$-plane *outside* the unit circle.

For continuous systems, in order that the Nyquist Path should not pass through any poles of $P(s)$, small semicircles along the imaginary axis or at the origin of $P(s)$ are required in the path if $P(s)$ has poles on the $j\omega$-axis or at the origin. The radii $\rho$ of these small circles are interpreted as approaching zero in the limit.

To enclose the RHP at infinity, and thus any poles in the interior of the RHP, a large semicircular path is drawn in the RHP and the radius $R$ of this semicircle is interpreted as being infinite in the limit.

The **generalized Nyquist Path in the s-plane** is illustrated by the s-plane contour in Fig. 11-12. It is apparent that *every pole and zero of $P(s)$ in the RHP is enclosed by the Nyquist Path* when it is mapped into the $P(s)$-plane.

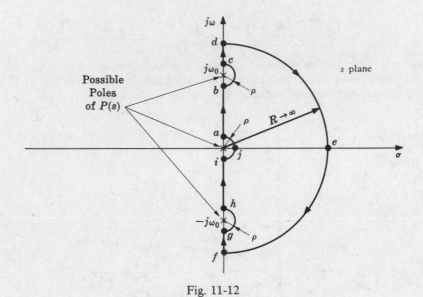

Fig. 11-12

The various portions of the Nyquist Path can be described analytically in the following manner.

Path $\overline{ab}$:          $s = j\omega$                                    $0 < \omega < \omega_0$                    $(11.5)$

Path $\overline{bc}$:          $s = \lim_{\rho \to 0} \left( j\omega_0 + \rho e^{j\theta} \right)$          $-90° \leq \theta \leq 90°$              $(11.6)$

Path $\overline{cd}$:          $s = j\omega$                                    $\omega_0 \leq \omega < \infty$             $(11.7)$

Path $\overline{def}$:         $s = \lim_{R \to \infty} R e^{j\theta}$          $+90° \leq \theta \leq -90°$             $(11.8)$

Path $\overline{fg}$:          $s = j\omega$                                    $-\infty < \omega < -\omega_0$             $(11.9)$

Path $\overline{gh}$:          $s = \lim_{\rho \to 0} \left( -j\omega_0 + \rho e^{j\theta} \right)$          $-90° \leq \theta \leq 90°$              $(11.10)$

Path $\overline{hi}$:          $s = j\omega$                                    $-\omega_0 < \omega < 0$                   $(11.11)$

Path $\overline{ija}$:         $s = \lim_{\rho \to 0} \rho e^{j\theta}$          $-90° \leq \theta \leq 90°$              $(11.12)$

The **generalized Nyquist Path in the z-plane** is given in Fig. 11-13. Every pole and zero of $P(z)$ outside the unit circle is enclosed by the Nyquist Path when it is mapped into the $P(z)$-plane. In traversing the unit circle as a function of increasing angular frequency $\omega$, any poles of $P(z)$ on the unit circle, which may include "integrators" at $z = 1$ (corresponding to $z \equiv e^{0 \cdot T} \equiv 1$ when $s = 0$), are excluded by infinitesimal circular arcs. For example, one pair of complex conjugate poles on the unit circle is shown in Fig. 11-13, circumvented by arcs of radius $\rho \to 0$. The remainder of the z-plane outside the unit circle is enclosed by the large circle of radius $R \to \infty$ shown in Fig. 11-13.

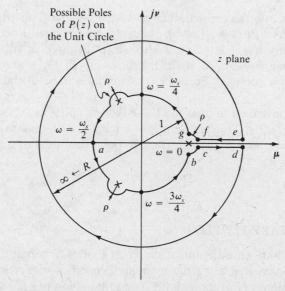

Fig. 11-13

The **unit circle in the z-plane** has a practical feature not shared by the Nyquist Path in the s-plane, one that facilitates drawing Polar Plots, as well as having other consequences in designing digital systems. First, we define the **angular sampling frequency** $\omega_s = 2\pi/T$ (radians per unit time). The advantage is that the unit circle repeats itself every angular sampling frequency $\omega_s$ as $\omega$ increases. This is shown in Fig. 11-14($a$), which illustrates that the portion of the $j\omega$-axis in the s-plane between

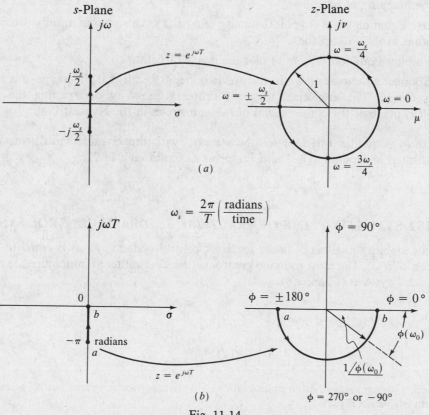

$$\omega_s = \frac{2\pi}{T}\left(\frac{\text{radians}}{\text{time}}\right)$$

Fig. 11-14

$-j\omega_s/2$ and $+j\omega_s/2$ maps into the entire unit circle in the $z$-plane. This property is useful in drawing Polar Plots of functions $P(z) = P(e^{j\omega T})$, because the same Polar Plot is obtained for $n\omega_s \leq \omega \leq (n+1)\omega_s$, for any $n = \pm 1, \pm 2, \ldots$. Also, since the circular arc from $\omega = 0$ to $\omega_s/2$ is the mirror image of that from $\omega = -\omega_s/2$ to $0$, the function $P(e^{j\omega T})$ need only be evaluated from $\omega = -\omega_s/2$ to $0$ to obtain a complete Polar Plot, taking advantage of the symmetry of the mapping (Property 2, Section 11.6).

It is sometimes also convenient to treat the Polar Plot mapping as a function of $\omega T$ rather than $\omega$. Then the strip $-(\omega_s/2)T \leq \omega T \leq 0$ is equivalent to $-\pi \leq \omega T \leq 0$ (in radians), because $\omega_s/2 = \pi/T$; this strip is mapped into the *lower half* of the unit circle in polar coordinates, from $-180°$ ($-\pi$ radians) to $0°$ or radians [Fig. 11-14(b)].

## 11.8   THE NYQUIST STABILITY PLOT

The Nyquist Stability Plot, an extension of the Polar Plot, is a mapping of the *entire* Nyquist Path into the $P$-plane. It is constructed using the mapping properties of Sections 11.4 and 11.6 and, for continuous systems, Equations (11.5) through (11.8) and Equation (11.12). A carefully drawn sketch is sufficient for most purposes.

A general construction procedure is outlined for continuous systems in the following steps.

**Step 1:**   Check $P(s)$ for poles on the $j\omega$-axis and at the origin.

**Step 2:**   Using Equation (11.5) through (11.7), sketch the image of path $\overline{ad}$ in the $P(s)$-plane. If there are no poles on the $j\omega$-axis, Equation (11.6) need not be employed. In this case, Step 2 should read: Sketch the Polar Plot of $P(j\omega)$.

**Step 3:**   Draw the mirror image about the real axis $\operatorname{Re} P$ of the sketch resulting from Step 2. This is the mapping of path $\overline{fi}$.

**Step 4:**   Use Equation (11.8) to plot the image of path $\overline{def}$. This path at infinity usually plots into a point in the $P(s)$-plane.

**Step 5:**   Employ Equation (11.12) to plot the image of path $\overline{ija}$.

**Step 6:**   Connect all curves drawn in the previous steps. Recall that the image of a closed contour is closed. The conformal mapping property helps by determining the image in the $P(s)$-plane of the corner angles of the semicircles in the Nyquist Path.

The procedure is similar for discrete-time systems, with the Nyquist Path given in Fig. 11-13 instead, as illustrated in Example 11.11 and Problems 11.65 through 11.72.

## 11.9   NYQUIST STABILITY PLOTS OF PRACTICAL FEEDBACK CONTROL SYSTEMS

For Nyquist stability analysis of linear feedback control systems, $P(\omega)$ is equal to the open-loop transfer function $GH(\omega)$. The most common control systems encountered in practice are those classified as type $0, 1, 2, \ldots, l$ systems (Chapter 9).

**EXAMPLE 11.8.**   *Type 0 continuous system*

$$GH(s) = \frac{1}{s+1}$$

By definition, a *type 0 system* has no poles at the origin. This particular system has no poles on the $j\omega$-axis. The Nyquist Path is given in Fig. 11-15.

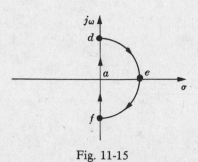

Fig. 11-15

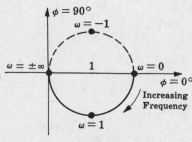

Fig. 11-16

The Polar Plot for this loop transfer function was constructed in Example 11.7, and it is shown in Fig. 11-16. This plot is the image of the $j\omega$-axis, or path $\overline{fad}$ of the Nyquist Path, in the $GH(s)$-plane. The semicircular path $\overline{def}$ at infinity is mapped into the $GH(s)$-plane in the following manner. Equation (11.8) implies substitution of $s = \lim_{R \to \infty} Re^{j\theta}$ into the expression for $GH(s)$, where $90° \leq \theta \leq -90°$. Hence

$$GH(s)\big|_{\text{path }\overline{def}} \equiv GH(\infty) = \frac{1}{\lim_{R \to \infty} Re^{j\theta} + 1}$$

By the elementary properties of limits,

$$GH(\infty) = \lim_{R \to \infty} \left[ \frac{1}{Re^{j\theta} + 1} \right]$$

But since $|a + b| \geq |\,|a| - |b|\,|$, then

$$|GH(\infty)| = \lim_{R \to \infty} \left| \frac{1}{Re^{j\theta} + 1} \right| \leq \lim_{R \to \infty} \left( \frac{1}{R - 1} \right) = 0$$

and the infinite semicircle plots into a point at the origin. Of course, this computation was unnecessary for this simple example because the Polar Plot produces a completely closed contour in the $GH(s)$-plane. In fact, Polar Plots of all *type 0 systems* exhibit this property. The Nyquist Stability Plot is a replica of the Polar Plot with the axes relabeled, and is given in Fig. 11-17.

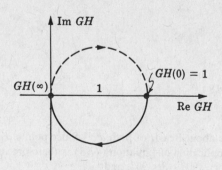

Fig. 11-17

**EXAMPLE 11.9.**  *Type 1 continuous system*

$$GH(s) = \frac{1}{s(s + 1)}$$

There is one pole at the origin. The Nyquist Path is given in Fig. 11-18.

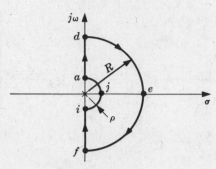

Fig. 11-18

Path $\overline{ad}$: $s = j\omega$ for $0 < \omega < \infty$, and

$$GH(j\omega) = \frac{1}{j\omega(j\omega + 1)} = \frac{1}{\omega\sqrt{\omega^2 + 1}}\underline{/-90° - \tan^{-1}\omega}$$

At extreme values of $\omega$ we have

$$\lim_{\omega \to 0} GH(j\omega) = \infty\underline{/-90°} \qquad \lim_{\omega \to \infty} GH(j\omega) = 0\underline{/-180°}$$

As $\omega$ increases in the interval $0 < \omega < \infty$, the magnitude of $GH$ decreases from $\infty$ to $0$ and the phase angle decreases steadily from $-90°$ to $-180°$. Therefore the contour does not cross the negative real axis, but approaches it from below as shown in Fig. 11-19.

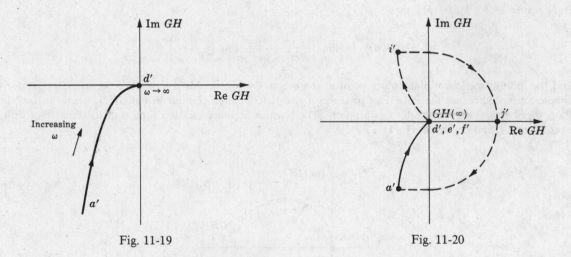

Fig. 11-19                                                          Fig. 11-20

Path $\overline{f'i'}$ is the mirror image about Re $GH$ of path $\overline{a'd'}$. Since points $d'$ and $f'$ meet at the origin, the origin is clearly the image of path $\overline{def}$. Application of Equation (11.8) is therefore unnecessary.

Path $\overline{ija}$: $s = \lim_{\rho \to 0}\rho e^{j\theta}$ for $-90° \le \theta \le 90°$, and

$$\lim_{\rho \to 0} GH(\rho e^{j\theta}) = \lim_{\rho \to 0}\left[\frac{1}{\rho e^{j\theta}(\rho e^{j\theta} + 1)}\right] = \lim_{\rho \to 0}\left[\frac{1}{\rho e^{j\theta}}\right] = \infty \cdot e^{-j\theta} = \infty\underline{/-\theta}$$

where we have used the fact that $(\rho e^{j\theta} + 1) \to 1$ as $\rho \to 0$. Hence path $\overline{ija}$ maps into a semicircle of infinite radius. For point $i$, $GH = \infty\underline{/90°}$; for point $j$, $GH = \infty\underline{/0°}$; and for point $a$, $GH = \infty\underline{/-90°}$. The resulting Nyquist Stability Plot is given in Fig. 11-20.

Path $\overline{i'j'a'}$ could also have been determined in the following manner. The Nyquist Path makes a 90° turn to the right at point $i$; hence by conformal mapping, a 90° right turn must be made at $i'$ in the $GH(s)$-plane. The same goes for point $a'$. Since both $i'$ and $a'$ are points at infinity, and since the Nyquist Stability Plot must be a closed contour, a CW semicircle of infinite radius must join point $i'$ to point $a'$.

### Type *l* Continuous Systems

The Nyquist Stability Plot of a type *l* system includes *l* infinite semicircles in its path. There are 180*l* degrees in the connecting arc at infinity of the *GH*(*s*)-plane.

**EXAMPLE 11.10.**  The type 3 system with

$$GH(s) = \frac{1}{s^3(s+1)}$$

has three infinite semicircles in its Nyquist Stability Plot (Fig. 11-21).

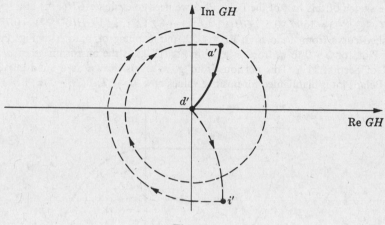

Fig. 11-21

### Discrete-Time Systems

Nyquist Stability Plots of discrete-time systems are drawn in the same manner as above, the only difference being that the Nyquist Path is that given in Fig. 11-13, instead of Fig. 11-12.

**EXAMPLE 11.11.**  Consider the type 1 digital control system with open-loop transfer function

$$GH(z) = \frac{K/4}{(z-1)\left(z-\frac{1}{2}\right)}$$

The Polar Plot of *GH* is determined by first mapping the lower half of the unit circle in the *z*-plane into the *GH*-plane. This is readily accomplished with the aid of the mapping illustrated in Fig. 11-14(*b*), that is, we evaluate $GH(e^{j\omega T})$ for increasing values of $\omega T$, from $-180°$ to $0°$ (or $-\pi$ to 0 radians). For given values of $K$ and $T$, say $K = 1$ and $T = 1$,

$$GH(e^{j\omega T}) = \frac{K/4}{\left(e^{j\omega T}-1\right)\left(e^{j\omega T}-\frac{1}{2}\right)} = \frac{1/4}{\left(e^{j\omega}-1\right)\left(e^{j\omega}-\frac{1}{2}\right)}$$

For hand calculations, a combination of the Polar Form, Euler Form, and Complex Form are useful in evaluating $GH(e^{j\omega T})$ at different values of $\omega$, because $e^{j\omega T} = 1\underline{/\omega T}\,(\text{rad}) = \cos(\omega T) + j\sin(\omega T) = \text{Re}(e^{j\omega T}) + j\,\text{Im}(e^{j\omega T})$. At $\omega = -\pi$ rad $(-180°)$, we have

$$GH(e^{-j\pi}) = GH\left(1\underline{/-180°}\right) = \frac{0.25\underline{/0°}}{\left(1\underline{/-180°} - 1\underline{/0°}\right)\left(1\underline{/-180°} - \frac{1}{2}\underline{/0°}\right)}$$

$$= \frac{0.25}{(-1+j0-1)(-1+j0-\frac{1}{2})} = \frac{0.25}{(-2)(-\frac{3}{2})}$$

$$= 0.083\underline{/0°}$$

Then, at $\omega = 270°$,

$$GH(e^{j3\pi/2}) = GH(e^{-j\pi/2}) = \frac{0.25\underline{/0°}}{\left(1\underline{/-90°} - 1\underline{/0°}\right)\left(1\underline{/90°} - \tfrac{1}{2}\underline{/0°}\right)}$$

$$= \frac{0.25}{(-j-1)\left(-j-\tfrac{1}{2}\right)} = \frac{0.25}{-\tfrac{1}{2}+j\tfrac{3}{2}}$$

$$= \frac{2(0.25)}{\sqrt{10}}\underline{/180° - \tan^{-1}(3)} = 0.158\underline{/-108.4°}$$

Similarly, we find that $GH(e^{j2\pi})$ does not exist, but $\lim_{\phi \to 360°}GH(e^{j\phi}) = \lim_{\omega \to 0}GH(e^{j\omega}) = \infty\underline{/90°}$.

To complete the sketch of this half of the Polar Plot, we need to evaluate $GH(e^{j\omega})$ at a few more values of $\omega$. We readily find $GH(e^{-j\pi/1000}) = 159\underline{/90.5°}$, $GH(e^{-j\omega/12}) = 1.8\underline{/127°}$, and $GH(e^{-j\pi/6}) = 0.779\underline{/159°}$. The result is shown as the dashed curve from $a'$ to $b'$ in Fig. 11-22, the mapping of $a$ to $b$ in Fig. 11-13. The remaining portion of the Polar Plot, for $\omega = 0$ to $\pi$, from $g'$ to $a'$ in Fig. 11-22, is the mirror image of $a'$ to $b'$ about the real axis, by Property 2 of Section 11.6. This portion, from $g'$ to $a'$, is drawn as a solid curve, keeping with the convention that the Polar Plot is highlighted for positive values of $\omega$, $0 < \omega T < (2n-1)\pi$, $n = 1, 2, \dots$.

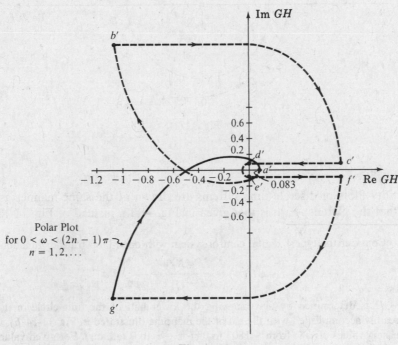

Fig. 11-22

The Nyquist Stability Plot is determined by completing the mapping of Fig. 11-13 segments $b$ to $c$, $c$ to $d$, $d$ to $e$, and $f$ to $g$, to the $GH$-plane. Using the mapping properties of Section 11.4 and limit calculations, $GH(e^{j\omega})$ makes a right turn at $b'$, from $\infty\underline{/90°}$ to $\infty\underline{/0°}$ at $c'$, then to $0\underline{/0°}$ at $d'$ and at $e'$, and $\infty\underline{/0°}$ at $f'$ to $\infty\underline{/-90°}$ at $g'$, using limit operations for radii $\rho$ and $R$ in Fig. 11-13. For example, $\lim_{\rho \to 0}GH(z = 1 + \rho e^{j\theta})$ for $-90° < \theta < 0°$, provides the mapping of the arc from $b$ to $c$ in Fig. 11-13 into the arc from $b'$ ($\infty\underline{/90°}$) to $c'$ ($\infty\underline{/0°}$) in Fig. 11-22.

## 11.10  THE NYQUIST STABILITY CRITERION

A linear closed-loop *continuous control system* is absolutely stable if the roots of the characteristic equation have negative real parts (Section 5.2). Equivalently, the poles of the closed-loop transfer function, or the *zeros* of the denominator, $1 + GH(s)$, of the closed-loop transfer function, must lie in the left-half plane (LHP). For continuous systems, the Nyquist Stability Criterion establishes the number of *zeros* of $1 + GH(s)$ *in the RHP* directly from the Nyquist Stability Plot of $GH(s)$. For *discrete-time control systems*, the Nyquist Stability Criterion establishes the number of *zeros* of $1 + GH(z)$ *outside the unit circle* of the $z$-plane, the region of instability for discrete systems.

For either class of systems, continuous or discrete-time, the Nyquist Stability Criterion may be stated as follows.

### Nyquist Stability Criterion

The closed-loop control system whose open-loop transfer function is $GH$ is stable if and only if

$$N = -P_0 \leq 0 \qquad (11.13)$$

where

$$P_0 \equiv \begin{cases} \textit{number of poles } (\geq 0) \textit{ of } GH \textit{ in the RHP for continuous systems} \\ \textit{number of poles } (\geq 0) \textit{ of } GH \textit{ outside the unit circle (of the } z\textit{-plane) for discrete-time systems} \end{cases}$$

$N \equiv$ total number of CW encirclements of the $(-1, 0)$ point (i.e., $GH = -1$) in the

$GH$-plane (continuous or discrete)

If $N > 0$, *the number of zeros $Z_0$ of $1 + GH$ in the RHP* for continuous systems, or *outside the unit circle* for discrete systems, is determined by

$$Z_0 = N + P_0 \qquad (11.14)$$

If $N \leq 0$, the $(-1, 0)$ point is not enclosed by the Nyquist Stability Plot. Therefore $N \leq 0$ if the region to the right of the contour in the prescribed direction does not include the $(-1, 0)$ point. Shading this region helps significantly in determining whether $N \leq 0$.

If $N \leq 0$ *and* $P_0 = 0$, then the system is absolutely stable if and only if $N = 0$; that is, if and only if the $(-1, 0)$ point *does not* lie in the shaded region.

**EXAMPLE 11.12.** The Nyquist Stability Plot for $GH(s) = 1/s(s + 1)$ was determined in Example 11.9 and is shown in Fig. 11-23. The region to the right of the contour has been shaded. Clearly, the $(-1, 0)$ point is not in the shaded region; therefore it is not enclosed by the contour and so $N \leq 0$. The poles of $GH(s)$ are at $s = 0$ and $s = -1$, neither of which are in the RHP; hence $P_0 = 0$. Thus

$$N = -P_0 = 0$$

and the system is absolutely stable.

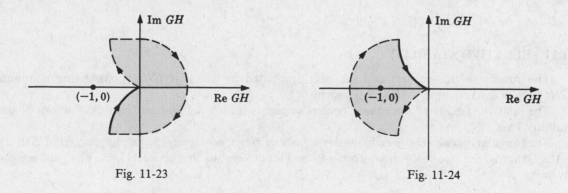

Fig. 11-23                                        Fig. 11-24

**EXAMPLE 11.13.** The Nyquist Stability Plot for $GH(s) = 1/s(s - 1)$ is given in Fig. 11-24. The region to the right of the contour has been shaded and the $(-1, 0)$ point is enclosed; then $N > 0$. (It is clear that $N = 1$.) The poles of $GH$ are at $s = 0$ and $s = +1$, the latter pole being in the RHP. Hence $P_0 = 1$.

$N \neq -P_0$ indicates that the system is *unstable*. From Equation (*11.14*) we have

$$Z_0 = N + P_0 = 2$$

zeros of $1 + GH$ in the RHP.

**EXAMPLE 11.14.**   The Nyquist Stability Plot for the discrete-time open-loop transfer function

$$GH(z) = \frac{K/4}{(z-1)(z-0.5)}$$

was determined in Example 11.11 and is repeated in Fig. 11-25 for $K = 1$. The region to the right of the contour has been shaded and the $(-1, 0)$ point is not enclosed for $K = 1$. Thus $N \leq 0$ and from Equation (*11.13*) there are no poles outside the unit circle of the $z$-plane, that is, $P_0 = 0$. Hence $N = -P_0 = 0$ and the system is therefore stable.

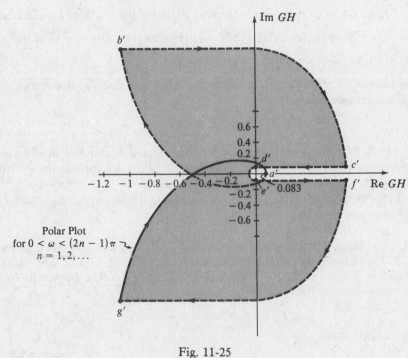

Fig. 11-25

## 11.11   RELATIVE STABILITY

The results in this section and the next are stated in terms of $GH(\omega)$, for either continuous $[GH(j\omega)]$ or discrete-time $[GH(e^{j\omega T})]$ systems.

The relative stability of a feedback control system is readily determined from the Polar or Nyquist Stability Plot.

The (angular) **phase crossover frequency** $\omega_\pi$ is that frequency at which the phase angle of $GH(\omega)$ is $-180°$, that is, the frequency at which the Polar Plot crosses the negative real axis. The **gain margin** is given by

$$\text{gain margin} = \frac{1}{|GH(\omega_\pi)|}$$

These quantities are illustrated in Fig. 11-26.

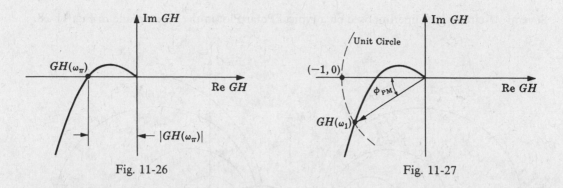

Fig. 11-26                          Fig. 11-27

The (angular) **gain crossover frequency** $\omega_1$ is that frequency at which $|GH(\omega)| = 1$. The **phase margin** $\phi_{PM}$ is the angle by which the Polar Plot must be rotated to cause it to pass through the $(-1, 0)$ point. It is given by

$$\phi_{PM} = [180 + \arg GH(\omega_1)] \text{ degrees}$$

These quantities are illustrated in Fig. 11-27.

## 11.12   M- AND N-CIRCLES*

The closed-loop frequency response of a unity feedback control system is given by

$$\frac{C}{R}(\omega) = \frac{G(\omega)}{1 + G(\omega)} = \left| \frac{G(\omega)}{1 + G(\omega)} \right| \Big/ \tan^{-1}\left[ \frac{\text{Im}(C/R)(\omega)}{\text{Re}(C/R)(\omega)} \right] \qquad (11.15)$$

The magnitude and phase angle characteristics of the closed-loop frequency response of a unity feedback control system can be determined directly from the Polar Plot of $G(\omega)$. This is accomplished by first drawing lines of constant magnitude, called **M-circles**, and lines of constant phase angle, called **N-circles**, directly onto the $G(\omega)$-plane, where

$$M \equiv \left| \frac{G(\omega)}{1 + G(\omega)} \right| \qquad (11.16)$$

$$N \equiv \frac{\text{Im}(C/R)(\omega)}{\text{Re}(C/R)(\omega)} \qquad (11.17)$$

The intersection of the Polar Plot with a particular M-circle yields the value of $M$ at the frequency $\omega$ of $G(\omega)$ at the point of intersection. The intersection of the Polar Plot with a particular N-circle yields the value of $N$ at the frequency $\omega$ of $G(\omega)$ at the intersection point. $M$ versus $\omega$ and $N$ versus $\omega$ are easily plotted from these points.

---

*The letter symbols $M, N$ used in this section for M- and N-circles are not equal to and should not be confused with the manipulated variable $M = M(s)$ defined in Chapter 2 and with the number of encirclements $N$ of the $(-1, 0)$ point of Section 11.10. It is unfortunate that the same symbols have been used to signify more than one quantity. But in the interest of being consistent with most other control system texts, we have maintained the terminology of the classical literature and have now pointed this out to the reader.

Several $M$-circles are superimposed on a typical Polar Plot in the $G(\omega)$-plane in Fig. 11-28.

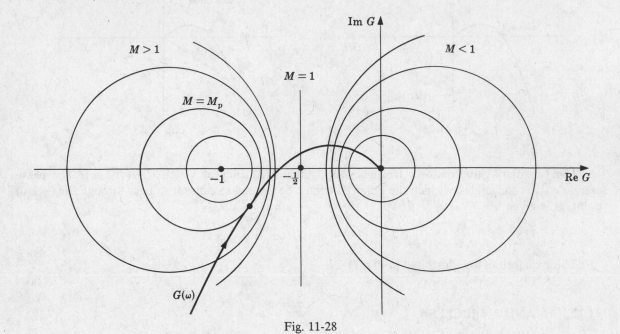

Fig. 11-28

The **radius of an $M$-circle** is given by

$$\text{radius of } M\text{-circle} = \left| \frac{M}{M^2 - 1} \right| \qquad (11.18)$$

The **center of an $M$-circle** always lies on the $\operatorname{Re} G(\omega)$-axis. The center point is given by

$$\text{center of } M\text{-circle} = \left( \frac{-M^2}{M^2 - 1}, 0 \right) \qquad (11.19)$$

The **resonance peak** $M_p$ is given by the largest value of $M$ of the $M$-circle(s) tangent to the Polar Plot. (There may be more than one tangency.)

The **damping ratio** $\zeta$ for a second-order continuous system with $0 \le \zeta \le 0.707$ is related to $M_p$ by

$$M_p = \frac{1}{2\zeta\sqrt{1 - \zeta^2}} \qquad (11.20)$$

Several $N$-circles are superimposed on the Polar Plot shown in Fig. 11-29. The **radius of an $N$-circle** is given by

$$\text{radius of } N\text{-circle} = \sqrt{\frac{1}{4} + \left( \frac{1}{2N} \right)^2} \qquad (11.21)$$

The **center of an $N$-circle** always falls on the line $\operatorname{Re} G(\omega) = -\frac{1}{2}$. The center point is given by

$$\text{center of } N\text{-circle} = \left( -\frac{1}{2}, \frac{1}{2N} \right) \qquad (11.22)$$

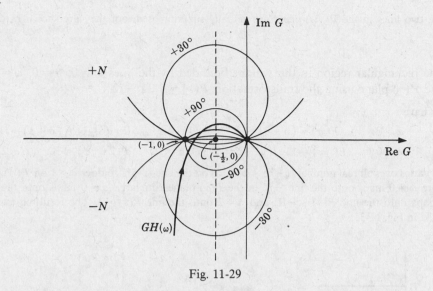

Fig. 11-29

# Solved Problems

### COMPLEX FUNCTIONS OF A COMPLEX VARIABLE

**11.1.** What are the values of $P(s) = 1/(s^2 + 1)$ for $s_1 = 2$, $s_2 = j4$, and $s_3 = 2 + j4$?

$$P(s_1) = P(2) = \frac{1}{(2)^2 + 1} = \frac{1}{5} + j0 \qquad P(s_2) = P(j4) = \frac{1}{(j4)^2 + 1} = -\frac{1}{15} + j0$$

$$P(s_3) = P(2 + j4) = \frac{1}{(2 + j4)^2 + 1} = \frac{1}{-11 + j16}$$

$$= \frac{1\underline{/0°}}{\sqrt{(11)^2 + (16)^2}\,\underline{/\tan^{-1}(16/-11)}} = \frac{1}{19.4}\underline{/0° - 124.6°}$$

$$= 0.0514\underline{/-124.6°} = -0.0514\underline{/55.4°} = -0.0292 - j0.0423$$

**11.2.** Map the imaginary axis in the $s$-plane onto the $P(s)$-plane, using the mapping function $P(s) = s^2$.

We have $s = j\omega$, $-\infty < \omega < \infty$. Therefore $P(j\omega) = (j\omega)^2 = -\omega^2$. Now when $\omega \to -\infty$, $P(j\omega) \to -\infty$ (or $-\infty^2$, if you prefer). When $\omega \to +\infty$, $P(j\omega) \to -\infty$; and when $\omega = 0$, $P(j0) = 0$. Thus as $j\omega$ increases along the negative *imaginary* axis from $-j\infty$ toward $j0$, $P(j\omega)$ increases along the negative *real* axis from $-\infty$ to 0. When $j\omega$ increases from $j0$ to $+j\infty$, $P(j\omega)$ decreases back to $-\infty$, again along the negative *real* axis. The mapping is plotted in the following manner (Fig. 11-30):

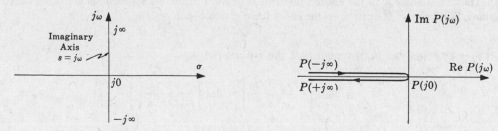

Fig. 11-30

The two lines in the $P(j\omega)$-plane are actually superimposed, but they are shown here separated for clarity.

**11.3.** Map the rectangular region in the $s$-plane bounded by the lines $\omega = 0$, $\sigma = 0$, $\omega = 1$, and $\sigma = 2$ onto the $P(s)$-plane using the transformation $P(s) = s + 1 - j2$.

We have

$$\omega = 0: \quad P(\sigma) = (\sigma + 1) - j2 \qquad \omega = 1: \quad P(\sigma + j1) = (\sigma + 1) - j1$$

$$\sigma = 0: \quad P(j\omega) = 1 + j(\omega - 2) \qquad \sigma = 2: \quad P(2 + j\omega) = 3 + j(\omega - 2)$$

Since $\sigma$ varies over all real numbers $(-\infty < \sigma < \infty)$ on the line $\omega = 0$, so does $\sigma + 1$ on $P(\sigma) = (\sigma + 1) - j2$. Therefore $\omega = 0$ maps onto the line $-j2$ in the $P(s)$-plane. Similarly, $\sigma = 0$ maps onto the line $P(s) = 1$, $\omega = 1$ maps onto the line $P(s) = -j1$, and $\sigma = 2$ onto the line $P(s) = 3$. The resulting transformation is illustrated in Fig. 11-31.

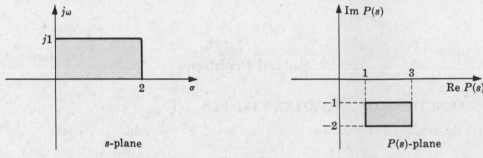

Fig. 11-31

This type of mapping is called a **translation** mapping. Note that the mapping would be exactly the same if $s = \sigma + j\omega$ were replaced by $z = \mu + j\nu$ in this example.

**11.4.** Find the derivative of $P(s) = s^2$ at the points $s = s_0$ and $s_0 = 1$.

$$\left.\frac{dP}{ds}\right|_{s=s_0} = \lim_{s \to s_0}\left[\frac{P(s) - P(s_0)}{s - s_0}\right] = \lim_{s \to s_0}\left[\frac{s^2 - s_0^2}{s - s_0}\right] = \lim_{s \to s_0}(s + s_0) = 2s_0$$

At $s_0 = 1$, we have $(dP/ds)|_{s=1} = 2$. Similarly, if $P(z) = z^2$, $(dP/dz)|_{z=1} = 2$.

## ANALYTIC FUNCTIONS AND SINGULARITIES

**11.5.** Is $P(s) = s^2$ an analytic function in any region of the $s$-plane? If so, which region?

From the preceding problem $(dP/ds)|_{s=s_0} = 2s_0$. Hence $s^2$ is analytic wherever $2s_0$ is finite (Definition 11.1). Thus $s^2$ is analytic in the entire finite region of the $s$-plane. Such functions are often called **entire functions**. Similarly, $z^2$ is analytic in the entire finite region of the $z$-plane.

**11.6.** Is $P(s) = 1/s$ analytic in any region of the $s$-plane?

$$\left.\frac{dP}{ds}\right|_{s=s_0} = \lim_{s \to s_0}\left[\frac{1/s - 1/s_0}{s - s_0}\right] = \lim_{s \to s_0}\left[\frac{-(s - s_0)}{ss_0(s - s_0)}\right] = \frac{-1}{s_0^2}$$

This derivative is unique and finite for all $s_0 \neq 0$. Hence $1/s$ is analytic at all points in the $s$-plane except the origin, $s = s_0 = 0$. The point $s = 0$ is a *singularity* (pole) of $1/s$. Singularities other than poles exist, but not in the transfer functions of ordinary control system components.

**11.7.** Is $P(s) = |s|^2$ analytic in any region of the $s$-plane?

First put $s = \sigma + j\omega$, $s_0 = \sigma_0 + j\omega_0$. Then

$$\frac{dP}{ds}\bigg|_{s=s_0} = \lim_{(s-s_0)\to 0}\left[\frac{|\sigma+j\omega|^2 - |\sigma_0+j\omega_0|^2}{(\sigma+j\omega)-(\sigma_0+j\omega_0)}\right]$$

$$= \lim_{[(\sigma-\sigma_0)+j(\omega-\omega_0)]\to 0}\left[\frac{(\sigma-\sigma_0)(\sigma+\sigma_0)+(\omega-\omega_0)(\omega+\omega_0)}{(\sigma-\sigma_0)+j(\omega-\omega_0)}\right]$$

If the limit exists it must be unique and should not depend on how $s$ approaches $s_0$, or equivalently how $[(\sigma-\sigma_0)+j(\omega-\omega_0)]$ approaches zero. So first let $s\to s_0$ along the $j\omega$-axis and obtain

$$\frac{dP}{ds}\bigg|_{s=s_0} = \lim_{\substack{\omega\to\omega_0\\\sigma=\sigma_0}}\left[\frac{(\omega-\omega_0)(\omega+\omega_0)}{j(\omega-\omega_0)}\right] = -j2\omega_0$$

Now let $s\to s_0$ along the $\sigma$-axis; that is,

$$\frac{dP}{ds}\bigg|_{s=s_0} = \lim_{\substack{\sigma\to\sigma_0\\\omega=\omega_0}}\left[\frac{(\sigma-\sigma_0)(\sigma+\sigma_0)}{\sigma-\sigma_0}\right] = 2\sigma_0$$

Hence the limit *does not* exist for arbitrary nonzero values of $\sigma_0$ and $\omega_0$, and therefore $|s|^2$ is not analytic anywhere in the $s$-plane except possibly at the origin. When $s_0 = 0$,

$$\frac{dP}{ds}\bigg|_{s=0} = \lim_{s\to 0}\left[\frac{|s|^2-0}{s}\right] = \lim_{s\to 0}\left[\frac{(\sigma+j\omega)(\sigma-j\omega)}{\sigma+j\omega}\right] = 0$$

Therefore $P(s) = |s|^2$ is analytic only at the origin, $s = 0$.

**11.8.** If $P(s)$ is analytic at $s_0$, prove that it must be continuous at $s_0$. That is, show that $\lim_{s\to s_0}P(s) = P(s_0)$.

Since

$$P(s) - P(s_0) = \frac{P(s)-P(s_0)}{(s-s_0)}\cdot(s-s_0)$$

for $s \neq s_0$, then

$$\lim_{s\to s_0}[P(s)-P(s_0)] = \lim_{s\to s_0}\left[\frac{P(s)-P(s_0)}{(s-s_0)}\right]\cdot\lim_{s\to s_0}(s-s_0) = \left[\frac{dP}{ds}\bigg|_{s=s_0}\right]\cdot 0 = 0$$

because $(dP/ds)|_{s=s_0}$ exists by hypothesis [i.e., $P(s)$ is analytic]. Therefore

$$\lim_{s\to s_0}[P(s)-P(s_0)] = 0 \quad\text{or}\quad \lim_{s\to s_0}P(s) = P(s_0)$$

**11.9. Polynomial functions** are defined by $Q(s) \equiv a_n s^n + a_{n-1}s^{n-1} + \cdots + a_1 s + a_0$, where $a_n \neq 0$, $n$ is a positive integer called the **degree of the polynomial**, and $a_0, a_1, \ldots, a_n$ are constants. Prove that $Q(s)$ is analytic in every bounded (finite) region of the $s$-plane.

First consider $s^n$:

$$\frac{d}{ds}[s^n]\bigg|_{s=s_0} = \lim_{s\to s_0}\left[\frac{s^n-s_0^n}{s-s_0}\right] = \lim_{s\to s_0}\left(s^{n-1}+s^{n-2}s_0+\cdots+ss_0^{n-2}+s_0^{n-1}\right) = ns_0^{n-1}$$

Thus $s^n$ is analytic in every finite region of the $s$-plane. Then, by mathematical induction, $s^{n-1}, s^{n-2},\ldots, s$ are also analytic. Hence, by the elementary theorems on limits of sums and products, we see that $Q(s)$ is analytic in every finite region of the $s$-plane.

**11.10. Rational algebraic functions** are defined by $P(s) \equiv N(s)/D(s)$, where $N(s)$ and $D(s)$ are polynomials. Show that $P(s)$ is analytic at every point $s$ where $D(s) \neq 0$; that is, prove that the transfer functions of control system elements that take the form of rational algebraic functions are analytic except at their poles.

The overwhelming majority of linear control system elements are in this category. The fundamental theorem of algebra, "a polynomial of degree $n$ has $n$ zeros and can be expressed as a product of $n$ linear factors," helps to put $P(s)$ in a form more recognizable as a control system transfer function; that is, $P(s)$ can be written in the familiar form

$$P(s) \equiv \frac{N(s)}{D(s)} = \frac{b_m s^m + b_{m-1} s^{m-1} + \cdots + b_0}{a_n s^n + a_{n-1} s^{n-1} + \cdots + a_0} = \frac{b_m (s+z_1)(s+z_2)\cdots(s+z_m)}{a_n (s+p_1)(s+p_2)\cdots(s+p_n)}$$

where $-z_1, -z_2, \ldots, -z_n$ are zeros, $-p_1, -p_2, \ldots, -p_n$ are poles, and $m \leq n$.

From the identity given by

$$\frac{N(s)}{D(s)} - \frac{N(s_0)}{D(s_0)} \equiv \frac{1}{D(s)D(s_0)} \big[ D(s_0)(N(s) - N(s_0)) - N(s_0)(D(s) - D(s_0)) \big]$$

where $D(s) \neq 0$, we get

$$\left. \frac{dP}{ds} \right|_{s=s_0} = \lim_{s \to s_0} \left[ \frac{\dfrac{N(s)}{D(s)} - \dfrac{N(s_0)}{D(s_0)}}{s - s_0} \right]$$

$$= \lim_{s \to s_0} \left[ \frac{1}{D(s)D(s_0)} \left( D(s_0) \left[ \frac{N(s) - N(s_0)}{s - s_0} \right] - N(s_0) \left[ \frac{D(s) - D(s_0)}{s - s_0} \right] \right) \right]$$

$$= \lim_{s \to s_0} \left[ \frac{1}{D(s)} \left( \frac{N(s) - N(s_0)}{s - s_0} \right) \right] - \lim_{s \to s_0} \left[ \frac{N(s_0)}{D(s)D(s_0)} \left( \frac{D(s) - D(s_0)}{s - s_0} \right) \right]$$

$$= \lim_{s \to s_0} \left[ \frac{1}{D(s)} \right] \cdot \lim_{s \to s_0} \left[ \frac{N(s) - N(s_0)}{s - s_0} \right] - \lim_{s \to s_0} \left[ \frac{N(s_0)}{D(s)D(s_0)} \right] \cdot \lim_{s \to s_0} \left[ \frac{D(s) - D(s_0)}{s - s_0} \right]$$

$$= \frac{1}{D(s_0)} \cdot \left. \frac{dN}{ds} \right|_{s=s_0} - \frac{N(s_0)}{D(s_0)^2} \cdot \left. \frac{dD}{ds} \right|_{s=s_0}$$

where we have used the results of Problems 11.8, 11.9, and Definition 11.1. Therefore the derivative of $P(s)$ exists ($P(s)$ is analytic) for all points $s$ where $D(s) \neq 0$.

Note that we have determined a formula for the derivative of a rational algebraic function (the last part of the above equation) in terms of the derivatives of its numerator and denominator, in addition to solving the required problem.

**11.11.** Prove that $e^{-sT}$ is analytic in every bounded region of the $s$-plane.

In complex variable theory $e^{-sT}$ is defined by the power series

$$e^{-sT} = \sum_{k=0}^{\infty} \frac{(-sT)^k}{k!}$$

By the ratio test, as $k \to \infty$ we have

$$\left| \frac{(-sT)^k / k!}{(-sT)^{k+1} / (k+1)!} \right| = \left| \frac{k+1}{-sT} \right| \to \infty$$

Hence the radius of convergence of this power series is infinite. The sum of a power series is analytic within its radius of convergence. Thus $e^{-sT}$ is analytic in every bounded region of the $s$-plane.

**11.12.** Prove that $e^{-sT}P(s)$ is analytic wherever $P(s)$ is analytic. Hence systems containing a combination of rational algebraic transfer functions and time-delay operators (i.e., $e^{-sT}$) are analytic except at the poles of the system.

By Problem 11.11, $e^{-sT}$ is analytic in every bounded region of the $s$-plane; and by Problem 11.10, $P(s)$ is analytic except at its poles. Now

$$\frac{d}{ds}\left[e^{-sT}P(s)\right]\bigg|_{s=s_0} = \lim_{s \to s_0}\left[\frac{e^{-sT}P(s) - e^{-s_0 T}P(s_0)}{s - s_0}\right]$$

$$= \lim_{s \to s_0}\left[e^{-sT}\left(\frac{P(s) - P(s_0)}{s - s_0}\right) + P(s_0)\left(\frac{e^{-sT} - e^{-s_0 T}}{s - s_0}\right)\right]$$

$$= e^{-s_0 T}\frac{dP}{ds}\bigg|_{s=s_0} + P(s_0)\frac{d}{ds}\left(e^{-sT}\right)\bigg|_{s=s_0}$$

Therefore $e^{-sT}P(s)$ is analytic wherever $P(s)$ is analytic.

**11.13.** Consider the function given by $P(s) = e^{-sT}(s^2 + 2s + 3)/(s^2 - 2s + 2)$. Where are the singularities of this function? Where is $P(s)$ analytic?

The singular points are at the poles of $P(s)$. Since $s^2 - 2s + 2 = (s - 1 + j1)(s - 1 - j1)$, the two poles are given by $-p_1 = 1 - j1$ and $-p_2 = 1 + j1$. $P(s)$ is analytic in every bounded region of the $s$-plane except at the points $s = -p_1$ and $s = -p_2$.

## CONTOURS AND ENCIRCLEMENTS

**11.14.** What points are *enclosed* by the following contours (Fig. 11-32)?

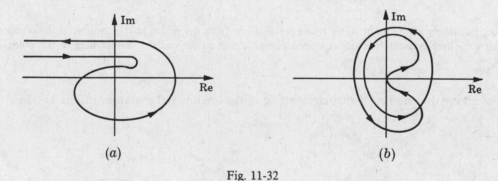

(a)                                            (b)

Fig. 11-32

By shading the region to the right of each contour as it is traversed in the prescribed direction, we get Fig. 11-33. All points in the shaded regions are enclosed.

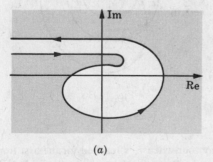
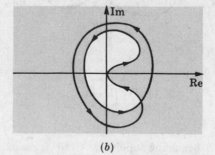

(a)                                            (b)

Fig. 11-33

**11.15.** What contours of Problem 11.14 are *closed*?

Clearly, the contour of part (*b*) is closed. The contour of part (*a*) may or may not close upon itself at infinity in the complex plane. This cannot be determined from the given graph.

**11.16.** What is the *direction* (positive or negative) of each contour in Problem 11.14(*a*) and (*b*)?

Using the origin as a base, each contour is directed in the counterclockwise, *negative* direction about the origin.

**11.17.** Determine the number of encirclements $N_0$ of the origin for the contour in Fig. 11-34.

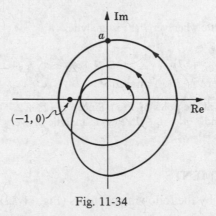

Fig. 11-34

Beginning at the point *a*, we rotate a radial line from the origin to the contour in the direction of the arrows. Three counterclockwise rotations of 360° result in the radial line returning to the point *a*. Hence $N_0 = -3$.

**11.18.** Determine the number of encirclements $N_0$ of the origin for the contour in Fig. 11-35.

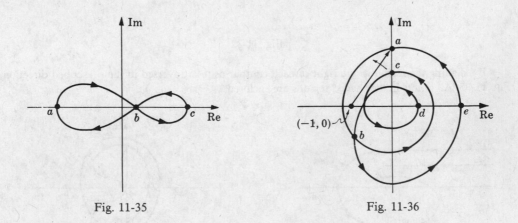

Fig. 11-35                                                        Fig. 11-36

Beginning at point *a*, +180° is swept out by the contour when *b* is reached for the first time. In going from *b* to *c* and back to *b*, the net angular gain is zero. Returning to *a* from *b* yields +180°. Thus $N_0 = +1$.

**11.19.** Determine the number of encirclements $N$ of the $(-1, 0)$ point (i.e., the $-1$ point on the real axis) for the contour of Problem 11.17.

Again beginning at point $a$, we rotate a radial line *from the* $(-1, 0)$ *point* to the contour in the direction of the arrows as shown in Fig. 11-36. In going from $a$ to $b$ to $c$, the radial line sweeps out somewhat less than $-360°$. But from $c$ to $d$ and back to $b$, the angle increases again toward the value reached in going only from $a$ to $b$. Then from $b$ to $e$ to $a$ the resultant angle is $-360°$. Thus $N = -1$.

## PROPERTIES OF THE MAPPING $P$

**11.20.** Are the following functions single-valued: (*a*) $P(s) = s^2$, (*b*) $P(s) = s^{1/2}$?

(*a*) Substitution of any complex number $s$ into $P(s) = s^2$ yields a unique value for $P(s)$. Hence $P(s) = s^2$ is a single-valued function.

(*b*) In polar form we have $s = |s|e^{j\theta}$, where $\theta = \arg(s)$. Therefore $s^{1/2} = |s|^{1/2}e^{j\theta/2}$. Now if we increase $\theta$ by $2\pi$ we return to the same point $s$. But

$$P(s) = |s|^{1/2}e^{j(\theta + 2\pi)/2} = |s|^{1/2}e^{j\theta/2}e^{j\pi} = P(s)e^{j\pi}$$

which is *another* point in the $P(s)$-plane. Hence $P(s) = s^{1/2}$ has two points in the $P(s)$-plane for every point in the $s$-plane. It is not a single-valued function; it is a **multiple-valued function** (with two values).

**11.21.** Prove that every closed contour containing no singular points of $P(s)$ in the $s$-plane maps into a closed contour in the $P(s)$-plane.

Suppose not. Then at some point $s_0$ where the $s$-plane contour closes upon itself the $P(s)$-plane contour is not closed. This means that one (nonsingular) point $s_0$ in the $s$-plane is mapped into more than one point in the $P(s)$-plane (the images of the point $s_0$). This contradicts the fact that $P(s)$ is a single-valued function (Property 1, Section 11.4).

**11.22.** Prove that $P$ is a conformal mapping wherever $P$ is analytic and $dP/ds \neq 0$.

Consider two curves: $C$ in the $s$-plane and $C'$, the image of $C$, in the $P(s)$-plane. Let the curve in the $s$-plane be described by a parameter $t$; that is, each $t$ corresponds to a point $s = s(t)$ along the curve $C$. Hence $C'$ is described by $P[s(t)]$ in the $P(s)$-plane. The derivatives $ds/dt$ and $dP/dt$ represent tangent vectors to corresponding points on $C$ and $C'$. Now

$$\left. \frac{dP[s(t)]}{dt} \right|_{P(s) = P(s_0)} = \frac{ds}{dt} \cdot \left. \frac{dP(s)}{ds} \right|_{s = s_0}$$

where we have used the fact that $P$ is analytic at some point $s_0 \equiv s(t_0)$. Put $dP/dt \equiv r_1 e^{j\phi}$, $dP/ds \equiv r_2 e^{j\alpha}$, and $ds/dt \equiv r_3 e^{j\theta}$. Then

$$r_1(s_0)e^{j\phi(s_0)} = r_2(s_0) \cdot r_3(s_0)e^{j[\theta(s_0) + \alpha(s_0)]}$$

Equating angles, we have $\phi(s_0) = \theta(s_0) + \alpha(s_0) = \theta(s_0) + \arg(dP/ds)|_{s = s_0}$, and we see that the tangent to $C$ at $s_0$ is rotated through an angle $\arg(dP/ds)|_{s = s_0}$ at $P(s_0)$ on $C'$ in the $P(s)$-plane.

Now consider two curves $C_1$ and $C_2$ intersecting at $s_0$, with images $C_1'$ and $C_2'$ in the $P(s)$-plane (Fig. 11-37).

Let $\theta_1$ be the angle of inclination of the tangent to $C_1$, and $\theta_2$ for $C_2$. Then the angles of inclination for $C_1'$ and $C_2'$ are $\theta_1 + \arg(dP/ds)|_{s = s_0}$, and $\theta_2 + \arg(dP/ds)|_{s = s_0}$. Therefore the angle $(\theta_1 - \theta_2)$ between $C_1$ and $C_2$ is equal in magnitude and sense to the angle between $C_1'$ and $C_2'$,

$$\theta_1 + \left. \arg\frac{dP}{ds} \right|_{s = s_0} - \theta_2 - \left. \arg\frac{dP}{ds} \right|_{s = s_0} = \theta_1 - \theta_2$$

Note that $\arg(dP/ds)|_{s = s_0}$ is indeterminate if $(dP/ds)|_{s = s_0} = 0$.

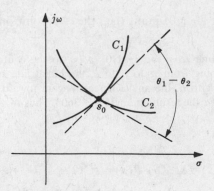

Fig. 11-37

**11.23.** Show that $P(s) = e^{-sT}$ is conformal in every bounded region of the $s$-plane.

$e^{-sT}$ is analytic (Problem 11.11). Moreover, $(d/ds)(e^{-sT}) = -Te^{-sT} \neq 0$ in any bounded (finite) region of the $s$-plane. Then by Problem 11.22, $P(s) = e^{-sT}$ is conformal.

**11.24.** Show that $P(s)e^{-sT}$ is conformal for rational $P(s)$ and $dP/ds \neq 0$.

By Problem 11.12, $Pe^{-sT}$ is analytic except at the poles of $P$. By Problem 11.12,

$$\frac{d}{ds}[Pe^{-sT}] = e^{-sT}\frac{dP}{ds} - PTe^{-sT} = e^{-sT}\left(\frac{dP}{ds} - TP\right)$$

Suppose $(d/ds)[Pe^{-sT}] = 0$. Then since $e^{-sT} \neq 0$ for any finite $s$, we have $dP/ds - TP = 0$ whose general solution is $P(s) = ke^{sT}$, $k$ constant. But $P$ is rational and $e^{sT}$ is not. Hence $(d/ds)[Pe^{-sT}] \neq 0$.

**11.25.** Two $s$-plane contours $C_1$ and $C_2$ intersect in a 90° angle in Fig. 11-38. The analytic function $P(s)$ maps these contours into the $P(s)$-plane and $dP/ds \neq 0$ at $s_0$. Sketch the image of contour $C_2$ in a neighborhood of $P(s_0)$. The image of $C_1$ is also given.

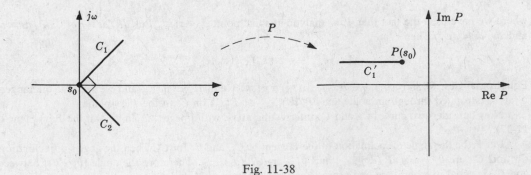

Fig. 11-38

By Problem 11.22, $P$ is conformal; hence the angle between $C_1'$ and $C_2'$ is 90°. Since $C_1$ makes a left turn onto $C_2$ at $s_0$, then $C_1'$ must also turn left at $P(s_0)$ (Fig. 11-39).

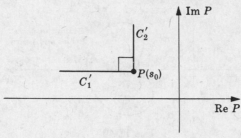

Fig. 11-39

**11.26.** Prove Equation (*11.1*):  $N_0 = Z_0 - P_0$.

The bulk of the proof is somewhat more involved than can be handled with the complex-variable theory presented in this book. So we assume knowledge of a well-known theorem of functions of a complex variable and continue from there. The theorem states that if $C$ is a closed contour in the $s$-plane, $P(s)$ an analytic function on $C$ and within $C$ except for possible poles, and $P(s) \neq 0$ on $C$, then

$$\frac{1}{2\pi j} \int_C \frac{P'(s)}{P(s)} \, ds = Z_0 - P_0$$

where $Z_0$ is the total number of zeros inside $C$, $P_0$ the total number of poles inside $C$, and $P' \equiv dP/ds$. Multiple poles and zeros are counted one for one; that is, a double pole at a point is two poles of the total, a triple zero is three zeros of the total.

Now since $d[\ln P(s)] = [P'(s)/P(s)] \, ds$ and $\ln P(s) \equiv \ln|P(s)| + j \arg P(s)$, we have

$$\frac{1}{2\pi j} \int_C \left[ \frac{P'(s)}{P(s)} \right] ds = \frac{1}{2\pi j} \int_C d[\ln P(s)] = \frac{1}{2\pi j} [\ln P(s)] \Big|_C = \frac{1}{2\pi j} \left[ \ln|P(s)| + j \arg P(s) \right] \Big|_C$$

$$= \frac{1}{2\pi j} \left[ \ln|P(s)| \right] \Big|_C + \frac{1}{2\pi j} \left[ j \arg P(s) \right] \Big|_C$$

Now since $\ln|P(s)|$ returns to its original value when we go once around $C$, the first term in the last equation is zero. Hence

$$Z_0 - P_0 = \frac{1}{2\pi} \left[ \arg P(s) \right] \Big|_C$$

Since $C$ is closed, the image of $C$ in the $P(s)$-plane is closed, and the net change in the angle $\arg P(s)$ around the $P(s)$ contour is $2\pi$ times the number of encirclements $N_0$ of the origin in the $P(s)$-plane. Then $Z_0 - P_0 = 2N_0\pi/2\pi = N_0$. This result is often called *the principle of the argument*. Note that this result would be the same if we replaced $s$ by $z$ in all of the above. Therefore Equation (*11.1*) is valid for discrete-time systems as well.

**11.27.** Determine the number $N_0$ of $P$-plane contour encirclements for the complex-plane contour mapped into the $P$-plane shown in Fig. 11-40.

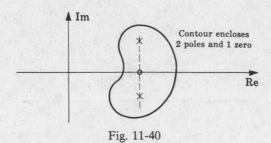

Fig. 11-40

$P_0 = 2$, $Z_0 = 1$. Therefore $N_0 = 1 - 2 = -1$.

**11.28.** Determine the number of zeros $Z_0$ enclosed by the complex-plane contour in Fig. 11-41, where $P_0 = 5$.

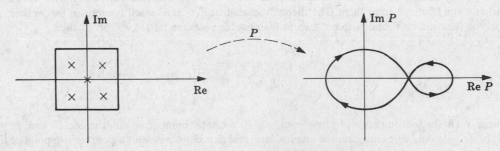

Fig. 11-41

$N_0 = 1$ was computed in Problem 11.18 for the given $P$-plane contour. Since $P_0 = 5$, then $Z_0 = N_0 + P_0 = 1 + 5 = 6$.

**11.29.** Determine the number of poles $P_0$ enclosed by the complex-plane contour in Fig. 11-42, where $Z_0 = 0$.

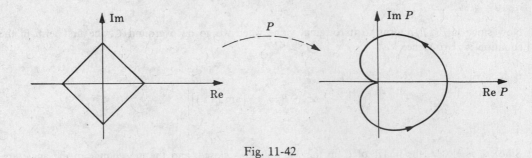

Fig. 11-42

Clearly, $N_0 = -1$. Hence $P_0 = Z_0 - N_0 = 0 + 1 = 1$.

**11.30.** Determine $N_0$ [Equation (*11.1*)] for the transfer function (transformation) and $s$-plane contour of Fig. 11-43.

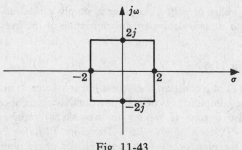

Fig. 11-43

The pole-zero map of $P(s)$ is given in Fig. 11-44. Hence three poles (two at $s = 0$ and one at $s = -1$) and no zeros are enclosed by the contour. Thus $P_0 = 3$, $Z_0 = 0$, and $N_0 = -3$.

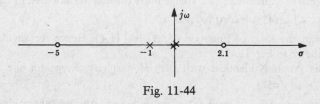

Fig. 11-44

**11.31.** Is the origin *enclosed* by the contour in Fig. 11-45?

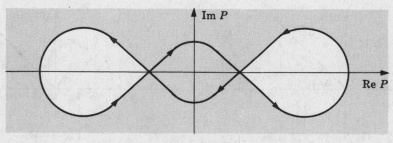

Fig. 11-45

The region to the right of the contour has been shaded. The origin falls in a shaded region and is therefore *enclosed* by the contour.

**11.32.** What is the sign of $N_0$ in Problem 11.31?

Since the origin is enclosed by the contour in a clockwise direction, $N_0 > 0$.

## POLAR PLOTS

**11.33.** Prove Property 1 of Section 11.6.

Let $P(\omega) \equiv P_1(\omega) + jP_2(\omega)$ and $a \equiv a_1 + ja_2$, where $P_1(\omega)$, $P_2(\omega)$, $a_1$, and $a_2$ are real. Then

$$P(\omega) + a = (P_1(\omega) + a_1) + j(P_2(\omega) + a_2)$$

and the image of any point $(P_1(\omega), P_2(\omega))$ in the $P(\omega)$-plane is $(P_1(\omega) + a_1, P_2(\omega) + a_2)$ in the

$(P(\omega) + |a)$-plane. Hence the image of a $P(\omega)$ contour is simply a *translation* (see Problem 11.3). Clearly, translation of the contour by $a$ units is equivalent to translation of the axes (origin) by $-a$ units.

**11.34.** Prove Property 2 of Section 11.6.

The transfer function $P(s)$ of a constant-coefficient linear system is, in general, a ratio of polynomials with constant coefficients. The complex roots of such polynomials occur in conjugate pairs; that is, if $a + jb$ is a root, then $a - jb$ is also a root. If we let an asterisk (*) represent complex conjugation, then $a + jb = (a - jb)^*$, and if $a = 0$, then $jb = (-jb)^*$. Therefore $P(j\omega) = P(-j\omega)^*$ or $P(-j\omega) = P(j\omega)^*$. Graphically this means that the plot for $P(-j\omega)$ is the mirror image about the real axis of the plot for $P(j\omega)$ since only the imaginary part of $P(j\omega)$ changes sign.

**11.35.** Sketch the Polar Plot of each of the following complex functions:

(a) $P(j\omega) = \omega^2 \underline{/45^\circ}$, (b) $P(j\omega) = \omega^2(\cos 45^\circ + j \sin 45^\circ)$, (c) $P(j\omega) = 0.707\omega^2 + 0.707j\omega^2$.

(a) $\omega^2 \underline{/45^\circ}$ is in the form of Equation (*11.2*). Hence polar coordinates are used in Fig. 11-46.

(b) $P(j\omega) = \omega^2(\cos 45^\circ + j \sin 45^\circ) = \omega^2(0.707 + 0.707j)$

That is, $P(j\omega)$ is in the form of Equation (*11.3*) or (*11.4*). Hence rectangular coordinates is the natural choice as shown in Fig. 11-47.

Note that this graph is identical with that of part (a) except for the coordinates. In fact, $\omega^2(0.707 + 0.707j) = \omega^2 \underline{/45^\circ}$.

(c) Clearly, (c) is identical with (b), and therefore with (a). Among other things, this problem has illustrated how a complex function of frequency $\omega$ can be written in three different but mathematically and graphically identical forms: the polar form, Equation (*11.2*); the trigonometric or *Euler form*, Equation (*11.3*); and the equivalent rectangular (complex) form, Equation (*11.4*).

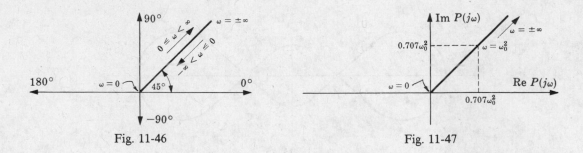

Fig. 11-46                                     Fig. 11-47

**11.36.** Sketch the Polar Plot of

$$P(j\omega) = 0.707\omega^2(1 + j) + 1$$

The Polar Plot of $0.707\omega^2(1 + j)$ was drawn in Problem 11.35(b). By Property 1 of Section 11.6, the required Polar Plot is given by that of Problem 11.35(b) with its origin shifted to $-a = -1$ as shown in Fig. 11-48.

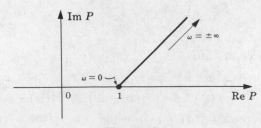

Fig. 11-48

**11.37.** Construct a Polar Plot from the set of graphs of the magnitude and phase angle of $P(j\omega)$ in Fig. 11-49, representing the frequency response of a linear constant-coefficient system.

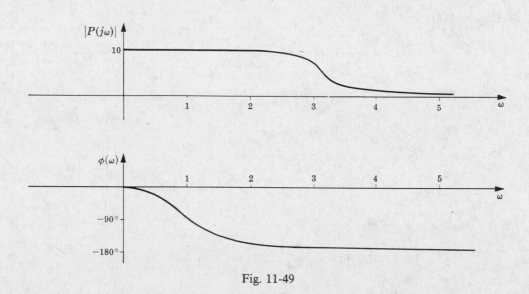

Fig. 11-49

The graphs shown above differ little from *Bode representations*, discussed in detail in Chapter 15. The Polar Plot is constructed by mapping this set of graphs into the $P(j\omega)$-plane. It is only necessary to choose values of $\omega$ and corresponding values of $|P(j\omega)|$ and $\phi(\omega)$ from the graphs and plot these points in the $P(j\omega)$-plane. For example at $\omega = 0$, $|P(j\omega)| = 10$ and $\phi(\omega) = 0$. The resulting Polar Plot is given in Fig. 11-50.

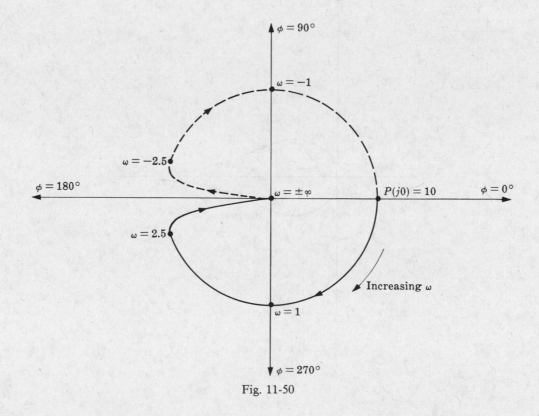

Fig. 11-50

The portion of the graph for $-\infty < \omega < 0$ has been drawn using the property of conjugate symmetry (Section 11.6).

**11.38.** Sketch the Polar Plot for

$$GH(s) = \frac{1}{s^4(s+p)} \qquad p > 0$$

Substituting $j\omega$ for $s$, and applying Equation (*11.2*), we obtain

$$GH(j\omega) = \frac{1}{j^4\omega^4(j\omega + p)}$$

$$= \frac{1}{\omega^4\sqrt{\omega^2 + p^2}} \underline{/-\tan^{-1}(\omega/p)}$$

For $\omega = 0$ and $\omega \to \infty$, we have

$$GH(j0) = \infty \underline{/0°} \qquad \lim_{\omega \to \infty} GH(j\omega) = 0 \underline{/-90°}$$

Clearly, as $\omega$ increases from zero to infinity, the phase angle remains negative and decreases to $-90°$, and the magnitude decreases monotonically to zero. Thus the Polar Plot may be sketched as shown in Fig. 11-51. The dashed line represents the mirror image of the plot for $0 < \omega < \infty$ (Section 11.6, Property 2). hence it is the Polar Plot for $-\infty < \omega < 0$.

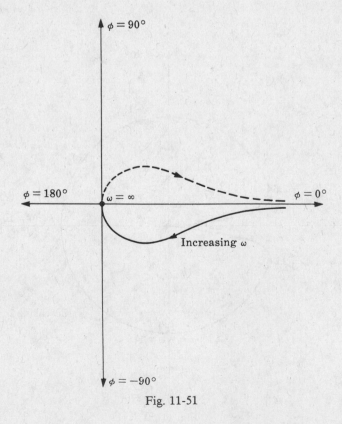

Fig. 11-51

## THE NYQUIST PATH

**11.39.** Prove that the infinite semicircle, portion $\overline{def}$ of the Nyquist Path, maps into the origin $P(s) = 0$ in the $P(s)$-plane for all transfer functions of the form:

$$P(s) = \frac{K}{\displaystyle\prod_{i=1}^{n}(s + p_i)}$$

where $n > 0$, $K$ is a constant, and $-p_i$ is any finite pole.

For $n > 0$,

$$\left| \lim_{R \to \infty} P(Re^{j\theta}) \right| \equiv |P(\infty)| = \lim_{R \to \infty} \left| \frac{K}{\displaystyle\prod_{i=1}^{n}(Re^{j\theta} + p_i)} \right|$$

$$= \lim_{R \to \infty} \frac{|K|}{\displaystyle\prod_{i=1}^{n}|Re^{j\theta} + p_i|} \leq \lim_{R \to \infty} \frac{|K|}{\displaystyle\prod_{i=1}^{n}|R - |p_i||} = 0$$

Since $|P(\infty)| \leq 0$, then clearly $|P(\infty)| \equiv 0$.

**11.40.** Prove that the infinite semicircle, portion $\overline{def}$ of the Nyquist Path, maps into the origin $P(s) = 0$ in the $P(s)$-plane for all transfer functions of the form:

$$P(s) = \frac{K\displaystyle\prod_{i=1}^{m}(s + z_i)}{\displaystyle\prod_{i=1}^{n}(s + p_i)}$$

where $m < n$, $K$ is a constant, and $-p_i$ and $-z_i$ are finite poles and zeros, respectively.

For $m < n$,

$$\left| \lim_{R \to \infty} P(Re^{j\theta}) \right| \equiv |P(\infty)| = \lim_{R \to \infty} \left| \frac{K\displaystyle\prod_{i=1}^{m}(Re^{j\theta} + z_i)}{\displaystyle\prod_{i=1}^{n}(Re^{j\theta} + p_i)} \right|$$

$$= \lim_{R \to \infty} \frac{|K|\displaystyle\prod_{i=1}^{m}|Re^{j\theta} + z_i|}{\displaystyle\prod_{i=1}^{n}|Re^{j\theta} + p_i|} \leq \lim_{R \to \infty} \frac{|K|\displaystyle\prod_{i=1}^{m}|R + |z_i||}{\displaystyle\prod_{i=1}^{n}|R - |p_i||} = 0$$

Since $|P(\infty)| \leq 0$, then $|P(\infty)| \equiv 0$.

## NYQUIST STABILITY PLOTS

**11.41.** Prove that a continuous type $l$ system includes $l$ infinite semicircles in the locus of its Nyquist Stability Plot. That is, show that portion $\overline{ija}$ of the Nyquist Path maps into an arc of $180l$ degrees at infinity in the $P(s)$-plane.

The transfer function of a continuous type $l$ system has the form:

$$P(s) = \frac{B_1(s)}{s^l B_2(s)}$$

where $B_1(0)$ and $B_2(0)$ are finite and nonzero. If we let $B_1(s)/B_2(s) \equiv F(s)$, then

$$P(s) = \frac{F(s)}{s^l}$$

where $F(0)$ is finite and nonzero. Now put $s = \rho e^{j\theta}$, as required by Equation (*11.12*). Clearly, $\lim_{\rho \to 0} F(\rho e^{j\theta}) = F(0)$. Then $P(\rho e^{j\theta}) = F(\rho e^{j\theta})/\rho^l e^{jl\theta}$ and

$$\lim_{\rho \to 0} P(\rho e^{j\theta}) = \infty \cdot e^{-jl\theta} \qquad -90° \leq \theta \leq +90°$$

At $\theta = -90°$, the limit is $\infty \cdot e^{j90l}$. At $\theta = +90°$, the limit is $\infty \cdot e^{-j90l}$. Hence the angle subtended in the $P(s)$-plane, by mapping the locus of the infinitesimal semicircle of the Nyquist Path in the neighborhood of the origin in the $s$-plane, is $90l - (-90l) = 180l$ degrees, which represents $l$ infinite semicircles in the $P(s)$-plane.

**11.42.** Sketch the Nyquist Stability Plot for the open-loop transfer function given by

$$GH(s) = \frac{1}{(s + p_1)(s + p_2)} \qquad p_1, p_2 > 0$$

The Nyquist Path for this type 0 system is shown in Fig. 11-52.

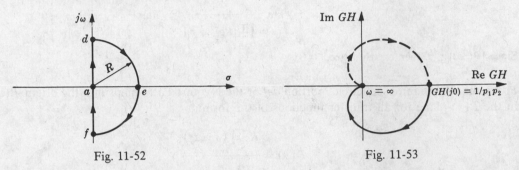

Fig. 11-52　　　　　　　　　　　　　　　　　　　Fig. 11-53

Since there are no poles on the $j\omega$-axis, Step 2 of Section 11.8 indicates that the Polar Plot of $GH(j\omega)$ yields the image of path $\overline{ad}$ (and hence $\overline{fad}$) in the $GH(s)$-plane. Letting $s = j\omega$ for $0 < \omega < \infty$, we get

$$GH(j\omega) = \frac{1}{(j\omega + p_1)(j\omega + p_2)} = \frac{1}{\sqrt{(\omega^2 + p_1^2)(\omega^2 + p_2^2)}} \Big/ -\tan^{-1}\left(\frac{\omega}{p_1}\right) - \tan^{-1}\left(\frac{\omega}{p_2}\right)$$

$$GH(j0) = \frac{1}{p_1 p_2} \Big/ 0° \qquad \lim_{\omega \to \infty} GH(j\omega) = 0 \Big/ 180°$$

For $0 < \omega < \infty$, the Polar Plot passes through the third and fourth quadrants because $\phi = -[\tan^{-1}(\omega/p_1) + \tan^{-1}(\omega/p_2)]$ varies from $0°$ to $180°$ when $\omega$ increases.

From Problem 11.39, path $\overline{def}$ plots into the origin $P(s) = 0$. Therefore the Nyquist Stability Plot is a replica of the Polar Plot. This is easily sketched from the above derivations, and is shown in Fig. 11-53.

**11.43.** Sketch the Nyquist Stability Plot for $GH(s) = 1/s$.

Mathcad

The Nyquist Path for this simple type 1 system is shown in Fig. 11-54.

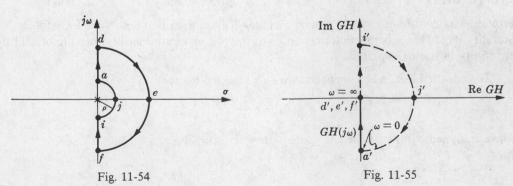

Fig. 11-54　　　　　　　　　　　　　　　　　　　Fig. 11-55

For path $\overline{ad}$, $s = j\omega$, $0 < \omega < \infty$, and

$$GH(j\omega) = \frac{1}{j\omega} = \frac{1}{\omega}\bigg/\!-90° \qquad \lim_{\omega \to 0} GH(j\omega) = \infty\bigg/\!-90° \qquad \lim_{\omega \to \infty} GH(j\omega) = 0\bigg/\!-90°$$

Path $\overline{def}$ maps into the origin (see Problem 11.39).

Path $\overline{f'i'}$ is the mirror image of $\overline{a'd'}$ about the real axis.

The image of path $\overline{ija}$ is determined from Equation *(11.12)*, by letting $s = \lim_{p \to 0}\rho e^{j\theta}$, where $-90° \le \theta \le 90°$:

$$\lim_{\rho \to 0} GH(\rho e^{j\theta}) = \lim_{\rho \to 0}\left[\frac{1}{\rho}e^{-i\theta}\right] = \infty \cdot e^{-i\theta} = \infty\bigg/\!-\theta$$

For point $i$, $\theta = -90°$; then $i$ maps into $i'$ at $\infty\big/90°$. At point $j$, $\theta = 0°$; then $j$ maps into $j'$ at $\infty\big/0°$. Similarly, $a$ maps into $a'$ at $\infty\big/\!-90°$. Path $\overline{i'j'a'}$ could also have been obtained from the conformal mapping property of the transformation as explained in Example 11.9 plus the statement proved in Problem 11.41.

The resulting Nyquist Stability Plot is shown in Fig. 11-55.

**11.44.** Sketch the Nyquist Stability Plot for $GH(s) = 1/s(s + p_1)(s + p_2)$, $p_1, p_2 > 0$.

The Nyquist Path for this type 1 system is the same as that for the preceding problem. For path $\overline{ad}$, $s = j\omega$, $0 < \omega < \infty$, and

$$GH(j\omega) = \frac{1}{j\omega(j\omega + p_1)(j\omega + p_2)} = \frac{1}{\omega\sqrt{(\omega^2 + p_1^2)(\omega^2 + p_2^2)}}\bigg/\!-90° - \tan^{-1}\!\left(\frac{\omega}{p_1}\right) - \tan^{-1}\!\left(\frac{\omega}{p_2}\right)$$

$$\lim_{\omega \to 0} GH(j\omega) = \infty\bigg/\!-90° \qquad \lim_{\omega \to \infty} GH(j\omega) = 0\bigg/\!-270° = 0\bigg/\!+90°$$

Since the phase angle changes sign as $\omega$ increases, the plot crosses the real axis. At intermediate values of frequency, the phase angle $\phi$ is within the range $-90° < \phi < -270°$. Hence the plot is in the second and third quadrants. An asymptote of $GH(j\omega)$ for $\omega \to 0$ is found by writing $GH(j\omega)$ as a real plus an imaginary part, and *then* taking the limit as $\omega \to 0$:

$$GH(j\omega) = \frac{-(p_1 + p_2)}{(\omega^2 + p_1^2)(\omega^2 + p_2^2)} - \frac{j(p_1 p_2 - \omega^2)}{\omega(\omega^2 + p_1^2)(\omega^2 + p_2^2)} \qquad \lim_{\omega \to 0} GH(j\omega) = \frac{-(p_1 + p_2)}{p_1^2 p_2^2} - j\infty$$

Hence the line $GH = -(p_1 + p_2)/p_1^2 p_2^2$ is an asymptote of the Polar Plot.

Path $\overline{def}$ maps into the origin (see Problem 11.39). Path $\overline{f'i'}$ is the mirror image of $\overline{a'd'}$ about the real axis. Path $\overline{i'j'a'}$ is most easily determined by the conformal mapping property and the fact that a type 1 system has *one* infinite semicircle in its path (Problem 11.41). The resulting Nyquist Stability Plot is shown in Fig. 11-56.

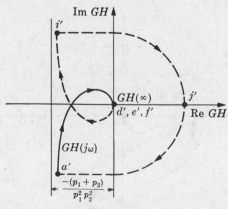

Fig. 11-56

**11.45.** Sketch the Nyquist Stability Plot for $GH(s) = 1/s^2$.

The Nyquist Path for this type 2 system is the same as that for the preceding problem, except there are two poles at the origin instead of one. For $\overline{ad}$,

$$GH(j\omega) = \frac{1}{j^2\omega^2} = \frac{1}{\omega^2}\bigg/\underline{180°} \qquad \lim_{\omega \to 0} GH(j\omega) = \infty\bigg/\underline{180°} \qquad \lim_{\omega \to \infty} GH(j\omega) = 0\bigg/\underline{180°}$$

The Polar Plot clearly lies along the negative real axis, increasing from $-\infty$ to $0$ as $\omega$ increases. Path $\overline{def}$ maps into the origin and path $\overline{ija}$ maps into *two* infinite semicircles at infinity (see Problem 11.41). Since the Nyquist Path makes right turns at $i$ and $a$, so does the Nyquist Stability Plot at $i'$ and $a'$. The resulting locus is shown in Fig. 11-57.

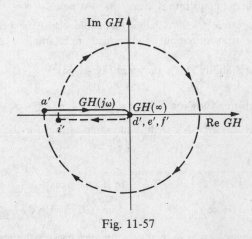

Fig. 11-57

**11.46.** Sketch the Nyquist Stability Plot for $GH(s) = 1/s^2(s + p)$, $p > 0$.

The Nyquist Path for this type 2 system is the same as that for the previous problem. For $\overline{ad}$,

$$GH(j\omega) = \frac{1}{j^2\omega^2(j\omega + p)} = \frac{1}{\omega^2\sqrt{\omega^2 + p^2}}\bigg/\underline{-180° - \tan^{-1}\left(\frac{\omega}{p}\right)}$$

$$\lim_{\omega \to 0} GH(j\omega) = \infty\bigg/\underline{-180°} \qquad \lim_{\omega \to \infty} GH(j\omega) = 0\bigg/\underline{-270°}$$

For $0 < \omega < \infty$ the phase angle varies continuously from $-180°$ to $-270°$; thus the plot lies in the second quadrant. The remainder of the Nyquist Path is mapped into the $GH$-plane as in the preceding problem. The resulting Nyquist Stability Plot is shown in Fig. 11-58.

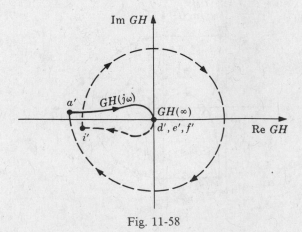

Fig. 11-58

**11.47.** Sketch the Nyquist Stability Plot for $GH(s) = 1/s^4(s+p)$, $p > 0$.

There are four poles at the origin in the $s$-plane, and the Nyquist Path is the same as that of the previous problem. The Polar Plot for this system was determined in Problem 11.38. The remainder of the Nyquist Path is mapped using the results of Problems 11.39 and 11.41, and the conformal mapping property. The resulting Nyquist Stability Plot is given in Fig. 11-59.

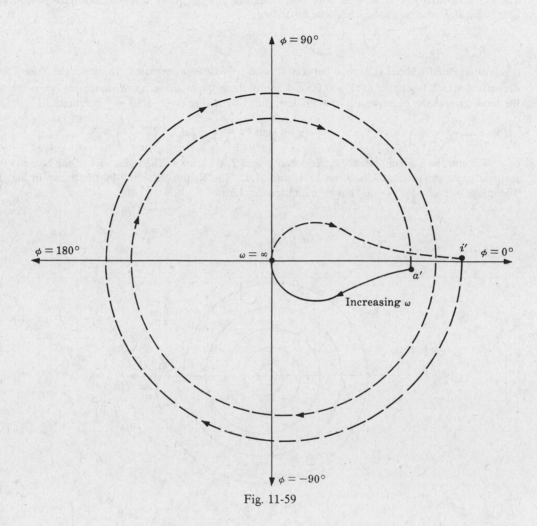

Fig. 11-59

**11.48.** Sketch the Nyquist Stability Plot for $GH(s) = e^{-Ts}/(s+p)$, $p > 0$.

The $e^{-Ts}$ term represents a time delay of $T$ seconds in the forward or feedback path. For example, a signal flow graph of such a system can be represented as in Fig. 11-60.

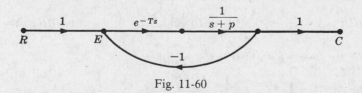

Fig. 11-60

The Nyquist Stability Plot for $1/(s + 1)$ was drawn in Example 11.8. The plot is modified by inclusion of the $e^{-Ts}$ term in the following manner. For path $\overline{ad}$,

$$GH(j\omega) = \frac{e^{-Tj\omega}}{j\omega + p} = \frac{1}{\sqrt{\omega^2 + p^2}} \bigg/ -\tan^{-1}\left(\frac{\omega}{p}\right) - T\omega \qquad GH(j0) = \frac{1}{p} \bigg/ 0°$$

The limit of $GH(j\omega)$ as $\omega \to \infty$ does not exist. But $\lim_{\omega \to \infty} |GH(j\omega)| = 0$ and $|GH(j\omega)|$ decreases monotonically as $\omega$ increases. The phase angle term

$$\phi(\omega) = -\tan^{-1}\left(\frac{\omega}{p}\right) - T\omega$$

revolves repeatedly about the origin between $0°$ and $-360°$ as $\omega$ increases. Therefore the Polar Plot is a decreasing spiral, beginning at $(1/p) \big/ 0°$ and approaching the origin in a CW direction. The points where the locus crosses the negative real axis are determined by letting $\phi = -180° = -\pi$ radians:

$$-\pi = -\tan^{-1}\left(\frac{\omega_\pi}{p}\right) - T\omega_\pi$$

or $\omega_\pi = p \tan(T\omega_\pi)$, which is easily solved when $p$ and $T$ are known. The remainder of the Nyquist Path is mapped using the results of Problems 11.41 and 11.42. The Nyquist Stability Plot is shown in Fig. 11-61. The image of path $\overline{fa}$ ($s = -j\omega$) has been omitted for clarity.

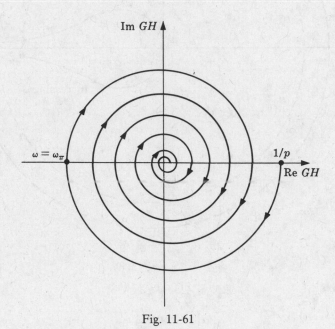

Fig. 11-61

**11.49.** Sketch the Nyquist Stability Plot for $GH(s) = 1/(s^2 + a^2)$.

The poles of $GH(s)$ are at $s = \pm ja \equiv \pm j\omega_0$. The Nyquist Path for this system is therefore as shown in Fig. 11-62.

For path $\overline{ab}$, $\omega < a$ and

$$GH(j\omega) = \frac{1}{a^2 - \omega^2} \bigg/ 0° \qquad GH(j0) = \frac{1}{a^2} \bigg/ 0° \qquad \lim_{\omega \to a} GH(j\omega) = \infty \bigg/ 0°$$

For path $\overline{bc}$, let $s \equiv ja + \rho e^{j\theta}$, $-90° \leq \theta \leq 90°$; then

$$\lim_{\rho \to 0} GH(ja + \rho e^{j\theta}) = \lim_{\rho \to 0}\left[\frac{1}{\rho e^{j\theta}(2ja + \rho e^{j\theta})}\right] = -j\infty \cdot e^{-j\theta} = \infty \bigg/ -\theta^0 - 90°$$

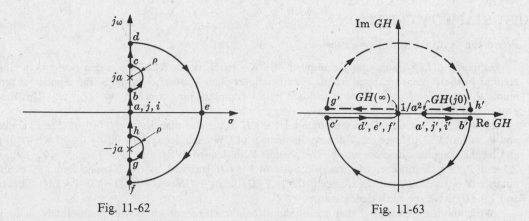

Fig. 11-62                                                          Fig. 11-63

At $\theta = -90°$ the limit is $\infty \underline{/0°}$; at $\theta = 0°$ it is $\infty \underline{/-90°}$; at $\theta = 90°$ it is $\infty \underline{/-180°}$.
For path $\overline{cd}$, $\omega > a$ and

$$\lim_{\omega \to a} GH(j\omega) = \infty \underline{/180°} \qquad \lim_{\omega \to \infty} GH(j\omega) = 0 \underline{/180°}$$

Path $\overline{def}$ maps into the origin by Problem 11.39, and $\overline{f'g'h'a'}$ is the mirror image of $\overline{a'b'c'd'}$ about the real axis. The resulting Nyquist Stability Plot is shown in Fig. 11-63.

**11.50.** Sketch the Nyquist Stability Plot for $GH(s) = (s - z_1)/s(s + p)$, $z_1, p > 0$.

The Nyquist Path for this type 1 system is the same as that for Problem 11.43. For path $\overline{ad}$,

$$GH(j\omega) = \frac{j\omega - z_1}{j\omega(j\omega + p)} = \frac{\sqrt{\omega^2 + z_1^2}}{\omega\sqrt{\omega^2 + p^2}} \underline{\bigg/ 90° - \tan^{-1}\bigg[\frac{\omega(p + z_1)}{pz_1 - \omega^2}\bigg]}$$

where we have used

$$\tan^{-1}x \pm \tan^{-1}y \equiv \tan^{-1}\bigg[\frac{x \pm y}{1 \mp xy}\bigg]$$

Now

$$\lim_{\omega \to 0} GH(j\omega) = \infty \underline{/+90°} \qquad GH(j\sqrt{pz_1}) = \frac{1}{p\sqrt{pz_1}} \underline{/0°} \qquad \lim_{\omega \to \infty} GH(j\omega) = 0 \underline{/-90°}$$

Thus the locus comes down in the first quadrant, crosses the positive real axis into the fourth quadrant, and approaches the origin from an angle of $-90°$.

Path $\overline{def}$ maps into the origin, and $\overline{ija}$ maps into one semicircle at infinity. The resulting plot is shown in Fig. 11-64.

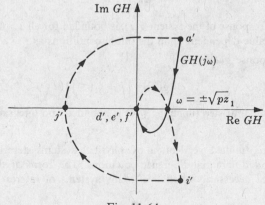

Fig. 11-64

## NYQUIST STABILITY CRITERION

**11.51.** Prove the Nyquist Stability Criterion.

Equation ($11.1$) states that the number of CW encirclements $N_0$ of the origin made by a closed $P$ contour in the $P$-plane, mapped from a closed complex-plane contour, is equal to the number of zeros $Z_0$ minus the number of poles $P_0$ of $P$ enclosed by the complex-plane contour: $N_0 = Z_0 - P_0$. This has been proven in Problem 11.26.

Now let $P \equiv 1 + GH$. Then the origin for $1 + GH$ in the $GH$-plane is at $GH = -1$. (See Example 11.6 and Problem 11.33.) Hence let $N$ denote the number of CW encirclements of this $-1 + j0 \equiv (-1, 0)$ point, and let the complex-plane contour be the Nyquist Path defined in Section 11.7. Then $N = Z_0 - P_0$, where $Z_0$ and $P_0$ are the number of zeros and poles of $1 + GH$ enclosed by the Nyquist Path. $P_0$ is also the number of poles of $GH$ enclosed, since if $GH \equiv N/D$, then $1 + GH = 1 + N/D = (D + N)/D$. That is, $GH$ and $1 + GH$ have the same denominator.

We know from Chapter 5 that a continuous (or discrete) feedback system is absolutely stable if and only if the zeros of the characteristic polynomial $1 + GH$ (the roots of the characteristic equation $1 + GH = 0$) are in the LHP (or unit circle), that is, $Z_0 = 0$. Therefore $N = -P_0$, and clearly $P_0 \geq 0$.

**11.52.** Extend the Nyquist Stability Criterion to a larger class of continuous linear systems than those already considered in this chapter.

The Nyquist Stability Criterion has been extended by Desoer [5]. The following statement is a modification of this generalization, found with its proof in the reference.

*A Generalized Nyquist Stability Criterion*: Consider the linear time-invariant system described by the block diagram in Fig. 11-65. If $g(t)$ satisfies the conditions given below and the Nyquist Stability Plot of $G(s)$ *does not* enclose the $(-1, 0)$ point, then the system is *stable*. If the $(-1, 0)$ point is enclosed, the system is unstable.

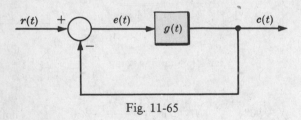

Fig. 11-65

1.  $G(s)$ represents a causal, linear time-invariant system element.

2.  The input-output relationship for $g(t)$ is

$$c(t) = c_a(t) + \int_0^t g(t - \tau) e(\tau) \, d\tau \qquad t \geq 0$$

where $c_a(t)$, the free response of the system $g(t)$, is bounded for all $t \geq 0$ and all initial conditions, and approaches a finite value dependent upon the initial conditions as $t \to \infty$.

3.  The unit impulse response $g(t)$ is

$$g(t) = [k + g_1(t)]\mathbf{1}(t)$$

where $k \geq 0$, $\mathbf{1}(t)$ is the unit step function, $g_1(t)$ is bounded and integrable for all $t \geq 0$, and $g_1(t) \to 0$ as $t \to \infty$.

These conditions are fulfilled very often by physical systems described by ordinary and partial differential equations, and differential-difference equations. The form of the closed-loop block diagram given in Fig. 11-65 is not necessarily restrictive. Many systems of interest can be transformed into this configuration.

**11.53.** Suppose the Nyquist Path for $GH(s) = 1/s(s + p)$ were modified so that the pole at the origin is enclosed as shown in Fig. 11-66. How does this modify application of the Nyquist Stability Criterion?

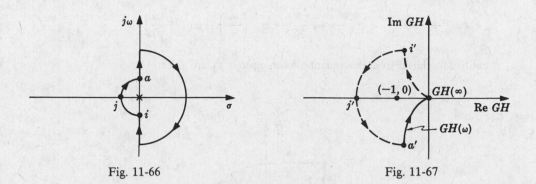

Fig. 11-66                                                         Fig. 11-67

The Polar Plot remains the same, but the image of path $\overline{ija}$ makes *left* instead of right turns at $i'$ and $a'$, just as in the Nyquist Path. The Nyquist Stability Plot is therefore given by Fig. 11-67. Clearly, $N = -1$. But since the pole of $GH$ at the origin is enclosed by the Nyquist Path, then $P_0 = 1$, and $Z_0 = N + P_0 = -1 + 1 = 0$. Therefore the system is stable. Application of the Nyquist Stability Criterion does not depend on the path chosen in the $s$-plane.

**11.54.** Is the system of Problem 11.42 stable or unstable?

Shading the region to the right of the contour in the prescribed direction yields Fig. 11-68. It is clear that $N = 0$. The $(-1, 0)$ point is not in the shaded region. Now, since $p_1 > 0$ and $p_2 > 0$, then $P_0 = 0$. Therefore $N = -P_0 = 0$, or $Z_0 = N + P_0 = 0$, and the system is stable.

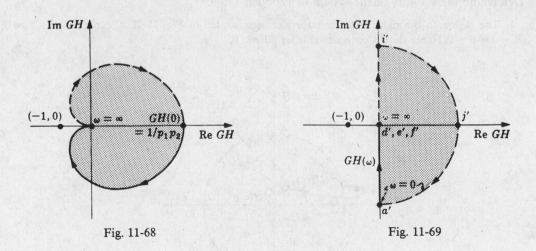

Fig. 11-68                                                         Fig. 11-69

**11.55.** Is the system of Problem 11.43 stable or unstable?

The region to the right of the contour has been shaded in Fig. 11-69. The $(-1, 0)$ point is not enclosed, and $N = 0$. Since $P_0 = 0$, then $Z_0 = P_0 + N = 0$, and the system is stable.

**11.56.** Determine the stability of the system of Problem 11.44.

The region to the right of the contour has been shaded in Fig. 11-70. If the $(-1,0)$ point lies to the left of point $k$, then $N = 0$; if it lies to the right, then $N = 1$. Since $P_0 = 0$, then $Z_0 = 0$ or 1. Hence the system is stable if and only if the $(-1,0)$ point lies to the left of point $k$. Point $k$ can be determined by solving for $GH(j\omega_\pi)$, where

$$-\pi = \frac{-\pi}{2} - \tan^{-1}\left(\frac{\omega_\pi}{p_1}\right) - \tan^{-1}\left(\frac{\omega_\pi}{p_2}\right)$$

$\omega_\pi$ is easily determined from this equation when $p_1$ and $p_2$ are given.

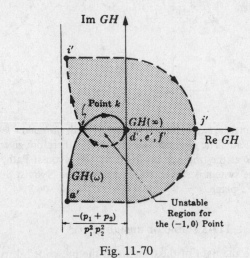

Fig. 11-70

**11.57.** Determine the stability of the system of Problem 11.46.

The region to the right of the contour has been shaded in Fig. 11-71. Clearly, $N = 1$, $P_0 = 0$, and $Z_0 = 1 + 0 = 1$. Hence the system is unstable for all $p > 0$.

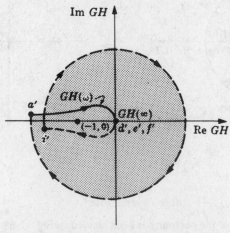

Fig. 11-71

**11.58.** Determine the stability of the system of Problem 11.47.

The region to the right of the contour has been shaded in Fig. 11-72.
It is clear that $N > 0$. Since $P_0 = 0$ for $p > 0$, then $N \neq -P_0$. Hence the system is unstable.

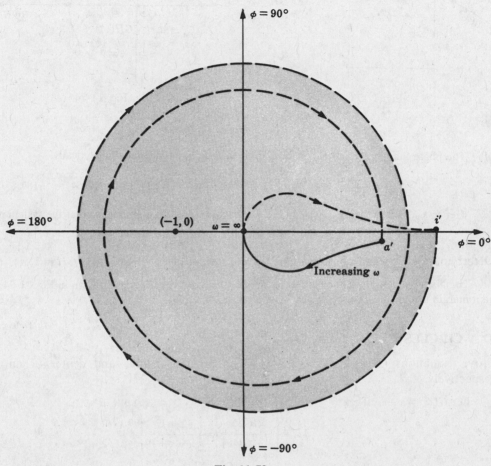

Fig. 11-72

## RELATIVE STABILITY

**11.59.** Determine: ($a$) the phase crossover frequency $\omega_\pi$, ($b$) the gain crossover frequency $\omega_1$, ($c$) the gain margin, and ($d$) the phase margin for the system of Problem 11.44 with $p_1 = 1$ and $p_2 = \frac{1}{2}$.

($a$)  Letting $\omega = \omega_\pi$, we have

$$\phi(\omega_\pi) = -\pi = \frac{-\pi}{2} - \tan^{-1}\omega_\pi - \tan^{-1}2\omega_\pi = \frac{-\pi}{2} - \tan^{-1}\left(\frac{3\omega_\pi}{1 - 2\omega_\pi^2}\right)$$

or $3\omega_\pi/(1 - 2\omega_\pi^2) = \tan(\pi/2) = \infty$. Hence $\omega_\pi = \sqrt{\frac{1}{2}} = 0.707$.

($b$)  From $|GH(\omega_1)| = 1$, we have $1/\omega_1\sqrt{(\omega_1^2 + 1)(\omega_1^2 + 0.25)} = 1$ or $\omega_1 = 0.82$.

($c$)  The gain margin $1/|GH(\omega_\pi)|$ is easily determined from the graph, as shown in Fig. 11-73. It can also be calculated analytically: $|GH(\omega_\pi)| = |GH(j0.707)| = 4/3$; hence gain margin $= 3/4$.

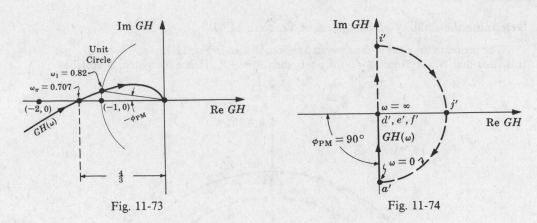

Fig. 11-73                                                    Fig. 11-74

(*d*)  The phase margin is easily determined from the graph, or calculated analytically:

$$\arg GH(\omega_1) = \arg GH(0.82) = -90° - \tan^{-1}(0.82) - \tan^{-1}(1.64) = -187.8°$$

Hence $\phi_{\mathrm{PM}} = 180° + \arg GH(\omega_1) = -7.8°$. Negative phase margin means that the system is unstable.

**11.60.** Determine the gain and phase margins for the system of Problem 11.43 ($GH = 1/s$).

The Nyquist Stability Plot of $1/s$ never crosses the negative real axis as shown in Fig. 11-74; hence the gain margin is undefined for this system. The phase margin is $\phi_{\mathrm{PM}} = 90°$.

## *M*- AND *N*-CIRCLES

**11.61.** Prove Equations (*11.18*) and (*11.19*), which give the radius and center of an *M*-circle, respectively.

Let $G(\omega) \equiv x + jy$. Then

$$M \equiv \left| \frac{G(\omega)}{1 + G(\omega)} \right| = \left| \frac{x + jy}{1 + x + jy} \right|$$

Squaring both sides and rearranging yields

$$\left[ x - \left( \frac{M^2}{1 - M^2} \right) \right]^2 + y^2 = \left( \frac{M}{1 - M^2} \right)^2 \qquad M < 1$$

$$\left[ x + \left( \frac{M^2}{M^2 - 1} \right) \right]^2 + y^2 = \left( \frac{M}{M^2 - 1} \right)^2 \qquad M > 1$$

For $M = \text{constant}$, these are equations of circles with radii $|M/(M^2 - 1)|$ and centers at $(-M^2/(M^2 - 1), 0)$.

**11.62.** Prove Equation (*11.20*).

The transfer function $G$ for the second-order continuous system whose signal flow graph is shown in Fig. 11-75 is $G = \omega_n^2/s(s + 2\zeta\omega_n)$. Now

$$M^2 = \left| \frac{G}{1 + G} \right|^2 = \frac{\omega_n^4}{\left( \omega_n^2 - \omega^2 \right)^2 + 4\zeta^2\omega_n^2\omega^2}$$

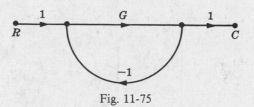

Fig. 11-75

To find $\omega_p$, we maximize the above expression:

$$\frac{d}{d\omega}(M^2) = \frac{\omega_n^4\left[2(\omega_n^2 - \omega^2)(-2\omega) + 8\zeta^2\omega_n^2\omega\right]}{\left[(\omega_n^2 - \omega^2)^2 + 4\zeta^2\omega_n^2\omega^2\right]^2} = 0$$

from which $\omega = \omega_p = \pm\omega_n\sqrt{1 - 2\zeta^2}$. Hence for $0 \le \zeta \le 0.707$,

$$M_p = \left[\frac{\omega_n^4}{\left[\omega_n^2 - \omega_n^2(1 - 2\zeta^2)\right]^2 + 4\zeta^2\omega_n^4(1 - 2\zeta^2)}\right]^{1/2} = \frac{1}{2\zeta\sqrt{1 - \zeta^2}}$$

**11.63.** Prove Equations ($11.21$) and ($11.22$), which give the radius and center of an $N$-circle.

Let $G(\omega) \equiv x + jy$. Then

$$\frac{C(\omega)}{R(\omega)} = \frac{x^2 + x + y^2 + jy}{(1 + x)^2 + y^2} \qquad \text{and} \qquad N \equiv \frac{\text{Im}(C/R)(\omega)}{\text{Re}(C/R)(\omega)} = \frac{y}{x^2 + x + y^2}$$

which yields

$$\left(x + \frac{1}{2}\right)^2 + \left(y - \frac{1}{2N}\right)^2 = \frac{1}{4}\left(1 + \frac{1}{N^2}\right)$$

For $N$ equal to a constant parameter, this is the equation of a circle with radius $\sqrt{\frac{1}{4} + (1/2N)^2}$ and center at $(-\frac{1}{2}, 1/2N)$.

**11.64.** Find $M_p$ and $\zeta$ for the unity feedback system given by $G = 1/s(s + 1)$.

The general open-loop transfer function for the second-order system is $G = \omega_n^2/s(s + 2\zeta\omega_n)$. Then $\omega_n = 1$, $\zeta = 0.5$, and $M_p = 1/(2\zeta\sqrt{1 - \zeta^2}) = 0.866$.

## MISCELLANEOUS PROBLEMS

**11.65.** Determine the Polar Plot for

$$P(z) = \frac{z}{z - 1}$$

for a sampling period $T = 1$.

The solution requires mapping the strip from $-j\omega_s/2$ to $j\omega_s/2$ on the $j\omega$-axis of the $s$-plane or, equivalently, $\omega = -\pi$ to $\omega = \pi$ radians on the unit circle of the $z$-plane, into the $P(e^{j\omega})$-plane. We have $P(e^{\pm j\pi}) = 0.5 \underline{/0^\circ}$, and $P(e^{j0}) = \infty \underline{/\pm 90^\circ}$. Evaluation of $P(e^{j\omega})$ for several values of $\omega$ between $-\pi$ and $0$ results in a straight line parallel to the imaginary axis in the $P$-plane, as shown in Fig. 11-76, where the segments $a$ to $b$ and $g$ to $a$ map the corresponding segments of the unit circle in Fig. 11-13.

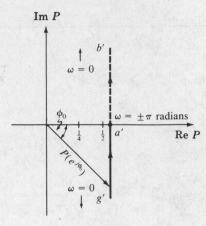

Fig. 11-76

**11.66.** Determine the Polar Plot of the type 0 discrete-time system open-loop transfer function

$$GH(z) = \frac{\frac{3}{8}(z+1)(z+\frac{1}{3})K}{z(z+\frac{1}{2})}$$

for $K = 1$ and $T = 1$.

In this case, the Polar Plot has been drawn by computer, as illustrated in Fig. 11-77. The computer program evaluates $GH(e^{j\omega})$ for values of $\omega T = \omega$ in the range $-\pi$ to $\pi$ radians, separates each result into real and imaginary parts (Complex Form), and then generates the rectangular plot from these coordinates.

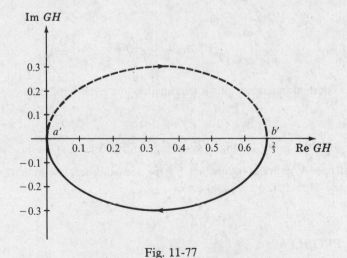

Fig. 11-77

**11.67.** Determine the Polar Plot of the type 1 discrete-time system open-loop transfer function

Mathcad

$$GH(z) = \frac{K(z+1)^2}{(z-1)(z+\frac{1}{3})(z+\frac{1}{2})}$$

for $K = 1$ and $T = 1$.

As in Problem 11.66, the Polar Plot given in Fig. 11-78 was generated by computer, exactly in the same manner as described in the previous problem.

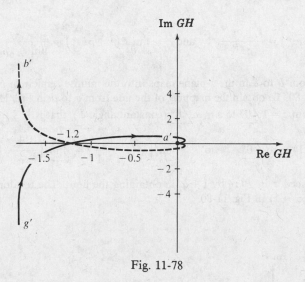

Fig. 11-78

**11.68.** Determine the absolute stability of the system given in Examples 11.11 and 11.14, for $K \geq 2$ and $T = 1$.

The Nyquist Stability Plot for $K = 2$ is given in Fig. 11-79. The region to the right has been shaded and the plot goes directly through $(-1, 0)$. Thus $N > 0$ and $N \neq -P_0$, which is zero for this problem. Therefore the system is marginally stable for $K = 2$. For $K > 2$, the $(-1, 0)$ point is completely enclosed, $N = 1$, and the closed-loop system is unstable.

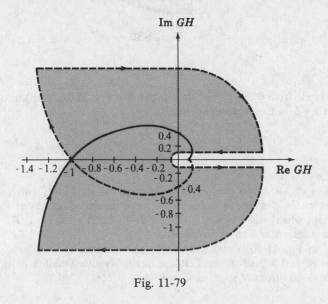

Fig. 11-79

**11.69.** Determine the Nyquist Stability Plot of the system given in Problem 11.65.

We note that $P(z) = z/(z - 1)$ has a pole at 1, so we must begin by mapping the segment $b$ to $c$ of the infinitesimal semicircle near $z = 1$ in Fig. 11-13 into the $P$-plane. Overall, we have a conformal mapping, so the plot must turn right at $b'$. Between $b$ and $c$, $z = 1 + \rho e^{j\phi}$, with $\phi$ increasing from $-90°$ to $0°$. Therefore

$$P(1 + \rho e^{j\phi}) = \frac{1 + \rho e^{j\phi}}{\rho e^{j\phi}} \quad \text{and} \quad \lim_{\rho \to 0} P(1 + \rho e^{j\phi}) = \frac{1}{\lim_{\rho \to 0} \rho e^{j\phi}} = \infty \underline{/-\phi}$$

Therefore the arc from $b$ to $c$ in the $z$-plane maps into the infinite semicircle $b'$ to $c'$, from $+90°$ back to $0°$, shown in Fig. 11-80. To obtain the mapping of the line from $c$ to $d$ in Fig. 11-13, we note that this is the mapping of $P(z)$ from $z = 1\underline{/0°}$ to $z = \infty\underline{/0°}$ (constant angle $\phi$), that is,

$$P(1) = \infty\underline{/0°} \quad \text{to} \quad \lim_{\alpha \to \infty} \left( \frac{1 + \alpha}{1 + \alpha - 1} \right) = \lim_{\alpha \to \infty} \left( \frac{1 + \alpha}{\alpha} \right) = 1\underline{/0°}$$

where we have replaced $z$ in $P(z)$ by $1 + \alpha$, in obtaining the limit. The resulting mapping is shown as the line from $c'$ to $d'$ ($\infty \to 1$) in Fig. 11-80.

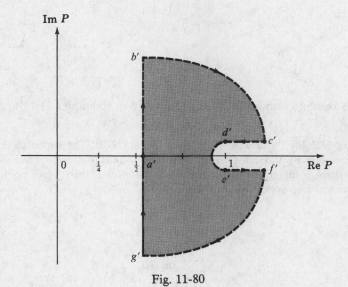

Fig. 11-80

The infinite circle from $0°$ to $-360°$, from $d$ to $e$ in Fig. 11-13, maps into an infinitesimal semicircle around the point $z = 1$ in the $P$-plane, because

$$P(Re^{j\phi}) = \frac{Re^{j\phi}}{Re^{j\phi} - 1} = \frac{e^{j\phi}}{e^{j\phi} - \frac{1}{R}}$$

and $P \to 1$ as $R \to \infty$ for any $\phi$, and a few evaluations of arg $P(Re^{j\phi})$ at values of $\phi$ between $0°$ and $-360°$ show that the limit is approached from values in the first quadrant of $P$ when $0 < \phi < -180°$, and the fourth quadrant when $-180° < \phi < -360°$, with $P(Re^{j\phi}) = 1/(1 + 1/R) < 1$ for $R > 0$ at $\phi = 180°$. The resulting arc is shown as $d'$ to $e'$ in Fig. 11-80.

Arc $e'$ to $f'$ in Fig. 11-80 is obtained in the same manner as that for $c'$ to $d'$, taking the limits of $(\alpha + 1)/\alpha$ as $\alpha \to \infty$ and 0. And the final closure of the Nyquist Stability Plot, arc $f'$ to $g'$, is obtained in the same manner as that from $b'$ to $c'$, as shown.

**11.70.** For $GH = P = z/(z-1)$ in Problem 11.69, is the closed-loop system stable?

The region to the right of the contour in Fig. 11-80 has been shaded and it does not enclose the $(-1,0)$ point. Therefore $N \le 0$. The only pole of $GH$ is at $z = 1$, which is not outside the unit circle. Thus $P_0 = 0$, $N = -P_0 = 0$, and the system is absolutely stable.

**11.71.** Determine the stability of the system given in Problem 11.66.

The open-loop transfer function is

$$GH = \frac{\frac{3}{8}(z+1)\left(z+\frac{1}{3}\right)}{z\left(z+\frac{1}{2}\right)}$$

The Polar Plot of $GH$ is given in Fig. 11-77, which is the mapping of arcs $a$ to $b$ and $g$ to $a$ of Fig. 11-13. There are no poles of $GH$ on the unit circle, so the infinitesimal arcs $b$ to $c$ and $f$ to $g$ in Fig. 11-13 are not needed. Setting $z = 1 + \alpha$ and using the same limiting procedures illustrated in Problem 11.70, the straight lines to and from infinity, $b$ to $d$ and $e$ to $f$ in Fig. 11-13, map into the lines from $b$ to $d$ and $e$ to $f$ between $\operatorname{Re} GH \frac{3}{8}$ and $\frac{2}{3}$. Similarly, with $z$ replaced by $1 + Re^{j\phi}$ and $R \to \infty$, the infinite arc from $d$ to $e$ maps into the infinitesimal semicircle about $\operatorname{Re} GH = \frac{3}{8}$, all as shown in Fig. 11-81.

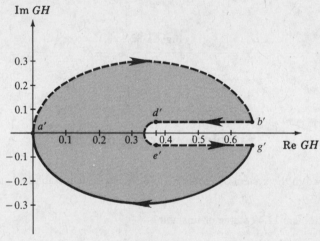

Fig. 11-81

The $(-1,0)$ point is not enclosed by this contour, as shown, $N = 0$, $P_0 = 0$, and the closed-loop system is absolutely stable.

**11.72.** Determine the stability of the system given in Problem 11.67.

The open-loop transfer function is

$$GH = \frac{(z+1)^2}{(z-1)\left(z+\frac{1}{3}\right)\left(z+\frac{1}{2}\right)}$$

The Polar Plot of $GH$ is given in Fig. 11-78. Completion of the closed contour mapping of the exterior of the unit circle in the $z$-plane (Fig. 11-13) closely parallels that described in Problem 11.69 and Example 11.11. In this case, the $(-1,0)$ point is enclosed once by the contour, that is, $N = 1$. Since $P_0 = 0$ and $Z_0 = N + P_0 = 1$, then one zero of $1 + GH$ is outside the unit circle of the $z$-plane and the closed-loop system is therefore unstable (Fig. 11-82).

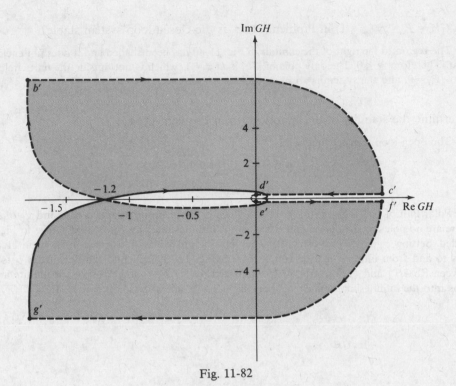

Fig. 11-82

# Supplementary Problems

**11.73.** Let $T = 2$ and $p = 5$ in the system of Problem 11.48. Is this system stable?

**11.74.** Is the system of Problem 11.49 stable or unstable?

**11.75.** Is the system of Problem 11.50 stable or unstable?

**11.76.** Sketch the Polar Plot for $GH = \dfrac{K(s + z_1)(s + z_2)}{s^3(s + p_1)(s + p_2)}$, $z_i, p_i > 0$.

**11.77.** Sketch the Polar Plot for $GH = \dfrac{K}{(s + p_1)(s + p_2)(s + p_3)}$, $p_i > 0$.

**11.78.** Find the closed-loop frequency response of the unity feedback system described by $G = \dfrac{10(s + 0.5)}{s^2(s + 1)(s + 10)}$, using $M$- and $N$-circles.

**11.79.** Sketch the Polar Plot for $GH = \dfrac{K(s + z_1)}{s^2(s + p_1)(s + p_2)(s + p_3)}$, $z_1, p_i > 0$.

**11.80.** Sketch the Nyquist Stability Plot for $GH = \dfrac{Ke^{-Ts}}{s(s + 1)}$.

Mathcad

**11.81.** Sketch the Polar Plot for $GH = \dfrac{s+z_1}{s(s+p_1)}$, $z_1, p_1 > 0$.

**11.82.** Sketch the Polar Plot for $GH = \dfrac{s+z_1}{s(s+p_1)(s+p_2)}$, $z_1, p_i > 0$.

**11.83.** Sketch the Polar Plot for $GH = \dfrac{K}{s^2(s+p_1)(s+p_2)}$, $p_i > 0$.

**11.84.** Sketch the Polar Plot for $GH = \dfrac{s+z_1}{s^2(s+p_1)}$, $z_1, p_1 > 0$.

**11.85.** Sketch the Polar Plot for $GH = \dfrac{s+z_1}{s^2(s+p_1)(s+p_2)}$, $z_1, p_i > 0$.

**11.86.** Sketch the Polar Plot for $GH = \dfrac{(s+z_1)(s+z_2)}{s^2(s+p_1)(s+p_2)(s+p_3)}$, $z_i, p_i > 0$.

**11.87.** Sketch the Polar Plot for $GH = \dfrac{K}{s^3(s+p_1)(s+p_2)}$, $p_i > 0$.

**11.88.** Sketch the Polar Plot for $GH = \dfrac{(s+z_1)}{s^3(s+p_1)(s+p_2)}$, $z_1, p_i > 0$.

**11.89.** Sketch the Polar Plot for $GH = \dfrac{s+z_1}{s^4(s+p_1)}$, $z_1, p_1 > 0$.

**11.90.** Sketch the Polar Plot for $GH = \dfrac{e^{-Ts}(s+z_1)}{s^2(s+p_1)}$, $z_1, p_1 > 0$.

**11.91.** Sketch the Polar Plot for $GH = \dfrac{e^{-Ts}(s+z_1)}{s^2(s^2+a)(s^2+b)}$, $z_1, a, b > 0$.

**11.92.** Sketch the Polar Plot for $GH = \dfrac{(s-z_1)}{s^2(s+p_1)}$, $z_1, p_1 > 0$.

**11.93.** Sketch the Polar Plot for $GH = \dfrac{s}{(s+p_1)(s-p_2)}$, $p_i > 0$.

**11.94.** The various portions of the Nyquist Path for continuous systems are illustrated in Fig. 11-12 and the different segments are defined mathematically by Equations (*11.5*) through (*11.12*). Write the corresponding equations for each segment of the Nyquist Path for the discrete-time systems given in Fig. 11-13. (One of these was given in Example 11.11. Also see Problems 11.69 and 11.70.)

# Answers to Some Supplementary Problems

**11.73.** Yes

**11.74.** Unstable

**11.75.** Unstable

**11.76.**

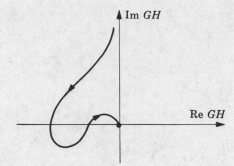

**11.77.**

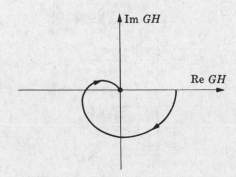

**11.79.**

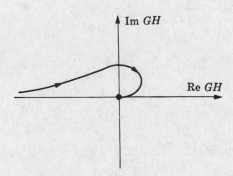

**11.80.**

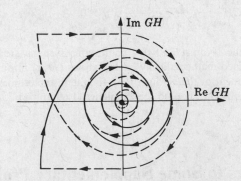

# Chapter 12

# Nyquist Design

## 12.1 DESIGN PHILOSOPHY

Design by analysis in the frequency domain using Nyquist techniques is performed in the same general manner as all other design methods in this book: appropriate compensation networks are introduced in the forward or feedback paths and the behavior of the resulting system is critically analyzed and reanalyzed. In this manner, the Polar Plot is shaped and reshaped until performance specifications are met. The procedure is greatly facilitated when computer programs for generating Polar Plots are used.

Since the Polar Plot is a graph of the open-loop frequency response function $GH(\omega)$, many types of compensation components can be used in either the forward or feedback path, becoming part of either $G$ or $H$. Often, compensation in only one path, or a combination of both cascade and feedback compensation, can be used to satisfy specifications. Cascade compensation is emphasized in this chapter.

## 12.2 GAIN FACTOR COMPENSATION

It was pointed out in Chapter 5 that an unstable feedback system can sometimes be stabilized, or a stable system destabilized, by appropriately adjusting the gain factor $K$ of $GH$. The root-locus method of Chapters 13 and 14 vividly illustrates this phenomenon, but it is also evidenced in Nyquist Stability Plots.

**EXAMPLE 12.1.** Figure 12-1 indicates an unstable *continuous* system when the gain factor is $K_1$, where

$$GH(s) = \frac{K_1}{s(s+p_1)(s+p_2)} \qquad p_1, p_2, K_1 > 0 \qquad P_0 = 0 \qquad N = 2$$

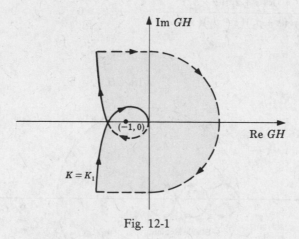

Fig. 12-1

A sufficient decrease in the gain factor to $K_2$ ($K_2 < K_1$) stabilizes the system, as illustrated in Fig. 12-2.

$$GH(s) = \frac{K_2}{s(s+p_1)(s+p_2)} \qquad 0 < K_2 < K_1 \qquad P_0 = 0 \qquad N = 0$$

Further decrease of $K$ does not alter stability.

299

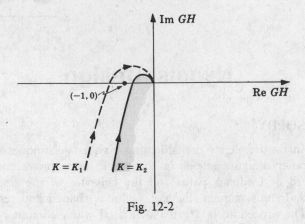

Fig. 12-2

**EXAMPLE 12.2.**   The type 1 *discrete-time* control system with

$$GH_1 = \frac{1}{(z-1)\left(z-\frac{1}{2}\right)}$$

is unstable, as shown in Fig. 11-79 and Problem 11.68. That is, the open-loop transfer function

$$GH = \frac{K/4}{(z-1)\left(z-\frac{1}{2}\right)}$$

was found to be unstable for $K \geq 2$. Therefore gain factor compensation can be used to stabilize $GH_1$, by attenuating the gain factor $K_1 = 1$ of $GH_1$ by a factor less than 0.5. For example, if the attenuator is given a value of 0.25, the resulting $GH \equiv GH_2$ would have the Nyquist Stability Plot in Fig. 11-25, shown in Example 11.14 to represent a stable system.

**EXAMPLE 12.3.**   The stable region for the $(-1, 0)$ point in Fig. 12-3 is indicated by the portion of the real axis in the unshaded area:

$$GH(s) = \frac{K(s+z_1)(s+z_2)}{s^2(s+p_1)(s+p_2)(s+p_3)} \qquad z_1, z_2 > 0 \qquad p_i > 0 \qquad P_0 = 0$$

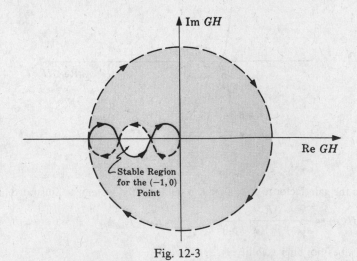

Fig. 12-3

If the $(-1, 0)$ point falls in the stable region, an increase or decrease in $K$ can cause enough shift in the $GH$ contour to the left or the right to destabilize the system. This can happen because a shaded (unstable) region appears both to the left and the right of the unshaded (stable) region. This phenomenon is called **conditional stability**.

Although absolute stability can often be altered by adjustment of the gain factor alone, other performance criteria such as those concerned with *relative stability* usually require additional compensators.

## 12.3   GAIN FACTOR COMPENSATION USING *M*-CIRCLES

The gain factor $K$ of $G$ for a *unity feedback* system can be determined for a specific resonant peak $M_p$ by the following procedure which entails drawing the Polar Plot once only.

**Step 1:**   Draw the Polar Plot of $G(\omega)$ for $K = 1$.

**Step 2:**   Calculate $\Psi_p$, given by

$$\Psi_p = \sin^{-1}\left(\frac{1}{M_p}\right) \tag{12.1}$$

**Step 3:**   Draw a radial line $\overline{AB}$ at an angle $\Psi_p$ below the negative real axis, as shown in Fig. 12-4.

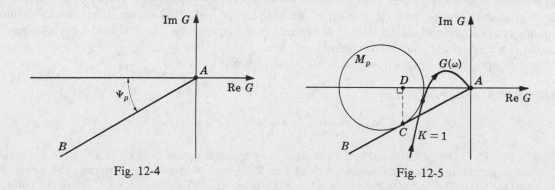

Fig. 12-4                                          Fig. 12-5

**Step 4:**   Draw the $M_p$ circle tangent to both $G(\omega)$ and line $\overline{AB}$ at $C$. Then draw a line $\overline{CD}$ perpendicular to the real axis shown in the example Polar Plot shown in Fig. 12-5.

**Step 5:**   Measure the length of line $\overline{AD}$ along the real axis. The required gain factor $K$ for the specified $M_p$ is given by

$$K_{M_p} = \frac{1}{\text{length of line } \overline{AD}} \tag{12.2}$$

If the Polar Plot of $G$ for a gain factor $K'$ other than $K = 1$ is already available, it is not necessary to repeat this plot for $K = 1$. Simply apply Steps 2 through 5 and use the following formula for the gain factor necessary to achieve the specified $M_p$:

$$K_{M_p} = \frac{K'}{\text{length of line } \overline{AD}} \tag{12.3}$$

## 12.4  LEAD COMPENSATION

The transfer function for a continuous system lead network, presented in Equation (6.2), is

$$P_{\text{lead}} = \frac{s + a}{s + b}$$

where $a < b$. The Polar Plot of $P_{\text{Lead}}$ for $0 \leq \omega < \infty$ is shown in Fig. 12-6.

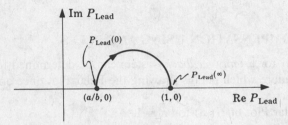

Fig. 12-6

For some systems in which lead compensation is applicable, appropriate choice of the zero at $-a$ and the pole at $-b$ permits an increase in the open-loop gain factor $K$, providing greater accuracy (and sometimes stability), without adversely affecting transient performance. Conversely, for a given $K$, transient performance can be improved. In some cases, both steady state and transient response can be favorably modified with lead compensation.

The lead network provides compensation by virtue of its phase lead property in the low-to-medium-frequency range and its negligible attenuation at high frequencies. The low-to-medium-frequency range is defined as the vicinity of the resonant frequency $\omega_p$. Several lead networks may be cascaded if a large phase lead is required.

Lead compensation generally increases the *bandwidth* of a system.

Mathcad  **EXAMPLE 12.4.**   The Polar Plot for

$$GH_1(s) = \frac{K_1}{s(s + p_1)(s + p_2)} \qquad K_1, p_1, p_2 > 0$$

is given in Fig. 12-7. The system is stable and the phase margin $\phi_{\text{PM}}$ is greater than 45°. For a given application, $\phi_{\text{PM}}$ is too large, causing a longer than desired delay time $T_d$ in the system transient response. The steady state error is also too large. That is, the velocity error constant $K_v$ is too small by a factor of $\lambda > 1$. We shall modify this system by a combination of gain factor compensation, to meet the steady state specification, and phase lead compensation, to improve the transient response. Assuming $H(s) = 1$, Equation (9.12) yields

$$K_{v1} = \lim_{s \to 0} \left[ sGH_1(s) \right] = \frac{K_1}{p_1 p_2}$$

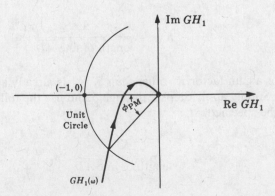

Fig. 12-7

and hence
$$\lambda K_{v1} = \frac{\lambda K_1}{p_1 p_2}$$

Putting $K_2 \equiv \lambda K_1$, the open-loop transfer function becomes
$$GH_2 = \frac{K_2}{s(s + p_1)(s + p_2)}$$

The system represented by $GH_2$ has the desired velocity constant $K_{v2} = \lambda K_{v1}$.

Let us now consider what would happen to $K_{v2}$ of $GH_2$ if a lead network were introduced. The lead network acts like an attenuator at low frequencies. That is,
$$\lim_{s \to 0} \left[ sGH_2(s) \cdot P_{\text{Lead}}(s) \right] = \frac{K_2 a}{p_1 p_2 b} < \lambda K_{v1}$$

since $a/b < 1$. Therefore if a lead network is used to modify the transient response, the gain factor $K_1$ of $GH_1$ must be increased $\lambda(b/a)$ times in order to meet the steady state requirement. The gain factor part of the total compensation should therefore be larger than that which would be called for if only the steady state specification has to be met. Hence we modify $GH_2$, yielding
$$GH_3 = \frac{\lambda K_1(b/a)}{s(s + p_1)(s + p_2)}$$

As is often the case, increasing the gain factor by an amount as large as $\lambda(b/a)$ times destabilizes the system, as shown in the Polar Plots of $GH_1$, $GH_2$, and $GH_3$ in Fig. 12-8.

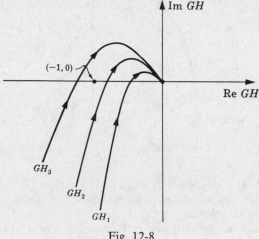

Fig. 12-8

Now let us insert the lead network and determine its effects. $GH_3$ becomes
$$GH_4 = \frac{\lambda K_1(b/a)(s + a)}{s(s + p_1)(s + p_2)(s + b)}$$

First, $\lim_{s \to 0}[sGH_4(s)] = \lambda K_{v1}$ convinces us that the steady state specification has been met. In fact, in the very low frequency region we have
$$GH_4(j\omega)\big|_{\omega \text{ very small}} \cong \frac{\lambda K_1}{j\omega(j\omega + p_1)(j\omega + p_2)}$$
$$= GH_2$$

Hence the $GH_4$ contour is almost coincident with the $GH_2$ contour in the very low frequency range.

In the very high frequency region,
$$GH_4(j\omega)\big|_{\omega \text{ very large}} \cong \frac{\lambda K_1(b/a)}{j\omega(j\omega + p_1)(j\omega + p_2)} = GH_3$$

Therefore the $GH_4$ contour is almost coincident with $GH_3$ for very high frequencies.

In the mid-frequency range, where the phase lead property of the lead network substantially alters the phase characteristic of $GH_4$, the $GH_4$ contour bends away from the $GH_2$ and toward the $GH_3$ locus as $\omega$ is increased. This effect is better understood if we write $GH_4$ in the following form:

$$GH_4(j\omega) = \left[ \frac{\lambda K_1(b/a)}{j\omega(j\omega + p_1)(j\omega + p_2)} \right] \cdot \left[ \frac{j\omega + a}{j\omega + b} \right]$$

$$= GH_3(j\omega) \cdot P_{\text{Lead}}(j\omega) = GH_3(j\omega) \cdot |P_{\text{Lead}}(j\omega)| \underline{/\phi(\omega)}$$

where $|P_{\text{Lead}}(j\omega)| = \sqrt{(\omega^2 + a^2)/(\omega^2 + b^2)}$, $\phi(\omega) \equiv \tan^{-1}(\omega/a) - \tan^{-1}(\omega/b)$, $a/b < |P_{\text{Lead}}(j\omega)| < 1$, $0° < \phi(\omega) < 90°$. Therefore the lead network modifies $GH_3$ as follows. $GH_3$ is shifted downwards beginning at $GH_3(j\infty)$ in a counterclockwise direction toward $GH_2$ due to the positive phase contribution of $P_{\text{Lead}}$ $[0° < \phi(\omega) < 90°]$. In addition, it is attenuated $[0 < |P_{\text{Lead}}(j\omega)| < 1]$. The resulting Polar Plot for $GH_4$ is illustrated in Fig. 12-9.

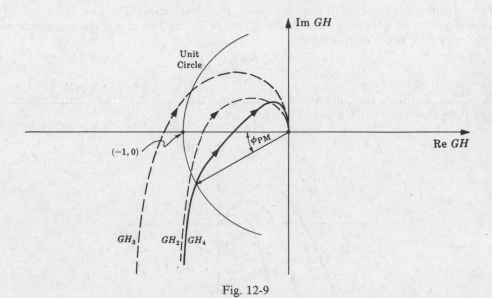

Fig. 12-9

The system represented by $GH_4$ is clearly stable, and $\phi_{\text{PM}}$ is less than $45°$, reducing the delay time $T_d$ of the original system represented by $GH_1$. By a trial-and-error procedure, the zero at $-a$ and the pole at $-b$ can be chosen such that a specific $M_p$ can be achieved.

A block diagram of the fully compensated system is shown in Fig. 12-10. Unity feedback is shown for convenience only.

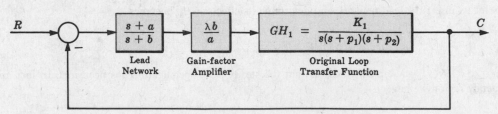

Fig. 12-10

## 12.5  LAG COMPENSATION

The transfer function for a continuous system lag network, presented in Equation (6.3), is

$$P_{\text{Lag}} = \frac{a}{b}\left[ \frac{s + b}{s + a} \right]$$

where $a < b$. The Polar Plot of $P_{\text{Lag}}$ for $0 \le \omega < \infty$ is shown in Fig. 12-11.

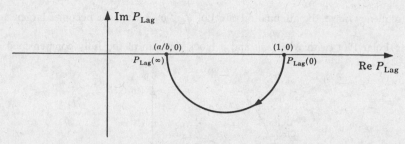

Fig. 12-11

The lag network usually provides compensation by virtue of its attenuation property in the high frequency portion of the Polar Plot, since $P_{\text{Lag}}(0) = 1$ and $P_{\text{Lag}}(\infty) = a/b < 1$. Several lag networks can be cascaded to provide even higher attenuation, if required. The phase lag contribution of the lag network is often restricted by design to the very low frequency range. Several general effects of lag compensation are:

1. The bandwidth of the system is usually decreased.
2. The dominant time constant $\tau$ of the system is usually increased, producing a more sluggish system.
3. For a given relative stability, the value of the error constant is increased.
4. For a given value of error constant, relative stability is improved.

The procedure for using lag compensation to improve system performance is essentially the same as that for lead compensation.

**EXAMPLE 12.5.** Let us redesign the system of Example 12.4 using gain factor plus *lag* compensation. The original open-loop transfer function is

$$GH_1 = \frac{K_1}{s(s+p_1)(s+p_2)}$$

The gain factor compensation transfer function is

$$GH_2 = \frac{\lambda K_1}{s(s+p_1)(s+p_2)}$$

Since $P_{\text{Lag}}(0) = 1$, introduction of the lag network after the steady state criterion has been met by gain factor compensation does not require an additional increase in gain factor.

Introducing the lag network, we get

$$GH_3' = \frac{\lambda K_1 (a/b)(s+b)}{s(s+p_1)(s+p_2)(s+a)}$$

Now

$$\lim_{s \to 0}\left[ sGH_3'(s)\right] = \lambda K_{v1}$$

where $K_{v1} = K_1/p_1 p_2$. Therefore the steady state specification is met by $GH_3'$.

In the very low frequency region,

$$GH_3'(j\omega)\big|_{\omega \text{ very small}} \cong \frac{\lambda K_1}{j\omega(j\omega+p_1)(j\omega+p_2)} = GH_2(j\omega)$$

Hence $GH_3'$ is almost coincident with $GH_2$ at very low frequencies, with the lag property of this network manifesting itself in this range.

In the very high frequency region,

$$GH_3'(j\omega)\big|_{\omega \text{ very large}} \cong \frac{\lambda(a/b)K_1}{j\omega(j\omega+p_1)(j\omega+p_2)} = \lambda(a/b)GH_1(j\omega)$$

Therefore, the $GH_3'$ contour lies above or below the $GH_1$ contour in the range, if $\lambda > b/a$ or $\lambda < b/a$, respectively. If $\lambda = b/a$, the $GH_3'$ and $GH_1$ contours coincide.

In the mid-frequency range, the attenuation effect of $P_{\text{Lag}}$ increases as $\omega$ becomes larger, and there is relatively small phase lag.

The resulting Polar Plot (with $\lambda = b/a$) and a block diagram of the fully compensated system are given in Figs. 12-12 and 12-13.

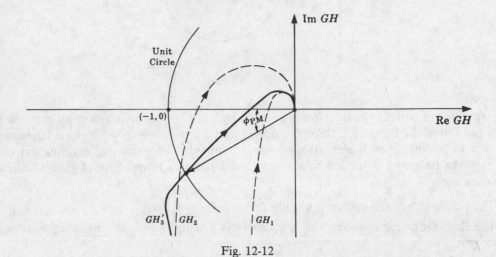

Fig. 12-12

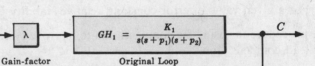

Fig. 12-13

## 12.6  LAG-LEAD COMPENSATION

The transfer function for a continuous system lag-lead network, presented in Equation (6.4), is

$$P_{\text{LL}} = \frac{(s + a_1)(s + b_2)}{(s + b_1)(s + a_2)}$$

where $a_1 b_2 / b_1 a_2 = 1$, $b_1/a_1 = b_2/a_2 > 1$, $a_i, b_i > 0$. The Polar Plot of $P_{\text{LL}}$ for $0 \leq \omega \leq \infty$ is shown in Fig. 12-14.

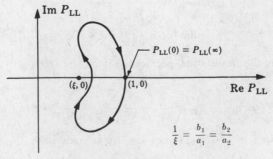

Fig. 12-14

Lag-lead compensation has all of the advantages of both lag compensation and lead compensation, and only a minimum of their usually undesirable characteristics. Satisfaction of many system specifications is possible without the burden of excessive bandwidth and small dominant time constants.

It is not easy to generalize about the application of lag-lead compensation or to prescribe a method for its employment, especially using Nyquist techniques. But, for illustrative purposes, we can describe how it alters the properties of a simple type 2 system in the following example.

**EXAMPLE 12.6.**   The Nyquist Stability Plot for

$$GH = \frac{K}{s^2(s + p_1)} \qquad p_1, K > 0$$

is given in Fig. 12-15. Clearly, the system is unstable, and no amount of gain factor compensation can stabilize it because the contour for $0 < \omega < \infty$ always lies above the negative real axis. Lag compensation is also inapplicable for basically the same reason.

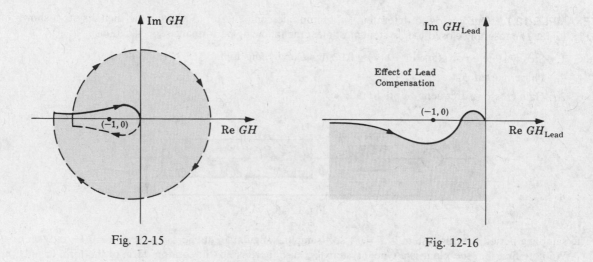

Fig. 12-15                                                          Fig. 12-16

Lead compensation may succeed in stabilizing the system, as shown in Fig. 12-16. But the desired application for the compensated system may call for a lower bandwidth than can be achieved with a lead network.

If a lag-lead network is used, the open-loop transfer function becomes

$$GH_{LL} = \frac{K(s + a_1)(s + b_2)}{s^2(s + p_1)(s + b_1)(s + a_2)}$$

and the Polar Plot is shown in Fig. 12-17. This system is conditionally stable if the $(-1, 0)$ point falls on the real axis in the unshaded region. By trial and error, the parameters of the lag-lead network can be chosen to yield good transient and steady state performance for this previously unstable system, and the bandwidth will be smaller than that of the lead-compensated system. A computer program control system design (CAD) package, or any program that readily generates Polar Plots, can be used to help accomplish this task quickly and effectively.

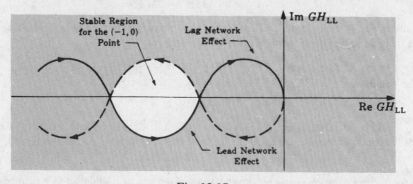

Fig. 12-17

## 12.7   OTHER COMPENSATION SCHEMES AND COMBINATIONS OF COMPENSATORS

Many other types of physical networks can be used to compensate feedback control systems. Compensation networks can also be implemented in software, as part of the control algorithm in a computer-controlled system. PID controllers are a popular class of such controllers (see Examples 2.14 and 6.7 and Section 10.5).

Combinations of gain factors and lead or lag networks were used as compensators in Examples 12.4 and 12.5, and a lag-lead compensator alone was used in Example 12.6 Other combinations are also feasible and effective, particularly where steady state error requirements cannot be met by gain factor compensation alone. This is often the case when the open-loop transfer function has too few "integrators," that is, denominator terms of the form $s^l$ for continuous systems, or $(z-1)^l$ for discrete-time systems as illustrated in the next example.

**EXAMPLE 12.7.** Our goal is to determine an appropriate compensator $G_1(z)$ for the digital system shown in Fig. 12-18. The resulting closed-loop system must meet the following performance specifications:

1.  Steady state error $e(\infty) = 1 - c(\infty) \leq 0.02$, for a unit *ramp* input.

2.  Phase margin $\phi_{PM} \geq 30°$.

3.  Gain crossover frequency $\omega_1 \geq 10$ rad/sec.*

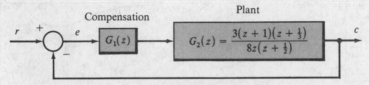

Fig. 12-18

The sampling period for this system is $T = 0.1$ sec (sampling angular frequency $\omega_s = 2\pi/0.1 = 20\pi$ rad/sec).

We note first that the plant is a type 0 system, because there is no "integrator" term of the form $(z-1)^l$ in the denominator of $G_2(z)$ for $l \geq 1$ (see Section 9.8). To meet the first performance specification, it is immediately clear that the overall open-loop system type must be increased by a factor of at least 1, that is, the compensated system must be at least type 1, to achieve a finite steady state error for a unit ramp input. Therefore we add a single pole at $z = 1$, as $G_1'$, as a first step in determining appropriate compensation:

$$G_1'G_2 = \frac{3(z+1)\left(z+\frac{1}{3}\right)}{8z(z-1)\left(z+\frac{1}{2}\right)}$$

Now, from the table in Section 9.9, the steady state error for a unit ramp input is $e(\infty) = T/K_v$, and the velocity error constant is $K_v = 3(2)(\frac{4}{3})/8(\frac{3}{2}) = \frac{2}{3}$. Therefore $e(\infty) = 0.15$, which is larger than the value of 0.02 required by performance specification 1.

The next obvious question is whether the addition of gain factor compensation would be sufficient to complete the design. This would require a gain increase by at least a factor of $\lambda = 0.1/(0.02)(\frac{2}{3}) = \frac{15}{2}$, yielding

$$G_1''G_2 = \frac{15}{2}G_1G_2' = \frac{45(z+1)\left(z+\frac{1}{3}\right)}{16z(z-1)\left(z+\frac{1}{2}\right)}$$

To check the remaining performance criteria (2 and 3) the gain crossover frequency $\omega_1$ and phase margin $\phi_{PM}$ can be evaluated from their defining equations in Section 11.11. We have

$$\phi_{PM} = \left[180 + \arg G_1''G_2(\omega_1)\right] \text{ degrees}$$

and $\omega_1$ satisfies the equation

$$|G_1''G_2(\omega_1)| = 1$$

Now, $\omega_1$ and $\phi_{PM}$ could be determined graphically from a Nyquist Stability Plot of $G_1''G_2$, as illustrated in Fig. 11-16. But a less difficult task is to solve for $\omega_1$ and $\phi_{PM}$ from their defining equations, preferably using a computer

---

*See Problem 12.16 for further discussion of this performance specification and its relationship to system *bandwidth* BW.

program capable of complex numerical calculations. This can be done by first substituting $e^{j\omega T}$ for $z$ in $G_1''G_2(z)$, using the Polar Form, Euler Form, and/or Complex Form substitutions [Equations $(11.2)$ through $(11.4)$], and then solving for $\omega_1 T$ such that $|G_1''G_2| = 1$. Trial-and-error solution for $\omega_1 T$ can be helpful in this regard, which we used to find $\omega_1 T = 2.54$ rad after several trials, resulting in $G_1''G_2(\omega_1) = -0.72 + j0.7$ and

$$\phi_{PM} = \left[ 180° - \tan^{-1}\left( \frac{0.7}{-0.72} \right) \right] = -44.4°$$

Clearly, $\omega_1 = 2.54/0.1 = 25.4 > 10$ rad/sec satisfies performance specification 3, but *not* the phase margin requirement 2, because $\phi_{PM} = -44.4 \ngeq 30°$, the negative phase margin also indicating that the closed-loop system with $G_1''G_2$ is unstable.

Introduction of a *lag* compensator might solve the remaining constraint, because it increases the phase margin without affecting the steady state error. The transfer function of a digital lag compensator was given in Example 6.12, Equation $(6.11)$, as

$$P_{Lag}(z) = \left( \frac{1 - p_c}{1 - z_c} \right) \left[ \frac{z - z_c}{z - p_c} \right] \tag{12.4}$$

where $z_c < p_c$. Note that $P_{Lag}(1) = P_{Lag}(e^{j0}) = 1$, which explains why the lag network does not affect the steady state response of this type 1 system. The Polar Plot of $P_{Lag}$ is shown in Fig. 12-26.

The problem now is to choose appropriate values of $z_c$ and $p_c$ to render $\phi_{PM} \geq 30°$ and $\omega_1 \geq 10$ rad/sec. Again, we accomplished this readily by trial and error, using a computer to evaluate the simultaneous solution for $z_c$ and $p_c$ of the two relations $|G_1''' G_2(10)| = 1$ and

$$\phi_{PM} = \left[ 180 + \arg G_1''' G_2(10) \right] \geq 30°$$

where $G_1''' G_2 = P_{Lag}(G_1''G_2)$. These equations have multiple solutions and, often, good choices for $p_c$ and $z_c$ are values close to 1, because $P_{Lag}$ then has minimal effect on the phase of $G_1''G_2$ at higher frequencies. The pole and zero of $P_{Lag}$ effectively cancel each other at high frequencies when their values are close to 1. After several trials, we obtained $a = 0.86$ and $b = 0.97$, and a final compensator:

$$G_1(z) \equiv G_1'''(z) = \frac{1.59(z - 0.86)}{(z - 1)(z - 0.97)}$$

The resulting Polar Plot (for $0 < \omega < \pi$) for the compensated system $G_1 G_2$ is shown with $\phi_{PM} > 30°$ in Fig. 12-19.

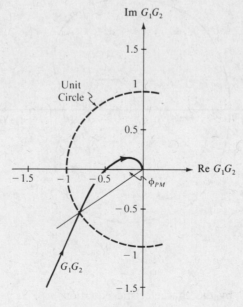

Fig. 12-19

Example 12.7 is reworked by root-locus techniques in Example 14.5, and also by Bode methods in Example 16.6, the latter solution using the $w$-transform introduced in Section 10.7.

# Solved Problems

## GAIN FACTOR COMPENSATION

**12.1.** Consider the open-loop transfer function $GH = -3/(s+1)(s+2)$. Is the system represented by
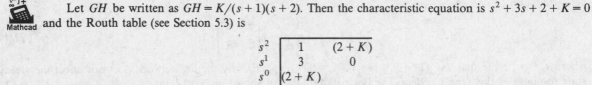 $GH$ stable or unstable?

Unstable. The characteristic equation is determined from $1 + GH = 0$ and is given by $s^2 + 3s - 1 = 0$. Since all the coefficients do not have the same sign, the system is unstable (see Problem 5.27).

**12.2.** Determine the minimum value of gain factor to stabilize the system of the previous problem.

Let $GH$ be written as $GH = K/(s+1)(s+2)$. Then the characteristic equation is $s^2 + 3s + 2 + K = 0$ and the Routh table (see Section 5.3) is

$$
\begin{array}{c|cc}
s^2 & 1 & (2+K) \\
s^1 & 3 & 0 \\
s^0 & (2+K) &
\end{array}
$$

Hence the minimum gain factor for stability is $K = -2 + \epsilon$, where $\epsilon$ is any small positive number.

**12.3.** The solution of the previous problem also tells us that the system of Problems 12.1 and 12.2 is stable for all $K > -2$. Sketch Polar Plots of this system, superimposed on the same coordinate axes, for $K_1 = -3$ and $K_2 = -1$. What general comments can you make about the transient response of the stable system? Assume it is a unity feedback system.

The required Polar Plots are shown in Fig. 12-20. The $M$-circle tangent to the plot for $K = -1$ has infinite radius; thus $M_p = 1$. This means that the peak overshoot is zero (no overshoot), and the system is either critically damped or overdamped.

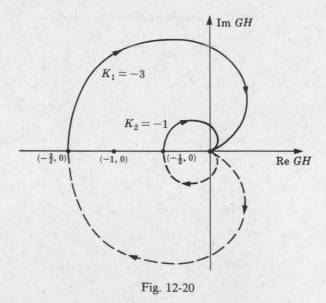

Fig. 12-20

**12.4.** Is the system represented by the characteristic equation $s^3 + 3s^2 + 3s + 1 + K = 0$ ever conditionally stable? Why?

Yes. The gain factor range for stability of this system was determined in Example 5.3 as $-1 < K < 8$. Since both limits are finite, an increase in the gain factor above 8 or a decrease below $-1$ destabilizes the system.

**12.5.** Determine the gain factor $K$ of a unity feedback system whose open-loop transfer function is given by $G = K/(s + 1)(s + 2)$ for a resonant peak specified by $M_p = 2$.

From Equation $(12.1)$ we have $\Psi_p = \sin^{-1}(\frac{1}{2}) = 30°$. The line $\overline{AB}$ drawn at an angle of 30° below the negative real axis is shown in Fig. 12-21, a replica of Fig. 12-20 for $K = -1$.

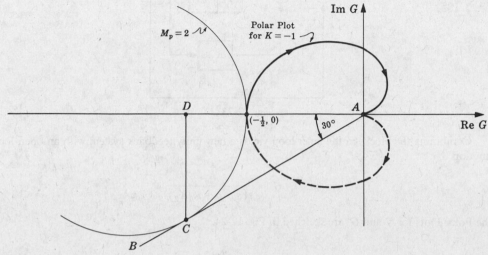

Fig. 12-21

The circle denoted by $M_p = 2$ has been drawn tangent to both $\overline{AB}$ and the Polar Plot of $K = -1$. Using the scale of this Polar Plot, line $\overline{AD}$ has a length equal to 0.76. Therefore Equation $(12.3)$ yields

$$K_{M_p} = \frac{K'}{\text{length of } \overline{AD}} = \frac{-1}{0.76} = -1.32$$

It is also possible to compute a *positive* value of gain for $M_p = 2$ from a Polar Plot of $G(s)$ for any positive value of $K$. The Polar Plot for $K = 1$ is the same as that in Fig. 12-21, but rotated by 180°.

## MISCELLANEOUS COMPENSATION

**12.6.** What kind of compensation is possible for a system whose Polar Plot is given by Fig. 12-22?

Lead, lag-lead, and simple gain factor compensation are capable of stabilizing the system and improving the relative stability.

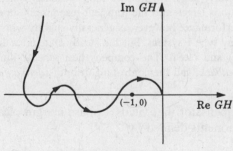

Fig. 12-22

**12.7.** Consider the unity feedback system whose open-loop transfer function is given by

$$G = \frac{K_1}{s(s + a)} \qquad a, K_1 > 0$$

How would the inclusion of a minor feedback loop with a transfer function $K_2s$ ($K_2 > 0$), as shown in the block diagram in Fig. 12-23, affect the transient and steady state performance of the system?

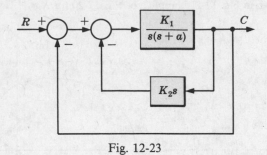

Fig. 12-23

Combining the blocks in the inner loop yields a new unity feedback system with an open-loop transfer function

$$G' = \frac{K_1}{s(s + a + K_1 K_2)}$$

The Polar Plots for $G$ and $G'$ are sketched in Fig. 12-24.

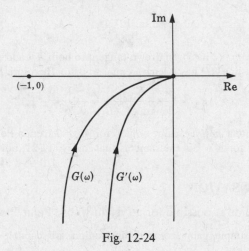

Fig. 12-24

The phase margin is clearly larger for the two-loop feedback system $G'$. Hence the peak overshoot is smaller, or the damping ratio is larger, and the transient response is superior to that of the uncompensated system. The steady state performance however, is generally slightly worse. For a unit step input the steady state error is zero, as for any type 1 system. But the steady state error for a unit ramp or velocity input is larger [see Equations (9.4) and (9.5)]. The compensation scheme illustrated by this problem is called *derivative* or *tachometric feedback*, and the control algorithm is *derivative* (*D*) control.

**12.8.** Determine a type of compensator that yields a phase margin of approximately 45° when added to the fixed system components defined by

$$GH = \frac{4}{s(s^2 + 3.2s + 64)}$$

An additional requirement is that the high-frequency response of the compensated system is to be approximately the same as that of the uncompensated system.

The Polar Plot for $GH$ is sketched in Fig. 12-25. It is very close to the negative imaginary axis for almost all values of $\omega$.

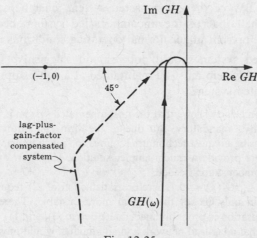

Fig. 12-25

The phase margin is almost 90°, and either an increase in gain factor and/or a lag compensator is capable of satisfying the phase margin requirement. But since the lag network may be designed to provide attenuation at high frequencies and lag in the low-frequency range, a combination of both would be ideal and sufficient (see Example 12.5), as shown in Fig. 12-25. Of course a lag-plus-gain factor compensator is not *necessary* for meeting the design requirements. There are probably an infinite number of different networks or transfer functions capable of satisfying these specifications. The lag network and amplifier, however, are *convenient* due to their standardization, availability, and ease of synthesis.

**12.9.** Outline the design of a servomechanism capable of following a constant velocity input with zero steady state error and approximately 25% maximum overshoot in the transient state. The fixed plant is given by $G_2 = 50/s^2(s + 5)$.

Since the plant is type 2, it is capable of following a constant velocity input with zero steady state error (see Chapter 9). However, the closed-loop system is unstable for any value of gain factor (see Example 12.6). Since no demands on bandwidth have been made, lead compensation should be sufficient (again see Example 12.6) to stabilize the system and meet the transient specification. But two lead networks in series are probably required because the phase margin of the unstable system is negative, and 25% overshoot is equivalent to about +45° phase margin. Most standard lead networks have a maximum phase lead of approximately 54° (see Fig. 16-2).

Detailed design would be very tedious using Nyquist analysis, if performed manually, because the Polar Plot usually must be drawn in some detail several times before converging to a satisfactory solution. If a computer is not available to facilitate this process, this problem may be solved much more easily using the design methods introduced in Chapters 14, 16, and 18. Actually, two compensating lead networks, each with a transfer function of approximately $P_{\text{Lead}} = (s + 3)/(s + 20)$, would satisfy the specifications. If the maximum steady state acceleration error were also specified, a preamplifier would be required with the lead networks. For example, if $K_a = 50$, then a preamplifier of gain $5(20/3)^2$ would be needed. This preamplifier should be placed *between* the two lead networks to prevent, or minimize, loading effects (see Section 8.7).

**12.10.** Outline a design for a unity feedback system with a plant given by

Mathcad

$$G_2 = \frac{2000}{s(s + 5)(s + 10)}$$

and the performance specifications:

(1)   $\phi_{\text{PM}} \cong 45°$.

(2)   $K_v = 50$.

(3)  The bandwidth BW of the compensated system must be approximately equal to or not much greater than that of the uncompensated system, because high-frequency "noise" disturbances are present under normal operating conditions.

(4)  The compensated system should not respond sluggishly; that is the predominant time constant $\tau$ of the system must be maintained at a value approximately the same as that of the uncompensated system.

A simple calculation clearly shows that the uncompensated system is unstable (e.g., try the Routh test). Therefore compensation is mandatory. But due to the stringent nature of the specifications, a detailed design for this system using Nyquist techniques requires too much effort, if done manually. The techniques of the next few chapters provide a much simpler solution. However, analysis of the problem statement indicates the kind of compensation needed.

For $G_2$, $K_v = \lim_{s \to 0} sG_2(s) = 40$. Therefore satisfaction of (2) requires a gain compensation of 5/4. But an increase in gain only makes the system more unstable. Therefore additional compensation is necessary. Lead compensation is probably inadequate due to (3), and lag compensation is not possible due to (4). Thus it appears that a lag-lead network and an amplifier would most likely satisfy all criteria. The lag portion of the lag-lead network would satisfy (3), and the lead portion (4) and (1).

**12.11.**  What is the effect on the Polar Plot of the system

$$GH = \frac{\prod_{i=1}^{m}(s + z_i)}{\prod_{i=1}^{n}(s + p_i)}$$

where $m \le n$, $0 < z_i < \infty$, $0 \le p_i < \infty$, when $k$ finite nonzero poles are included in $GH$, in addition to the original $n$ poles?

For low frequencies the Polar Plot is modified in magnitude only, since

$$\lim_{s \to 0} GH' = \lim_{s \to 0} \left[ \frac{\prod_{i=1}^{m}(s + z_i)}{\prod_{i=1}^{n+k}(s + p_i)} \right] = \frac{\prod_{i=1}^{m} z_i}{\prod_{i=1}^{n+k} p_i} = \left( \frac{1}{\prod_{i=1}^{k} p_i} \right) \lim_{s \to 0} GH$$

For high frequencies addition of $k$ poles reduces the phase angle of $GH$ by $k\pi/2$ radians, since

$$\lim_{\omega \to \infty} \arg GH'(\omega) = \lim_{\omega \to \infty} \left[ \sum_{i=1}^{m} \tan^{-1}\left( \frac{\omega}{z_i} \right) - \sum_{i=1}^{n+k} \tan^{-1}\left( \frac{\omega}{p_i} \right) \right]$$

$$= \frac{m\pi}{2} - \frac{(n+k)\pi}{2} = \lim_{\omega \to \infty} \arg GH - \frac{k\pi}{2}$$

Therefore the portion of the Polar Plot near the origin is rotated clockwise by $k\pi/2$ degrees when $k$ poles are added.

**12.12.**  Draw the Polar Plot of the digital lag compensator given by Equation ($12.4$):

Mathcad

$$P_{\text{Lag}}(z) = \left( \frac{1 - p_c}{1 - z_c} \right) \left[ \frac{z - z_c}{z - p_c} \right] \qquad z_c < p_c$$

Let $z_c = 0.86$ and $p_c = 0.97$, to simplify the task.

At $\omega = 0$, $P_{\text{Lag}} = P_{\text{Lag}}(e^{j0T}) = P_{\text{Lag}}(1) = 1$. At $\omega T = \pi$,

$$P_{\text{Lag}}(e^{j\pi}) = \left( \frac{1 - p_c}{1 - z_c} \right) \left[ \frac{-1 - z_c}{-1 - p_c} \right] = \frac{1 - z_c p_c - (p_c - z_c)}{1 - z_c p_c + (p_c - z_c)} \equiv c = 0.2$$

At a few intermediate values, $P_{\text{Lag}}(e^{j\pi/4}) \approx 0.02 - j0.03$ and $P_{\text{Lag}}(e^{j\pi/2}) \approx 0.2 - j0.012$. The resulting Polar Plot, for $0 \le \omega T \le \pi$ radians is shown in Fig. 12-26. It is instructive to compare this Polar Plot of the digital lag compensator with its continuous-time equivalent in Fig. 12-11.

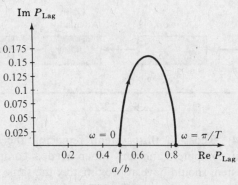

Fig. 12-26

**12.13.** Draw the Polar Plot of the particular *digital lead* compensator:

$$P_{\text{Lead}}(z) = \left(\frac{a}{b}\right)\left[\frac{1 - e^{-bT}}{1 - e^{-aT}}\right]\left[\frac{z - e^{-aT}}{z - e^{-bT}}\right]$$

where $a < b$.

We have

$$P_{\text{Lead}}(e^{j0T}) = P_{\text{Lead}}(1) = \left(\frac{a}{b}\right)\left(\frac{1 - e^{-bT}}{1 - e^{-aT}}\right)\left(\frac{1 - e^{-aT}}{1 - e^{-bT}}\right) = \frac{a}{b} < 1$$

The remainder of the plot has been drawn by computer, by evaluating $P_{\text{Lead}}(1\underline{/\phi})$ for values of the angle $\phi$ in the range $0 < \phi \leq \pi$ radians, for specific values $a = 1$ and $b = 2$. The result is given in Fig. 12-27, which should be compared with Fig. 12-6, the Polar Plot of a continuous system lead network.

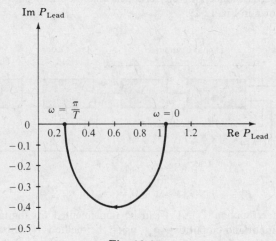

Fig. 12-27

This form of the general digital lead compensator, given in Equation (*6.9*), has a gain factor

$$K_{\text{Lead}} = \frac{a}{b}\left[\frac{1 - e^{-bT}}{1 - e^{-aT}}\right]$$

This compensator is a direct digital analog of the continuous lead compensator $P_{\text{Lead}} = (s + a)/(s + b)$, in which the zeros and poles at $-a$ and $-b$ in the $s$-plane have been transformed directly into zeros and poles in $z$-plane $z_c = e^{-aT}$ and $p_c = e^{-bT}$, and the steady state gain (at $\omega = 0$) has been preserved as $a/b$.

**12.14.** The closed-loop continuous system with both gain factor and lead compensation shown in Fig. 12-28 is stable, with a damping ratio $\zeta \approx 0.7$ and dominant time constant $\tau \approx 4.5$ sec (see

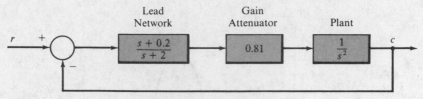

Fig. 12-28

Sections 4.13 and 10.4). Redesign this system, replacing the controller (including summing junction) with a digital computer, and any other needed components for analog-digital data conversion. The new system should have approximately the same dynamic characteristics.

The sampling rate of the digital components must be sufficiently fast to reproduce the signals accurately. The natural frequency $\omega_n$ is estimated from Equation ($10.7$) as $\omega_n \approx 1/\xi\tau = 1/(0.7)(4.5) = 0.317$ rad/sec. For a continuous system with this $\omega_n$, a safe angular sampling frequency $\omega_s \approx 20\omega_n = 6.35 \approx 2\pi$ rad/sec, equivalent to $f_s = 1$ Hz, because $\omega_s = 2\pi f_s$. Therefore we choose $T = 1$ sec.

We now replace the continuous lead compensator by the digital lead compensator given in Problem 12.13:

$$P_{\text{Lead}}(z) = \left(\frac{a}{b}\right)\left(\frac{1 - e^{-bT}}{1 - e^{-aT}}\right)\left[\frac{z - e^{-aT}}{z - e^{-bT}}\right]$$

$$\simeq 0.55\left[\frac{z - 0.82}{z - 0.14}\right]$$

where $a = 0.2$ and $b = 2$ from Fig. 12-28. The factor of 0.55 can be obtained with the gain factor compensator for the continuous system, $K = 0.81$, yielding an overall factor of 0.55 (0.81) = 0.45. The resulting design also needs samplers in the feedback and the input paths, and a zero-order hold in the forward path, all as shown in Fig. 12-29.

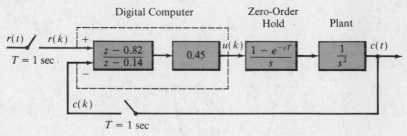

Fig. 12-29

The digital transfer function $P_{\text{Lead}}(z)$ can be implemented for digital computation as a difference equation between the input and output of $P_{\text{Lead}}$, using the methods described in Section 4.9. That is, write $P_{\text{Lead}}(z)$ as a function of $z^{-1}$ instead of $z$, and treat $z^{-1}$ as a unit time-shift operator. Combining the gain factor 0.45 with $P_{\text{Lead}}$, we obtain

$$0.48P_{\text{Lead}} = \frac{0.45 - 0.39z^{-1}}{1 - 0.14z^{-1}} \equiv \frac{u(k)}{r(k) - c(k)}$$

Then, cross-multiplying terms and letting $z^{-1}u(k) = u(k-1)$, etc., we obtain the desired difference equation:

$$u(k) = 0.14u(k-1) + 0.45[r(k) - c(k)] - 0.39[r(k-1) - c(k-1)]$$

**12.15.** Digitize the remaining continuous components in Fig. 12-29 and compare the Polar Plot of: ($a$) the original continuous plant without compensation, $G_2(s) = 1/s^2$, ($b$) the compensated system of Fig. 12-28, $G_1G_2(s)$, and ($c$) the digital system of Fig. 12-30, $G_1G_2(z)$.

The combination of the zero-order hold and the plant $G_2(s) = 1/s^2$ can be digitized using Equation (6.9):

$$G_2'(z) = \left(\frac{z-1}{z}\right)\mathbb{Z}\left\{\mathscr{L}^{-1}\left(\frac{1}{s^2}\right)\bigg|_{t=kT}\right\}$$

$$= \frac{T^2}{2}\left(\frac{z+1}{(z-1)^2}\right) = \frac{0.5(z+1)}{(z-1)^2}$$

The closed-loop discrete-time equivalent system is shown in Fig. 12-30.

Nyquist Stability Plots (not shown) would indicate that the compensated systems are absolutely stable. To check relative stability, the Polar Plots of the three systems are shown superimposed in Fig. 12-31, for $\omega > 0$ only. The phase margin of $G_1G_2(s)$ is $\phi_{\mathrm{PM}} \simeq 53°$, a substantial improvement over that of $G_2(s)$. The Polar Plots for $G_1G_2(s)$ and $G_1G_2(z)$ are quite similar, over a wide range of $\omega$, and the phase margin for $G_1G_2(z)$ is still quite good, $\phi_{\mathrm{PM}} \simeq 37°$.

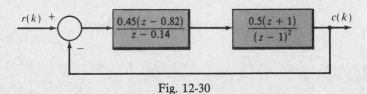

Fig. 12-30

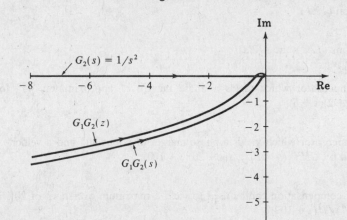

Fig. 12-31

**12.16.** Determine the closed-loop system bandwidth BW of the compensated system designed in Example 12.7.

Performance specification 3 was given in terms of the gain crossover frequency $\omega_1$, as $\omega_1 \geq 10$ rad/sec. This may appear somewhat unrealistic, or artificial, given that a specific phase margin $\phi_{\mathrm{PM}} = [180 + \arg GH(\omega_1)]$ degrees was also given in performance specification 2. Actually, the *bandwidth* (BW) of the closed-loop system would be the more likely frequency of interest in control system design. (These design criteria are discussed in Chapter 10.) However, as noted in Section 10.4, it is often the case that $\omega_1$ is a good approximation of the closed-loop system bandwidth BW, when it is given its common interpretation as the range of frequencies over which the magnitude ratio of the system, which in this case means $|C/R|$, does not fall more than 3 db from its steady state value, at $\omega = 0$ ($z = 1$). For this problem

$$G_1 = \frac{1.59(z-0.86)}{(z-1)(z-0.97)}$$

$$G_2 = \frac{3(z+1)\left(z+\frac{1}{3}\right)}{8z\left(z+\frac{1}{2}\right)}$$

$$\frac{C}{R} = \frac{G_1G_2}{1+G_1G_2}$$

We easily find that

$$\lim_{\omega \to 0}\left(\frac{C}{R}\right) = \lim_{z \to 1}\left(\frac{C}{R}\right) = 1$$

Now, 3 db down from 1 is 0.707 [see Equation (*10.5*)]. Therefore the BW is the frequency $\omega_{BW}$ that satisfies the equation:

$$\left|\frac{C}{R}(\omega_{BW})\right| = 0.707$$

We quickly obtain the solution $\omega_{BW} = 10.724$ rad/sec by trial and error using a computer to evaluate the magnitude ratio at a few values of $\omega$ in the vicinity of $\omega_1 = 10$. Thus the approximation $\omega_1 \simeq \omega_{BW}$ is confirmed as a good one for the problem solved in Example 12.7.

# Supplementary Problems

**12.17.** Determine a positive value of gain factor $K$ when $M_p = 2$ for the system of Problem 12.5.

**12.18.** Prove Equation (*12.1*).

**12.19.** Prove Equations (*12.2*) and (*12.3*).

**12.20.** Design a compensator which yields a phase margin of approximately 45° for the system defined by $GH = 84/s(s + 2)(s + 6)$.

**12.21.** Design a compensator which yields a phase margin of about 40° and a velocity constant $K_v = 40$ for the system defined by $GH = (4 \times 10^5)/s(s + 20)(s + 100)$.

**12.22.** What kind of compensation can be used to yield a maximum overshoot of 20% for the system defined by $GH = (4 \times 10^4)/s^2(s + 100)$?

**12.23.** Show that the addition of $k$ finite zeros ($z_i \neq 0$) to the system of Problem 12.11 rotates the high-frequency portion of the Polar Plot by $k\pi/2$ radians in the counterclockwise direction.

# Answers to Some Supplementary Problems

**12.17.** $K = 31.2$

**12.18.** $P_{\text{Lead}} = \dfrac{s + 30}{s + 120}$

**12.21.** $P_{\text{Lead}} = \dfrac{s + 20}{s + 100}$, no preamplifier required

**12.22.** Lag-lead, and possibly lead plus gain factor compensation.

# Chapter 13

## Root-Locus Analysis

### 13.1 INTRODUCTION

It was shown in Chapters 4 and 6 that the poles of a transfer function can be displayed graphically in the $s$-plane or $z$-plane by means of a pole-zero map. An analytical method is presented in this chapter for displaying the location of the poles of the closed-loop transfer function

$$\frac{G}{1+GH}$$

as a function of the gain factor $K$ (see Sections 6.2 and 6.6) of the open-loop transfer function $GH$. This method, called *root-locus analysis*, requires that only the location of the poles and zeros of $GH$ be known, and does not require factorization of the characteristic polynomial.

Root-locus techniques permit accurate computation of the time-domain response in addition to yielding readily available frequency response information.

The following discussion of root-locus analysis applies identically to continuous systems in the $s$-plane and discrete-time systems in the $z$-plane.

### 13.2 VARIATION OF CLOSED-LOOP SYSTEM POLES: THE ROOT-LOCUS

Consider the canonical feedback control system given in Fig. 13-1. The closed-loop transfer function is

$$\frac{C}{R} = \frac{G}{1+GH}$$

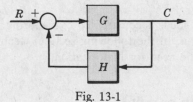

Fig. 13-1

Let the open-loop transfer function $GH$ be represented by

$$GH \equiv \frac{KN}{D}$$

where $N$ and $D$ are finite polynomials in the complex variable $s$ or $z$ and $K$ is the open-loop gain factor. The closed-loop transfer function then becomes

$$\frac{C}{R} = \frac{G}{1+KN/D} = \frac{GD}{D+KN}$$

The closed-loop poles are roots of the characteristic equation

$$D + KN = 0 \qquad\qquad (13.1)$$

In general the location of these roots in the $s$-plane or $z$-plane changes as the open-loop gain factor $K$ is varied. A locus of these roots plotted in the $s$-plane or $z$-plane as a function of $K$ is called a **root-locus**.

For $K$ equal to zero, the roots of Equation ($13.1$) are the roots of the polynomial $D$, which are the same as the poles of the open-loop transfer function $GH$. If $K$ becomes very large, the roots approach those of the polynomial $N$, the open-loop zeros. Thus, as $K$ is increased from zero to infinity, the loci of the closed-loop poles originate from the open-loop poles and terminate at the open-loop zeros.

**EXAMPLE 13.1.**   Consider the continuous system open-loop transfer function

$$GH = \frac{KN(s)}{D(s)} = \frac{K(s+1)}{s^2 + 2s} = \frac{K(s+1)}{s(s+2)}$$

For $H = 1$, the closed-loop transfer function is

$$\frac{C}{R} = \frac{K(s+1)}{s^2 + 2s + K(s+1)}$$

The closed-loop poles of this system are easily determined by factoring the denominator polynomial:

$$p_1 = -\tfrac{1}{2}(2+K) + \sqrt{1 + \tfrac{1}{4}K^2}$$

$$p_2 = -\tfrac{1}{2}(2+K) - \sqrt{1 + \tfrac{1}{4}K^2}$$

The locus of these roots plotted as a function of $K$ (for $K > 0$) is shown in the $s$-plane in Fig. 13-2. As observed in the figure, this root-locus has two *branches*: one for a closed-loop pole which moves from the open-loop pole at the origin to the open-loop zero at $-1$, and from the open-loop pole at $-2$ to the open-loop zero at $-\infty$.

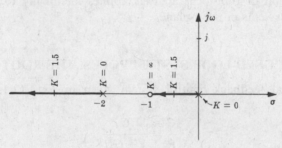

Fig. 13-2

In the example above, the root-locus is constructed by factoring the denominator polynomial of the system closed-loop transfer function. In the following sections, techniques are described which permit construction of root-loci without the need for factorization.

## 13.3   ANGLE AND MAGNITUDE CRITERIA

In order for a branch of a root-locus to pass through a particular point $p_1$ in the complex plane, it is necessary that $p_1$ be a root of the characteristic Equation (*13.1*) for some real value of $K$. That is,

$$D(p_1) + KN(p_1) = 0 \qquad\qquad (13.2)$$

or, equivalently,

$$GH = \frac{KN(p_1)}{D(p_1)} = -1 \qquad\qquad (13.3)$$

Therefore the complex number $GH(p_1)$ must have a phase angle of $180° + 360l°$, where $l$ is an arbitrary integer. Thus we have the **angle criterion**

$$\arg GH(p_1) = 180° + 360l° = (2l+1)\pi \text{ radians} \qquad l = 0, \pm 1, \pm 2, \dots \qquad (13.4a)$$

which can also be written as

$$\arg\left[\frac{N(p_1)}{D(p_1)}\right] = \begin{cases} (2l+1)\pi \text{ radians} & \text{for } K > 0 \\ 2l\pi \text{ radians} & \text{for } K < 0 \end{cases} \qquad l = 0, \pm 1, \pm 2, \dots \qquad (13.4b)$$

In order for $p_1$ to be a closed-loop pole of the system, on the root-locus, it is necessary that Equation (*13.3*) be satisfied with regard to *magnitude* in addition to phase angle. That is, $K$ must have the particular value that satisfies the **magnitude criterion**: $|GH(p_1)| = 1$, or

$$|K| = \left| \frac{D(p_1)}{N(p_1)} \right| \tag{13.5}$$

The angle and magnitude of $GH$ at any point in the complex $s$- or $z$-plane can be determined graphically as described in Sections 4.12 and 6.5. In this way, it is possible to construct the root-locus manually by a trial-and-error procedure of testing points in the complex plane. That is, the root-locus is drawn through all points which satisfy the angle criterion, Equation (*13.4b*), and the magnitude criterion is used to determine the values of $K$ at points along the loci. Digital computer programs for routinely plotting root-loci are widely available. However, manual construction is simplified considerably, using certain shortcuts or construction rules as described in the following sections.

## 13.4  NUMBER OF LOCI

The number of loci, that is, the number of branches of the root-locus, is equal to the number of poles of the open-loop transfer function $GH$ (for $n \geq m$).

**EXAMPLE 13.2.**  The open-loop transfer function of the discrete-time system $GH(z) = K(z + \frac{1}{2})/z^2(z + \frac{1}{4})$ has three poles. Hence there are three loci in the root-locus plot.

## 13.5  REAL AXIS LOCI

Those sections of the root-locus on the real axis in the complex plane are determined by counting the total number of finite poles and zeros of $GH$ to the right of the points in question. The following rule depends on whether the open-loop gain factor $K$ is positive or negative.

*Rule for $K > 0$*
Points of the root-locus on the real axis lie to the left of an *odd* number of finite poles and zeros.

*Rule for $K < 0$*
Points of the root-locus on the real axis lie to the left of an *even* number of finite poles and zeros.

If no points on the real axis lie to the left of an odd number of finite poles and zeros, then no portion of the root-locus for $K > 0$ lies on the real axis. A similar statement is true for $K < 0$.

**EXAMPLE 13.3.**  Consider the pole-zero map of an open-loop transfer function $GH$ shown in Fig. 13-3. Since all the points on the real axis between 0 and $-1$ and between $-1$ and $-2$ lie to the left of an odd number of finite poles and zeros, these points are on the root-locus for $K > 0$. The portion of the real axis between $-\infty$ and $-4$ lies to the left of an odd number of finite poles and zeros; hence these points are also on the root-locus for $K > 0$. All portions of the root-locus for $K > 0$ on the real axis are illustrated in Fig. 13-4. All remaining portions of the real axis, that is, between $-2$ and $-4$ and between 0 and $\infty$, lie on the root-locus for $K < 0$.

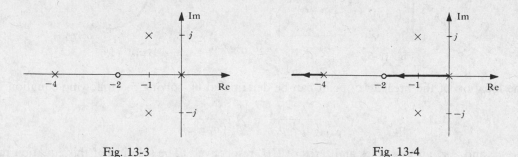

Fig. 13-3                                    Fig. 13-4

322       ROOT-LOCUS ANALYSIS       [CHAP. 13

## 13.6  ASYMPTOTES

For large distances from the origin in the complex plane, the branches of a root-locus approach a set of straight-line asymptotes. These asymptotes emanate from a point in the complex plane on the real axis called the **center of asymptotes** $\sigma_c$ given by

$$\sigma_c = -\frac{\sum_{i=1}^{n} p_i - \sum_{i=1}^{m} z_i}{n - m} \tag{13.6}$$

where $-p_i$ are the poles, $-z_i$ are the zeros, $n$ is the number of poles, and $m$ the number of zeros of $GH$.

The angles between the asymptotes and the real axis are given by

$$\beta = \begin{cases} \dfrac{(2l+1)180}{n-m} \text{ degrees} & \text{for } K > 0 \\ \dfrac{(2l)180}{n-m} \text{ degrees} & \text{for } K < 0 \end{cases} \tag{13.7}$$

for $l = 0, 1, 2, \ldots, n - m - 1$. This results in a number of asymptotes equal to $n - m$.

**EXAMPLE 13.4.**  The center of asymptotes for $GH = K(s + 2)/s^2(s + 4)$ is located at

$$\sigma_c = -\frac{4 - 2}{2} = -1$$

Since $n - m = 3 - 1 = 2$, there are two asymptotes. Their angles with the real axis are 90° and 270°, for $K > 0$, as shown in Fig. 13-5.

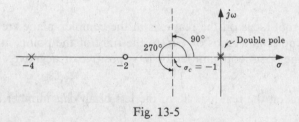

Fig. 13-5

## 13.7  BREAKAWAY POINTS

A **breakaway point** $\sigma_b$ is a point on the real axis where two or more branches of the root-locus depart from or arrive at the real axis. Two branches leaving the real axis are illustrated in the root-locus plot in Fig. 13-6. Two branches coming onto the real axis are illustrated in Fig. 13-7.

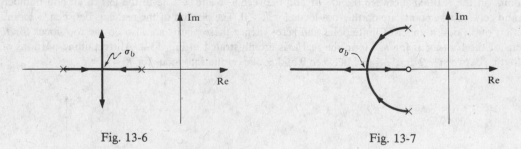

Fig. 13-6                     Fig. 13-7

The location of the breakaway point can be determined by solving the following equation for $\sigma_b$:

$$\sum_{i=1}^{n} \frac{1}{(\sigma_b + p_i)} = \sum_{i=1}^{m} \frac{1}{(\sigma_b + z_i)} \tag{13.8}$$

where $-p_i$ and $-z_i$ are the poles and zeros of $GH$, respectively. The solution of this equation requires factorization of an $(n + m - 1)$-order polynomial in $\sigma_b$. Consequently, the breakaway point can only be easily determined analytically for relatively simple $GH$. However, an approximate location can often be determined intuitively; then an iterative process can be used to solve the equation more exactly (see Problem 13.20). Computer programs for factorization of polynomials could also be applied.

**EXAMPLE 13.5.**   To determine the breakaway points for $GH = K/s(s + 1)(s + 2)$, the following equation must be solved for $\sigma_b$:

$$\frac{1}{\sigma_b} + \frac{1}{\sigma_b + 1} + \frac{1}{\sigma_b + 2} = 0$$

$$(\sigma_b + 1)(\sigma_b + 2) + \sigma_b(\sigma_b + 2) + \sigma_b(\sigma_b + 1) = 0$$

which reduces to $3\sigma_b^2 + 6\sigma_b + 2 = 0$ whose roots are $\sigma_b = -0.423, -1.577$.

Applying the real axis rule of Section 13.5 for $K > 0$ indicates that there are branches of the root-locus between 0 and $-1$ and between $-\infty$ and $-2$. Therefore the root at $-0.423$ is a breakaway point, as shown in Fig. 13-8. The value $\sigma_b = -1.577$ represents a breakaway on the root-locus for negative values of $K$ since the portion of the real axis between $-1$ and $-2$ is on the root-locus for $K < 0$.

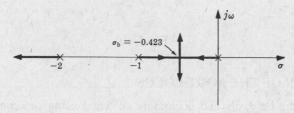

Fig. 13-8

## 13.8   DEPARTURE AND ARRIVAL ANGLES

The **departure angle** of the root-locus from a *complex pole* is given by

$$\theta_D = 180° + \arg GH' \tag{13.9}$$

where $\arg GH'$ is the phase angle of $GH$ computed at the complex pole, but ignoring the contribution of that particular pole.

**EXAMPLE 13.6.**   Consider the continuous system open-loop transfer function

$$GH = \frac{K(s + 2)}{(s + 1 + j)(s + 1 - j)} \qquad K > 0$$

The departure angle of the root-locus from the complex pole at $s = -1 + j$ is determined as follows. The angle of $GH$ for $s = -1 + j$, ignoring the contribution of the pole at $s = -1 + j$, is $-45°$. Therefore the departure angle is

$$\theta_D = 180° - 45° = 135°$$

and is illustrated in Fig. 13-9.

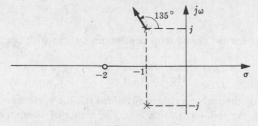

Fig. 13-9

The **angle of arrival** of the root-locus at a *complex zero* is given by

$$\theta_A = 180° - \arg GH''  \qquad (13.10)$$

where $\arg GH''$ is the phase angle of $GH$ at the complex zero, ignoring the effect of that zero.

**EXAMPLE 13.7.** Consider the discrete-time system open-loop transfer function

$$\frac{K(z+j)(z-j)}{z(z+1)} \qquad K > 0$$

The arrival angle of the root-locus for the complex zero at $z = j$ is $\theta_A = 180° - (-45°) = 225°$ as shown in Fig. 13-10.

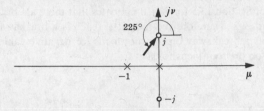

Fig. 13-10

## 13.9   CONSTRUCTION OF THE ROOT-LOCUS

A root-locus plot may be easily and accurately sketched using the construction rules of Sections 13.4 through 13.8. An efficient procedure is the following. First, determine the portions of the root-locus on the real axis. Second, compute the center and angles of the asymptotes and draw the asymptotes on the plot. Then determine the departure and arrival angles at complex poles and zeros (if any) and indicate them on the plot. Next, make a rough sketch of the branches of the root-locus so that each branch of the locus either terminates at a zero or approaches infinity along one of the asymptotes. The accuracy of this last step should of course improve with experience.

The accuracy of the plot may be improved by applying the angle criterion in the vicinity of the estimated branch locations. The rule of Section 13.7 can also be applied to determine the exact location of breakaway points.

The magnitude criterion of Section 13.3 is used to determine the values of $K$ along the branches of the root-locus.

Since complex poles must occur in complex conjugate pairs (assuming real coefficients for the numerator and denominator polynomials of $GH$), the root-locus is symmetric about the real axis. Thus it is sufficient to plot only the upper half of the root-locus. However, it must be remembered that, in doing this, the lower halves of open-loop complex poles and zeros must be included when applying the magnitude and angle criteria.

Often, for analysis or design purposes, an accurate plot of the root-locus is required only in certain regions of the complex plane. In this case, the angle and magnitude criteria need only be applied in those regions of interest after a rough sketch has established the general shape of the plot. Of course, if a computer and appropriate software are available, plotting of even very complex root-loci can be a simple matter.

**EXAMPLE 13.8.** The root-locus for the closed-loop continuous system with open-loop transfer function

$$GH = \frac{K}{s(s+2)(s+4)} \qquad K > 0$$

is constructed as follows. Applying the real axis rule of Section 13.5, the portions of the real axis between 0 and $-2$ and between $-4$ and $-\infty$ lie on the root-locus for $K > 0$. The center of asymptotes is determined from Equation (13.6) to be $\sigma_c = -(2+4)/3 = -2$, and there are three asymptotes located at angles of $\beta = 60°$, 180°, and 300°.

Since two branches of the root-locus for $K > 0$ come together on the real axis between 0 and $-2$, a breakaway point exists on that portion of the real axis. Hence the root-locus for $K > 0$ may be sketched by estimating the location of the breakaway point and continuing the branches of the root-locus to the asymptotes, as shown in Fig. 13-11. To improve the accuracy of this plot, the exact location of the breakaway point is determined from Equation (*13.8*):

$$\frac{1}{\sigma_b} + \frac{1}{\sigma_b + 2} + \frac{1}{\sigma_b + 4} = 0$$

which simplifies to $3\sigma_b^2 + 12\sigma_b + 8 = 0$. The appropriate solution of this equation is $\sigma_b = -0.845$.

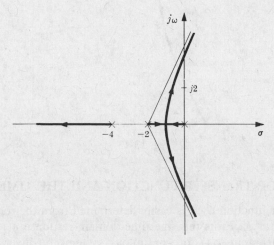

Fig. 13-11

The angle criterion is applied to points in the vicinity of the approximate root-locus to improve the accuracy of the location of the branches in the complex part of the $s$-plane; the magnitude criterion is used to determine the values of $K$ along the root-locus. The resulting root-locus plot for $K > 0$ is shown in Fig. 13-12.

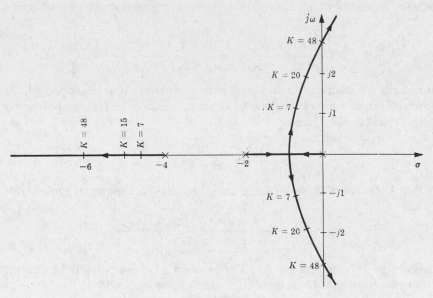

Fig. 13-12

The root-locus for $K < 0$ is constructed in a similar manner. In this case, however, the portions of the real axis between 0 and $\infty$ and between $-2$ and $-4$ lie on the root-locus; the breakaway point is located at $-3.155$; and the asymptotes have angles of $0°$, $120°$, and $240°$. The root-locus for $K < 0$ is shown in Fig. 13-13.

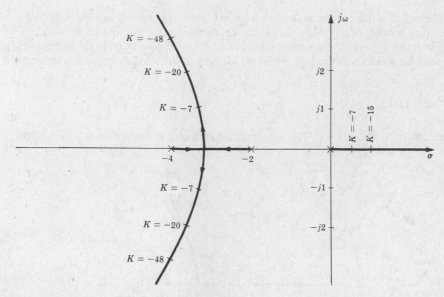

Fig. 13-13

## 13.10   THE CLOSED-LOOP TRANSFER FUNCTION AND THE TIME-DOMAIN RESPONSE

The closed-loop transfer function $C/R$ is easily determined from the root-locus plot for a specified value of open-loop gain factor $K$. From this, the time-domain response $c(t)$ may be determined for a given Laplace transformable input $r(t)$ for continuous systems by inversion of $C(s)$. For discrete systems, $c(k)$ can be similarly determined by inversion of $C(z)$.

Consider the closed-loop transfer function $C/R$ for the canonical *unity* (*negative*) *feedback* system

$$\frac{C}{R} = \frac{G}{1+G} \tag{13.11}$$

Open-loop transfer functions which are rational algebraic expressions can be written (for continuous systems) as

$$G = \frac{KN}{D} = \frac{K(s+z_1)(s+z_2)\cdots(s+z_m)}{(s+p_1)(s+p_2)\cdots(s+p_n)} \tag{13.12}$$

$G$ has the same form for discrete-time systems, with $z$ replacing $s$ in Equation (13.12). In Equation (13.12), $-z_i$ are the zeros, $-p_i$ are the poles of $G$, $m \le n$, and $N$ and $D$ are polynomials whose roots are $-z_i$ and $-p_i$, respectively. Then

$$\frac{C}{R} = \frac{KN}{D+KN} \tag{13.13}$$

and it is clear that $C/R$ and $G$ have the same zeros but not the same poles (unless $K=0$). Hence

$$\frac{C}{R} = \frac{K(s+z_1)(s+z_2)\cdots(s+z_m)}{(s+\alpha_1)(s+\alpha_2)\cdots(s+\alpha_n)}$$

where $-\alpha_i$ denote the $n$ closed-loop poles. The location of these poles is by definition determined directly from the root-locus plot for a specified value of open-loop gain $K$.

**EXAMPLE 13.9.**   Consider the continuous system whose open-loop transfer function is

$$G = \frac{K(s+2)}{(s+1)^2} \qquad K > 0$$

The root-locus plot is given in Fig. 13-14.

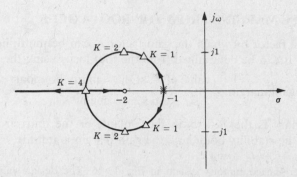

Fig. 13-14

Several values of gain factor $K$ are shown at points on the loci denoted by *small triangles*. These points are the *closed-loop poles* corresponding to the specified values of $K$. For $K = 2$, the closed-loop poles are $-\alpha_1 = -2 + j$ and $-\alpha_2 = -2 - j$. Therefore

$$\frac{C}{R} = \frac{2(s+2)}{(s+2+j)(s+2-j)}$$

When the system is not unity feedback, then

$$\frac{C}{R} = \frac{G}{1+GH} \qquad (13.14)$$

and

$$GH = \frac{KN}{D} \qquad (13.15)$$

The closed-loop poles may be determined directly from the root-locus for a given $K$, but the closed-loop zeros are not equal to the open-loop zeros. The open-loop zeros must be computed separately by clearing fractions in Equation (13.14).

**EXAMPLE 13.10.**   Consider the continuous system described by

$$G = \frac{K(s+2)}{s+1} \qquad H = \frac{1}{s+1} \qquad GH = \frac{K(s+2)}{(s+1)^2} \qquad K > 0$$

and

$$\frac{C}{R} = \frac{K(s+1)(s+2)}{(s+1)^2 + K(s+2)} = \frac{K(s+1)(s+2)}{(s+\alpha_1)(s+\alpha_2)}$$

The root-locus plot for this example is the same as that for Example 13.9. Hence for $K = 2$, $\alpha_1 = 2 + j$ and $\alpha_2 = 2 - j$. Thus

$$\frac{C}{R} = \frac{2(s+1)(s+2)}{(s+2+j)(s+2-j)}$$

**EXAMPLE 13.11.**   For the discrete-time system with $GH(z) = K/z(z-1)$, the root-locus for $K > 0$ is shown in Fig. 13-15. For $K = 0.25$, the roots are at $z = 0.5$ and the closed-loop transfer function is

$$\frac{C}{R} = \frac{0.25}{(z-0.5)^2}$$

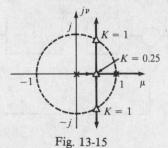

Fig. 13-15

### 13.11   GAIN AND PHASE MARGINS FROM THE ROOT-LOCUS

The **gain margin** is the factor by which the gain factor $K$ can be multiplied before the closed-loop system becomes unstable. It can be determined from the root-locus using the following formula:

$$\text{gain margin} = \frac{\text{value of } K \text{ at the stability boundary}}{\text{design value of } K} \qquad (13.16)$$

where the stability boundary is the $j\omega$-axis in the $s$-plane, or the unit circle in the $z$-plane. If the root-locus does not cross the stability boundary, the gain margin is infinite.

**EXAMPLE 13.12.**   Consider the continuous system in Fig. 13-16. The design value for the gain factor is 8, producing the closed-loop poles (denoted by small triangles) shown in the root-locus of Fig. 13-17. The gain factor at the $j\omega$-axis crossing is 64; hence the gain margin for this system is $64/8 = 8$.

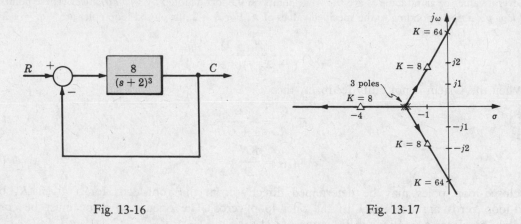

Fig. 13-16                                    Fig. 13-17

**EXAMPLE 13.13.**   The root-locus for the discrete-time system of Example 13.11 crosses the stability boundary (unit circle) for $K = 1$. For a design value of $K = 0.25$, the gain margin is $1/0.25 = 4$.

The **phase margin** can also be determined from the root-locus. In this case it is necessary to find the point $\omega_1$ on the stability boundary for which $|GH| = 1$ for the design value of $K$; that is,

$$|D(\omega_1)/N(\omega_1)| = K_{\text{design}}$$

It is usually necessary to use a trial-and-error procedure to locate $\omega_1$. The phase margin is then computed from $\arg GH(\omega_1)$ as

$$\phi_{\text{PM}} = \left[180° + \arg GH(\omega_1)\right] \text{ degrees} \qquad (13.17)$$

**EXAMPLE 13.14.**   For the system of Example 13.12, $|GH(\omega_1)| = |8/(j\omega_1 + 2)^3| \equiv 1$ when $\omega_1 = 0$; the phase angle of $GH(0)$ is $0°$. The phase margin is therefore $180°$.

**EXAMPLE 13.15.**   For the continuous system of Fig. 13-18, the root-locus is shown in Fig. 13-19. The point on the $j\omega$-axis for which $|GH(\omega_1)| = |24/j\omega_1(j\omega_1 + 4)^2| \equiv 1$ is at $\omega_1 = 1.35$; the angle of $GH(1.35)$ is $-129.6°$. Therefore the phase margin is $\phi_{\text{PM}} = 180° - 129.6° = 50.4°$.

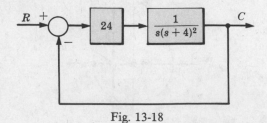

Fig. 13-18

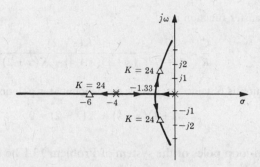

Fig. 13-19

## 13.12 DAMPING RATIO FROM THE ROOT-LOCUS FOR CONTINUOUS SYSTEMS

The gain factor $K$ required to give a specified damping ratio $\zeta$ (or vice versa) for the second-order continuous system

$$GH = \frac{K}{(s+p_1)(s+p_2)} \qquad K, p_1, p_2 > 0$$

is easily determined from the root-locus. Simply draw a line from the origin at an angle of plus or minus $\theta$ with the negative real axis, where

$$\theta = \cos^{-1}\zeta \qquad\qquad\qquad (13.18)$$

(See Section 4.13.) The gain factor at the point of intersection with the root-locus is the required value of $K$. This procedure can be applied to any pair of complex conjugate poles, for systems of second or higher order. For higher-order systems, the damping ratio determined by this procedure for a *specific pair* of complex poles does not necessarily determine the damping (predominant time constant) of the system.

**EXAMPLE 13.16.** Consider the third-order system of Example 13.15. The damping ratio $\zeta$ of the *complex poles* for $K = 24$ is easily determined by drawing a line from the origin to the point on the root-locus where $K = 24$, as shown in Fig. 13-20. The angle $\theta$ is measured as 60°; hence

$$\zeta = \cos\theta = 0.5$$

This value of $\zeta$ is a good approximation for the damping of the third-order system with $K = 24$ because the complex poles dominate the response.

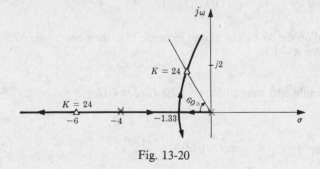

Fig. 13-20

# Solved Problems

## VARIATION OF SYSTEM CLOSED-LOOP POLES

**13.1.** Determine the closed-loop transfer function and the characteristic equation of the unity negative feedback control system whose open-loop transfer function is $G = K(s + 2)/(s + 1)(s + 4)$.

The closed-loop transfer function is

$$\frac{C}{R} = \frac{G}{1+G} = \frac{K(s+2)}{(s+1)(s+4) + K(s+2)}$$

The characteristic equation is obtained by setting the denominator polynomial equal to zero:

$$(s+1)(s+4) + K(s+2) = 0$$

**13.2.** How would the closed-loop poles of the system of Problem 13.1 be determined for $K = 2$ from its root-locus plot?

The root-locus is a plot of the closed-loop poles of the feedback system as a function of $K$. Therefore the closed-loop poles for $K = 2$ are determined by the points on the root-locus which correspond to $K = 2$ (one point on each branch of the locus).

**13.3.** How can a root-locus be employed to factor the polynomial $s^2 + 6s + 18$?

Since the root-locus is a plot of the roots of the characteristic equation of a system, Equation (*13.1*), as a function of its open-loop gain factor, the roots of the above polynomial can be determined from the root-locus of any system whose characteristic polynomial is equivalent to it for some value of $K$. For example, the root-locus for $GH = K/s(s + 6)$ factors the characteristic polynomial $s^2 + 6s + K$. For $K = 18$ this polynomial is equivalent to the one we desire to factor. Thus the desired roots are located on this root-locus at the points corresponding to $K = 18$.

Note that other forms for $GH$ could be chosen, such as $GH = K/(s + 2)(s + 4)$ whose closed-loop characteristic polynomial corresponds to the one we wish to factor, but now for $K = 10$.

## ANGLE AND MAGNITUDE CRITERIA

**13.4.** Show that the point $p_1 = -0.5$ satisfies the angle criterion, Equation (*13.4*), and the magnitude criterion, Equation (*13.5*), when $K = 1.5$ in the open-loop transfer function of Example 13.1.

$$\arg GH(p_1) = \arg \frac{K(p_1 + 1)}{p_1(p_1 + 2)} = \arg \frac{1.5(0.5)}{-0.5(1.5)} = 180° \qquad |GH(p_1)| = \left| \frac{1.5(0.5)}{-0.5(1.5)} \right| = 1$$

or

$$\left| \frac{D(p_1)}{N(p_1)} \right| = \left| \frac{-0.5(1.5)}{0.5} \right| = 1.5 = K$$

Thus as illustrated on the root-locus plot of Example 13.1, the point $p_1 = -0.5$ is on the root-locus and is a closed-loop pole for $K = 1.5$.

**13.5.** Determine the angle and magnitude of $GH(j2)$ for $GH = K/s(s + 2)^2$. What value of $K$ satisfies $|GH(j2)| = 1$?

$$GH(j2) = \frac{K}{j2(j2 + 2)^2} \qquad \arg GH(j2) = \begin{cases} -180° & \text{for} \quad K > 0 \\ 0° & \text{for} \quad K < 0 \end{cases} \qquad |GH(j2)| = \frac{|K|}{2(8)} = \frac{|K|}{16}$$

and for $|GH(j2)| = 1$ it is necessary that $|K| = 16$.

**13.6.** Illustrate the graphical composition of $\arg GH(j2)$ and $|GH(j2)|$ in Problem 13.5.

$$\arg GH(j2) = -90° - 45° - 45° = -180° \qquad |GH(j2)| = \frac{|K|}{2(2\sqrt{2})^2} = \frac{|K|}{16}$$

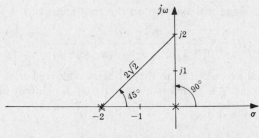

Fig. 13-21

**13.7.** Show that the point $p_1 = -1 + j\sqrt{3}$ is on the root-locus for

$$GH(s) = \frac{K}{(s+1)(s+2)(s+4)} \qquad K > 0$$

and determine $K$ at this point.

$$\arg \frac{N(p_1)}{D(p_1)} = \arg \frac{1}{j\sqrt{3}\left(1+j\sqrt{3}\right)\left(3+j\sqrt{3}\right)} = -90° - 60° - 30° = -180°$$

The angle criterion, Equation (13.4b), is thus satisfied for $K > 0$ and the point $p_1 = -1 + j\sqrt{3}$ is on the root-locus. From Equation (13.5),

$$K = \left| \frac{j\sqrt{3}\left(1+j\sqrt{3}\right)\left(3+j\sqrt{3}\right)}{1} \right| = \sqrt{3(4)12} = 12$$

## NUMBER OF LOCI

**13.8.** Why must the number of loci equal the number of open-loop poles for $m \le n$?

Each branch of the root-locus represents the locus of one closed-loop pole. Consequently there must be as many branches or loci as there are closed-loop poles. Since the number of closed-loop poles is equal to the number of open-loop poles for $m \le n$, the number of loci must equal the number of open-loop poles.

**13.9.** How many loci are in the root-locus for

$$GH(z) = \frac{K\left(z + \frac{1}{3}\right)\left(z + \frac{1}{2}\right)}{z\left(z + \frac{1}{2} + j/2\right)\left(z - \frac{1}{2} - j/2\right)}$$

Since the number of open-loop poles is three, there are three loci in the root-locus plot.

## REAL AXIS LOCI

**13.10.** Prove the real axis loci rules.

For any point on the real axis, the angle contributed to $\arg GH$ by any real axis pole or zero is either $0°$ or $180°$, depending on whether or not the point is to the right or to the left of the pole or zero. The total angle contributed to $\arg GH(s)$ by a pair of complex poles or zeros is zero because

$$\arg\left(s + \sigma_1 + j\omega_1\right) + \arg\left(s + \sigma_1 - j\omega_1\right) = 0$$

for all real values of $s$. Thus $\arg GH(s)$ for real values of $s$ $(s = \sigma)$ may be written as

$$\arg GH(\sigma) = 180 n_r + \arg K$$

where $n_r$ is the total number of finite poles and zeros to the right of $\sigma$. In order to satisfy the angle criterion, $n_r$ must be odd for positive $K$ and even for negative $K$. Thus for $K > 0$, points of the root-locus on the real axis lie to the left of an odd number of finite poles and zeros; and for $K < 0$, points of the root-locus on the real axis lie to the left of an even number of finite poles and zeros.

**13.11.** Determine which parts of the real axis are on the root-locus of

$$GH = \frac{K(s+2)}{(s+1)(s+3+j)(s+3-j)} \qquad K > 0$$

The points on the real axis which lie to the left of an odd number of finite poles and zeros are only those points between $-1$ and $-2$. Therefore by the rule for $K > 0$, only the portion of real axis between $-1$ and $-2$ lies on the root-locus.

**13.12.** Which parts of the real axis are on the root-locus for

$$GH = \frac{K}{s(s+1)^2(s+2)} \qquad K > 0$$

Points on the real axis between $0$ and $-1$ and between $-1$ and $-2$ lie to the left of an odd number of poles and zeros and therefore are on the root-locus for $K > 0$.

## ASYMPTOTES

**13.13.** Prove that the angles of the asymptotes are given by

$$\beta = \begin{cases} \dfrac{(2l+1)180}{n-m} \text{ degrees} & \text{for} \quad K > 0 \\[2ex] \dfrac{(2l)180}{n-m} \text{ degrees} & \text{for} \quad K < 0 \end{cases} \qquad (13.7)$$

For points $s$ far from the origin in the $s$-plane, the angle contributed to $\arg GH$ by each of $m$ zeros is

$$\arg(s + z_i)\big|_{|s| \gg |z_i|} \cong \arg(s)$$

Similarly, the angle contributed to $\arg GH$ by each of $n$ poles is approximately equal to $-\arg(s)$. Therefore

$$\arg\left[\frac{N(s)}{D(s)}\right] \cong -(n-m) \cdot \arg(s) = -(n-m)\beta$$

where $\beta \equiv \arg(s)$. In order for $s$ to be on the root-locus the angle criterion, Equation $(13.4b)$, must be satisfied. Thus

$$\arg\left[\frac{N(s_1)}{D(s_1)}\right] = -(n-m)\beta = \begin{cases} (2l+1)\pi & \text{for} \quad K > 0 \\ (2l)\pi & \text{for} \quad K < 0 \end{cases}$$

and, since $\pm\pi$ radians ($\pm 180°$) are the same angle in the $s$-plane, then

$$\beta = \begin{cases} \dfrac{(2l+1)180}{n-m} \text{ degrees} & \text{for} \quad K > 0 \\[2ex] \dfrac{(2l)180}{n-m} \text{ degrees} & \text{for} \quad K < 0 \end{cases}$$

The proof is similar for the $z$-plane.

**13.14.** Show that the center of asymptotes is given by

$$\sigma_c = -\frac{\displaystyle\sum_{i=1}^{n} p_i - \sum_{i=1}^{m} z_i}{n-m} \qquad (13.6)$$

The points on the root-locus satisfy the characteristic equation $D + KN = 0$, or

$$s^n + b_{n-1}s^{n-1} + \cdots + b_0 + K\left(s^m + a_{m-1}s^{m-1} + \cdots + a_0\right) = 0$$

Dividing by the numerator polynomial $N(s)$, this becomes

$$s^{n-m} + \left(b_{n-1} - a_{m-1}\right)s^{n-m-1} + \cdots + K = 0$$

(same for the $z$-plane, with $z$ replacing $s$). When the first coefficient of a polynomial is unity, the second coefficient is equal to minus the sum of the roots (see Problem 5.26). Thus from $D(s) = 0$, $b_{n-1} = \sum_{i=1}^{n} p_i$. From $N(s) = 0$, $a_{m-1} = \sum_{i=1}^{m} z_i$; and $-(b_{n-1} - a_{m-1})$ is equal to the sum of $n - m$ roots of the characteristic equation.

Now for large values of $K$ and correspondingly large distances from the origin these $n - m$ roots approach the straight-line asymptotes and, along the asymptotes, the sum of the $n - m$ roots is equal to $-(b_{n-1} - a_{m-1})$. Since $b_{n-1} - a_{m-1}$ is a real number, the asymptotes must intersect at a point on the real axis. The center of asymptotes is therefore given by the point on the real axis where $n - m$ equal roots add up to $-(b_{n-1} - a_{m-1})$. Thus

$$\sigma_c = -\frac{b_{n-1} - a_{m-1}}{n - m} = -\frac{\sum_{i=1}^{n} p_i - \sum_{i=1}^{m} z_i}{n - m}$$

For a more detailed proof, see reference [6].

**13.15.** Find the angles and center of, and sketch the asymptotes for

$$GH = \frac{K(s + 2)}{(s + 1)(s + 3 + j)(s + 3 - j)(s + 4)} \qquad K > 0$$

The center of asymptotes is

$$\sigma_c = -\frac{1 + 3 + j + 3 - j + 4 - 2}{4 - 1} = -3$$

There are three asymptotes located at angles of $\beta = 60°$, $180°$, and $300°$ as shown in Fig. 13-22.

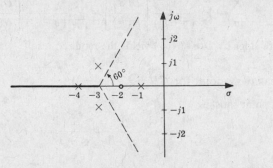

Fig. 13-22

**13.16.** Sketch the asymptotes for $K > 0$ and $K < 0$ for

$$GH = \frac{K}{s(s + 2)(s + 1 + j)(s + 1 - j)}$$

The center of asymptotes is $\sigma_c = -(0 + 2 + 1 + j + 1 - j)/4 = -1$.
For $K > 0$, the angles of the asymptotes are $\beta = 45°$, $135°$, $225°$, and $315°$ as shown in Fig. 13-23.
For $K < 0$, the angles of the asymptotes are $\beta = 0°$, $90°$, $180°$, and $270°$ as shown in Fig. 13-24.

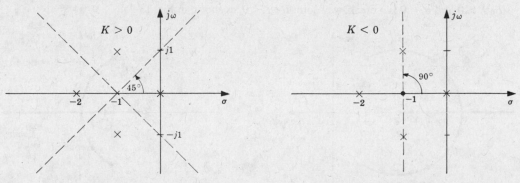

Fig. 13-23                                         Fig. 13-24

## BREAKAWAY POINTS

**13.17.** Show that a breakaway point $\sigma_b$ satisfies

$$\sum_{i=1}^{n} \frac{1}{(\sigma_b + p_i)} = \sum_{i=1}^{m} \frac{1}{(\sigma_b + z_i)} \qquad (13.8)$$

A breakaway point is a point on the real axis where the gain factor $K$ along the real axis portion of the root-locus is a maximum for poles leaving the real axis, or a minimum for poles coming onto the real axis, (see Section 13.2). The gain factor along the root-locus is given by

$$|K| = \left| \frac{D}{N} \right| \qquad (13.5)$$

On the real axis, $s = \sigma$ (or $z = \mu$) and the magnitude signs may be dropped because $D(\sigma)$ and $N(\sigma)$ are both real. Then

$$K = \frac{D(\sigma)}{N(\sigma)}$$

To find the value of $\sigma$ for which $K$ is a maximum or minimum, the derivative of $K$ with respect to $\sigma$ is set equal to zero:

$$\frac{dK}{d\sigma} = \frac{d}{d\sigma} \left[ \frac{(\sigma + p_1) \cdots (\sigma + p_n)}{(\sigma + z_1) \cdots (\sigma + z_m)} \right] = 0$$

By repeated differentiation and factorization, this can be written as

$$\frac{dK}{d\sigma} = \sum_{i=1}^{n} \frac{1}{(\sigma + p_i)} \left[ \frac{D(\sigma)}{N(\sigma)} \right] - \sum_{i=1}^{m} \frac{1}{(\sigma + z_i)} \left[ \frac{D(\sigma)}{N(\sigma)} \right] = 0$$

Finally, dividing both sides by $D(\sigma)/N(\sigma)$ yields the required result.

**13.18.** Determine the breakaway point for $GH = K/s(s + 3)^2$.

The breakaway point satisfies

$$\frac{1}{\sigma_b} + \frac{1}{\sigma_b + 3} + \frac{1}{\sigma_b + 3} = 0$$

from which $\sigma_b = -1$.

**13.19.** Find the breakaway point for

$$GH = \frac{K(s + 2)}{(s + 1 + j\sqrt{3})(s + 1 - j\sqrt{3})}$$

From Equation (13.8),

$$\frac{1}{\sigma_b + 1 + j\sqrt{3}} + \frac{1}{\sigma_b + 1 - j\sqrt{3}} = \frac{1}{\sigma_b + 2}$$

which gives $\sigma_b^2 + 4\sigma_b = 0$. This equation has the solution $\sigma_b = 0$ and $\sigma_b = -4$; $\sigma_b = -4$ is a breakaway point for $K > 0$ and $\sigma_b = 0$ is a breakaway point for $K < 0$, as shown in Fig. 13-25.

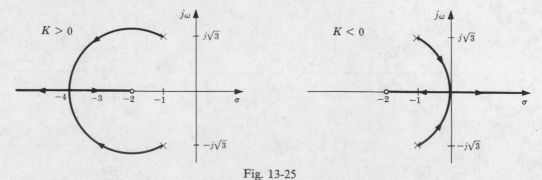

Fig. 13-25

**13.20.** Find the breakaway point between 0 and $-1$ for

$$GH = \frac{K}{s(s+1)(s+3)(s+4)}$$

The breakaway point must satisfy

$$\frac{1}{\sigma_b} + \frac{1}{(\sigma_b+1)} + \frac{1}{(\sigma_b+3)} + \frac{1}{(\sigma_b+4)} = 0$$

If this equation were simplified, a third-order polynomial would be obtained. To avoid solving a third-order polynomial, the following procedure may be used. As a first guess, assume $\sigma_b = -0.5$ and use this value in the two terms for the poles furthest from the breakaway point. Then

$$\frac{1}{\sigma_b} + \frac{1}{\sigma_b+1} + \frac{1}{2.5} + \frac{1}{3.5} = 0$$

which simplifies to $\sigma_b^2 + 3.92\sigma_b + 1.46 = 0$ and has the root $\sigma_b = -0.43$ between 0 and $-1$. This value is used to obtain a better approximation as follows:

$$\frac{1}{\sigma_b} + \frac{1}{\sigma_b+1} + \frac{1}{2.57} + \frac{1}{3.57} = 0 \qquad \sigma_b^2 + 3.99\sigma_b + 1.496 = 0 \qquad \sigma_b = -0.424$$

The second computation did not result in a value much different from the first. A reasonable first guess can often result in a fairly accurate approximation with only one computation.

## DEPARTURE AND ARRIVAL ANGLES

**13.21.** Show that the departure angle of the root-locus from a complex pole is given by

$$\theta_D = 180° + \arg GH' \qquad\qquad (13.9)$$

Consider a circle of infinitesimally small radius around the complex pole. Clearly, the phase angle $\arg GH'$ of $GH$, neglecting the contribution of the complex pole, is constant around this circle. If $\theta_D$ represents the departure angle, the total phase angle of $GH$ at the point on the circle where the root-locus crosses it is

$$\arg GH = \arg GH' - \theta_D$$

since $-\theta_D$ is the phase angle contributed to $\arg GH$ by the complex pole. In order to satisfy the angle criterion, $\arg GH = \arg GH' - \theta_D = 180°$ or $\theta_D = 180° + \arg GH'$ since $+180°$ and $-180°$ are equivalent.

**13.22.** Determine the relationship between the departure angle from a complex pole for $K > 0$ with that for $K < 0$.

Since $\arg GH'$ changes by 180° if $K$ changes from a positive number to a negative one, the departure angle for $K < 0$ is 180° different from the departure angle for $K > 0$.

**13.23.** Show that the arrival angle at a complex zero satisfies

$$\theta_A = 180° - \arg GH'' \qquad\qquad (13.10)$$

In the same manner as in the solution to Problem 13.21, the phase angle of $GH$ in the vicinity of the complex zero is given by $\arg GH = \arg GH'' + \theta_A$ since $\theta_A$ is the phase angle contributed to $\arg GH$ by the complex zero. Then applying the angle criterion yields $\theta_A = 180° - \arg GH''$.

**13.24.** Graphically determine $\arg GH'$ and compute the departure angle of the root-locus from the complex pole at $s = -2 + j$ for

$$GH = \frac{K}{(s+1)(s+2-j)(s+2+j)} \qquad K > 0$$

From Fig. 13-26, $\arg GH' = -135° - 90° = -225°$; and $\theta_D = 180° - 225° = -45°$ as shown in Fig. 13-27.

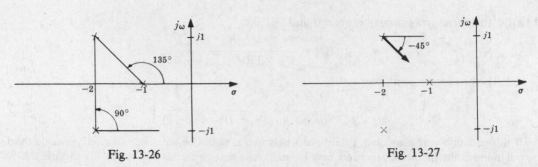

Fig. 13-26                                        Fig. 13-27

**13.25.** Determine the departure angles from the complex poles and the arrival angles at the complex zeros for the open-loop transfer function

$$GH = \frac{K(s+1+j)(s+1-j)}{s(s+2j)(s-2j)} \qquad K > 0$$

For the complex pole at $s = 2j$,

$$\arg GH' = 45° + 71.6° - 90° - 90° = -63.4° \qquad \text{and} \qquad \theta_D = 180° - 63.4° = 116.6°$$

Since the root-locus is symmetric about the real axis, the departure angle from the pole at $s = -2j$ is $-116.6°$. For the complex zero $s = -1 + j$,

$$\arg GH'' = 90° - 108.4° - 135° - 225° = -18.4° \qquad \text{and} \qquad \theta_A = 180° - (-18.4°) = 198.4°$$

Thus the arrival angle at the complex zero $s = -1 - j$ is $\theta_A = -198.4°$.

## CONSTRUCTION OF THE ROOT-LOCUS

**13.26.** Construct the root-locus for

$$GH = \frac{K}{(s+1)(s+2-j)(s+2+j)} \qquad K > 0$$

The real axis from $-1$ to $-\infty$ is on the root-locus. The center of asymptotes is at

$$\sigma_c = \frac{-1 - 2 + j - 2 - j}{3} = -1.67$$

There are three asymptotes ($n - m = 3$), located at angles of 60°, 180°, and 300°. The departure angle from the complex pole at $s = -2 + j$ computed in Problem 13.24 is $-45°$. A sketch of the resulting root-locus is shown in Fig. 13-28. An accurate root-locus plot is obtained by checking the angle criterion at points along the sketched branches, adjusting the location of the branches if necessary, and then applying the magnitude criterion to determine the values of $K$ at selected points along the branches. The completed root-locus is shown in Fig. 13-29.

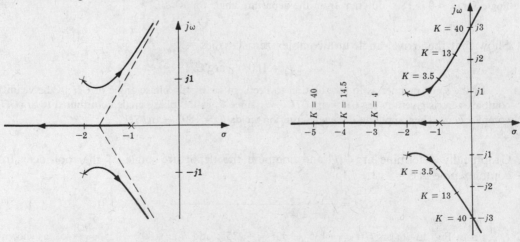

Fig. 13-28                                        Fig. 13-29

**13.27.** Sketch the branches of the root-locus for the transfer function

$$GH = \frac{K(s+2)}{(s+1)(s+3+j)(s+3-j)} \qquad K > 0$$

The real axis between $-1$ and $-2$ is on the root-locus (Problem 13.11). There are two asymptotes with angles of 90° and 270°. The center of asymptotes is easily computed as $\sigma_c = -2.5$ and the departure angle from the complex pole at $s = -3 + j$ as 72°. By symmetry, the departure angle from the pole at $-3 - j$ is $-72°$. The branches of the root-locus may therefore be sketched as shown in Fig. 13-30.

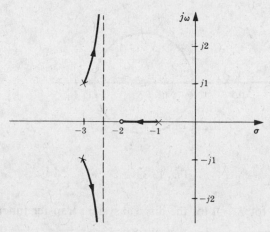

Fig. 13-30

**13.28.** Construct the root-locus for $K > 0$ and $K < 0$ for the transfer function

Mathcad

$$GH = \frac{K}{s(s+1)(s+3)(s+4)}$$

For this transfer function the center of asymptotes is simply $\sigma_c = -2$; and $n - m = 4$. Therefore for $K > 0$ the asymptotes have angles of 45°, 135°, 225°, and 315°. The real axis sections between 0 and $-1$ and between $-3$ and $-4$ lie on the root-locus for $K > 0$ and it was determined in Problem 13.20 that a breakaway point is located at $\sigma_b = -0.424$. From the symmetry of the pole locations, another breakaway point is located at $-3.576$. This can be verified by substituting this value into the relation for the breakaway point, Equation (*13.8*). The completed root-locus for $K > 0$ is shown in Fig. 13-31.

For $K < 0$, the asymptotes have angles of 0°, 90°, 180°, and 270°. In this case the real axis portions between $\infty$ and 0, between $-1$ and $-3$, and between $-4$ and $-\infty$ are on the root-locus. There is only one breakaway point, located at $-2$. The completed root-locus for $K < 0$ is shown in Fig. 13-32.

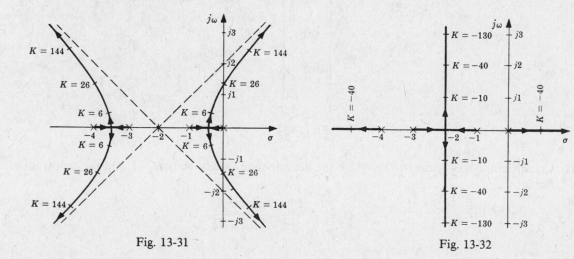

Fig. 13-31                                          Fig. 13-32

**13.29.** Construct the root-locus for $K > 0$ for the discrete system transfer function

$$GH(z) = \frac{K(z - 0.5)}{(z - 1)^2}$$

This root-locus has two loci and one asymptote. The root-locus lies on the real axis for $z < 0.5$. The breakaway points are at $z = 0$ and $z = 1$. The completed root-locus is shown in Fig. 13-33.

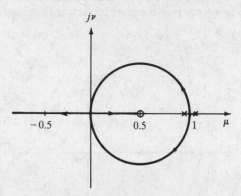

Fig. 13-33

**13.30.** Construct the root-locus for $K > 0$ for the discrete system transfer function

$$GH(z) = \frac{K}{(z + 0.5)(z - 1.5)}$$

This root-locus has two branches and two asymptotes. The breakaway point and the center of asymptotes are at $z = 0.5$. The root-locus is shown in Fig. 13-34.

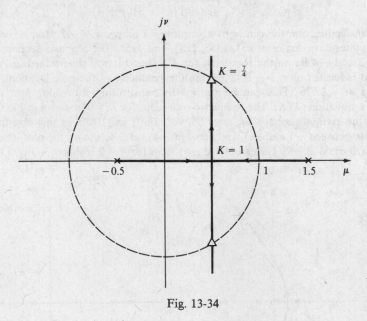

Fig. 13-34

**13.31.** Construct the root-locus for $K > 0$ for the discrete-time system with $H = 1$ and forward transfer function

$$G(z) = \frac{K(z + \frac{1}{3})(z + 1)}{z(z + \frac{1}{2})(z - 1)}$$

The system has one more pole than zero, so the root-locus has only one asymptote, along the negative real axis. The root-locus is on the real axis between 0 and 1, between $-\frac{1}{3}$ and $-\frac{1}{2}$, and to the left of $-1$. Breakaway points are located between 0 and 1 and to the left of $-1$. By trial and error (or computer solution), breakaway points are found at $z = 0.383$ and $z = -2.22$.

The root-locus is an ellipse between the breakaway points at $z = 0.383$ and $z = -2.22$. The point on the $j\nu$-axis, where $\arg G(z) = -180°$ is found by trial and error to be $z = j0.85$. Similarly, the point on the line $z = -1 + j\nu$, where $\arg G(z) = -180°$ is $z = -1 + j1.26$. The root-locus is drawn in Fig. 13-35. The gain factor along the root-locus is determined graphically from the pole-zero map or analytically by evaluating $G(z)$.

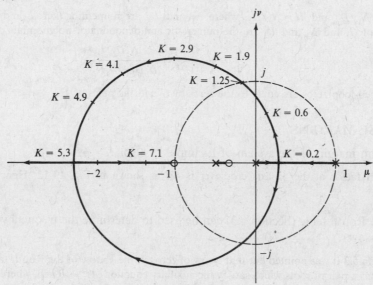

Fig. 13-35

## THE CLOSED-LOOP TRANSFER FUNCTION AND THE TIME-DOMAIN RESPONSE

**13.32.** Determine the closed-loop transfer function of the continuous system of Example 13.8 for $K = 48$, given the following transfer functions for $H$: (a) $H = 1$, (b) $H = 4/(s + 1)$, (c) $H = (s + 1)/(s + 2)$.

From the root-locus plot of Example 13.8, the closed-loop poles for $K = 48$ are located at $s = -6$, $j2.83$, and $-j2.83$. For $H = 1$,

$$G = \frac{48}{s(s+2)(s+4)} \quad \text{and} \quad \frac{C}{R} = \frac{GH}{1+GH} = \frac{48}{(s+6)(s-j2.83)(s+j2.83)}$$

For $H = 4/(s + 1)$,

$$G = \frac{12(s+1)}{s(s+2)(s+4)} \quad \text{and} \quad \frac{C}{R} = \frac{1}{H}\left(\frac{GH}{1+GH}\right) = \frac{12(s+1)}{(s+6)(s-j2.83)(s+j2.83)}$$

For $H = (s + 1)/(s + 2)$,

$$G = \frac{48}{s(s+1)(s+4)} \quad \text{and} \quad \frac{C}{R} = \frac{48(s+2)}{(s+1)(s+6)(s-j2.83)(s+j2.83)}$$

Note that in this last case there are four closed-loop poles, while $GH$ has only three poles. This is due to the cancellation of a pole of $G$ by a zero of $H$.

**13.33.** Determine the unit step response of the system of Example 13.1 with $K = 1.5$.

The closed-loop transfer function of this system is

$$\frac{C}{R} = \frac{1.5(s+1)}{(s+0.5)(s+3)}$$

For $R = 1/s$,

$$C = \frac{1.5(s+1)}{s(s+0.5)(s+3)} = \frac{1}{s} + \frac{-0.6}{s+0.5} + \frac{-0.4}{s+3}$$

and the unit step response is $\mathscr{L}^{-1}[C(s)] = c(t) = 1 - 0.6e^{-0.5t} - 0.4e^{-3t}$.

**13.34.** Determine the relationship between the closed-loop zeros and the poles and zeros of $G$ and $H$, assuming there are no cancellations.

Let $G = N_1/D_1$ and $H = N_2/D_2$, where $N_1$ and $D_1$ are numerator (zeros) and denominator (poles) polynomials of $G$, and $N_2$ and $D_2$ are the numerator and denominator polynomials of $H$. Then

$$\frac{C}{R} = \frac{G}{1 + GH} = \frac{N_1 D_2}{D_1 D_2 + N_1 N_2}$$

Thus the closed-loop zeros are equal to the zeros of $G$ and the poles of $H$.

## GAIN AND PHASE MARGINS

**13.35.** Find the gain margin of the system of Example 13.8 for $K = 6$.

The gain factor at the $j\omega$-axis crossover is 48, as shown in Fig. 13-12. Hence the gain margin is $48/6 = 8$.

**13.36.** Show how a Routh table (Section 5.3) can be used to determine the frequency and the gain at the $j\omega$-axis crossover.

In Section 5.3 it was pointed out that a row of zeros in the $s^1$ row of the Routh table indicates that the polynomial has a pair of roots which satisfy the auxiliary equation $As^2 + B = 0$, where $A$ and $B$ are the first and second elements of the $s^2$ row. If $A$ and $B$ have the same sign, the roots of the auxiliary equation are imaginary (on the $j\omega$-axis). Thus if a Routh table is constructed for the characteristic equation of a system, the values of $K$ and $\omega$ corresponding to $j\omega$-axis crossovers can be determined. For example, consider the system with the open-loop transfer function

$$GH = \frac{K}{s(s+2)^2}$$

The characteristic equation for this system is

$$s^3 + 4s^2 + 4s + K = 0$$

The Routh table for the characteristic polynomial is

$$
\begin{array}{c|cc}
s^3 & 1 & 4 \\
s^2 & 4 & K \\
s^1 & (16-K)/4 & \\
s^0 & K &
\end{array}
$$

The $s^1$ row is zero for $K = 16$. The auxiliary equation then becomes

$$4s^2 + 16 = 0$$

Thus for $K = 16$ the characteristic equation has solutions (closed-loop poles) at $s = \pm j2$, and the root-locus crosses the $j\omega$-axis at $j2$.

**13.37.** Determine the phase margin for the system of Example 13.8 (Figure 13-12) for $K = 6$.

First, the point on the $j\omega$-axis for which $|GH(j\omega)| = 1$ is found by trial and error to be $j0.7$. Then $\arg GH(j0.7)$ is computed as $-120°$. Hence the phase margin is $180° - 120° = 60°$.

**13.38.** Is it necessary to construct the entire root-locus in order to determine the gain and phase margins of a system?

No. Only one point on the root-locus is required to determine the gain margin. This point, at $\omega_\pi$, where the root-locus crosses the stability boundary, can be determined by trial and error or by the use of a Routh table as described in Problem 13.36. To determine the phase margin, it is only necessary to determine the point on the stability boundary where $|GH(j\omega)| = 1$. Although the entire root-locus plot is not necessary, it can often be helpful, especially in the case of multiple stability boundary crossings.

## DAMPING RATIO FROM THE ROOT-LOCUS FOR CONTINUOUS SYSTEMS

**13.39.** Prove Equation (*13.18*).

The roots of $s^2 + 2\zeta\omega_n s + \omega_n^2$ are $s_{1,2} = -\zeta\omega_n \pm j\omega_n\sqrt{1-\zeta^2}$. Then

$$|s_1| = |s_2| = \sqrt{\zeta^2\omega_n^2 + \omega_n^2(1-\zeta^2)} = \omega_n$$

and

$$\arg s_{1,2} = \mp\tan^{-1}\left(\sqrt{1-\zeta^2}\,/\zeta\right) \equiv 180° \pm \theta$$

or $s_{1,2} = \omega_n \underline{/180° \pm \theta}$. Thus $\cos\theta = \zeta\omega_n/\omega_n = \zeta$.

**13.40.** Determine the positive value of gain which results in a damping ratio of 0.55 for the complex poles on the root-locus shown in Fig. 13-12.

The angle of the desired poles is $\theta = \cos^{-1}0.55 = 56.6°$. A line drawn from the origin at an angle of $55.6°$ with the negative real axis intersects the root-locus of Fig. 13-12 at $K = 7$.

**13.41.** Find the damping ratio of the complex poles of Problem 13.26 for $K = 3.5$.

Mathcad

A line drawn from the root-locus at $K = 3.5$ to the origin makes an angle of $53°$ with the negative real axis. Hence the damping ratio of the complex poles is $\zeta = \cos 53° = 0.6$.

# Supplementary Problems

**13.42.** Determine the angle and magnitude of

$$GH = \frac{16(s+1)}{s(s+2)(s+4)}$$

at the following points in the s-plane: (*a*) $s = j2$, (*b*) $s = -2+j2$, (*c*) $s = -4+j2$, (*d*) $s = -6$, (*e*) $s = -3$.

**13.43.** Determine the angle and magnitude of

$$GH = \frac{20(s+10+j10)(s+10-j10)}{(s+10)(s+15)(s+25)}$$

at the following points in the s-plane: (*a*) $s = j10$, (*b*) $s = j20$, (*c*) $s = -10+j20$, (*d*) $s = -20+j20$, (*e*) $s = -15+j5$.

**13.44.** For each transfer function, find the breakaway points on the root-locus:

(*a*) $\quad GH = \dfrac{K}{s(s+6)(s+8)}$, $\qquad$ (*b*) $\quad GH = \dfrac{K(s+5)}{(s+2)(s+4)}$, $\qquad$ (*c*) $\quad GH = \dfrac{K(s+1)}{s^2(s+9)}$.

**13.45.** Find the departure angle of the root-locus from the pole at $s = -10+j10$ for

$$GH = \frac{K(s+8)}{(s+14)(s+10+j10)(s+10-j10)} \qquad K > 0$$

**13.46.** Find the departure angle of the root-locus from the pole at $s = -15 + j9$ for

$$GH = \frac{K}{(s+5)(s+10)(s+15+j9)(s+15-j9)} \qquad K > 0$$

**13.47.** Find the arrival angle of the root-locus to the zero at $s = -7 + j5$ for

$$GH = \frac{K(s+7+j5)(s+7-j5)}{(s+3)(s+5)(s+10)} \qquad K > 0$$

**13.48.** Construct the root-locus for $K > 0$ for the transfer function of Problem 13.44($a$).

**13.49.** Construct the root-locus for $K > 0$ for the transfer function of Problem 13.44($c$).

**13.50.** Construct the root-locus for $K > 0$ for the transfer function of Problem 13.45.

**13.51.** Construct the root-locus for $K > 0$ for the transfer function of Problem 13.46.

**13.52.** Determine the gain and phase margins for the system with the open-loop transfer function of Problem 13.46 if the gain factor $K$ is set equal to 20,000.

# Answers to Some Supplementary Problems

**13.42.** ($a$) $\arg GH = -99°$, $|GH| = 1.5$; ($b$) $\arg GH = -153°$, $|GH| = 2.3$; ($c$) $\arg GH = -232°$, $|GH| = 1.8$; ($d$) $\arg GH = 0°$, $|GH| = 1.7$; ($e$) $\arg GH = -180°$, $|GH| = 10.7$

**13.43.** ($a$) $\arg GH = -38°$, $|GH| = 0.68$; ($b$) $\arg GH = -40°$, $|GH| = 0.37$; ($c$) $\arg GH = -41°$, $|GH| = 0.60$; ($d$) $\arg GH = -56°$, $|GH| = 0.95$; ($e$) $\arg GH = +80°$, $|GH| = 6.3$

**13.44.** ($a$) $\sigma_b = -2.25, -7.07$; ($b$) $\sigma_b = -3.27, -6.73$; ($c$) $\sigma_b = 0, -3$

**13.45.** $\theta_D = 124°$

**13.46.** $\theta_D = 193°$

**13.47.** $\theta_A = 28°$

**13.52.** Gain margin = 3.7; phase margin = 102°

# Chapter 14

# Root-Locus Design

## 14.1 THE DESIGN PROBLEM

The root-locus method can be quite effective in the design of either continuous or discrete-time feedback control systems, because it graphically illustrates the variation of the system closed-loop poles as a function of the open-loop gain factor $K$. In its simplest form, design is accomplished by choosing a value of $K$ which results in satisfactory closed-loop behavior. This is called *gain factor compensation* (also see Section 12.2). Specifications on allowable steady state errors usually take the form of a minimum value of $K$, expressed in terms of error constants, for example, $K_p$, $K_v$, and $K_a$ (Chapter 9). If it is not possible to meet all system specifications using gain factor compensation alone, other forms of compensation can be added to the system to alter the root-locus as needed, for example, lag, lead, lag-lead networks, or PID controllers.

In order to accomplish system design in the $s$-plane or the z-plane using root-locus techniques, it is necessary to interpret the system specifications in terms of desired pole-zero configurations.

Digital computer programs for constructing root-loci can be very helpful in system design, as well as analysis as indicated in Chapter 13.

**EXAMPLE 14.1.** Consider the design of a continuous unity feedback system with the plant $G = K/(s+1)(s+3)$ and the following specifications: (1) Overshoot less than 20%, (2) $K_p \geq 4$, (3) 10 to 90% rise time less than 1 sec.

The root-locus for this system is shown in Fig. 14-1. The system closed-loop transfer function may be written as

$$\frac{C}{R} = \frac{K}{s^2 + 2\zeta\omega_n s + \omega_n^2}$$

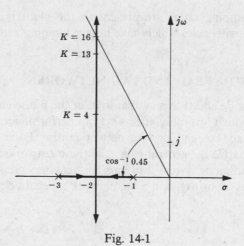

Fig. 14-1

where $\zeta$ and $\omega_n$ can be determined from the root-locus for a given value of $K$. In order to satisfy the first specification, $\zeta$ must be greater than 0.45 (see Fig. 3-4). Then from the root-locus we see that $K$ must be less than 16 (see Section 13.12). For this system, $K_p$ is given by $K/3$. Thus in order to satisfy the second specification, $K$ must be greater than 12. The rise time is a function of both $\zeta$ and $\omega_n$. Suppose a trial value of $K = 13$ is chosen. In this case, $\zeta = 0.5$, $\omega_n = 4$, and the rise time is 0.5 sec. Hence all the specifications can be met by setting $K = 13$. Note that if the specification on $K_p$ was greater than 5.33, or the specification on rise time was less than 0.34 sec, all the specifications could not be met by simply adjusting the open-loop gain factor.

## 14.2 CANCELLATION COMPENSATION

If the pole-zero configuration of the plant is such that the system specifications cannot be met by an adjustment of the open-loop gain factor, a more complicated cascade compensator, as shown in Fig. 14-2, can be added to the system to cancel some or all of the poles and zeros of the plant. Due to realizability considerations, the compensator must have no more zeros than poles. Consequently, when poles of the plant are cancelled by zeros of the compensator, the compensator also adds new poles to the forward-loop transfer function. The philosophy of this compensation technique is then to replace undesirable with desirable poles.

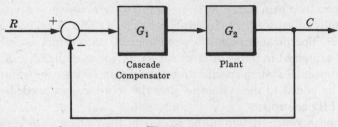

Fig. 14-2

The difficulty encountered in applying this scheme is that it is not always apparent what open-loop pole-zero configuration is desirable from the standpoint of meeting specifications on closed-loop system performance.

Some situations where cancellation compensation can be used to advantage are the following:

1. If the specifications on system rise time or bandwidth cannot be met without compensation, cancellation of low-frequency poles and replacement with high-frequency poles is helpful.

2. If the specifications on allowable steady state errors cannot be met, a low-frequency pole can be cancelled and replaced with a lower-frequency pole, yielding a larger forward-loop gain at low frequencies.

3. If poles with small damping ratios are present in the plant transfer function, they may be cancelled and replaced with poles which have larger damping ratios.

## 14.3 PHASE COMPENSATION: LEAD AND LAG NETWORKS

A cascade compensator can be added to a system to alter the phase characteristics of the open-loop transfer function in a manner which favorably affects system performance. These effects were illustrated in the frequency domain for lead, lag, and lag-lead networks using Polar Plots in Chapter 12, Sections 12.4 through 12.7, which summarize the general effects of these networks.

The pole-zero maps of continuous system lead and lag networks are shown in Figs. 14-3 and 14-4. Note that a lead network makes a positive, and a lag network a negative phase contribution. A lag-lead

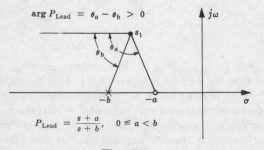

$$P_{\text{Lead}} = \frac{s+a}{s+b}, \quad 0 \le a < b$$

Fig. 14-3

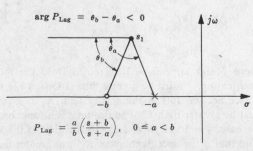

$$P_{\text{Lag}} = \frac{a}{b}\left(\frac{s+b}{s+a}\right), \quad 0 \le a < b$$

Fig. 14-4

network may be obtained by appropriately combining a lag and a lead network in series, or from the implementation described in Problem 6.14.

Since the compensated system root-locus is determined by the points in the complex plane for which the phase angle of $G = G_1 G_2$ is equal to $-180°$, the branches of the locus can be moved by proper selection of the phase angle contributed by the compensator. In general, lead compensation has the effect of moving the loci to the left.

**EXAMPLE 14.2.** The phase lead compensator $G_1 = (s + 2)/(s + 8)$ alters the root-locus of the system with the plant $G_2 = K/(s + 1)^2$, as illustrated in Fig. 14-5.

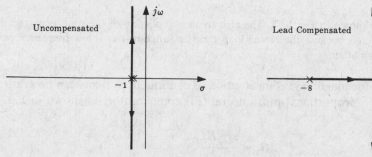

Fig. 14-5

**EXAMPLE 14.3.** The use of *simple lag compensation* (one pole at $-1$, no zero) to alter the breakaway angles of a root-locus from a pair of complex poles is illustrated in Fig. 14-6.

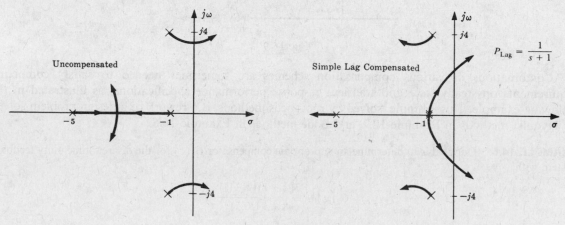

Fig. 14-6

## 14.4 MAGNITUDE COMPENSATION AND COMBINATIONS OF COMPENSATORS

Compensation networks may be employed to alter the closed-loop magnitude characteristic $(|(C/R)(\omega)|)$ of a feedback control system. The low-frequency characteristic can be modified by addition of a low-frequency pole-zero pair, or **dipole**, in such a manner that high-frequency behavior is essentially unaltered.

**EXAMPLE 14.4.** The continuous system root-locus for $GH = K/s(s + 2)^2$ is shown in Fig. 14-7.

Let us assume that this system has a satisfactory transient response with $K = 3$, but the resulting velocity error constant, $K_v = 0.75$, is too small. We can increase $K_v$ to 5 without seriously affecting the transient response by adding the compensator $G_1 = (s + 0.1)/(s + 0.015)$ since

$$K_v' = K_v G_1(0) = \frac{0.75(0.1)}{0.015} = 5$$

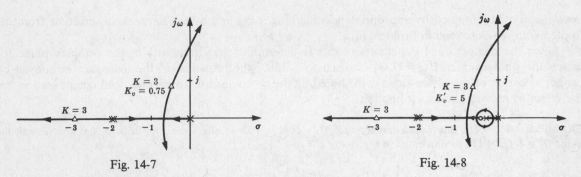

Fig. 14-7                                                    Fig. 14-8

The resulting root-locus is shown in Fig. 14-8. The high-frequency portion of the root-locus and the transient response are essentially unaffected because the closed-loop transfer function has a low-frequency pole-zero pair which approximately cancel each other.

A low-frequency dipole for magnitude compensation of continuous systems can be synthesized with the pole at the origin using a proportional plus integral (PI) compensator, as shown in Fig. 14-9, with transfer function

$$G_1 = \frac{s + K_I}{s}$$

Fig. 14-9

Combinations of various compensation schemes are sometimes needed to satisfy competing requirements on steady state and transient response performance specifications, as illustrated in the following example. This example, solved by root-locus methods, is a rework of a design problem solved by Nyquist methods in Example 12.7, and Bode methods in Example 16.6.

**EXAMPLE 14.5.**   Our goal is to determine an appropriate compensator $G_1(z)$ for the discrete-time unity feedback system with

$$G_2(z) = \frac{3(z+1)\left(z+\frac{1}{3}\right)}{8z(z+0.5)}$$

The resulting closed-loop system must satisfy the following performance specifications:

1.   Steady state error less than or equal to 0.02 for a unit ramp input
2.   Phase margin $= \phi_{PM} \geq 30°$
3.   Gain crossover frequency $\omega_1 \geq 10$ rad/sec, where $T = 0.1$ sec.

In order to have a finite steady state error with a ramp input, the system must be type 1. The compensation must therefore provide a pole at $z = 1$. Consider the compensator

$$G_1' = \frac{K_1}{z - 1}$$

The forward-loop transfer function then becomes

$$G_1'G_2(z) = \frac{3K_1(z+1)\left(z+\frac{1}{3}\right)}{8(z-1)z(z+0.5)}$$

From Section 9.9, the velocity error coefficient is

$$K_v = \frac{3K_1(1+1)\left(1+\frac{1}{3}\right)}{8(1)(1+0.5)} = 0.667K_1$$

Now, in order for the system to have a steady state error of less than 0.02 with a ramp input, we must have $K_v \geq 5$, or $K_1 \geq 7.5$. To investigate the effects of added gain, we consider the root-locus for

$$G_1'G_2(z) = \frac{K(z+1)\left(z+\frac{1}{3}\right)}{z(z+0.5)(z-1)}$$

where $K = 3K_1/8$. This root-locus was constructed in Problem 13.31 and is repeated in Fig. 14-10.

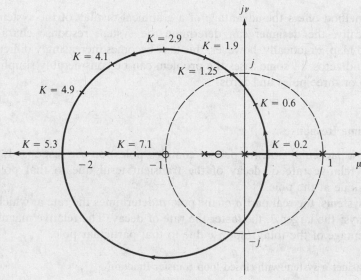

Fig. 14-10

At the point $z = -0.18 + j0.98$ where the root-locus crosses the unit circle, $\omega_\pi T = 1.75$ rad and $K = 1.25$ ($K_1 = 8K/3 = 3.33$). Since this is less than the gain $K_1 = 7.5$ needed to make $K_v = 5$, simple gain factor compensation is insufficient.

The next step is to evaluate the magnitude and phase of $G_1'G_2(z)$ at the required minimum gain crossover frequency, $\omega_1 = 10$, or $\omega_1 T = 1$ rad. This is the point $z = e^{j\omega T} = e^j$ on the unit circle. At this point, $|G_1'G_2(e^j)| = 1.66K$ and $\arg G_1'G_2(e^j) = -142.5°$. If the gain $K$ were adjusted so that $|G_1'G_2(e^j)| = 1$, that is, $K = 0.6$, the phase margin would be $(180 - 142.5)° = 37.5°$ and the $30°$ requirement would be met. This requires that $K_1 = 8K/3 = 1.6$, and the velocity constant becomes $K_v = 0.667K_1 = 1.067$.

To complete the design, additional gain must be added to increase the velocity constant to the required value of 5 at low frequencies, without significantly altering the desired high-frequency characteristics obtained so far. This requires an additional gain of $5/1.067 = 4.69$, which can be supplied by a lag compensator. The lag compensator should have a gain at $z = 1$ that is 4.69 times as large as the gain at $\omega T = 1$, without adding more than $7.5°$ phase lag at $\omega T = 1$, to satisfy the requirement for $\phi_{PM} \geq 30°$. If a value of 0.97 is chosen for the pole of the lag compensator, the zero should be located so that

$$P_{Lag} = \frac{1 - z_1}{1 - 0.97} \geq 4.69$$

or, $z_1 \leq 0.86$. If we set $z_1 = 0.86$, then

$$|P_{Lag}| = \left| \frac{z - 0.86}{z - 0.97} \right| = \begin{cases} 4.7 & \text{for} \quad z = 1 \\ 0.95 & \text{for} \quad z = e^j \end{cases} \quad (\omega T = 1)$$

and

$$\arg P_{Lag} = \arg\left( \frac{e^j - 0.86}{e^j - 0.97} \right) = -6.25° \quad \text{for} \quad z = e^j$$

The compensator then becomes

$$G_1 = \frac{K_1(z - 0.86)}{(z - 0.97)(z - 1)}$$

Finally, for $\omega_1 T = 1$, we need $|G_1 G_2(e^j)| = 1$, so $K_1 = 1.60/0.95 = 1.68$, to account for the gain of the lag compensator at $\omega T = 1$. The completed compensator is

$$G_1 = \frac{1.68(z - 0.86)}{(z - 0.97)(z - 1)}$$

which is nearly the same design obtained by Nyquist methods in Example 12.7.

## 14.5  DOMINANT POLE-ZERO APPROXIMATIONS

The root-locus method offers the advantage of a graphical display of the system closed-loop poles and zeros. Theoretically, the designer can determine the system response characteristics from the closed-loop pole-zero map. Practically, however, this task becomes increasingly difficult for systems with four or more poles and zeros. In some cases the problem can be considerably simplified if the response is dominated by two or three poles and zeros.

### Effects on System Time Responses

The influence of a particular pole (or pair of complex poles) on the response is mainly determined by two factors: the relative rate of decay of the transient term due to that pole, and the relative magnitude of the residue at the pole.

For **continuous systems**, the real part $\sigma$ of the pole $p$ determines the rate at which the transient term due to that pole decays; the larger $\sigma$, the faster the rate of decay. The relative magnitude of the residue determines the percentage of the total response due to that particular pole.

**EXAMPLE 14.6.**   Consider a system with closed-loop transfer function

$$\frac{C}{R} = \frac{5}{(s + 1)(s + 5)}$$

The step response of this system is

$$c(t) = 1 - 1.25e^{-t} + 0.25e^{-5t}$$

The term in the response due to the pole at $s_1 = \sigma_1 = -5$ decays five times as fast as the term due to the pole at $s_2 = \sigma_2 = -1$. Furthermore, the residue at the pole at $s_1 = -5$ is only $\frac{1}{5}$ that of the one at $s_2 = -1$. Therefore for most practical purposes the effect of the pole at $s_1 = -5$ can be ignored and the system approximated by

$$\frac{C}{R} \cong \frac{1}{s + 1}$$

The pole at $s_1 = -5$ has been removed from the transfer function and the numerator has been adjusted to maintain the same steady state gain $((C/R)(0) = 1)$. The response of the approximate system is $c(t) = 1 - e^{-t}$.

**EXAMPLE 14.7.**   The system with the closed-loop transfer function

$$\frac{C}{R} = \frac{5.5(s + 0.91)}{(s + 1)(s + 5)}$$

has the step response

$$c(t) = 1 + 0.125e^{-t} - 1.125e^{-5t}$$

In this case, the presence of a zero close to the pole at $-1$ significantly reduces the magnitude of the residue at that pole. Consequently, it is the pole at $-5$ which now dominates the response of the system. The closed-loop pole and zero effectively cancel each other and $(C/R)(0) = 1$ so that an approximate transfer function is

$$\frac{C}{R} \cong \frac{5}{s + 5}$$

and the corresponding approximate step response is $c \cong 1 - e^{-5t}$.

For **discrete-time systems** with distinct (nonrepeated) poles $p_1, p_2, \ldots,$ the transient portion $y_T(k)$ of the response due to a pole $p$ has the form $y_T(k) = p^k$, $k = 0, 1, 2, \ldots$ (see Table 4.2). Therefore each successive time sample is equal to the previous sample multiplied by $p$, that is,

$$y_T(k+1) = py_T(k)$$

The magnitude of a distinct pole therefore determines the decay rate of the transient response, with the decay rate inversely proportional to $|p|$: the smaller the magnitude, the faster the rate of decay. For example, poles near the unit circle decay more slowly than poles near the origin, since their magnitudes are smaller.

For systems with repeated poles, the analysis is more complicated and approximations may not be appropriate.

**EXAMPLE 14.8.** The discrete system with closed-loop transfer function

$$\frac{C}{R} = \frac{0.45z}{(z-0.1)(z-0.5)}$$

has the step response

$$c(k) = 1 - 1.125(0.5)^k + 0.125(0.1)^k \qquad k = 0,1,2,\ldots$$

For the term in the response due to the pole at $z = 0.1$, the sample value at time $k$ is only 10% of the sample value at time $k - 1$, and it therefore decays five times faster than the term due to the pole at $z = 0.5$. The magnitude of the residue at $z = 0.1$ is 0.125, which is one-ninth as large as the magnitude of the residue 1.125 at $z = 0.5$. Consequently, for many practical purposes, the pole at $z = 0.1$ can often be ignored and the system approximated by

$$\frac{C}{R} \cong \frac{0.5}{z - 0.5}$$

where the numerator has been adjusted to maintain the same steady state gain

$$\frac{C}{R}(1) = 1$$

and the zero at $z = 0$ was deleted to maintain one more pole than zeros in the approximate system. This is necessary to give the same initial delay (one sample time) in the approximate system as in the original system. The step response of the approximate system is $c(k) = 1 - (0.5)^k$, $k = 0,1,2,\ldots$.

**Effects on Other System Characteristics**

The effect of a closed-loop real axis *pole* at $-p_r < 0$ on the overshoot and rise time $T_r$ of a continuous system also having complex poles $-p_c$, $-p_c^*$ is illustrated in Figs. 14-11 and 14-12. For

$$\frac{p_r}{\zeta\omega_n} > 5 \tag{14.1}$$

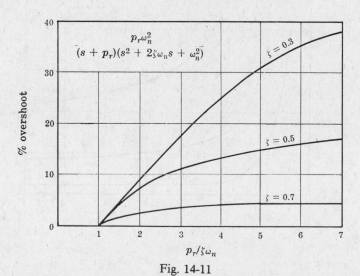

Fig. 14-11

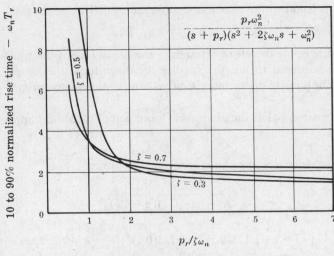

Fig. 14-12

the overshoot and rise time approach that of a second-order system containing only complex poles (see Fig. 3-4). Therefore $p_r$ can be neglected in determining overshoot and rise time if $\zeta > 0.5$ and

$$p_r > 5|\text{Re } p_c| = 5\zeta\omega_n \tag{14.2}$$

There is no overshoot if

$$p_r \leq |\text{Re } p_c| = \zeta\omega_n \tag{14.3}$$

and the rise time approaches that of a first-order system containing only the real axis pole.

The effect of a closed-loop real axis *zero* at $-z_r < 0$ on the overshoot and rise time $T_r$ of a continuous system also having complex poles $-p_c$, $-p_c^*$ is illustrated in Figs. 14-13 and 14-14. These graphs show that $z_r$ can be neglected in determining overshoot and rise time if $\zeta > 0.5$ and

$$z_r > 5|\text{Re } p_c| = 5\zeta\omega_n \tag{14.4}$$

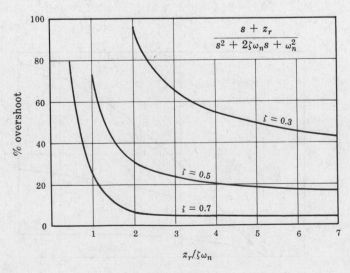

Fig. 14-13

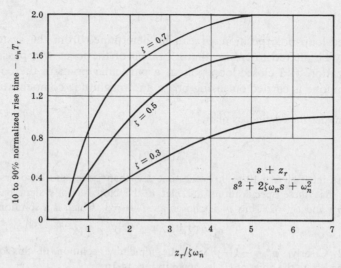

Fig. 14-14

**EXAMPLE 14.9.** The closed-loop transfer function of a particular continuous system is represented by the pole-zero map shown in Fig. 14-15. Given that the steady state gain $(C/R)(j0) = 1$, a dominant pole-zero approximation is

$$\frac{C}{R} \cong \frac{4}{s^2 + 2s + 4}$$

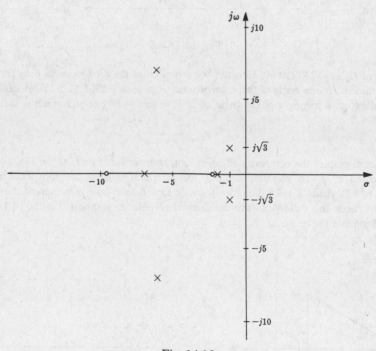

Fig. 14-15

This is a reasonable approximation because the pole and zero near $s = -2$ effectively cancel each other and all other poles and zeros satisfy Equations (*14.2*) and (*14.4*) with $-p_c = -1 + j\sqrt{3}$ and $\zeta = 0.5$.

## 14.6 POINT DESIGN

If a desired closed-loop pole position $p_1$ can be determined from the system specifications, the system root-locus may be altered to ensure that a branch of the locus will pass through the required point $p_1$. The specification of a closed-loop pole at a particular point in the complex plane is called **point design**. The technique is carried out using phase and magnitude compensation.

**EXAMPLE 14.10.** Consider the continuous plant

$$G_2 = \frac{K}{s(s+2)^2}$$

The closed-loop response must have a 10 to 90% rise time less than 1 sec, and an overshoot less than 20%. We observe from Fig. 3-4 that these specifications are met if the closed-loop system has a dominant two-pole configuration with $\zeta = 0.5$ and $\omega_n = 2$. Thus $p_1$ is chosen at $-1 + j\sqrt{3}$, which is a solution of

$$p_1^2 + 2\zeta\omega_n p_1 + \omega_n^2 = 0$$

for $\zeta = 0.5$ and $\omega_n = 2$. Clearly, $p_1^* = -1 - j\sqrt{3}$ is the remaining solution of this quadratic equation. The orientation of $p_1$ with respect to the poles of $G_2$ is shown in Fig. 14-16.

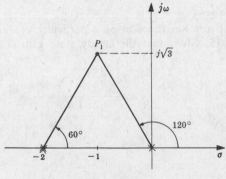

Fig. 14-16

The phase angle of $G_2$ is $-240°$ at $p_1$. In order for a branch of the root-locus to pass through $p_1$, the system must be modified so that the phase angle of the compensated system is $-180°$ at $p_1$. This can be accomplished by adding a cascade lead network having a phase angle of $240° - 180° = 60°$ at $p_1$, which is satisfied by

$$G_1 = P_{\text{Lead}} = \frac{s+1}{s+4}$$

as shown in the pole-zero map of the compensated open-loop transfer function $G_1 G_2$ in Fig. 14-17. The closed-loop pole can now be located at $p_1$ by choosing a value for $K$ which satisfies the root-locus magnitude criterion. Solution of Equation (13.5) yields $K = 16$. The root-locus or closed-loop pole-zero map of the compensated system should be sketched to check the validity of the dominant two-pole assumption. Figure 14-18 illustrates that the poles at $p_1$ and $p_1^*$ dominate the response.

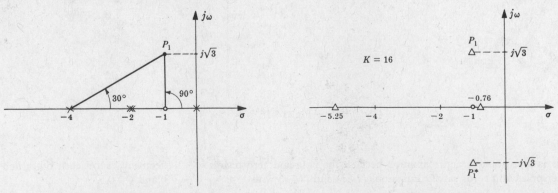

Fig. 14-17                                                    Fig. 14-18

## 14.7 FEEDBACK COMPENSATION

Addition of compensation elements to a feedback path of a control system can be employed in root-locus design in a manner similar to that discussed in the preceding sections. The compensation elements affect the root-locus of the open-loop transfer function in the same manner. But, although the root-locus is the same when the compensator is in either the forward or feedback path, the closed-loop transfer function may be significantly different. It was shown in Problem 13.34 that feedback *zeros* do not appear in the closed-loop transfer function, while feedback *poles* become zeros of the closed-loop transfer function (assuming no cancellations).

**EXAMPLE 14.11.** Suppose a feedback compensator were added to a continuous system with the forward transfer function

$$G = \frac{K}{(s+1)(s+4)(s+5)}$$

in an attempt to cancel the pole at $-1$ and replace it with a pole at $-6$. Then the compensator would be $H = (s+1)/(s+6)$, $GH$ would be given by $GH = K/(s+4)(s+5)(s+6)$ and the closed-loop transfer function would become

$$\frac{C}{R} = \frac{K(s+6)}{(s+1)[(s+4)(s+5)(s+6)+K]}$$

Although the pole at $-1$ is cancelled from $GH$, it reappears as a *closed-loop* pole. Furthermore, the feedback pole at $-6$ becomes a closed-loop zero. Consequently, *the cancellation technique does not work with a compensator in the feedback path*.

**EXAMPLE 14.12.** The continuous system block diagram in Fig. 14-19 contains two feedback paths.

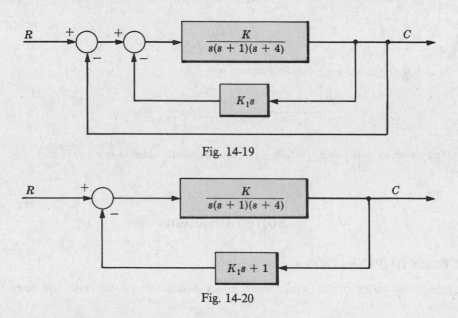

Fig. 14-19

Fig. 14-20

These two paths may be combined, as shown in Fig. 14-20.

In this representation the feedback path contains a zero at $s = -1/K_1$. This zero appears in $GH$ and consequently affects the root-locus. However, it does not appear in the closed-loop transfer function, which contains three poles no matter where the zero is located.

The fact that feedback zeros do not appear in the closed-loop transfer function may be used to advantage in the following manner. If closed-loop poles are desired at certain locations in the complex plane, feedback zeros can be placed at these points. Since branches of the root-locus will terminate on these zeros, the desired closed-loop pole locations can be obtained by setting the open-loop gain factor sufficiently high.

**EXAMPLE 14.13.** The continuous system feedback compensator

$$H = \frac{s^2 + 2s + 4}{(s+6)^2}$$

is added to the system with the forward-loop transfer function

$$G = \frac{K}{s(s+2)}$$

in order to guarantee that the dominant closed-loop poles will be near $s = -1 \pm j\sqrt{3}$. The resulting root-locus is shown in Fig. 14-21.

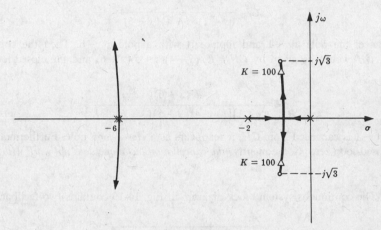

Fig. 14-21

If $K$ is set at 100, the closed-loop transfer function is

$$\frac{C}{R} = \frac{100(s+6)^2}{(s^2 + 1.72s + 2.96)(s^2 + 12.3s + 135)}$$

and the dominant complex pole pair $s_{1,2} = 0.86 \pm j1.5$ are sufficiently close to $-1 \pm j\sqrt{3}$.

# Solved Problems

## GAIN FACTOR COMPENSATION

**14.1.** Determine the value of the gain factor $K$ for which the system with the open-loop transfer function

$$GH = \frac{K}{s(s+2)(s+4)}$$

has closed-loop poles with a damping ratio $\zeta = 0.5$.

The closed-loop poles will have a damping ratio of 0.5 when they make an angle of 60° with the negative real axis [Equation (13.18)]. The desired value of $K$ is determined at the point where the root-locus crosses the $\zeta = 0.5$ line in the $s$-plane. A sketch of the root-locus is shown in Fig. 14-22. The desired value of $K$ is 8.3.

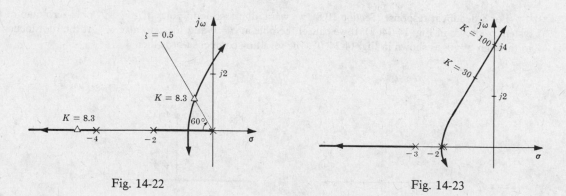

Fig. 14-22                                    Fig. 14-23

**14.2.** Determine a value of $K$ for which the system with the open-loop transfer function

$$GH = \frac{K}{(s+2)^2(s+3)}$$

satisfies the following specifications: ($a$) $K_p \geq 2$, ($b$) gain margin $\geq 3$.

For this system, $K_p$ is equal to $K/12$. Hence, in order to satisfy the first specification, $K$ must be greater than 24. The value of $K$ at the $j\omega$-axis crossover of the root-locus is equal to 100, as shown in Fig. 14-23. Then, in order to satisfy the second specification, $K$ must be less than $100/3 = 33.3$. A value of $K$ that will satisfy both specifications is 30.

**14.3.** Determine a gain factor $K$ for which the system in Example 13.11 has a gain margin of 2.

As shown in Fig. 13-15, the gain at the stability boundary is $K = 1$. Therefore, in order to have a gain margin of 2, $K$ must be 0.5.

## CANCELLATION COMPENSATION

**14.4.** Can right-half $s$-plane poles of a plant be effectively cancelled by a compensator with a right-half $s$-plane zero?

No. For example, suppose a particular plant has the transfer function

$$G_2 = \frac{K}{s-1} \qquad K > 0$$

and a cascade compensator is added with the transfer function $G_1 = (s - 1 + \epsilon)/(s + 1)$. The $\epsilon$ term in the transfer function represents any small error between the desired zero location at $+1$ and the actual location. The closed-loop transfer function is then

$$\frac{C}{R} = \frac{K(s - 1 + \epsilon)}{s^2 + Ks + K\epsilon - K - 1}$$

By applying the Hurwitz or Routh Stability Criterion (Chapter 5) to the denominator of this transfer function, it can be seen that the system is unstable for any value of $K$ if $\epsilon$ is less than $(1 + K)/K$, which is usually the case because $\epsilon$ represents the error in the desired zero location.

**14.5.** For the discrete-time unity feedback system with forward-loop transfer function

$$G_2 = \frac{z+1}{z(z-1)}$$

determine a compensator $G_1$ that provides a deadbeat response for the closed-loop system.

For a deadbeat response (Section 10.8), we want all closed-loop poles at $z = 0$. A pole-zero map of the system is shown in Fig. 14-24($a$). If we cancel the pole at $z = 0$ and the zero at $z = -1$, the root-locus will go through $z = 0$, as shown in Fig. 14-24($b$). The resulting compensator is then

$$G_1 = \frac{z}{z + 1}$$

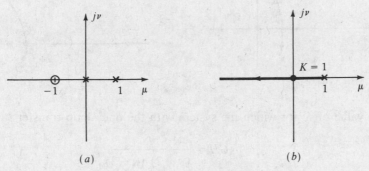

$$(a) \qquad\qquad\qquad\qquad (b)$$

Fig. 14-24

and the closed-loop transfer function is

$$\frac{C}{R} = \frac{G_1 G_2}{1 + G_1 G_2} = \frac{1}{z}$$

## PHASE COMPENSATION

**14.6.** It is desired to add to a system a compensator with a *zero* at $s = -1$ to produce 60° phase lead at $s = -2 + j3$. How can the proper location of the *pole* be determined?

With reference to Fig. 14-3, we want the phase contribution of the network to be $\theta_a - \theta_b = 60°$. From Fig. 14-25, $\theta_a = 108°$. Hence $\theta_b = \theta_a - 60° = 48°$ and the pole should be located at $s = -4.7$, as shown in Fig. 14-25.

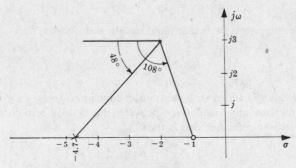

Fig. 14-25

**14.7.** Determine a compensator that will change the departure angle of the root-locus from the pole at $s = -0.5 + j$ to $-135°$ for the plant transfer function

$$G_2 = \frac{K}{s(s^2 + s + 1.25)}$$

The departure angle of the uncompensated system is $-27°$. To change this to $-135°$, a lag compensator with 108° phase lag at $s = -0.5 + j$ can be employed. The required amount of phase lag could be supplied by a *simple* lag compensator (one pole, no zero) with a pole at $s = -0.18$, as shown in Fig. 14-26($a$), or by two simple lags in cascade with two poles at $s = -1.22$, as shown in Fig. 14-26($b$).

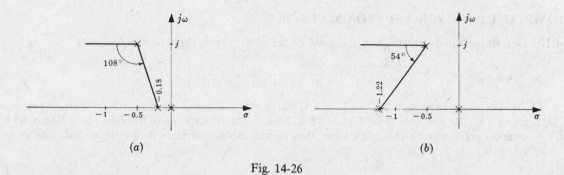

Fig. 14-26

**14.8.** Determine a compensator for the discrete-time system with

$$GH(z) = \frac{K}{z(z-1)}$$

that provides a phase crossover frequency $\omega_\pi$ such that $\omega_\pi T = \pi/2$ rad.

Arg $GH$ at $z = e^{j\pi/2} = j$ is determined from the pole-zero map in Fig. 14-27 as $-225°$. In order for the root-locus to pass through this point, we need to add $45°$ of phase lead, so that arg $GH = \pm 180°$. This can be provided by the compensator

$$P_{\text{Lead}}(z) = \frac{z}{z+1}$$

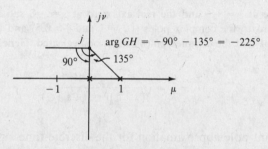

Fig. 14-27

The zero at $z = 0$ provides $90°$ of phase lead and the pole at $z = -1$ provides $45°$ of lag, resulting in a net lead of $45°$.

## MAGNITUDE COMPENSATION

**14.9.** In Example 14.4, the velocity error constant $K_v$ was increased by a factor of $6\frac{2}{3}$ without increasing the gain factor. How was this accomplished?

It was assumed that the compensator $G_1$ had a high-frequency gain of 1 and a low-frequency (d.c.) gain of $6\frac{2}{3}$. This compensator cannot be mechanized passively because a passive lag compensator has a d.c. gain of 1. Consequently, $G_1$ must include an amplifier. An alternative method would be to let $G_1$ be the passive lag compensator

$$G_1' = \frac{0.015}{0.1} \left( \frac{s + 0.1}{s + 0.015} \right)$$

and then amplify the gain factor by $6\frac{2}{3}$. However, when root-locus techniques are employed it is usually more convenient to assume the compensator just adds a pole and zero, as was done in Example 14.4. Appropriate adjustments can be made in the final stages of design to achieve the simplest and/or least expensive compensator mechanization.

## DOMINANT POLE-ZERO APPROXIMATIONS

**14.10.** Determine the overshoot and rise time of the system with the transfer function

$$\frac{C}{R} = \frac{1}{(s+1)(s^2+s+1)}$$

For this system, $\omega_n = 1$, $\zeta = 0.5$, $p_r = 1$, and $p_r/\zeta\omega_n = 2$. From Fig. 14-11 the percentage overshoot is about 8%. The rise time from Fig. 14-12 is 2.4 sec. The corresponding numbers for a system with the complex poles only are 18% and 1.6 sec. Thus the real axis pole reduces the overshoot and slows down the response.

**14.11.** Determine the overshoot and rise time of the system with the transfer function

$$\frac{C}{R} = \frac{s+1}{s^2+s+1}$$

For this system $\omega_n = 1$, $\zeta = 0.5$, $z_r = 1$, and $z_r/\zeta\omega_n = 2$. From Fig. 14-13 the percentage overshoot is 31%. From Fig. 14-14 the 10 to 90% rise time is 1.0 sec. The corresponding numbers for a system without the zero are 18% and 1.6 sec. The real axis zero thus increases the overshoot and decreases the rise time, that is, speeds up the response.

**14.12.** What is a suitable dominant pole-zero approximation for the following system?

Mathcad

$$\frac{C}{R} = \frac{2(s+8)}{(s+1)(s^2+2s+3)(s+6)}$$

The real axis pole at $s = -6$ and the real axis zero at $s = -8$ satisfy Equations (14.2) and (14.4), respectively, with regard to the complex poles ($\zeta\omega_n = 1$ and $\zeta > 0.5$) and therefore may be neglected. The real axis pole at $s = -1$ and the complex poles cannot be neglected. Hence a suitable approximation (with the same d.c. gain) is

$$\frac{C}{R} = \frac{8}{3(s+1)(s^2+2s+3)}$$

**14.13.** Determine a dominant pole approximation for the discrete-time system with transfer function

$$\frac{C}{R} = \frac{0.16}{(z-0.2)(z-0.8)}$$

The step response is given by

$$c(k) = 1 - 1.33(0.8)^k + 0.33(0.2)^k \qquad k = 0, 1, 2, \ldots$$

The magnitude 0.33 of the residue at $z = 0.2$ is four times smaller than the magnitude 1.33 of the residue at $z = 0.8$. Also, the transient response due to the pole at $z = 0.2$ decays $0.8/0.2 = 4$ times faster than that for the pole at $z = 0.8$. Thus the approximate closed-loop system should only have a pole at $z = 0.8$. However, to maintain a system response delay of two samples (the original system has two more poles than zeros), it is necessary to add a pole at $z = 0$ to the approximation. Then

$$\frac{C}{R} \cong \frac{0.2}{z(z-0.8)}$$

The step response of the approximate system is

$$c(k) = \begin{cases} 0 & \text{for } k = 0 \\ 1 - 1.25(0.8)^k & \text{for } k > 0 \end{cases}$$

Note that the only effect of the pole at $z = 0$ on the response is to delay it by one sample.

## POINT DESIGN

**14.14.** Determine $K$, $a$, and $b$ so that the system with open-loop transfer function

$$GH = \frac{K(s+a)}{(s+b)(s+2)^2(s+4)}$$

has a closed-loop pole at $p_1 = -2 + j3$.

The angle contributed to $\arg GH(s_1)$ by the poles at $s = -2$ and $s = -4$ is $-237°$. To satisfy the angle criterion, the angle contributions of the zero at $s = -a$ and the pole $s = -b$ must total $-180° - (-237°)$ $= 57°$. Since this is a positive angle, the zero must be farther to the right than the pole ($b > a$). Either $a$ or $b$ may be chosen arbitrarily as long as the remaining one can be fixed in the finite left-half $s$-plane to give a total contribution of $57°$. Let $a$ be set equal to 2, resulting in a $90°$ phase contribution. Then $b$ must be placed where the contribution of the pole is $-33°$. A line drawn from $p_1$ at $33°$ intercepts the real axis at $6.6 = b$, as shown in Fig. 14-28.

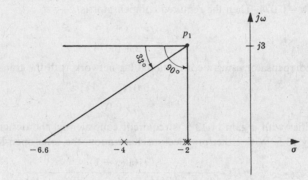

Fig. 14-28

The necessary value of $K$ required to satisfy the magnitude criterion at $p_1$ can now be computed using the chosen values of $a$ and $b$. From the following calculation, the required value of $K$ is

$$\left| \frac{(p_1 + 6.6)(p_1 + 2)^2(p_1 + 4)}{(p_1 + 2)} \right|_{p_1 = -2+j3} = 60$$

**14.15.** Determine the required compensation for a system with the plant transfer function

$$G_2 = \frac{K}{(s+8)(s+14)(s+20)}$$

to satisfy the following specifications: ($a$) overshoot $\leq 5\%$, ($b$) 10 to 90% rise time $T_r \leq 150$ msec, ($c$) $K_p > 6$.

The first specification may be satisfied with a closed-loop transfer function whose response is dominated by two complex poles with $\zeta \geq 0.7$, as seen from Fig. 3-4. A wide variety of dominant pole-zero configurations can satisfy the overshoot specification; but the two-pole configuration is usually the simplest obtainable form. We also see from Fig. 3-4 that, if $\zeta = 0.7$, the normalized 10 to 90% rise time is about $\omega_n T_r = 2.2$. Thus, in order to satisfy the second specification with $\zeta = 0.7$, we have $T_r = 2.2/\omega_n \leq 0.15$ sec or $\omega_n \geq 14.7$ rad/sec.

But let us choose $\omega_n = 17$ so as to achieve some margin with respect to the rise time specification. Other closed-loop poles, which may appear in the final design, may slow down the response. Thus, in order to satisfy the first two specifications, we shall design the system to have a dominant two-pole response with $\zeta = 0.7$ and $\omega_n = 17$. An $s$-plane evaluation of $\arg G_2(p_1)$, where $p_1 = -12 + j12$ (corresponding to $\zeta = 0.7$, $\omega_n = 17$), yields $\arg G_2(p_1) = -245°$. Then, to satisfy the angle criterion at $p_1$, we must compensate the system with phase lead so that the total angle becomes $-180°$. Hence we add a cascade lead compensator with $245° - 180° = 65°$ phase lead at $p_1$. Arbitrarily placing the zero of the lead compensator at $s = -8$

results in $\theta_a = 108°$ (see Fig. 14-3). Then, since we want $\theta_a - \theta_b = 65°$, $\theta_b = 108° - 65° = 43°$. Drawing a line from $p_1$ to the real axis at the required $\theta_b$ determines the pole location at $s = -25$. Addition of the lead compensator with $a = 8$ and $b = 25$ yields an open-loop transfer function

$$G_2 G_{\text{Lead}} = \frac{K}{(s+14)(s+20)(s+25)}$$

The value of $K$ necessary to satisfy the magnitude criterion at $p_1$ is $K = 3100$. The resulting positional error constant for this design is $K_p = 3100/(14)(20)(25) = 0.444$, which is substantially less than the specified value of 6 or more. $K_p$ could be increased slightly by trying other design points (higher $\omega_n$); but the required $K_p$ cannot be achieved without some form of low-frequency magnitude compensation. The required increase is $6/0.444 = 13.5$ and may be obtained with a low-frequency lag compensator with $b/a = 13.5$. The only other requirement is that $a$ and $b$ for the lag compensator must be small enough so as not to affect the high-frequency design accomplished with the lead network. That is,

$$\arg P_{\text{Lag}}(p_1) \cong 0$$

Let $b = 1$ and $a = 0.074$. Then the required compensator is

$$G_{\text{Lag}} = \frac{s+1}{s+0.074}$$

To synthesize this compensator using a conventional lag network with the transfer function

$$P_{\text{Lag}} = \frac{0.074(s+1)}{s+0.074}$$

an additional amplifier with a gain of 13.5 is required; equivalently, the design value of $K$ chosen above may be increased by 13.5. With either practical mechanization, the total open-loop transfer function is

$$GH = \frac{3100(s+1)}{(s+0.075)(s+14)(s+20)(s+25)}$$

The closed-loop poles and zeros are shown in Fig. 14-29. The low-frequency pole and zero effectively cancel each other. The real axis pole at $s = -35$ will slightly affect the response of the system because $p_r/\zeta\omega_n$ for this pole is only about 3 [Equation (*14.2*)]. However, reference to Figs. 14-11 and 14-12 verify that the overshoot and rise time are still well within the specifications. If the system had been designed to barely meet the required rise time specification with the dominant two-pole approximation, the presence of the additional pole in the closed-loop transfer function may have slowed the response enough to dissatisfy the specification.

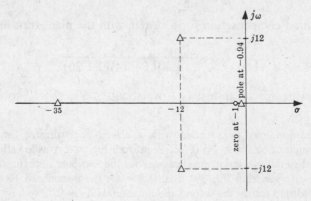

Fig. 14-29

## FEEDBACK COMPENSATION

**14.16.** A positional control system with a tachometer feedback path has the block diagram shown in Fig. 14-30. Determine values of $K_1$ and $K_2$ which result in a system design which yields a 10 to 90% rise time of less than 1 sec and an overshoot of less than 20%.

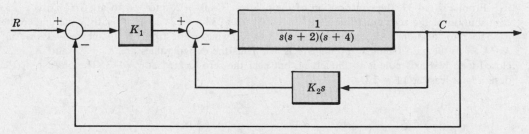

Fig. 14-30

A straightforward way to accomplish this design is to determine a suitable design point in the s-plane and use the point design technique. If the two feedback paths are combined, the block diagram shown in Fig. 14-31 is obtained.

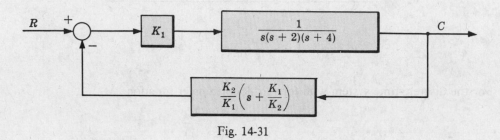

Fig. 14-31

For this configuration

$$GH = \frac{K_2(s + K_1/K_2)}{s(s+2)(s+4)}$$

The zero location at $s = -K_1/K_2$ appears in the feedback path and the gain factor is $K_2$. Thus for a fixed zero location (ratio of $K_1/K_2$), a root-locus for the system may be constructed as a function of $K_2$. The closed-loop transfer function will then contain three poles, but no zeros. Rough sketches of the root-locus (Fig. 14-32) reveal that if the ratio $K_1/K_2$ is set anywhere between 0 and 4, the closed-loop transfer function will probably contain two complex poles (if $K_2$ is large enough) and a real axis pole near the value of $-K_1/K_2$.

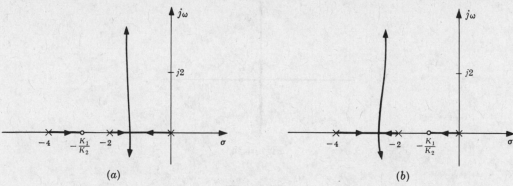

Fig. 14-32

A three-pole dominant configuration may then be appropriate for the design. A value of $\zeta = 0.5$ for the complex poles will satisfy the overshoot requirement. For $\zeta = 0.5$ and $p_r/\zeta\omega_n = 2$, Fig. 14-12 shows a normalized rise time $\omega_n T_r = 2.3$. Thus $T_r = 2.3/\omega_n < 1$ sec or $\omega_n > 2.3$ rad/sec. If $p_r/\zeta\omega_n$ turns out to be greater than 2, the rise time will be faster, and vice versa. In order to have a little margin in case $p_r/\zeta\omega_n$ is smaller than 2, let us choose $\omega_n = 2.6$. The design point in the s-plane is therefore $p_1 = -1.3 + j2.3$, corresponding to $\zeta = 0.5$ and $\omega_n = 2.6$.

From Fig. 14-33, the contribution of the poles at $s = 0$, $-2$, and $-4$ to $\arg GH(p_1)$ is $-233°$. The contribution of the zero must therefore be $-180° - (-233°) = 53°$ at $p_1$ to satisfy the angle criterion at $p_1$. The zero location is determined at $s = -3$ by drawing a line from $p_1$ to the real axis at $53°$. With $K_1/K_2 = 3$, the gain factor at $p_1$ for $GH$ is 7.5. Thus the design values are $K_2 = 7.5$ and $K_1 = 22.5$. The closed-loop real axis pole is to the left of, but near the zero located at $s = -3$. Therefore $p_r/\zeta\omega_n$ for this design is at least $3/1.3 = 2.3$.

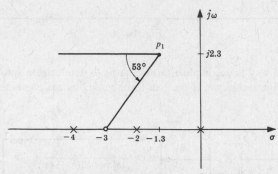

Fig. 14-33

Mathcad

**14.17.** For the discrete-time system with forward-loop transfer function

$$G_2 = \frac{K}{z(z-1)}$$

determine a feedback compensator that yields a closed-loop system with a deadbeat response.

For a deadbeat response (Section 10.8), the closed-loop transfer function must have all its poles at $z = 0$. Since poles cancelled by feedback zeros appear in the closed-loop transfer function, let $H$ have a zero at $z = 0$. This eliminates the pole at $z = 0$ from the root-locus but it remains in the closed-loop transfer function.

For realizability, $H$ must also have at least one pole. If we place the pole of $H$ at $z = -1$, the resulting root-locus goes through $z = 0$, as shown in Fig. 14-34. Then, by setting $K = 1$, all the closed-loop poles are located at $z = 0$ and the system has a deadbeat response.

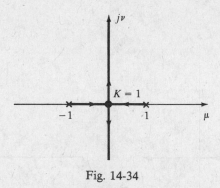

Fig. 14-34

# Supplementary Problems

**14.18.** For the system with the open-loop transfer function $GH = K(s + a)/(s^2 - 1)(s + 5)$ determine $K$ and $a$ such that the closed-loop system has dominant poles with $\zeta = 0.5$ and $\omega_n = 2$. What is the percentage overshoot of the closed-loop system with these values of $K$ and $a$?

**14.19.** Determine a suitable compensator for the system with the plant transfer function

$$G_2 = \frac{1}{s(s+1)(s+4)}$$

to satisfy the following specifications: (1) overshoot < 20%, (2) 10 to 90% rise time $\leq 1$ sec, (3) gain margin $\geq 5$.

**14.20.** Determine suitable compensation for the system with the plant transfer function $G_2 = 1/s(s+4)^2$ to satisfy the following specifications: (1) overshoot < 20%, (2) velocity error constant $K_v \geq 10$.

**14.21.** For the system shown in the block diagram of Fig. 14-35, determine $K_1$ and $K_2$ such that the system has closed-loop poles at $s = -2 \pm j2$.

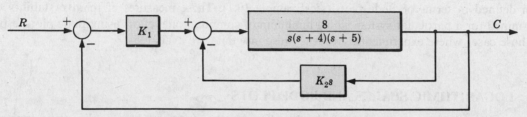

Fig. 14-35

**14.22.** Determine a value of $K$ for the system with the open-loop transfer function $GH = K/s(s^2 + 6s + 25)$ such that the velocity error constant $K_v > 1$, the closed-loop step response has no overshoot, and the gain margin > 5.

**14.23.** Design a compensator for the system with the plant transfer function $G_2 = 63/s(s+7)(s+9)$ such that the velocity error constant $K_v > 30$, the overshoot is less than 20%, and the 10 to 90% rise time is less than 0.5 sec.

# Answers to Supplementary Problems

**14.18.** $K = 11.25$, $a = 1.6$, overshoot = 38%; note that the system has a closed-loop zero at $s = -a = -1.6$.

**14.19.** $G_1 = 24(s+1)/(s+4)$

**14.20.** $G_1 = 24(s+0.2)/(s+0.03)$

**14.21.** $K_2 = 1$, $K_1 = 5$

**14.22.** $K = 28$

**14.23.** $G_1 = 3(s+0.5)/(s+0.05)$

# Chapter 15

# Bode Analysis

## 15.1 INTRODUCTION

The analysis of feedback control systems using the Bode method is equivalent to Nyquist analysis in that both techniques employ graphical representations of the open-loop frequency response function $GH(\omega)$, where $GH(\omega)$ refers to either a discrete-time or a continuous-time system. However, **Bode plots** consist of two graphs: the magnitude of $GH(\omega)$, and the phase angle of $GH(\omega)$, both plotted as a function of frequency $\omega$. Logarithmic scales are usually used for the frequency axes and for $|GH(\omega)|$.

Bode plots clearly illustrate the relative stability of a system. In fact, gain and phase margins are often defined in terms of Bode plots (see Example 10.1). These measures of relative stability can be determined for a particular system with a minimum of computational effort using Bode plots, especially for those cases where experimental frequency response data are available.

## 15.2 LOGARITHMIC SCALES AND BODE PLOTS

Logarithmic scales are used for Bode plots because they considerably simplify their construction, manipulation, and interpretation.

A logarithmic scale is used for the $\omega$-axes (abscissas) because the magnitude and phase angle may be graphed over a greater range of frequencies than with linear frequency axes, all frequencies being equally emphasized, and such graphs for continuous-time systems often result in straight lines (Section 15.4).

The magnitude $|P(\omega)|$ of any frequency response function $P(\omega)$ for any value of $\omega$ is plotted on a logarithmic scale in decibel (db) units, where

$$\text{db} = 20\log_{10}|P(\omega)| \qquad (15.1)$$

[Also see Equation (10.4).]

**EXAMPLE 15.1.** If $|P(2)| \equiv |GH(2)| = 10$, the magnitude is $20\log_{10}10 = 20$ db.

Since the decibel is a logarithmic unit, the **db magnitude** of a frequency response function composed of a *product* of terms is equal to the *sum* of the db magnitudes of the individual terms. Thus, when the logarithmic scale is employed, the magnitude plot of a frequency response function expressible as a product of more than one term can be obtained by adding the individual db magnitude plots for each product term.

The *db magnitude versus log $\omega$* plot is called the **Bode magnitude plot**, and the *phase angle versus log $\omega$* plot is the **Bode phase angle plot**. The Bode magnitude plot is sometimes called the *log-modulus plot* in the literature.

**EXAMPLE 15.2.** The Bode magnitude plot for the continuous-time frequency response function

$$P(j\omega) = \frac{100[1+j(\omega/10)]}{1+j\omega}$$

may be obtained by adding the Bode magnitude plots for: 100, $1+j(\omega/10)$, and $1/(1+j\omega)$.

## 15.3    THE BODE FORM AND THE BODE GAIN FOR CONTINUOUS-TIME SYSTEMS

It is convenient to use the so-called *Bode form* of a continuous-time frequency response function when using Bode plots for analysis and design because of the asymptotic approximations in Section 15.4.

The **Bode form** for the function

$$\frac{K(j\omega + z_1)(j\omega + z_2)\cdots(j\omega + z_m)}{(j\omega)^l(j\omega + p_1)(j\omega + p_2)\cdots(j\omega + p_n)}$$

where $l$ is a nonnegative integer, is obtained by factoring out all $z_i$ and $p_i$ and rearranging it in the form

$$\frac{\left[K\prod_{i=1}^{m}z_i \middle/ \prod_{i=1}^{n}p_i\right](1 + j\omega/z_1)(1 + j\omega/z_2)\cdots(1 + j\omega/z_m)}{(j\omega)^l(1 + j\omega/p_1)(1 + j\omega/p_2)\cdots(1 + j\omega/p_n)} \qquad (15.2)$$

The **Bode gain** $K_B$ is defined as the coefficient of the numerator in Equation $(15.2)$:

$$K_B \equiv \frac{K\prod_{i=1}^{m}z_i}{\prod_{i=1}^{n}p_i} \qquad (15.3)$$

## 15.4    BODE PLOTS OF SIMPLE CONTINUOUS-TIME FREQUENCY RESPONSE FUNCTIONS AND THEIR ASYMPTOTIC APPROXIMATIONS

The constant $K_B$ has a magnitude $|K_B|$, a phase angle of $0°$ if $K_B$ is positive, and $-180°$ if $K_B$ is negative. Therefore the Bode plots for $K_B$ are simply horizontal straight lines as shown in Figs. 15-1 and 15-2.

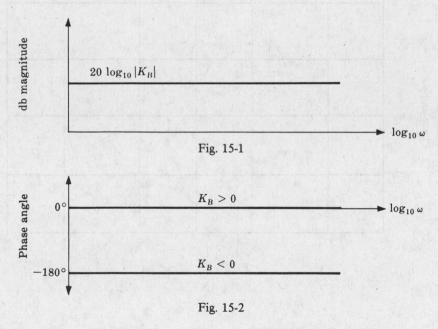

Fig. 15-1

Fig. 15-2

The frequency response function (or sinusoidal transfer function) for a *pole of order $l$ at the origin* is

$$\frac{1}{(j\omega)^l} \qquad (15.4)$$

The bode plots for this function are straight lines, as shown in Figs. 15-3 and 15-4.

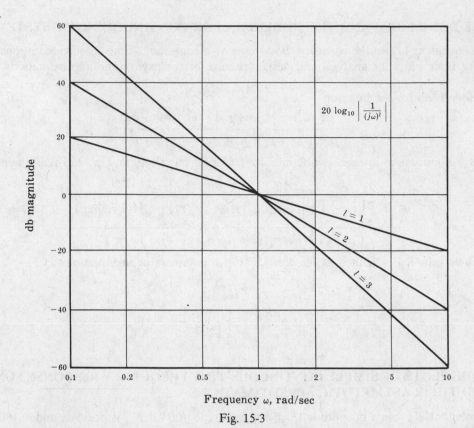

Fig. 15-3

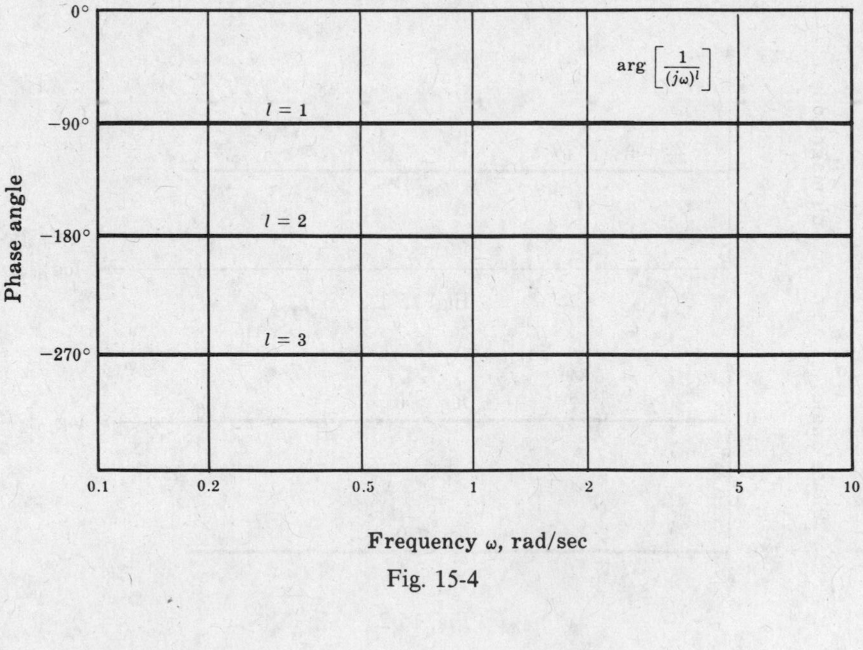

Fig. 15-4

For a *zero of order l at the origin*,

$$(j\omega)^l. \tag{15.5}$$

the Bode plots are the reflections about the 0-db and 0° lines of Figs. 15-3 and 15-4, as shown in Figs. 15-5 and 15-6.

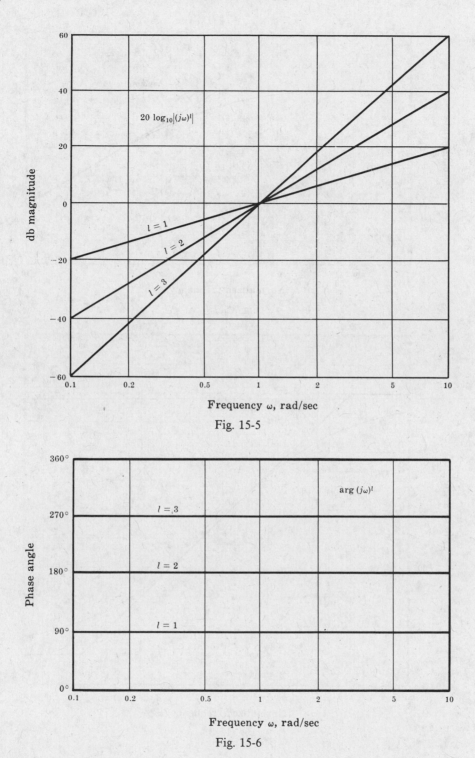

Fig. 15-5

Fig. 15-6

Consider the *single-pole* transfer function $p/(s+p)$, $p > 0$. The Bode plots for its frequency response function

$$\frac{1}{1 + j\omega/p} \qquad (15.6)$$

are given in Figs. 15-7 and 15-8. Note that the logarithmic frequency scale is normalized in terms of $p$.

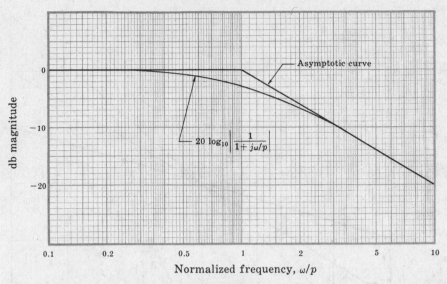

Fig. 15-7

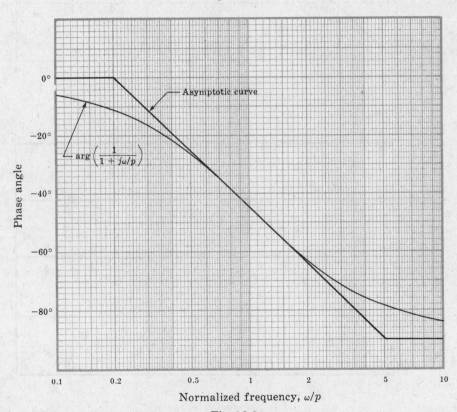

Fig. 15-8

To determine the *asymptotic approximations* for these Bode plots, we see that for $\omega/p \ll 1$, or $\omega \ll p$,

$$20 \log_{10}\left|\frac{1}{1 + j\omega/p}\right| \cong 20 \log_{10} 1 = 0 \text{ db}$$

and for $\omega/p \gg 1$, or $\omega \gg p$,

$$20 \log_{10}\left|\frac{1}{1 + j\omega/p}\right| \cong 20 \log_{10}\left|\frac{1}{j\omega/p}\right| = -20 \log_{10}\left(\frac{\omega}{p}\right)$$

Therefore the Bode magnitude plot asymptotically approaches a horizontal straight line at 0 db as $\omega/p$ approaches zero and $-20\log_{10}(\omega/p)$ as $\omega/p$ approaches infinity (Fig. 15-7). Note that this high-frequency asymptote is a straight line with a slope of $-20$ db/decade, or $-6$ db/octave when plotted on a logarithmic frequency scale as shown in Fig. 15-7. The two asymptotes intersect at the **corner frequency** $\omega = p$ rad/sec. To determine the phase angle asymptote, we see that for $\omega/p \ll 1$, or $\omega \ll p$,

$$\arg\left(\frac{1}{1+j\omega/p}\right) = -\tan^{-1}\left(\frac{\omega}{p}\right)\Bigg|_{\omega \ll p} \cong 0°$$

and for $\omega/p \gg 1$, or $\omega \gg p$,

$$\arg\left(\frac{1}{1+j\omega/p}\right) = -\tan^{-1}\left(\frac{\omega}{p}\right)\Bigg|_{\omega \gg p} \cong -90°$$

Thus the Bode phase angle plot asymptotically approaches $0°$ as $\omega/p$ approaches zero, and $-90°$ as $\omega/p$ approaches infinity, as shown in Fig. 15-8. A negative-slope straight-line asymptote can be used to join the $0°$ asymptote and the $-90°$ asymptote by drawing a line from the $0°$ asymptote at $\omega = p/5$ to the $-90°$ asymptote at $\omega = 5p$. Note that it is tangent to the exact curves at $\omega = p$.

The *errors* introduced by these asymptotic approximations are shown in Table 15-1 for the single-pole transfer function at various frequencies.

Table 15-1.  Asymptotic Errors for $\dfrac{1}{1+j\omega/p}$

| $\omega$ | $p/5$ | $p/2$ | $p$ | $2p$ | $5p$ |
|---|---|---|---|---|---|
| Magnitude error (db) | $-0.17$ | $-0.96$ | $-3$ | $-0.96$ | $-0.17$ |
| Phase angle error | $-11.3°$ | $-0.8°$ | $0°$ | $+0.8°$ | $+11.3°$ |

The Bode plots and their asymptotic approximations for the *single-zero* frequency response function

$$1 + \frac{j\omega}{z_1} \tag{15.7}$$

are shown in Figs. 15-9 and 15-10.

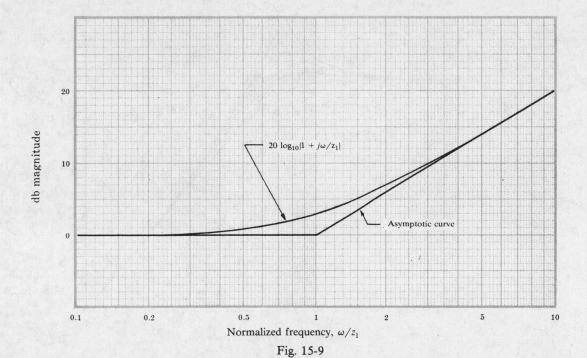

Fig. 15-9

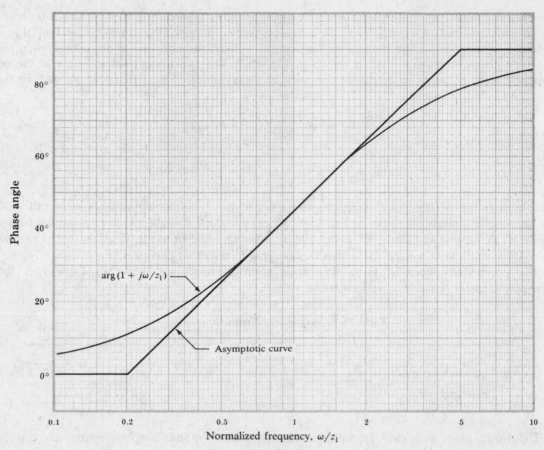

Fig. 15-10

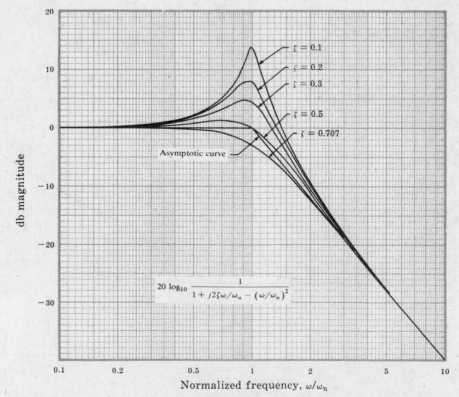

Fig. 15-11

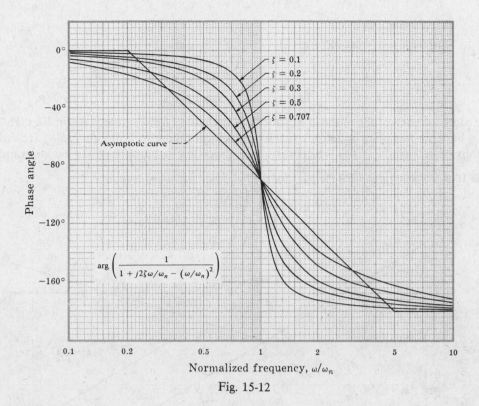

Fig. 15-12

The Bode plots and their asymptotic approximations for the second-order frequency response function with *complex poles*,

$$\frac{1}{1 + j2\zeta\omega/\omega_n - (\omega/\omega_n)^2} \qquad 0 \le \zeta \le 1 \qquad (15.8)$$

are shown in Figs. 15-11 and 15-12. Note that the damping ratio $\zeta$ is a parameter on these graphs.

The magnitude asymptote shown in Fig. 15-11 has a corner frequency at $\omega = \omega_n$ and a high-frequency slope twice that of the asymptote for the single-pole case of Fig. 15-7. The phase angle asymptote is similar to that of Fig. 15-8 except that the high-frequency portion is at $-180°$ instead of $-90°$ and the point of tangency, or inflection, is at $-90°$.

The Bode plots for a pair of *complex zeros* are the reflections about the 0 db and $0°$ lines of those for the complex poles.

## 15.5   CONSTRUCTION OF BODE PLOTS FOR CONTINUOUS-TIME SYSTEMS

Bode plots of continuous-time frequency response functions can be constructed by summing the magnitude and phase angle contributions of each pole and zero (or pairs of complex poles and zeros). The asymptotic approximations of these plots are often sufficient. If more accurate plots are desired, many software packages are available for rapidly accomplishing this task.

For the general open-loop frequency response function

$$GH(j\omega) = \frac{K_B(1 + j\omega/z_1)(1 + j\omega/z_2)\cdots(1 + j\omega/z_m)}{(j\omega)^l(1 + j\omega/p_1)(1 + j\omega/p_2)\cdots(1 + j\omega/p_n)} \qquad (15.9)$$

where $l$ is a positive integer or zero, the magnitude and phase angle are given by

$$20\log_{10}|GH(j\omega)| = 20\log_{10}|K_B| + 20\log_{10}\left|1 + \frac{j\omega}{z_1}\right| + \cdots + 20\log_{10}\left|1 + \frac{j\omega}{z_m}\right|$$

$$+ 20\log_{10}\frac{1}{|(j\omega)^l|} + 20\log_{10}\frac{1}{|1 + j\omega/p_1|} + \cdots + 20\log_{10}\frac{1}{|1 + j\omega/p_n|} \qquad (15.10)$$

and

$$\arg GH(j\omega) = \arg K_B + \arg\left(1 + \frac{j\omega}{z_1}\right) + \cdots + \arg\left(1 + \frac{j\omega}{z_m}\right)$$

$$+ \arg\left(\frac{1}{(j\omega)^l}\right) + \arg\left(\frac{1}{1 + j\omega/p_1}\right) + \cdots + \arg\left(\frac{1}{1 + j\omega/p_n}\right) \qquad (15.11)$$

The Bode plots for each of the terms in Equations (15.10) and (15.11) were given in Figs. 15-1 to 15-12. If $GH(j\omega)$ has complex poles or zeros, terms having a form similar to Equation (15.8) are simply added to Equations (15.10) and (15.11). The construction procedure is best illustrated by an example.

**EXAMPLE 15.3.**   The asymptotic Bode plots for the frequency response function

$$GH(j\omega) = \frac{10(1 + j\omega)}{(j\omega)^2\left[1 + j\omega/4 - (\omega/4)^2\right]}$$

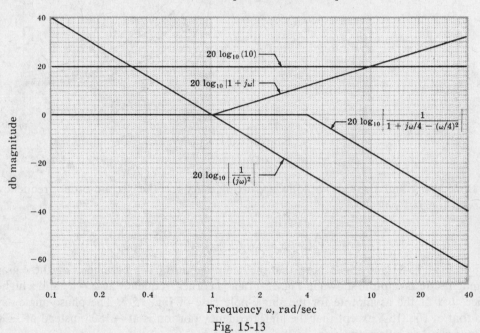

Fig. 15-13

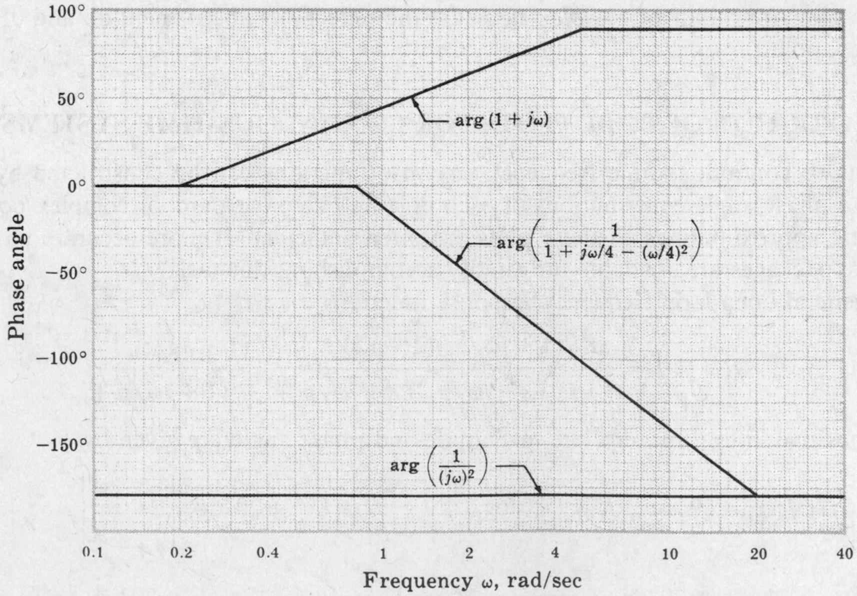

Fig. 15-14

are constructed using Equations ($15.10$) and ($15.11$):

$$20 \log_{10}|GH(j\omega)| = 20 \log_{10} 10 + 20 \log_{10}|1 + j\omega| + 20 \log_{10}\left|\frac{1}{(j\omega)^2}\right| + 20 \log_{10}\left|\frac{1}{1 + j\omega/4 - (\omega/4)^2}\right|$$

$$\arg GH(j\omega) = \arg(1 + j\omega) + \arg(1/(j\omega)^2) + \arg\left(\frac{1}{1 + j\omega/4 - (\omega/4)^2}\right)$$

The graphs for each of the terms in these equations are obtained from Figs. 15-1 to 15-12 and are shown in Figs. 15-13 and 15-14. The asymptotic Bode plots for $GH(j\omega)$ are obtained by adding these curves, as shown in Figs. 15-15 and 15-16, where computer-generated Bode plots for the frequency response function are also given for comparison with the asymptotic approximations.

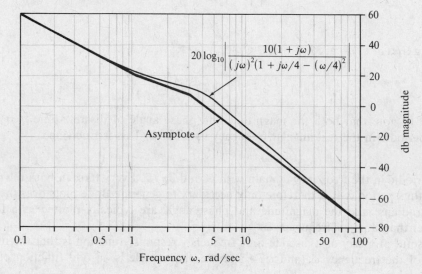

$$20 \log_{10}\left|\frac{10(1 + j\omega)}{(j\omega)^2(1 + j\omega/4 - (\omega/4)^2)}\right|$$

Asymptote

Frequency $\omega$, rad/sec

Fig. 15-15

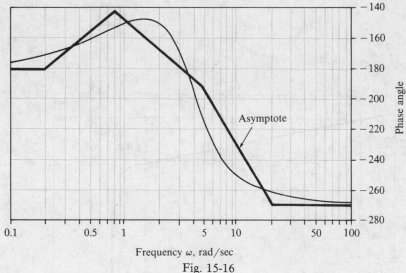

Asymptote

Frequency $\omega$, rad/sec

Fig. 15-16

## 15.6  BODE PLOTS OF DISCRETE-TIME FREQUENCY RESPONSE FUNCTIONS

The factored form for the general open-loop discrete-time frequency response function is

$$GH(e^{j\omega T}) = \frac{K(e^{j\omega T} + z_1)(e^{j\omega T} + z_2)\cdots(e^{j\omega T} + z_m)}{(e^{j\omega T} + p_1)(e^{j\omega T} + p_2)\cdots(e^{j\omega T} + p_n)} \qquad (15.12)$$

Simple asymptotic approximations, similar to those in Section 15.4, do not exist for the individual terms in Equation (*15.12*). Thus there is no particular advantage to a *Bode form* of the type in Equation (*15.2*) for discrete-time systems. In general, computers provide the most convenient way to generate Bode plots for discrete-time systems and several software packages exist to accomplish this task.

For the general open-loop frequency response function Equation (*15.12*), the magnitude and phase angle are given by

$$
\begin{aligned}
20\log_{10}\left|GH(e^{j\omega T})\right| = {} & 20\log_{10}|K| \\
& + 20\log_{10}|e^{j\omega T} + z_1| + \cdots + 20\log_{10}|e^{j\omega T} + z_m| \\
& + 20\log_{10}\frac{1}{|e^{j\omega T} + p_1|} + \cdots + 20\log_{10}\frac{1}{|e^{j\omega T} + p_n|}
\end{aligned}
\tag{15.13}
$$

and

$$
\begin{aligned}
\arg GH(e^{j\omega T}) = {} & \arg K + \arg\!\left(e^{j\omega T} + z_1\right) + \cdots \\
& + \arg\!\left(e^{j\omega T} + z_m\right) + \arg\frac{1}{\left(e^{j\omega T} + p_1\right)} + \cdots + \arg\!\left(\frac{1}{e^{j\omega T} + p_n}\right)
\end{aligned}
\tag{15.14}
$$

It is important to note that both the magnitude and phase angle of discrete-time frequency response functions are periodic in the real angular frequency variable $\omega$. This is true since

$$
e^{j\omega T} = e^{j(\omega + 2k\pi/T)T} = e^{j\omega T}e^{j2k\pi}
$$

thus $e^{j\omega T}$ is periodic in the frequency domain with period $2\pi/T$. Every term in both the magnitude and phase angle is thus periodic. It is therefore only necessary to generate Bode plots over the angular range $-\pi \le \omega T \le \pi$ radians; and the magnitude and phase angle are typically plotted as a function of the angle $\omega T$ rather than angular frequency $\omega$.

Another useful property of a discrete-time frequency response function is that the magnitude is an even function of the frequency $\omega$ (and $\omega T$) and the phase angle is an odd function of $\omega$ (and $\omega T$).

**EXAMPLE 15.4.**   The Bode plots for the discrete-time frequency response function

$$
GH(e^{j\omega T}) = \frac{\frac{1}{100}\left(e^{j\omega T} + 1\right)^2}{\left(e^{j\omega T} - 1\right)\left(e^{j\omega T} + \frac{1}{3}\right)\left(e^{j\omega T} + \frac{1}{2}\right)}
$$

are shown in Figs. 15-17 and 15-18.

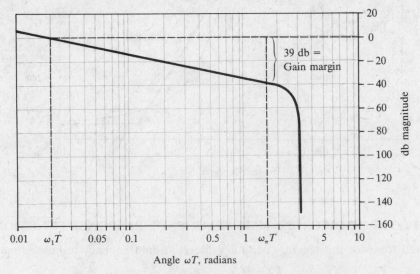

Fig. 15-17

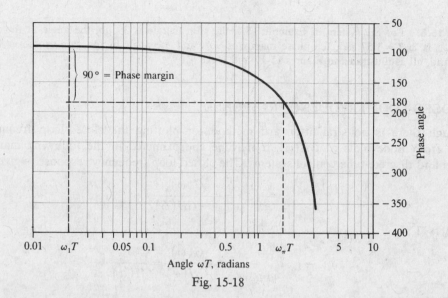

Fig. 15-18

## 15.7  RELATIVE STABILITY

The relative stability indicators "gain margin" and "phase margin" for both discrete-time and continuous-time systems are defined in terms of the system open-loop frequency response in Section 10.4. Consequently these parameters are easily determined from the Bode plots of $GH(\omega)$ as illustrated in Example 10.1, and in Example 15.4 above. Since 0 db corresponds to a magnitude of 1, the **gain margin** is the number of decibels that $|GH(\omega)|$ is *below* 0 db at the phase crossover frequency $\omega_\pi$ ($\arg GH(\omega_\pi) = -180°$). The **phase margin** is the number of degrees $\arg GH(\omega)$ is above $-180°$ at the gain crossover frequency $\omega_1$ ($|GH(\omega_1)| = 1$). Computer-generated Bode plots should be used to accurately determine $\omega_\pi$, $\omega_1$ and the gain and phase margins.

In most cases positive gain and phase margins, as defined above, will ensure stability of the closed-loop system. However, a Nyquist Stability Plot (Chapter 11) may be sketched, or one of the methods of Chapter 5 can be used to verify the absolute stability of the system.

**EXAMPLE 15.5.**  The continuous-time system whose Bode plots are shown in Fig. 15-19 has a gain margin of 8 db and a phase margin of 40°.

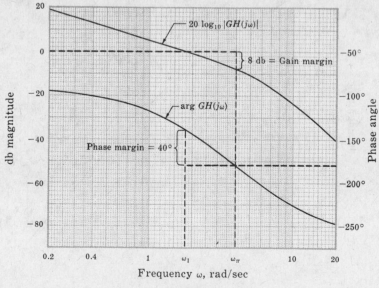

Fig. 15-19

**EXAMPLE 15.6.**  For the system in Example 15.4, the gain margin is 39 db, the angle at the phase crossover frequency $\omega_\pi$ is $\omega_\pi T = 1.57$ rad, the phase margin is 90°, and the angle at the gain crossover frequency $\omega_1$ is $\omega_1 T = 0.02$ rad, all as illustrated in Figures 15-17 and 15-18.

## 15.8   CLOSED-LOOP FREQUENCY RESPONSE

Although there is no straightforward method for plotting the closed-loop frequency response $(C/R)(\omega)$ from Bode plots of $GH(\omega)$, it may be approximated in the following manner, for both continuous and discrete-time control systems. The closed-loop frequency response is given by

$$\frac{C}{R}(\omega) = \frac{G(\omega)}{1 + GH(\omega)}$$

If $|GH(\omega)| \gg 1$,

$$\left.\frac{C}{R}(\omega)\right|_{|GH(\omega)| \gg 1} \cong \frac{G(\omega)}{GH(\omega)} = \frac{1}{H(\omega)}$$

If $|GH(\omega)| \ll 1$,

$$\left.\frac{C}{R}(\omega)\right|_{|GH(\omega)| \ll 1} \cong G(\omega)$$

The open-loop frequency response of most systems is characterized by high gain for low frequencies and decreasing gain for higher frequencies, due to the usual excess of poles over zeros. Thus the closed-loop frequency response *for a unity feedback system* ($H = 1$) is approximated by a magnitude of 1 (0 db) and phase angle of 0° for frequencies below the gain crossover frequency $\omega_1$. For frequencies above $\omega_1$, the closed-loop frequency response may be approximated by the magnitude and phase angle of $G(\omega)$. *An approximate closed-loop bandwidth for many systems is the gain crossover frequency $\omega_1$* (See Example 12.7.)

**EXAMPLE 15.7.**  The open-loop Bode magnitude plot and approximate closed-loop Bode magnitude plot for the continuous-time unity feedback system represented by $G(j\omega) = 10/j\omega(1 + j\omega)$ are shown in Fig. 15-20.

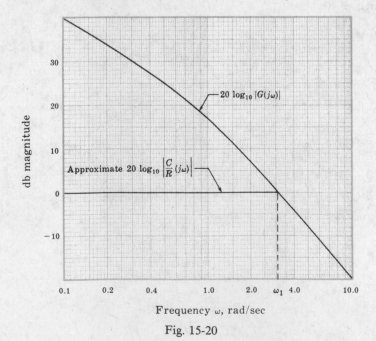

Fig. 15-20

## 15.9   BODE ANALYSIS OF DISCRETE-TIME SYSTEMS USING THE $w$-TRANSFORM

The $w$-transform discussed in Section 10.7 can be used in the Bode analysis of discrete-time systems. The algorithm for Bode analysis using the $w$-transform is:

1.   Substitute $(1 + w)/(1 - w)$ for $z$ in the open-loop transfer function $GH(z)$:

$$GH(z)\big|_{z = \frac{1+w}{1-w}} \equiv GH'(w)$$

2.   Let $w = j\omega_w$ and generate Bode plots for $GH'(j\omega_w)$, using the methods of Sections 15.3 through 15.5.

3.   Analyze the relative stability of the system in the $w$-plane by determining the gain and phase margins, the gain and phase crossover frequencies, the closed-loop frequency response, the bandwidth, and/or any other frequency-related characteristics of interest.

4.   Transform the critical frequencies determined in step 3 to the frequency domain of the $z$-plane using the transformation $\omega T = 2\tan^{-1}\omega_w$.

**EXAMPLE 15.8.**   The open-loop transfer function

$$GH(z) = \frac{\frac{1}{100}(z + 1)^2}{(z - 1)\left(z + \frac{1}{3}\right)\left(z + \frac{1}{2}\right)}$$

is transformed into the $w$-domain by letting

$$z = \frac{1 + w}{1 - w}$$

which yields

$$GH'(w) = \frac{-\frac{6}{100}(w - 1)}{w(w + 2)(w + 3)}$$

Note, in particular, that the minus sign contributes $-180°$ of phase angle, and the zero at $+1$ contributes $+90°$ at $\omega_w = 0°$. The Bode plots of $GH'(j\omega_w)$ are shown in Figs. 15-21 and 15-22.

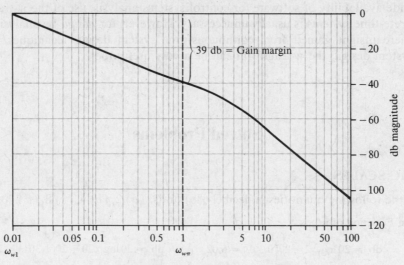

Fig. 15-21

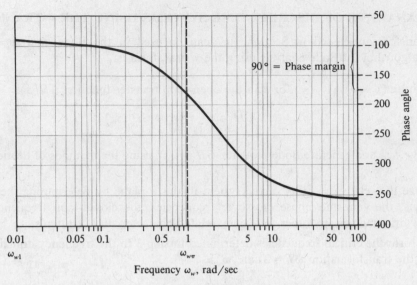

Fig. 15-22

**EXAMPLE 15.9.** From the Bode plots of Example 15.8, the gain margin in the $w$-domain is 39 db and the phase crossover frequency is $\omega_{w\pi} = 1$ rad/sec. Transforming back to the $z$-domain, the phase crossover frequency $\omega_\pi$ is obtained from

$$\omega_\pi T = 2\tan^{-1}\omega_{w\pi} = 1.57 \text{ rad}$$

Compare these results with those of Example 15.6, which are the same.

**EXAMPLE 15.10.** From the Bode plots of Example 15.8, the phase margin is 90° and the gain crossover frequency is $\omega_{w1} = 0.01$ rad/sec. Transforming to the $z$-domain, the gain crossover frequency $\omega_1$ is obtained from

$$\omega_1 T = 2\tan^{-1}\omega_{w1} = 0.02 \text{ rad}$$

Compare these results with those of Example 15.6, which are the same.

With the wide availability of software for control systems analysis, use of the $w$-transform for Bode *analysis* of discrete-time systems is usually unnecessary. However, for *design* by analysis, as discussed in Chapter 16 where insight gained from continuous-time system design techniques is transferred to discrete-time system design, the $w$-transform can be a very useful tool.

# Solved Problems

## LOGARITHMIC SCALES

**15.1.** Express the following quantities in decibel (db) units: $(a)$ 2, $(b)$ 4, $(c)$ 8, $(d)$ 20, $(e)$ 25, $(f)$ 140.

From Equation $(15.1)$,

$$\text{db}_a = 20\log_{10}2 = 20(0.301) = 6.02 \qquad \text{db}_d = 20\log_{10}20 = 20(1.301) = 26.02$$

$$\text{db}_b = 20\log_{10}4 = 20(0.602) = 12.04 \qquad \text{db}_e = 20\log_{10}25 = 20(1.398) = 27.96$$

$$\text{db}_c = 20\log_{10}8 = 20(0.903) = 18.06 \qquad \text{db}_f = 20\log_{10}140 = 20(2.146) = 42.92$$

Note that since $4 = 2 \times 2$, then for part ($b$) we have

$$20\log_{10}4 = 20\log_{10}2 + 20\log_{10}2 = 12.04$$

and since $8 = 2 \times 4$, then for part ($c$) we have

$$20\log_{10}8 = 20\log_{10}2 + 20\log_{10}4 = 6.02 + 12.04 = 18.06$$

## THE BODE FORM AND THE BODE GAIN FOR CONTINUOUS-TIME SYSTEMS

**15.2.** Determine the Bode form and the Bode gain for the transfer function

Mathcad

$$GH = \frac{K(s+2)}{s^2(s+4)(s+6)}$$

Factoring 2 from the numerator, 4 and 6 from the denominator and putting $s = j\omega$ results in the Bode form

$$GH(j\omega) = \frac{(K/12)(1 + j\omega/2)}{(j\omega)^2(1 + j\omega/4)(1 + j\omega/6)}$$

The Bode gain is $K_B = K/12$.

**15.3.** When is the Bode gain equal to the d.c. gain (zero frequency magnitude) of a transfer function?

The Bode gain is equal to the d.c. gain of any transfer function with no poles or zeros at the origin [$l = 0$ in Equation (*15.2*)].

## BODE PLOTS OF SIMPLE FREQUENCY RESPONSE FUNCTIONS

**15.4.** Prove that the Bode Magnitude plot for $(j\omega)^l$ is a straight line.

The Bode magnitude plot for $(j\omega)^l$ is a plot of $20\log_{10}\omega^l$ versus $\log_{10}\omega$. Thus

$$\text{slope} = \frac{d(20\log_{10}\omega^l)}{d(\log_{10}\omega)} = \frac{20l\,d(\log_{10}\omega)}{d(\log_{10}\omega)} = 20l$$

Since the slope is constant for any $l$, the Bode magnitude plot is a straight line.

**15.5.** Determine: (1) the conditions under which the Bode magnitude plot for a pair of complex poles has a peak at a nonzero, finite value of $\omega$; and (2) the frequency at which the peak occurs.

Mathcad

The Bode magnitude is given by

$$20\log_{10}\left|\frac{1}{1 + j2\zeta\omega/\omega_n - (\omega/\omega_n)^2}\right|$$

Since the logarithm is a monotonically increasing function, the magnitude in decibels has a peak (maximum) if and only if the magnitude itself is maximum. The magnitude squared, which is maximum when the magnitude is maximum, is

$$\frac{1}{\left[1 - (\omega/\omega_n)^2\right]^2 + 4(\zeta\omega/\omega_n)^2}$$

Taking the derivative of this function and setting it equal to zero yields

$$\frac{\left(4\omega/\omega_n^2\right)\left[1-\left(\omega/\omega_n\right)^2\right]-8\zeta^2\omega/\omega_n^2}{\left\{\left[1-\left(\omega/\omega_n\right)^2\right]^2+4\left(\zeta\omega/\omega_n\right)^2\right\}^2}=0$$

or

$$1-\left(\frac{\omega}{\omega_n}\right)^2-2\zeta^2=0$$

and the frequency at the peak is $\omega = \omega_n\sqrt{1-2\zeta^2}$. Since $\omega$ must be real, by definition, the magnitude has a peak at a nonzero value $\omega$ only if $1-2\zeta^2 > 0$ or $\zeta < 1/\sqrt{2} = 0.707$. For $\zeta \geq 0.707$, the Bode magnitude is monotonically decreasing.

## CONSTRUCTION OF BODE PLOTS FOR CONTINUOUS-TIME SYSTEMS

**15.6.** Construct the asymptotic Bode plots for the frequency response function

$$GH(j\omega) = \frac{1+j\omega/2-\left(\omega/2\right)^2}{j\omega\left(1+j\omega/0.5\right)\left(1+j\omega/4\right)}$$

The asymptotic Bode plots are determined by summing the graphs of the asymptotic representations for each of the terms of $GH(j\omega)$, as in equations (15.10) and (15.11). The asymptotes for each of these terms are shown in Figs. 15-23 and 15-24 and the asymptotic Bode plots for $GH(j\omega)$ in Figs. 15-25 and 15-26. The exact Bode plots generated by computer are shown for comparison.

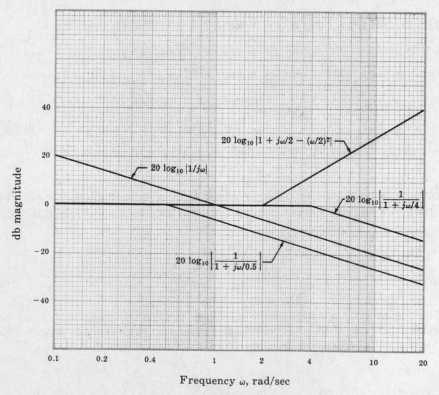

Fig. 15-23

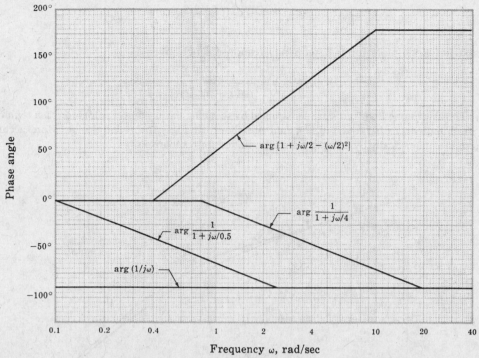

Fig. 15-24

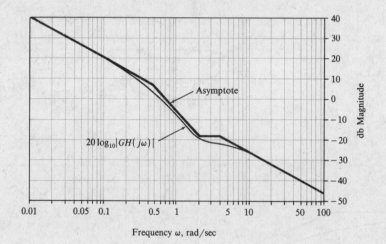

Fig. 15-25

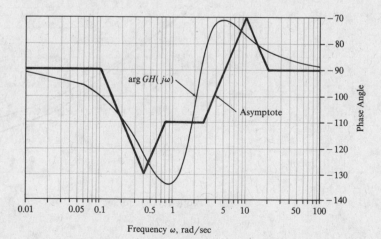

Fig. 15-26

**15.7.** Construct Bode plots for the frequency response function

$$GH(j\omega) = \frac{2}{j\omega(1 + j\omega/2)(1 + j\omega/5)}$$

The asymptotic Bode plots are constructed by summing the asymptotic plots for each term of $GH(j\omega)$, as in Equation (*15.10*) and (*15.11*), and are shown in Figs. 15-27 and 15-28. More accurate curves determined numerically by computer are also plotted for comparison.

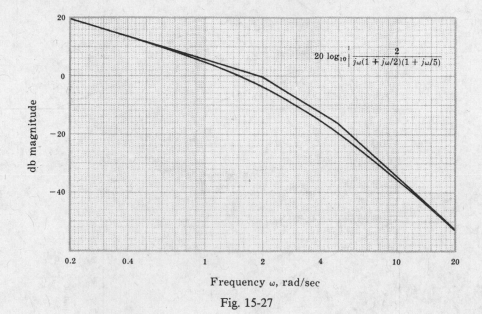

Fig. 15-27

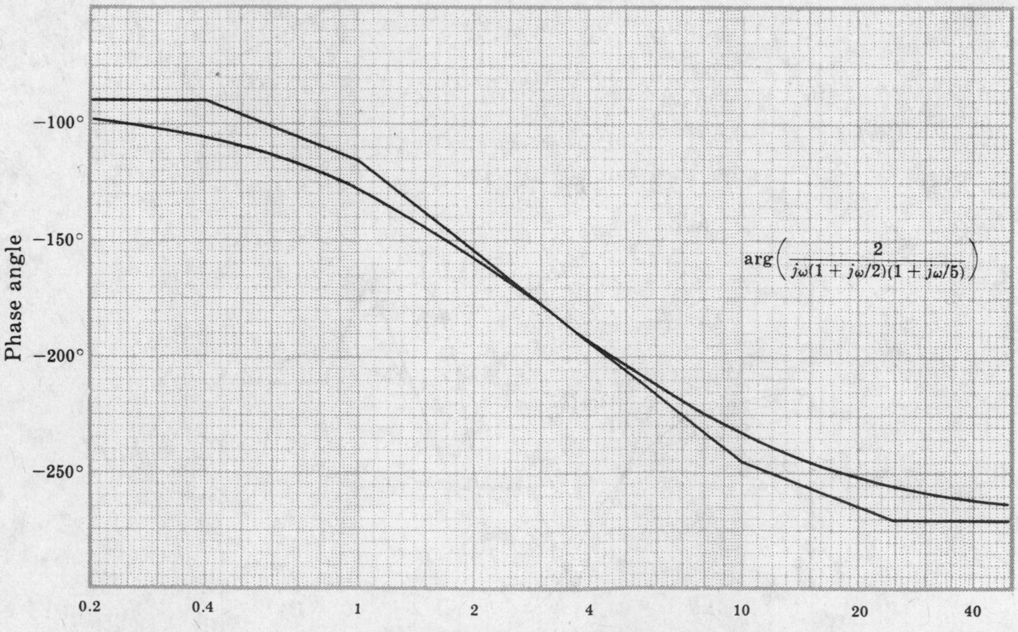

Fig. 15-28

**15.8.** Construct the Bode plots for the open-loop transfer function $GH = 2(s+2)/(s^2-1)$.

Mathcad

With $s = j\omega$, the Bode form for this transfer function is

$$GH(j\omega) = \frac{-4(1 + j\omega/2)}{(1 + j\omega)(1 - j\omega)}$$

This function has a right-half plane pole [due to the term $1/(1 - j\omega)$] which is not one of the standard functions introduced in Section 15.4. However, this function has the same magnitude as $1/(1 + j\omega)$ and the same phase angle as $1 + j\omega$. Thus for a function of the form $1/(1 - j\omega/p)$, the magnitude can be determined from Fig. 15-7 and the phase angle from Fig. 15-10. For this problem the phase angle contributions from the terms $1/(1 + j\omega)$ and $1/(1 - j\omega)$ cancel each other. The asymptotes for the Bode magnitude plot are shown in Fig. 15-29 along with a more accurate Bode magnitude plot. The Bode phase angle is determined solely from $\arg K_B = \arg(-4) - 180°$ and the zero at $\omega = 2$, as shown in Fig. 15-30.

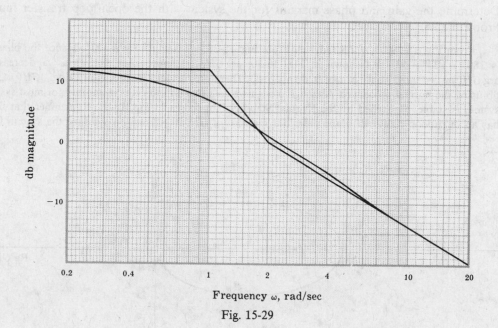

Frequency $\omega$, rad/sec

Fig. 15-29

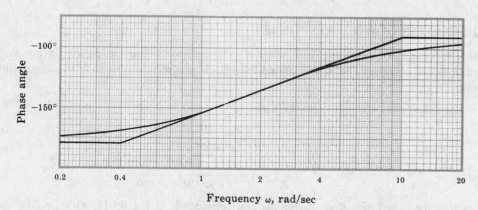

Frequency $\omega$, rad/sec

Fig. 15-30

## RELATIVE STABILITY

**15.9.** For the system with the open-loop transfer function of Problem 15.6, find $\omega_1$, $\omega_\pi$, the gain margin, and the phase margin.

Using the exact magnitude curve shown in Fig. 15-25, the gain crossover frequency is $\omega_1 = 0.62$. The phase crossover frequency $\omega_\pi$ is indeterminate because $\arg GH(j\omega)$ never crosses $-180°$. (See Fig. 15-26.) Arg $GH(j\omega_1) = \arg GH(j0.62)$ is $-129°$. Hence the phase margin is $-129° + 180° = 51°$. Since $\omega_\pi$ is indeterminate, the gain margin is also indeterminate.

**Mathcad**

**15.10.** Determine the gain and phase margins for the systems with the open-loop frequency response function of Problem 15.7.

From Fig. 15-27, $\omega_1 = 1.5$; and from Fig. 15-28, $\arg GH(j\omega_1) = -144°$. Therefore the phase margin is $180° - 144° = 36°$. From Fig. 15-28, $\omega_\pi = 3.2$; and the gain margin is read from Fig. 15-27 as $-20\log_{10}|GH(j\omega_\pi)| = 11$ db.

**Mathcad**

**15.11.** Determine the gain and phase margins for the system with the open-loop transfer function of Problem 15.8.

From Fig. 15-29, $\omega_1 = 2.3$ rad/sec. From Fig. 15-30, $\arg GH(j\omega_1) = -127°$. Hence the phase margin is $180° - 127° = 53°$. As shown in Fig. 15-30, $\arg GH(j\omega)$ approaches $-180°$ as $\omega$ decreases. Since $\arg GH(j\omega) = -180°$ only at $\omega = 0$, then $\omega_\pi = 0$. Therefore the gain margin is $-20\log_{10}|GH(j\omega_\pi)| = -12$ db using the normal procedure. Although a negative gain margin indicates instability for most systems, this system is stable, as verified by the Nyquist Stability Plot shown in Fig. 15-31. Remember that the system has an open-loop right-half plane pole; but the zero of $GH$ at $-2$ acts to stabilize the system for $K = 2$.

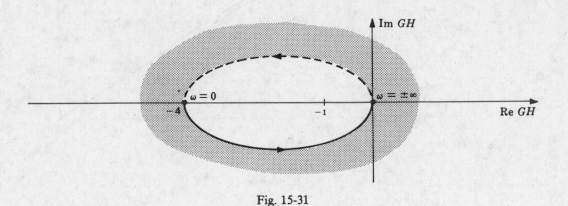

Fig. 15-31

## CLOSED-LOOP FREQUENCY RESPONSE

**15.12.** For the system of Example 15.7 with $H = 1$, determine the closed-loop frequency response function and compare the actual closed-loop Bode magnitude plot with the approximate one of Example 15.7.

For this system, $GH = 10/s(s + 1)$. Then

$$\frac{C}{R} = \frac{10}{s^2 + s + 10}$$

and

$$\frac{C}{R}(j\omega) = \frac{1}{1 + j\omega/10 - \omega^2/10}$$

Therefore the closed-loop Bode magnitude plot corresponds to Fig. 15-11, with $\zeta = 0.18$ and $\omega_n = 3.16$. From this plot the actual 3-db bandwidth is $\omega/\omega_n = 1.5$ in normalized form; hence, since $\omega_n = 3.16$, BW $= 1.5(3.16) = 4.74$ rad/sec. The approximate 3-db bandwidth determined from Fig. 15-20 of Example

15.7 is 3.7 rad/sec. Note that $\omega_n = 3.16$ rad/sec for the closed-loop system corresponds very well with $\omega_1 = 3.1$ rad/sec from Fig. 15-20. Thus the gain crossover frequency of the open-loop system corresponds very well with $\omega_n$ of the closed-loop system, although the approximate 3-db bandwidth determined above is not very accurate. The reason for this is that the approximate Bode magnitude plot of Fig. 15-20 does not show the peaking that occurs in the exact curve.

**15.13.** For the discrete-time system with open-loop frequency response function

$$GH(z) = \frac{3(z+1)\left(z+\frac{1}{3}\right)}{8z(z-1)\left(z+\frac{1}{2}\right)} \qquad H = 1$$

find the gain margin, phase margin, phase crossover angle, and gain crossover angle.

The Bode plots for this system are given in Figs. 15-32 and 15-33. The phase crossover angle $\omega_\pi T$ is determined from Fig. 15-33 as 1.74 rad. The corresponding gain margin is found on Fig. 15-32 as 11 db. The gain crossover angle $\omega_1 T$ is determined from Fig. 15-32 as 0.63 rad. The corresponding phase margin is found on Fig. 15-33 as 57°.

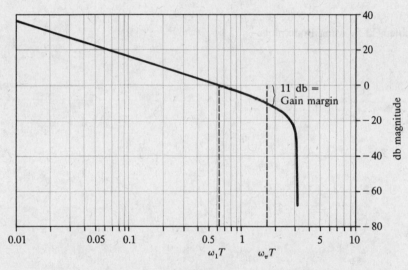

Angle $\omega T$, radians

Fig. 15-32

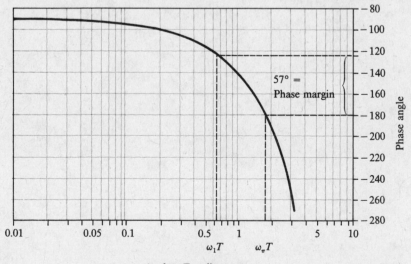

Angle $\omega T$, radians

Fig. 15-33

# Supplementary Problems

**15.14.** Construct the Bode plots for the open-loop frequency response function

$$GH(j\omega) = \frac{4(1 + j\omega/2)}{(j\omega)^2(1 + j\omega/8)(1 + j\omega/10)}$$

**15.15.** Construct the Bode plots and determine the gain and phase margins for the system with the open-loop frequency response function

$$GH(j\omega) = \frac{4}{(1 + j\omega)(1 + j\omega/3)^2}$$

**15.16.** Solve Problems 13.35 and 13.37 by constructing the Bode plots.

**15.17.** Work Problem 13.52 using Bode plots.

**15.18.** Work Problem 11.59 using Bode plots.

# Chapter 16

## Bode Design

### 16.1 DESIGN PHILOSOPHY

Design of a feedback control system using Bode techniques entails shaping and reshaping the Bode magnitude and phase angle plots until the system specifications are satisfied. These specifications are most conveniently expressed in terms of frequency-domain figures of merit such as gain and phase margin for the transient performance and the error constants (Chapter 9) for the steady state time-domain response.

Shaping the asymptotic Bode plots of continuous-time systems by adding cascade or feedback compensation is a relatively simple procedure. Bode plots for several common continuous-time compensation networks are presented in Sections 16.3, 16.4, and 16.5. With these graphs, the magnitude and phase angle contributions of a particular compensator can be added directly to the uncompensated system Bode plots. It is usually necessary to correct the asymptotic Bode plots in the final stages of design to accurately verify satisfaction of the performance specifications.

Since simple asymptotic Bode plots do not exist for discrete-time systems, the shaping and reshaping of Bode plots for discrete-time systems is usually not as simple and intuitive as for continuous-time systems. However, by transforming the discrete-time transfer function into the $w$-plane, design of discrete-time systems can be accomplished by continuous-time techniques.

### 16.2 GAIN FACTOR COMPENSATION

It is possible in some cases to satisfy all system specifications by simply adjusting the open-loop gain factor $K$. Adjustment of the gain factor $K$ does not affect the phase angle plot. It only shifts the magnitude plot up or down to correspond to the increase or decrease in $K$. The simplest procedure is to alter the db scale of the magnitude plot in accordance with the change in $K$ instead of replotting the curve. For example, if $K$ is doubled, the db scale should be shifted down by $20\log_{10}2 = 6.02$ db.

When working with continuous-time Bode plots, it is more convenient to use the Bode gain:

$$K_B = \frac{K\displaystyle\prod_{i=1}^{m} z_i}{\displaystyle\prod_{i=1}^{n} p_i}$$

where $-p_i$ and $-z_i$ are the finite poles and zeros of $GH$.

**EXAMPLE 16.1.** The Bode plots for

$$GH(j\omega) = \frac{K_B}{j\omega(1 + j\omega/2)}$$

are shown in Fig. 16-1 for $K_B = 1$.

The maximum amount $K_B$ may be increased to improve the system steady state performance without decreasing the phase margin below 45° is determined as follows. In Fig. 16-1, the phase margin is 45° if the gain crossover frequency $\omega_1$ is 2 rad/sec and the magnitude plot can be raised by as much as 9 db before $\omega_1$ becomes 2 rad/sec. Thus $K_B$ can be increased by 9 db without decreasing the phase margin below 45°.

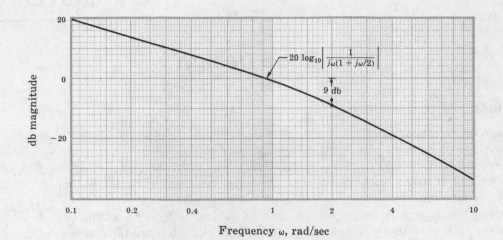

Fig. 16-1

 **16.3  LEAD COMPENSATION FOR CONTINUOUS-TIME SYSTEMS**

The lead compensator, presented in Sections 6.3 and 12.4, has the following Bode form frequency response function:

$$P_{\text{Lead}}(j\omega) = \frac{(a/b)(1+j\omega/a)}{1+j\omega/b} \qquad (16.1)$$

The Bode plots for this compensator, for various *lead ratios* $a/b$, are shown in Fig. 16-2. These graphs illustrate that addition of a cascade lead compensator to a system lowers the overall magnitude curve in the low-frequency region and raises the overall phase angle curve in the low-to-mid-frequency region. Other properties of the lead compensator are discussed in Section 12.4.

The amount of low-frequency attenuation and phase lead produced by the lead compensator depends on the lead ratio $a/b$. Maximum phase lead occurs at the frequency $\omega_m = \sqrt{ab}$ and is equal to

$$\phi_{\max} = \left(90 - 2\tan^{-1}\sqrt{a/b}\,\right) \text{ degrees} \qquad (16.2)$$

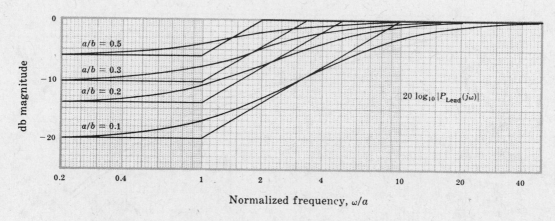

Normalized frequency, $\omega/a$

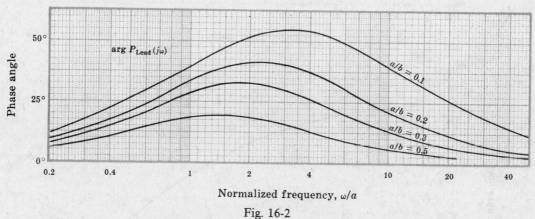

Normalized frequency, $\omega/a$

Fig. 16-2

Lead compensation is normally used to increase the gain and/or phase margins of a system or increase its bandwidth. An additional modification of the Bode gain $K_B$ is usually required with lead networks, as described in Section 12.4.

**EXAMPLE 16.2.** An uncompensated continuous-time system whose open-loop transfer function is

$$GH = \frac{24}{s(s+2)(s+6)} \qquad H = 1$$

is to be designed to meet the following performance specifications:

1.  when the input is a ramp with slope (velocity) $2\pi$ rad/sec, the steady state position error must be less than or equal to $\pi/10$ radians.

2.  $\phi_{PM} = 45° \pm 5°$.

3.  gain crossover frequency $\omega_1 \geq 1$ rad/sec.*

Lead compensation is appropriate, as previously outlined in detail in Example 12.4. Transforming $GH(j\omega)$ into Bode form,

$$GH(j\omega) = \frac{2}{j\omega(1+j\omega/2)(1+j\omega/6)}$$

we note that the Bode gain $K_B$ is equal to the velocity error constant $K_{v1} = 2$. The Bode plots for this system are shown in Fig. 16-3.

---

*When using Bode techniques, closed-loop system *bandwidth* specifications are often interpreted in terms of the gain crossover frequency $\omega_1$, which is easily determined from the Bode magnitude plot. The bandwidth and $\omega_1$ are not generally equivalent; but, when one increases or decreases, the other usually follows. As noted in Sections 10.4, and 15.8 and Problem 12.16, $\omega_1$ is often a reasonable approximation for the bandwidth.

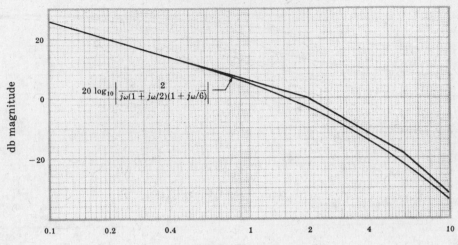

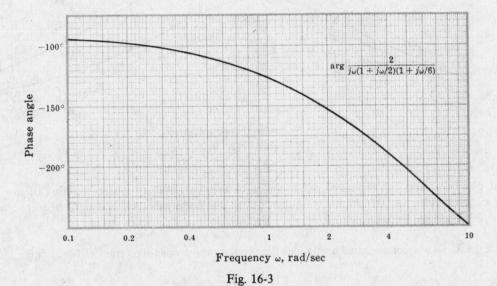

Fig. 16-3

The steady state error $e(\infty)$ is given by Equation (9.13) as $1/K_v$ for a unit ramp function input. Therefore, if $e(\infty) \leq \pi/10$ radians and the ramp has a slope of $2\pi$ instead of 1, then the required velocity error constant is

$$K_{v2} \geq \frac{2\pi}{\pi/10} = 20 \text{ sec}^{-1}$$

Thus a cascade amplifier with a gain of $\lambda = 10$, or 20 db, will satisfy the steady state specification. But this gain must be further increased after the lead network parameters are chosen, as described in Example 12.4. When the Bode gain is increased by 20 db, the gain margin is $-8$ db and the phase margin $-28°$, as read directly from the plots of Fig. 16-3. Therefore the lead compensator must be chosen to bring the phase margin to 45°. This requires a large amount of phase lead. Furthermore, since addition of the lead compensator must be accompanied by an increase in gain of $b/a$, the net effect is to increase the gain at mid and high frequencies, thus raising the gain crossover frequency. Hence a phase margin of 45° has to be established at a higher frequency, requiring even more phase lead. For these reasons we add two cascaded lead networks (with the necessary isolation to reduce loading effects, if required).

To determine the parameters of the lead compensator, assume that the Bode gain has been increased by 20 db so that the 0-db line is effectively lowered by 20 db. If we choose $b/a = 10$, then the lead compensator plus an additional Bode gain increase of $(b/a)^2$ for the two networks has the following combined form:

$$[10P_{\text{Lead}}(j\omega)]^2 = G_c(j\omega) = \frac{(1+j\omega/a)^2}{(1+j\omega/10a)^2}$$

Now we must choose an appropriate value for $a$. A useful method for improving system stability is to try to cross the 0-db line at a slope of $-6$ db/octave. Crossing at a slope of $-12$ db/octave usually results in too low a value for the phase margin. If $a$ is set equal to 2, a sketch of the asymptotes reveals that the 0-db line is crossed at $-12$ db/octave. If $a = 4$, the 0-db line is crossed at a slope of $-6$ db/octave. The Bode magnitude and phase angle plots for the system with $a = 4$ rad/sec are shown in Fig. 16-4. The gain margin is 14 db and the phase margin is $50°$. Thus the second specification is satisfied. The gain crossover frequency $\omega_1 = 14$ rad/sec is substantially higher than the value specified, indicating that the system will respond a good deal faster than required by the third specification. The compensated system block diagram is shown in Fig. 16-5. A properly designed amplifier may additionally serve the purpose of load-effect isolation if it is placed *between* the two lead networks.

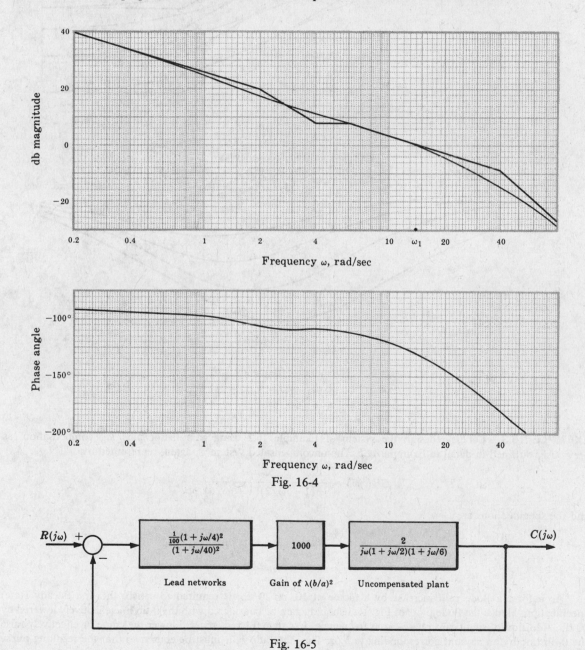

Fig. 16-4

Fig. 16-5

## 16.4 LAG COMPENSATION FOR CONTINUOUS-TIME SYSTEMS

The lag compensator, presented in Sections 6.3 and 12.5, has the following Bode form frequency response function:

$$P_{\text{Lag}}(j\omega) = \frac{1 + j\omega/b}{1 + j\omega/a} \qquad (16.3)$$

The Bode plots for the lag compensator, for several *lag ratios* $b/a$, are shown in Figure 16-6. The properties of this compensator are discussed in Section 12.5.

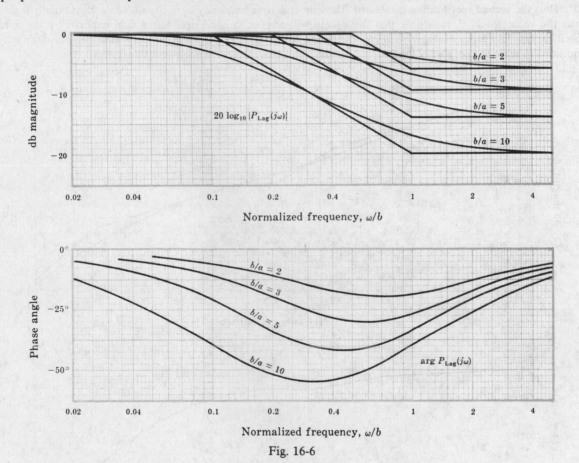

Fig. 16-6

**EXAMPLE 16.3.** Let us redesign the system of Example 16.2 using gain factor plus *lag* compensation, as previously outlined in detail in Example 12.5. The uncompensated system is, again, represented by

$$GH(j\omega) = \frac{2}{j\omega(1 + j\omega/2)(1 + j\omega/6)}$$

and the specifications are

1. $K_v \geq 20 \text{ sec}^{-1}$
2. $\phi_{\text{PM}} = 45° \pm 5°$
3. $\omega_1 \geq 1 \text{ rad/sec}$

As before, a Bode gain increase by a factor of 10, or 20 db, is required to satisfy the first (steady state) specification. Hence the Bode plots of Fig. 16-3 should again be considered with the 0-db line effectively lowered by 20 db. Addition of significant phase-lag at frequencies less than 0.1 rad/sec will lower the curve or effectively raise the 0-db line by an amount corresponding to $b/a$. Thus the ratio $b/a$ must be chosen so that the resulting phase margin is 45°. From the Bode phase angle plot (Fig. 16-3) we see that a 45° phase margin is obtained if the gain crossover frequency is $\omega_1 = 1.3$ rad/sec. From the Bode magnitude plot, this requires that the magnitude curve be lowered by 2 + 20 = 22 db. Thus a gain decrease of 22 db, or a factor of 14, is needed. This can be obtained using a lag compensator with $b/a = 14$. The actual location of the compensator is arbitrary as long as the phase shift produced at $\omega_1$ is negligible. Values of $a = 0.01$ and $b = 0.14$ rad/sec are adequate. The compensated system block diagram is shown in Fig. 16-7.

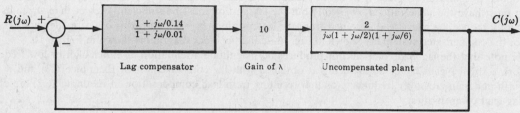

Fig. 16-7

## 16.5  LAG-LEAD COMPENSATION FOR CONTINUOUS-TIME SYSTEMS

It is sometimes desirable, as discussed in Section 12.6, to simultaneously employ both lead and lag compensation. Although one each of these two networks can be connected in series to achieve the desired effect, it is usually more convenient to mechanize the combined lag-lead compensator described in Example 6.6. This compensator can be constructed with a single $R$-$C$ network, as shown in Problem 6.14.

The Bode form of the frequency response function for the lag-lead compensator is

$$P_{LL}(j\omega) = \frac{(1 + j\omega/a_1)(1 + j\omega/b_2)}{(1 + j\omega/b_1)(1 + j\omega/a_2)}$$

with $b_1 > a_1$, $b_2 > a_2$ and $a_1 b_2 = b_1 a_2$. A typical Bode magnitude plot in which $a_1 > b_2$ is shown in Fig. 16-8. The Bode plots for a specific lag-lead compensator can be determined by combining the Bode plots for the lag portion from Fig. 16-6 with those for the lead portion from Fig. 16-2. Additional properties of the lag-lead compensator are discussed in Section 12.6.

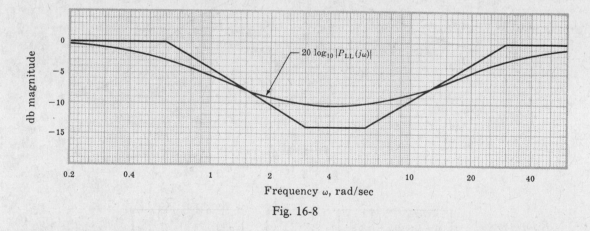

Fig. 16-8

**EXAMPLE 16.4.**  Let us redesign the system of Example 16.2 using lag-lead compensation. Suppose, for example, that we want the gain crossover frequency $\omega_1$ (approximate closed-loop bandwidth) to be greater than 2 rad/sec but less than 5 rad/sec, with all the other specifications the same as Example 16.2. For this application, we shall see that the lag-lead compensator has advantages over either lag or lead compensation. The uncompensated system is, again, represented by

$$GH(j\omega) = \frac{2}{j\omega(1 + j\omega/2)(1 + j\omega/6)}$$

The Bode plots are shown in Fig. 16-3. As in Example 16.2, a Bode gain increase of 20 db is required to satisfy the specification on steady state performance. Once again referring to Fig. 16-3 with the 0-db line shifted down by 20 db to correspond to the Bode gain increase, the parameters of the lag-lead compensator must be chosen to result in a gain crossover frequency between 2 and 5 rad/sec, with a phase margin of about 45°. The phase angle plot of Fig. 16-3 shows about $-188°$ phase angle at approximately 4 rad/sec. Thus we need about 53° phase lead to establish a 45° phase margin in that frequency range. Let us choose a lead ratio of $a_1/b_1 = 0.1$ to make sure we have enough phase lead. To place it in about the right frequency range, let $a_1 = 0.8$ and $b_1 = 8$ rad/sec. The lag portion must have the same ratio $a_2/b_2 = 0.1$; but the lag portion must be sufficiently lower than $a_1$ so as not to significantly reduce the phase lead obtained from the lead portion; $b_2 = 0.2$ and $a_2 = 0.02$ are adequate. The Bode plots for the compensated system are shown in Fig. 16-9; and the block diagram is shown in Fig. 16-10.

We note that the lag-lead compensator produces no magnitude attenuation at either high or low frequencies. Therefore a smaller gain factor adjustment (as obtained with lag compensation in Example 16.3) and a smaller bandwidth and gain crossover frequency (as that resulting from lead compensation in Example 16.2) are obtained using lag-lead compensation.

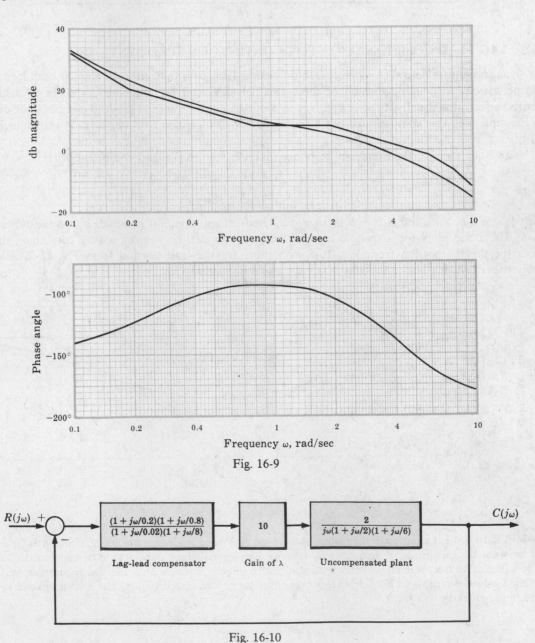

Fig. 16-9

Fig. 16-10

## 16.6  BODE DESIGN OF DISCRETE-TIME SYSTEMS

Bode design of discrete-time systems is based on the same philosophy as Bode design of continuous-time systems in that it entails shaping and reshaping the Bode magnitude and phase angle plots until the system specifications are met. But the effort involved can be substantially greater.

It is sometimes possible to satisfy specifications by simply adjusting the open-loop gain factor $K$, as described in Section 16.2 for continuous-time systems.

**EXAMPLE 16.5.**  Consider the discrete-time system of Example 15.4, with open-loop frequency response function

$$GH\left(e^{j\omega T}\right) = \frac{\frac{1}{100}\left(e^{j\omega T}+1\right)^2}{\left(e^{j\omega T}-1\right)\left(e^{j\omega T}+\frac{1}{3}\right)\left(e^{j\omega T}+\frac{1}{2}\right)}$$

and $H = 1$. Figures 16-11 and 16-12 are the Bode plots of $GH$, drawn by computer, which illustrate the gain and

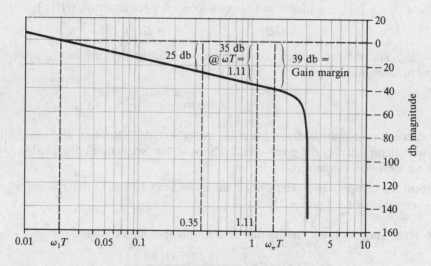

Fig. 16-11

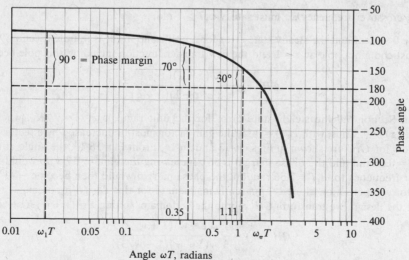

Fig. 16-12

phase margins and the gain and phase crossover frequencies. We now show that gain factor compensation alone can be used to satisfy the following system specifications:

1.  $\phi_{PM} \geq 30°$.
2.  10 db $\leq$ gain margin $\leq$ 15 db.

From Fig. 16-12 we see if $\omega_1 T$ can be increased to 1.11 rad, then $\phi_{PM} = 30°$. To accomplish this, the gain must be increased by 35 db, as shown in Fig. 16-11, resulting in a gain margin of $39 - 35 = 4$ db, which is too small. If we increase the gain by only 25 db (increase $K$ by a factor of 18), then $\omega_1 T = 0.35$ rad and the phase margin is 70°. Note that changing $K$ does not change $\omega_\pi T$.

For discrete-time system design specifications which cannot be satisfied by gain factor compensation alone, Bode design in the $z$-domain is not as straightforward as in the $s$-domain. Continuous-time system design methods can, however, be transferred to the design of discrete-time systems using the *w-transform*. Based on developments in Sections 10.7 and 15.9, the design algorithm is as follows:

1.  Substitute $(1 + w)/(1 - w)$ for $z$ in the open-loop transfer function $GH(z)$:

$$GH(z)|_{z=(1+w)/(1-w)} \equiv GH'(w)$$

2.  Set $w = j\omega_w$, and then transform critical frequencies in the performance specifications from the $z$- to the $w$-domain, using:

$$\omega_w = \tan \frac{\omega T}{2}$$

3.  Develop continuous-time compensation (as in Sections 16.3 through 16.5) such that the system in the $w$-domain satisfies the given specifications at the frequencies obtained in Step 2 (as if the $w$-domain were the $s$-domain).

4.  Transform the compensation elements obtained in Step 3 back to the $z$-domain to complete the design, using $w = (z - 1)/(z + 1)$.

**EXAMPLE 16.6.**  The unity feedback discrete-time system with open-loop transfer function

$$G(z) = GH(z) = \frac{3}{8} \frac{(z+1)\left(z+\frac{1}{3}\right)}{z\left(z+\frac{1}{2}\right)}$$

and sampling period $T = 0.1$ sec is to be compensated so that it meets the following specifications:

1.  The steady state error must be less than or equal to 0.02 for a unit ramp input.
2.  $\phi_{PM} \geq 30°$.
3.  The gain crossover frequency $\omega_1$ must satisfy $\omega_1 T \geq 1$ rad.

This is a type 0 system and the steady state error for a unit ramp input is infinite (Section 9.9). Therefore the compensation must contain a pole at $z = 1$ and the new transfer function including this pole becomes

$$GH'(z) = \frac{3}{8} \frac{(z+1)\left(z+\frac{1}{3}\right)}{z(z-1)\left(z+\frac{1}{2}\right)}$$

From the table in Section 9.9 the steady state error for the unit ramp is $e(\infty) = T/K_v$, where $K_v = GH(1) = \lim_{z \to 1}(z - 1)GH'(z) = \frac{2}{3}$. Thus, with $e(\infty) = 0.15$, the gain factor must be increased by a factor of 15/2 (17.5 db).

The Bode plots for $GH'$ are shown in Figs. 16-13 and 16-14. From Fig. 16-13, the angle at the gain crossover frequency is $\omega_1 T = 0.68$ rad and the phase margin is 56°. Increasing the gain by 17.5 db would move the angle at the gain crossover frequency to $\omega_1 T = 2.56$ rad, but the phase margin would then become $-41°$, destabilizing the system. Gain factor compensation alone is apparently inadequate for this design problem.

To complete the design, we transform $GH(z)$ into the $w$-domain, setting $z = (1 + w)/(1 - w)$ and forming

$$GH''(w) = \frac{1}{3} \frac{(1-w)(1+w/2)}{w(1+w)(1+w/3)}$$

The Bode plots for $GH''$ are shown in Figs. 16-15 and 16-16.

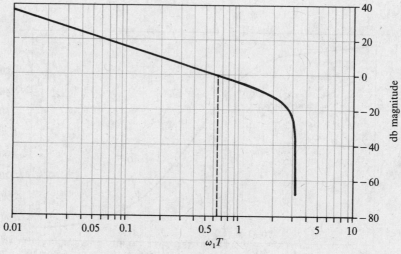

Angle $\omega T$, radians

Fig. 16-13

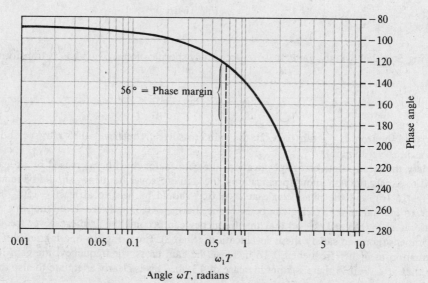

$56° = $ Phase margin

Angle $\omega T$, radians

Fig. 16-14

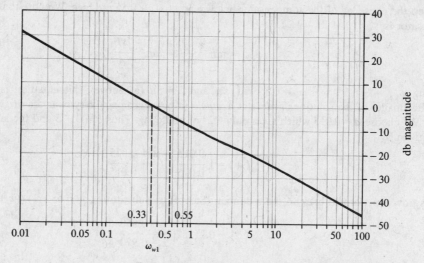

0.33    0.55

Frequency $\omega_w$, rad/sec

Fig. 16-15

397

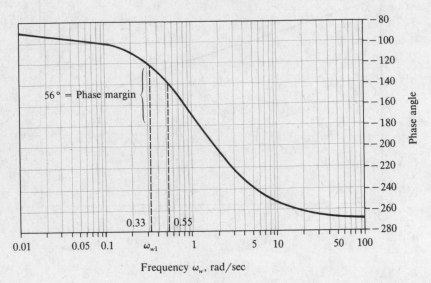

Fig. 16-16

Following Step 2 above, the gain crossover frequency specification $\omega_1 T \geq 1$ rad is transformed into the $w$-plane using

$$\omega_{w1} = \tan \frac{\omega_1 T}{2} \geq \tan \frac{1}{2} = 0.55 \text{ rad/sec}$$

From Fig. 16-15 [or from $\omega_{w1} = \tan(0.68/2)$] the gain crossover frequency is 0.35 rad/sec and the phase margin is 56° (as it was in the $z$-domain).

To satisfy the steady state error specification, the gain factor must be increased by at least 17.5 db (as noted earlier), and to satisfy the remaining specifications, the gain crossover frequency should be increased to at least 0.55 rad/sec (Fig. 16-16), and the phase angle at $\omega_w = 0.55$ should be held to at least $-150°$. This last requirement implies that no more than 6.5° of lag can be introduced at $\omega_w = 0.55$ rad/sec. Note that this requires about 4.3-db gain increase at $\omega_w = 0.55$ rad/sec so that this frequency can become the gain crossover frequency.

Lag compensation can satisfy these specifications (Step 3). From Fig. 16-6, a lag ratio of $b/a = 5$ provides 14 db of attenuation at higher frequencies. To increase the gain crossover frequency, the gain factor is increased by 18.3 db, so that at $\omega_w = 0.55$ there is a net increase of 4.3 db. This is clearly adequate to also satisfy the steady state error specification (17.5 db is needed).

Now the parameter $a$ in the lag ratio can be chosen to satisfy the phase margin requirement. As noted above, we must keep the phase lag of the compensator below 6.5° at $\omega_w = 0.55$ rad/sec. We note that the phase lag of the lag compensator is

$$\phi_{\text{Lag}} = \tan^{-1} \frac{\omega T}{b} - \tan^{-1} \frac{\omega T}{a}$$

Thus, setting $\phi_{\text{Lag}} = -6.5°$, $\omega = \omega_w = 0.55$ rad/sec and $b = 5a$ (as above), this equation is easily solved for $a$. Choosing the smaller of the solutions generates a *dipole* (a pole-zero pair) very near the origin of the $w$-plane, for $a = 0.0157$. We choose $a = 0.015$ which gives only 6.2° of phase lag. Thus $b = 0.075$ and the lag compensator in the $w$-plane is given by

$$P_{\text{Lag}}(w) = \left( \frac{0.015}{0.075} \right) \left( \frac{w + 0.075}{w + 0.015} \right)$$

$P_{\text{Lag}}$ is now transformed back to the $z$-domain (Step 4) by substituting $w = (z - 1)/(z + 1)$. The result is

$$P_{\text{Lag}}(z) = 0.21182 \left( \frac{z - 0.86046}{z - 0.97044} \right)$$

Combining this with the pole at $z = 1$ and the gain factor increase of 18.3 db (a gain factor ratio increase of 8.22), the complete compensation element $G_1(z)$ is

$$G_1(z) = 1.7417\left[\frac{z - 0.86046}{(z - 1)(z - 0.97044)}\right]$$

The compensated control system is shown in Fig. 16-17. Note that this design is quite similar to those developed for this same system and specifications in Examples 12.7 and 14.5.

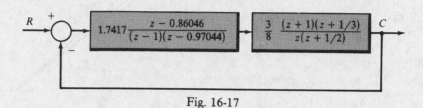

Fig. 16-17

# Solved Problems

## GAIN FACTOR COMPENSATION

**16.1.** Determine the maximum value for the Bode gain $K_B$ which will result in a gain margin of 6 db or more and a phase margin of 45° or more for the system with the open-loop frequency response function

$$GH(j\omega) = \frac{K_B}{j\omega(1 + j\omega/5)^2}$$

The Bode plots for this system with $K_B = 1$ are shown in Fig. 16-18.

The gain margin, measured at $\omega_\pi = 5$ rad/sec, is 20 db. Thus the Bode gain can be raised by as much as $20 - 6 = 14$ db and still satisfy the gain margin requirement. However, the Bode phase angle plot indicates that, for $\phi_{PM} \geq 45°$, the gain crossover frequency $\omega_1$ must be less than about 2 rad/sec. The magnitude curve can be raised by as much as 7.5 db before $\omega_1$ exceeds 2 rad/sec. Thus the maximum value of $K_B$ satisfying *both* specifications is 7.5 db, or 2.37.

**16.2.** Design the system of Problem 15.7, to have a phase margin of 55°.

The Bode phase angle plot in Fig. 15-28 indicates that the gain crossover frequency $\omega_1$ must be 0.9 rad/sec for 55° phase margin. From the Bode magnitude of Fig. 15-27, $K_B$ must be reduced by 6 db, or a factor of 2, to achieve $\omega_1 = 0.9$ rad/sec and hence $\phi_{PM} = 55°$.

## LEAD COMPENSATION

**16.3.** Show that the maximum phase lead of the lead compensator [Equation $(16.1)$] occurs at $\omega_m = \sqrt{ab}$ and prove Equation $(16.2)$.

The phase angle of the lead compensator is $\phi = \arg P_{Lead}(j\omega) = \tan^{-1}\omega/a - \tan^{-1}\omega/b$. Then

$$\frac{d\phi}{d\omega} = \frac{1}{a\left[1 + (\omega/a)^2\right]} - \frac{1}{b\left[1 + (\omega/b)^2\right]}$$

Setting $d\phi/d\omega$ equal to zero yields $\omega^2 = ab$. Thus the maximum phase lead occurs at $\omega_m = \sqrt{ab}$. Hence $\phi_{max} = \tan^{-1}\sqrt{b/a} - \tan^{-1}\sqrt{a/b}$. But since $\tan^{-1}\sqrt{b/a} = \pi/2 - \tan^{-1}\sqrt{a/b}$, we have $\phi_{max} = (90 - 2\tan^{-1}\sqrt{a/b})$ degrees.

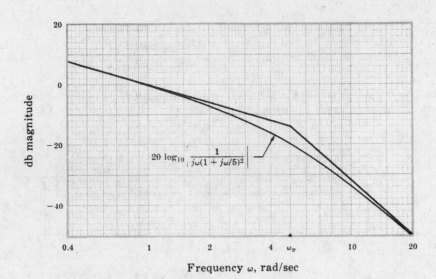

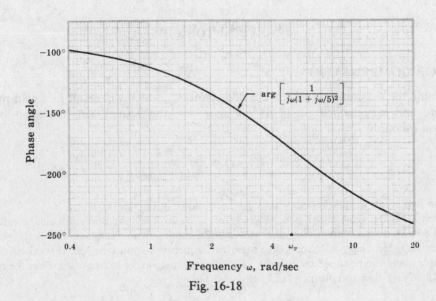

Fig. 16-18

**16.4.** What attenuation (magnitude) is produced by a lead compensator at the frequency of maximum phase lead $\omega_m = \sqrt{ab}$ ?

The attenuation factor is given by

$$\left| P_{\text{Lead}}\left( j\sqrt{ab} \right) \right| = \left| \frac{(a/b)\left(1 + j\sqrt{b/a}\right)}{\left(1 + j\sqrt{a/b}\right)} \right| = \frac{a}{b}\sqrt{\frac{1 + b/a}{1 + a/b}} = \sqrt{\frac{a}{b}}$$

**16.5.** Design compensation for the system

$$GH(j\omega) = \frac{8}{(1 + j\omega)(1 + j\omega/3)^2}$$

which will yield an overall phase margin of 45° and the same gain crossover frequency $\omega_1$ as the uncompensated system. The latter is essentially the same as designing for the same bandwidth, as discussed in Section 15.8.

The Bode plots for the uncompensated system are shown in Fig. 16-19.

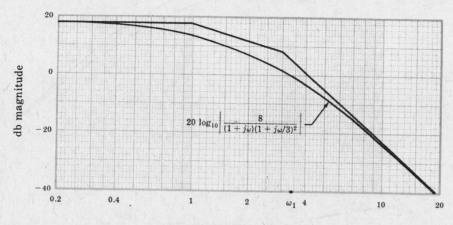

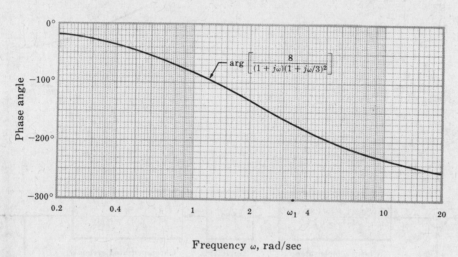

Fig. 16-19

The gain crossover frequency $\omega_1$ is 3.4 rad/sec and the phase margin is 10°. The specifications can be met with a cascade lead compensator and gain factor amplifier. Choosing $a$ and $b$ for the lead compensator is somewhat arbitrary, as long as the phase lead at $\omega_1 = 3.4$ is sufficient to raise the phase margin from 10° to 45°. However, it is often desirable, for economic reasons, to minimize the low-frequency attenuation obtained from the lead network by choosing the largest lead ratio $a/b < 1$ that will supply the required amount of phase lead. Assuming this is the case, the maximum lead ratio that will yield $45° - 10° = 35°$ phase lead is about 0.3 from Fig. 16-2. Solution of Equation (16.2) yields a value of $a/b = 0.27$. But we shall use $a/b = 0.3$ because we have the curves available for this value in Fig. 16-2. We want to choose $a$ and $b$ such that the maximum phase lead, which occurs at $\omega_m = \sqrt{ab}$, is obtained at $\omega_1 = 3.4$ rad/sec. Thus $\sqrt{ab} = 3.4$. Substituting $a = 0.3b$ into this equation and solving for $b$, we find $b = 6.2$ and $a = 1.86$. But this compensator produces $20 \log_{10} \sqrt{6.2/1.86} = 5.2$ db attenuation at $\omega_1 = 3.4$ rad/sec (see Problem 16.4). Thus an amplifier with a gain of 5.2 db, or 1.82, is required, in addition to the lead compensator, to maintain $\omega_1$ at 3.4 rad/sec. The Bode plots for the compensated system are shown in Fig. 16-20 and the block diagram in Fig. 16-21.

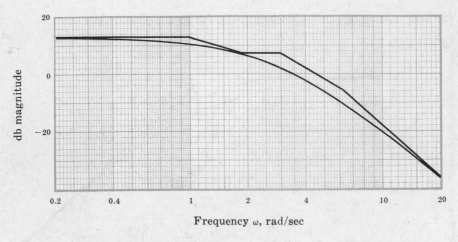

Fig. 16-20

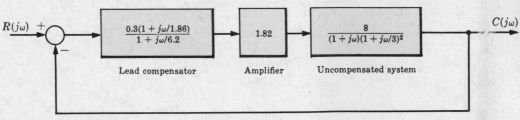

Fig. 16-21

## LAG COMPENSATION

**16.6.** What is the maximum phase lag produced by the lag compensator [Equation (*16.*

The phase angle of the lag compensator is

$$\arg P_{\text{Lag}}(j\omega) = \tan^{-1}\frac{\omega}{b} - \tan^{-1}\frac{\omega}{a} = -\arg P_{\text{Lead}}(j\omega)$$

Thus the maximum phase lag (negative phase angle) of the lag compensator is the same as the maximum phase lead of the lead compensator with the same values of $a$ and $b$. Hence the maximum also occurs at $\omega_m = \sqrt{ab}$ and, from Equation (*16.2*), we get

$$\phi_{\max} = \left(90 - 2\tan^{-1}\sqrt{\frac{a}{b}}\right) \text{ degrees}$$

Expressed in terms of the lag ratio $b/a$, this equation becomes

$$\phi_{max} = \left(2\tan^{-1}\sqrt{\frac{b}{a}} - 90\right) \text{ degrees}$$

**16.7.** Design compensation for the system of Problem 16.1 to satisfy the same specifications and, in addition, to have a gain crossover frequency $\omega_1$ less than or equal to 1 rad/sec and a velocity error constant $K_v > 5$.

The Bode plots for this system, shown in Fig. 16-18, indicate that $\omega_1 = 1$ rad/sec for $K_B = 1$. Hence $K_v = K_B = 1$ for $\omega_1 = 1$. The gain and phase margin requirements are easily met with any $K_B < 2.37$; but the steady state specification requires $K_v = K_B > 5$. Therefore a low-frequency cascade lag compensator with $b/a = 5$ can be used to increase $K_v$ to 5, while maintaining the crossover frequency and the gain and phase margins at their previous values. A lag compensator with $b = 0.5$ and $a = 0.1$ satisfies this requirements, as shown in Fig. 16-22.

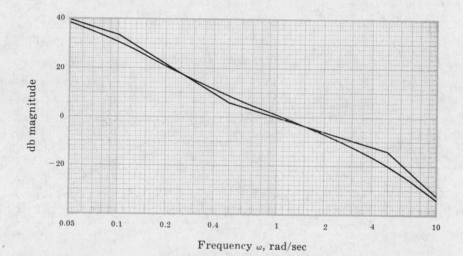

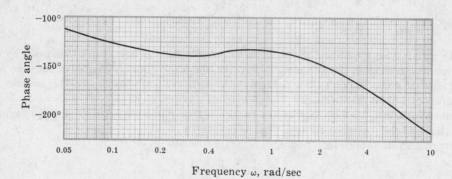

Fig. 16-22

The compensated open-loop frequency response function is $\dfrac{5(1 + j\omega/0.5)}{j\omega(1 + j\omega/0.1)(1 + j\omega/5)^2}$.

**16.8.** Design a discrete-time unity feedback system, with the fixed plant

$$G_2(z) = \frac{27}{64} \frac{(z + 1)^3}{\left(z + \frac{1}{2}\right)^3}$$

satisfying the specifications: (1) $K_p \geq 4$, (2) gain margin $\geq 12$ db, (3) phase margin $\geq 45°$.

The specification on the position error constant $K_p$ requires a gain factor increase of 4. This transfer function is transformed into the $w$-plane by letting $z = (1 + w)/(1 - w)$ thus forming

$$G_2'(w) = \frac{1}{(1 + w/3)^3}$$

The Bode plots for this system, with the gain factor increased by $20\log_{10}4 = 12$ db, are shown in Fig. 16-23.

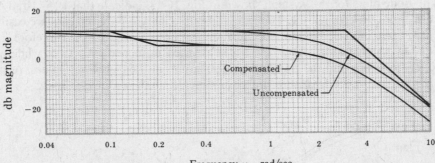

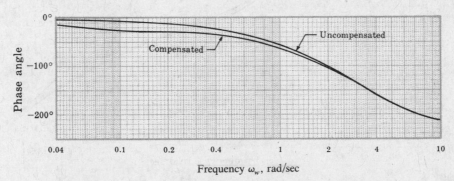

Fig. 16-23

The gain margin is 6 db and the phase margin is 30°. These margins can be increased by adding a lag compensator. To increase the gain margin by 12 db, the high-frequency magnitude must be reduced by 6 db. To raise the phase margin to 45°, $\omega_{w1}$ must be lowered to 3.0 rad/sec or less. This requires a magnitude attenuation of 3 db at that frequency. Therefore let us choose a lag ratio $b/a = 2$ to yield a high-frequency attenuation of $20\log_{10}2 = 6$ db. For $a = 0.1$ and $b = 0.2$ the phase margin is 65° and the gain margin is 12 db, as shown in the compensated Bode plots of Fig. 16-23.

The compensated open-loop frequency response function is

$$\frac{4(1 + j\omega_w/0.2)}{(1 + j\omega_w/0.1)(1 + j\omega_w)^3}$$

The compensation element

$$G_1'(w) = \frac{4(1 + w/0.2)}{1 + w/0.1}$$

is transformed back to the $z$-domain by letting $w = (z - 1)/(z + 1)$ thus forming

$$G_1(z) = \frac{24}{11}\frac{\left(z - \frac{2}{3}\right)}{\left(z - \frac{9}{11}\right)}$$

## LAG-LEAD COMPENSATION

**16.9.** Determine compensation for the system of Problem 16.5 that will result in a position error constant $K_p \geq 10$, $\phi_{PM} \geq 45°$ and the same gain crossover frequency $\omega_1$ as the uncompensated system.

The compensation determined in Problem 16.5 satisfies all the specifications except that $K_p$ is only 4.4 The lead compensator chosen in that problem has a low-frequency attenuation of 10.4 db, or a factor of 3.33. Let us replace the lead network with a lag-lead compensator, choosing $a_1 = 1.86$, $b_1 = 6.2$, and $a_2/b_2 = 0.3$. The low-frequency magnitude becomes $a_1 b_2/b_1 a_2 = 1$, or 0 db, and the attenuation produced by the lead network is erased, effectively raising $K_p$ for the system by a factor of 3.33 to 14.5. The lag portion of the compensator should be placed at frequencies sufficiently low so that the phase margin is not reduced below the specified value of 45°. This can be accomplished with $a_2 = 0.09$ and $b_2 = 0.3$. The compensated system block diagram is shown in Fig. 16-24. Note that an amplifier with a gain of 1.82 is included, as in Problem 16.5, to maintain $\omega_1 = 3.4$.

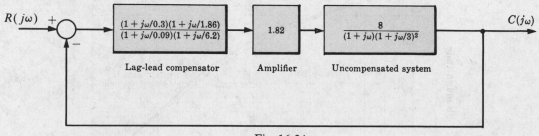

Fig. 16-24

The compensated Bode plots are shown in Fig. 16-25.

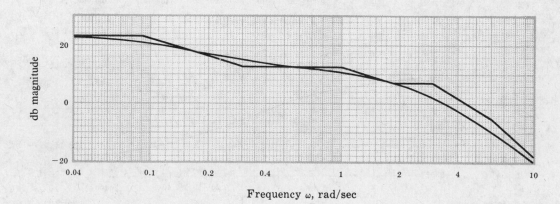

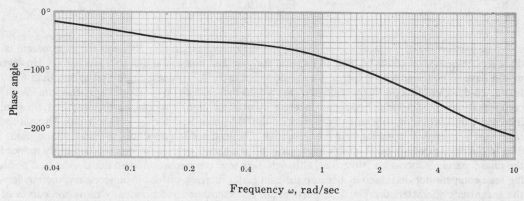

Fig. 16-25

**16.10.** Design cascade compensation for a unity feedback control system, with the plant

$$G_2(j\omega) = \frac{1}{j\omega(1 + j\omega/8)(1 + j\omega/20)}$$

to meet the following specifications:

| | | | |
|---|---|---|---|
| (1) | $K_v \geq 100$ | (3) | gain margin $\geq 10$ db |
| (2) | $\omega_1 \geq 10$ rad/sec | (4) | phase margin $\phi_{PM} \geq 45°$ |

To satisfy the first specification, a Bode gain increase by a factor of 100 is required since the uncompensated $K_v = 1$. The Bode plots for this system, with the gain increased to 100, are shown in Fig. 16-26.

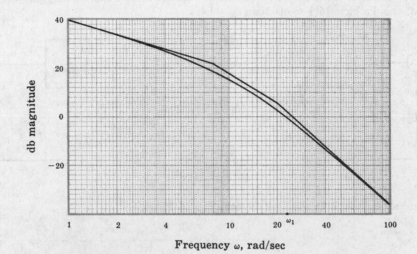

Fig. 16-26

The gain crossover frequency $\omega_1 = 23$ rad/sec, the phase margin is $-30°$, and the gain margin is $-12$ db. Lag compensation could be used to increase the gain and phase margins by reducing $\omega_1$. However, $\omega_1$ would have to be lowered to less than 8 rad/sec to achieve a 45° phase margin and to less than 6 rad/sec for a 10-db gain margin. Consequently, we would not satisfy the second specification. With lead compensation, an additional Bode gain increase by a factor of $b/a$ would be required and $\omega_1$ would be increased, thus requiring substantially more than the 75° phase lead for $\omega_1 = 23$ rad/sec. These disadvantages can be overcome using lag-lead compensation. The lead portion produces attenuation and phase lead. The frequencies at which these effects occur must be positioned near $\omega_1$ so that $\omega_1$ is slightly reduced and the phase margin is increased. Note that, although *pure* lead compensation increases $\omega_1$, the lead portion of lag-lead compensator decreases $\omega_1$ because the gain factor increase of $b/a$ is unnecessary, thereby lowering the magnitude characteristic. The lead portion can be determined independently using the curves of Fig. 16-2; but it must be kept in mind that, when the lag portion is included, the attenuation and phase lead may be somewhat reduced. Let us try a lead ratio of $a_1/b_1 = 0.1$, with $a_1 = 5$ and $b_1 = 50$. The maximum phase lead then occurs at 15.8 rad/sec. This enables the magnitude asymptote to cross the 0-db line with a slope of $-6$ db/octave (see Example 16.2). The compensated Bode plots are shown in Fig. 16-27 with $a_2$ and $b_2$ chosen as 0.1 and 1.0 rad/sec, respectively. The resulting parameters are $\omega_1 = 12$ rad/sec, gain margin $= 14$ db, and $\phi_{PM} = 52°$, as shown on the graphs. The compensated open-loop frequency response function is

$$\frac{100(1 + j\omega)(1 + j\omega/5)}{j\omega(1 + j\omega/0.1)(1 + j\omega/8)(1 + j\omega/20)(1 + j\omega/50)}$$

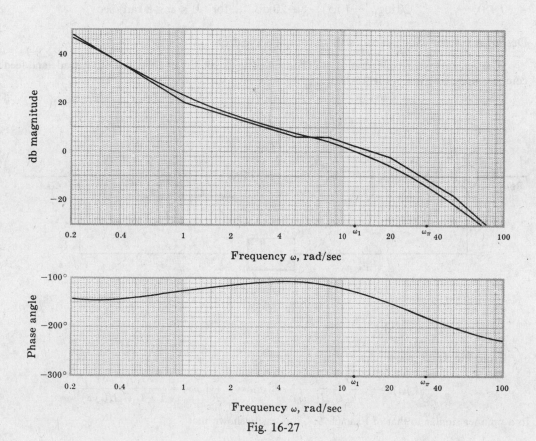

Fig. 16-27

## MISCELLANEOUS PROBLEM

**16.11.** The nominal frequency response function of a certain plant is

$$G_2(j\omega) = \frac{1}{j\omega(1 + j\omega/8)(1 + j\omega/20)}$$

A feedback control system must be designed to control the output of this plant for a certain application and it must satisfy the following frequency domain specifications:

(1)                               gain margin $\geq 6$ db

(2)                               phase margin $(\phi_{PM}) \geq 30°$

In addition, it is known that the "fixed" parameters of the plant may vary slightly during operation of the system. The effects of this variation on the system response must be minimized over the frequency range of interest, which is $0 \leq \omega \leq 8$ rad/sec, and the actual requirement can be interpreted as a specification on the sensitivity of $(C/R)(j\omega)$ with respect to $|G_2(j\omega)|$, that is,

$$(3) \qquad 20\left|\log_{10} S^{(C/R)(j\omega)}_{|G_2(j\omega)|}\right| \le -10 \text{ db} \qquad \text{for} \quad 0 \le \omega \le 8 \text{ rad/sec}$$

It is also known that the plant will be subjected to an uncontrollable, additive disturbance input, represented in the frequency domain by $U(j\omega)$. For this application, the system response to this disturbance input must be suppressed in the frequency range $0 \le \omega \le 8$ rad/sec. Therefore the design problem includes the additional constraint on the magnitude ratio of the output to the disturbance input given by

$$(4) \qquad 20\log_{10}\left|\frac{C}{U}(j\omega)\right| \le -20 \text{ db} \qquad \text{for} \quad 0 \le \omega \le 8 \text{ rad/sec}$$

Design a system which satisfies these four specifications.

The general system configuration, which includes the possibility of either or both cascade and feedback compensators, is shown in Fig. 16-28.

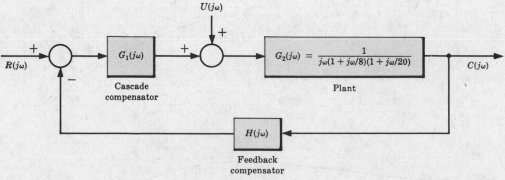

Fig. 16-28

From Fig. 16-28, we see that

$$\frac{C}{U}(j\omega) = \frac{G_2(j\omega)}{1 + G_1 G_2 H(j\omega)} \qquad \text{and} \qquad \frac{C}{R}(j\omega) = \frac{G_1 G_2(j\omega)}{1 + G_1 G_2 H(j\omega)}$$

In a manner similar to that of Example 9.7, it is easily shown that

$$S^{(C/R)(j\omega)}_{|G_2(j\omega)|} = \frac{1}{1 + G_1 G_2 H(j\omega)}$$

If we assume that $|G_1 G_2 H(j\omega)| \gg 1$ in the frequency range $0 \le \omega \le 8$ rad/sec (this inequality must be checked upon completion of the design and, if it is not satisfied, the compensation may have to be recomputed), then specification $(3)$ may be approximated by

$$20\log_{10}\left|S^{(C/R)(j\omega)}_{|G_2(j\omega)|}\right| \cong 20\log_{10}\left|\frac{1}{G_1 G_2 H(j\omega)}\right|$$

$$= -20\log_{10}|G_1 G_2 H(j\omega)| \le -10 \text{ db}$$

or $\qquad\qquad\qquad\qquad 20\log_{10}|G_1 G_2 H(j\omega)| \ge 10 \text{ db}$

Similarly, specification $(4)$ can be approximated by

$$20\log_{10}\left|\frac{C}{U}(j\omega)\right| \cong 20\log_{10}\frac{|G_2(j\omega)|}{|G_1 G_2 H(j\omega)|}$$

$$= 20\log_{10}|G_2(j\omega)| - 20\log_{10}|G_1 G_2 H(j\omega)| \le -20 \text{ db}$$

or $\qquad\qquad\qquad 20\log_{10}|G_1 G_2 H(j\omega)| \ge \left[20 + 20\log_{10}|G_2(j\omega)|\right] \text{ db}$

Specifications (3) and (4) can therefore be translated into the following combined form. We require that the open-loop frequency response, $G_1G_2H(j\omega)$, lie in a region on a Bode magnitude plot which simultaneously satisfies the two inequalities:

$$20 \log_{10}\left|G_1G_2H(j\omega)\right| \geq 10 \text{ db}$$

$$20 \log_{10}\left|G_1G_2H(j\omega)\right| \geq \left[20 + 20\log_{10}\left|G_2(j\omega)\right|\right] \text{ db}$$

This region lies above the broken line shown in the Bode magnitude plot in Fig. 16-29, which also includes Bode plots of $G_2(j\omega)$. The design may be completed by determining compensation which satisfies the gain and phase margin requirements, (1) and (2), subject to this magnitude constraint.

A 32-db increase in Bode gain, which is necessary at $\omega = 8$ rad/sec, would satisfy specifications (3) and (4), but not (1) and (2). Therefore a more complicated compensation is required. For a second trial, we find that the lag-lead compensation:

$$G_1H'(j\omega) = \frac{100(1+j\omega/2.5)(1+j\omega/0.25)}{(1+j\omega/25)(1+j\omega/0.025)}$$

results in a system with a gain margin of 6 db and $\phi_{PM} \cong 26°$, as shown in Fig. 16-29. We see from the figure that 10° to 15° more phase lead is necessary near $\omega = 25$ rad/sec and $|G_1H'(j\omega)|$ must be increased by at least 2 db in the neighborhood of $\omega = 8$ rad/sec to satisfy the magnitude constraint. If we introduce an additional lead network and increase the Bode gain to compensate for the low-frequency attenuation of the lead network, the compensation becomes

$$G_1H''(j\omega) = 300\left(\frac{1+j\omega/10}{1+j\omega/30}\right)\left[\frac{(1+j\omega/2.5)(1+j\omega/0.25)}{(1+j\omega/25)(1+j\omega/0.025)}\right]$$

This results in a gain margin of 7 db, $\phi_{PM} \cong 30°$, and satisfaction of specifications (3) and (4), as shown in Fig. 16-29. The assumption that $|G_1G_2H(j\omega)| \gg 1$ for $0 \leq \omega \leq 8$ rad/sec is easily shown to be justified by

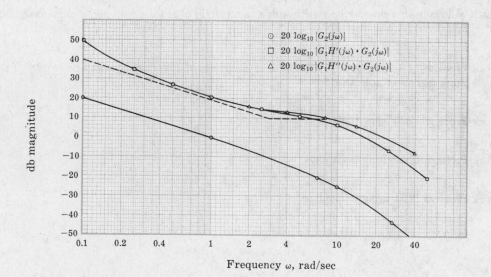

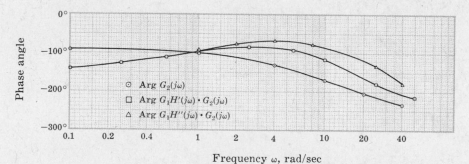

Fig. 16-29

calculating the actual values of the db magnitudes of

$$\left| S_{|G_2(j\omega)|}^{(C/R)(j\omega)} \right| \quad \text{and} \quad \left| \frac{C}{U}(j\omega) \right|$$

The compensator $G_1 H''(j\omega)$ can be divided between the forward and feedback paths, or put all in one path, depending on the form desired for $(C/R)(j\omega)$ if such a form is specified by the application.

# Supplementary Problems

**16.12.** Design a compensator for the system with the open-loop frequency response function

$$GH(j\omega) = \frac{20}{j\omega(1 + j\omega/10)(1 + j\omega/25)(1 + j\omega/40)}$$

to result in a closed-loop system with a gain margin of at least 10 db and a phase margin of at least 45°.

**16.13.** Determine a compensator for the system of Problem 16.1 which will result in the same gain and phase margins but with a crossover frequency $\omega_1$ of at least 4 rad/sec.

**16.14.** Design a compensator for the system with the open-loop frequency response function

$$GH(j\omega) = \frac{2}{(1 + j\omega)\left[1 + j\omega/10 - (\omega/4)^2\right]}$$

which will result in a closed-loop system with a gain margin of at least 6 db and a phase margin of at least 40°.

**16.15.** Work Problem 12.9 using Bode plots. Assume a maximum of 25% overshoot will be ensured if the system has a phase margin of at least 45°.

**16.16.** Work Problem 12.10 using Bode plots.

**16.17.** Work Problem 12.20 using Bode plots.

**16.18.** Work Problem 12.21 using Bode plots.

# Chapter 17

# Nichols Chart Analysis

## 17.1  INTRODUCTION

Nichols chart analysis, a frequency response method, is a modification of the Nyquist and Bode methods. The *Nichols chart* is essentially a transformation of the $M$- and $N$-circles on the Polar Plot (Section 11.12) into noncircular $M$ and $N$ contours on a db magnitude versus phase angle plot in rectangular coordinates. If $GH(\omega)$ represents the open-loop frequency response function of either a continuous-time or discrete-time system, then $GH(\omega)$ plotted on a Nichols chart is called a *Nichols chart plot* of $GH(\omega)$. The relative stability of the closed-loop system is easily obtained from this graph. The determination of absolute stability, however, is generally impractical with this method and either the techniques of Chapter 5 or the Nyquist Stability Criterion (Section 11.10) are preferred.

The reasons for using Nichols chart analysis are the same as those for the other frequency response methods, the Nyquist and Bode techniques, and are discussed in Chapters 11 and 15. The Nichols chart plot has at least two advantages over the Polar Plot: (1) a much wider range of magnitudes can be graphed because $|GH(\omega)|$ is plotted on a logarithmic scale; and (2) the graph of $GH(\omega)$ is obtained by algebraic summation of the individual magnitude and phase angle contributions of its poles and zeros. While both of these properties are also shared by Bode plots, $|GH(\omega)|$ and $\arg GH(\omega)$ are included on a single Nichols chart plot rather than on two Bode plots.

Nichols chart techniques are useful for directly plotting $(C/R)(\omega)$ and are especially applicable in system design, as shown in the next chapter.

## 17.2  db MAGNITUDE-PHASE ANGLE PLOTS

The polar form of both continuous-time and discrete-time open-loop frequency response functions is

$$GH(\omega) = |GH(\omega)| \big/\!\underline{\arg GH(\omega)} \tag{17.1}$$

***Definition 17.1:***  The **db magnitude-phase angle plot** of $GH(\omega)$ is a graph of $|GH(\omega)|$, in decibels, versus $\arg GH(\omega)$, in degrees, on rectangular coordinates with $\omega$ as a parameter.

**EXAMPLE 17.1.**  The db magnitude-phase angle plot of the continuous-time open-loop frequency response function

$$GH(j\omega) = 1 + j\omega = \sqrt{1 + \omega^2} \big/\!\underline{\tan^{-1}\omega}$$

is shown in Fig. 17-1.

## 17.3  CONSTRUCTION OF db MAGNITUDE-PHASE ANGLE PLOTS

The db magnitude-phase angle plot for either a continuous-time or discrete-time system can be constructed directly by evaluating $20\log_{10}|GH(\omega)|$ and $\arg GH(\omega)$ in degrees, for a sufficient number of values of $\omega$ (or $\omega T$) and plotting the results in rectangular coordinates with the log magnitude as the ordinate and the phase angle as the abscissa. Available software makes this a relatively simple process.

**EXAMPLE 17.2.**  The db magnitude-phase angle plot of the open-loop frequency response function

$$GH(e^{j\omega T}) = \frac{\frac{1}{100}(e^{j\omega T} + 1)^2}{(e^{j\omega T} - 1)(e^{j\omega T} + \frac{1}{3})(e^{j\omega T} + \frac{1}{2})}$$

is shown in Fig. 17-2. Note that $\omega T$ is the parameter along the curve.

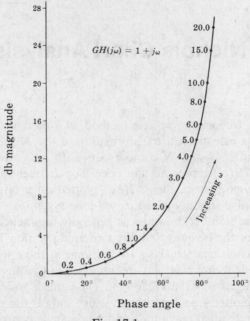

$$GH(j\omega) = 1 + j\omega$$

Fig. 17-1

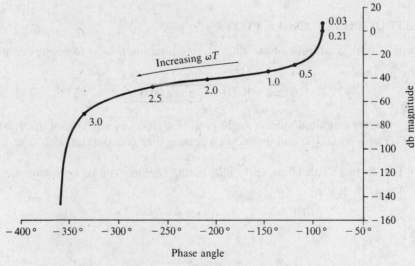

Fig. 17-2

A graphical approach to construction of db magnitude-phase angle plots is illustrated by examining the technique for continuous-time systems.

First write $GH(j\omega)$ in the *Bode form* (Section 15.3):

$$GH(j\omega) = \frac{K_B(1 + j\omega/z_1) \cdots (1 + j\omega/z_m)}{(j\omega)^l(1 + j\omega/p_1) \cdots (1 + j\omega/p_n)}$$

where $l$ is a nonnegative integer. For $K_B > 0$ [if $K_B < 0$, add $-180°$ to $\arg GH(j\omega)$],

$$20 \log_{10}|GH(j\omega)| = 20 \log_{10} K_B + 20 \log_{10}\left|1 + \frac{j\omega}{z_1}\right| + \cdots + 20 \log_{10}\left|1 + \frac{j\omega}{z_m}\right|$$

$$+ 20 \log_{10}\left|\frac{1}{(j\omega)^l}\right| + 20 \log_{10}\left|\frac{1}{1 + \frac{j\omega}{p_1}}\right| + \cdots + 20 \log_{10}\left|\frac{1}{1 + \frac{j\omega}{p_n}}\right| \quad (17.2)$$

$$\arg GH(j\omega) = \arg\left(1 + \frac{j\omega}{z_1}\right) + \cdots + \arg\left(1 + \frac{j\omega}{z_m}\right) + \arg\left[\frac{1}{(j\omega)^l}\right]$$

$$+ \arg\frac{1}{1 + \frac{j\omega}{p_1}} + \cdots + \arg\frac{1}{1 + \frac{j\omega}{p_n}} \quad (17.3)$$

Using Equations ($17.2$) and ($17.3$), the db magnitude-phase angle plot of $GH(j\omega)$ is generated by summing the db magnitudes and phase angles of the poles and zeros, or pairs of poles and zeros when they are complex conjugates.

The db magnitude-phase angle plot of $K_B$ is a straight line parallel to the phase angle axis. The ordinate of the straight line is $20 \log_{10} K_B$.

The db magnitude-phase angle plot for a *pole of order l at the origin*,

$$\frac{1}{(j\omega)^l} \quad (17.4)$$

is a straight line parallel to the db magnitude axis with an abscissa $-90l°$ as shown in Fig. 17-3. Note that the parameter along the curve is $\omega^l$.

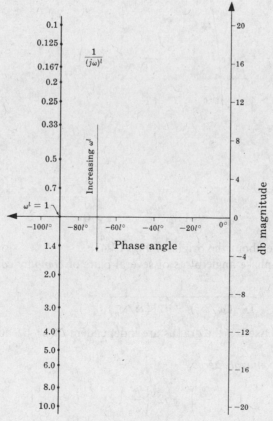

Fig. 17-3

The plot for a *zero of order l at the origin*,

$$(j\omega)^l \qquad\qquad\qquad\qquad (17.5)$$

is a straight line parallel to the db magnitude axis with an abscissa of $90l°$. The plot for $(j\omega)^l$ is the diagonal mirror image about the origin of the plot for $1/(j\omega)^l$. That is, for fixed $\omega$ the db magnitude and phase angle of $1/(j\omega)^l$ are the negatives of those for $(j\omega)^l$.

The db magnitude-phase angle plot for a *real pole*,

$$\frac{1}{1 + j\omega/p} \qquad p > 0 \qquad\qquad\qquad (17.6)$$

is shown in Fig. 17-4. The shape of the graph is independent of $p$ because the frequency parameter along the curve is normalized to $\omega/p$.

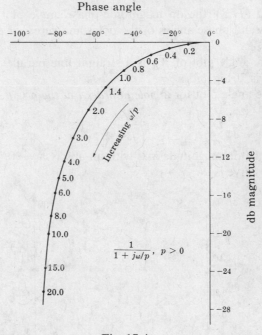

Fig. 17-4

The plot for a *real zero*,

$$1 + \frac{j\omega}{z} \qquad z > 0 \qquad\qquad\qquad (17.7)$$

is the diagonal mirror image about the origin of Fig. 17-4.

A set of db magnitude-phase angle plots of several pairs of *complex conjugate poles*,

$$\frac{1}{1 - (\omega/\omega_n)^2 + j2\zeta(\omega/\omega_n)} \qquad 0 < \zeta < 1 \qquad\qquad (17.8)$$

are shown in Fig. 17-5. For fixed $\zeta$, the graphs are independent of $\omega_n$ because the frequency parameter is normalized to $\omega/\omega_n$.

The plots for *complex conjugate zeros*,

$$1 - \left(\frac{\omega}{\omega_n}\right)^2 + j2\zeta\left(\frac{\omega}{\omega_n}\right) \qquad 0 < \zeta < 1 \qquad\qquad (17.9)$$

are diagonal mirror images about the origin of Fig. 17-5.

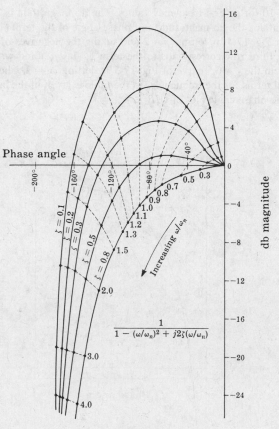

Fig. 17-5

**EXAMPLE 17.3.**   The db magnitude-phase angle plot of

$$GH(j\omega) = \frac{10(1 + j\omega/2)}{(1 + j\omega)\left[1 - (\omega/2)^2 + j\omega/2\right]}$$

is constructed by adding the db magnitudes and phase angles of the individual factors:

$$10 \qquad 1 + \frac{j\omega}{2} \qquad \frac{1}{1 + j\omega} \qquad \frac{1}{1 - (\omega/2)^2 + j\omega/2}$$

Tabulation of these factors is helpful, as in Table 17.1. The first row contains the db magnitude and phase

**Table 17.1**

| Term \ Frequency $\omega$ | 0 | 0.4 | 0.8 | 1.2 | 1.6 | 2 | 2.8 | 4 | 6 | 8 |
|---|---|---|---|---|---|---|---|---|---|---|
| 10 | 20 db 0° | 20 0° | 20 0° | 20 0° | 20 0° | 20 0° | 20 0° | 20 0° | 20 0° | 20 0° |
| $1 + \dfrac{j\omega}{2}$ | 0 db 0° | 0.2 11° | 0.6 21° | 1.3 31° | 2.2 39° | 3.0 45° | 4.7 54° | 7 63° | 10 71° | 12.3 76° |
| $\dfrac{1}{1 + j\omega}$ | 0 db 0° | −0.6 −21° | −2.2 −39° | −3.8 −50° | −5.4 −57° | −7.0 −63° | −9.4 −70° | −12.3 −76° | −15.7 −81° | −18.1 −83° |
| $\dfrac{1}{1 - (\omega/2)^2 + j\omega/2}$ | 0 db 0° | 0.3 −12° | 0.6 −26° | 0.9 −46° | 1.0 −68° | 0 −90° | −4.8 −126° | −12 −148° | −19.5 −160° | −24.5 −166° |
| Sum = $GH(j\omega)$ | 20 db 0° | 19.9 −22° | 19.0 −44° | 18.4 −65° | 17.8 −86° | 16 −108° | 10.5 −142° | 2.7 −161° | −5.2 −170° | −10.3 −173° |

angle of the Bode gain $K_B = 10$ for several frequency values. The db magnitude is 20 db and the phase angle is $0°$ for all $\omega$. The second row contains the db magnitude and phase angle of the term $(1 + j\omega/2)$ for the same values of $\omega$. These were obtained from Fig. 17-4 by letting $p = 2$ and taking the negatives of the values on the curve for the frequencies in the table. The third row corresponds to the term $1/(1 + j\omega)$ and was also obtained from Fig. 17-4. The fourth row was taken from the $\zeta = 0.5$ curve of Fig. 17-5 by letting $\omega_n = 2$. The sum of the db magnitudes and phase angles of the individual terms for the frequencies in the table is given in the last row. These values are plotted in Fig. 17-6, the db magnitude-phase angle plot of $GH(j\omega)$.

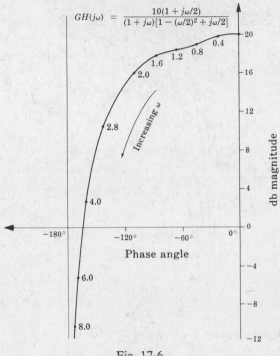

$$GH(j\omega) = \frac{10(1 + j\omega/2)}{(1 + j\omega)[1 - (\omega/2)^2 + j\omega/2]}$$

Fig. 17-6

## 17.4  RELATIVE STABILITY

The gain and phase margins for both continuous-time and discrete-time systems are readily determined from the db magnitude-phase angle plot of $GH(\omega)$.

The *phase crossover* frequency $\omega_\pi$ is the frequency at which the graph of $GH(\omega)$ intersects the $-180°$ line on the db magnitude-phase angle plot. The *gain margin in db* is given by

$$\text{gain margin} = -20\log_{10}|GH(\omega_\pi)| \text{ db} \qquad (17.10)$$

and is read directly from the db magnitude-phase angle plot.

The *gain crossover* frequency $\omega_1$ is the frequency at which the graph of $GH(\omega)$ intersects the 0-db line on the db magnitude-phase angle plot. The phase margin is given by

$$\text{phase margin} = \left[180 + \arg GH(\omega_1)\right] \text{ degrees}$$

and can be read directly from the db magnitude-phase angle plot.

In most cases, positive gain and phase margins will ensure stability of the closed-loop system; however, absolute stability should be established by some other means (for example, see Chapters 5 and 11) to guarantee that this is true.

**EXAMPLE 17.4.**  For a stable system, the db magnitude-phase angle plot of $GH(\omega)$ is shown in Fig. 17-7. The gain margin is 15 db and the phase margin is $35°$, as indicated.

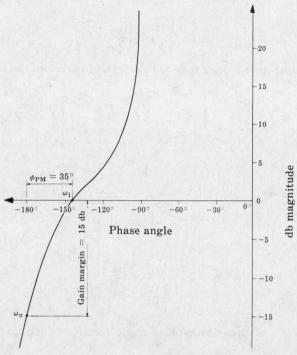

Fig. 17-7

## 17.5  THE NICHOLS CHART

The remaining discussion is restricted to either continuous-time or discrete-time unity feedback systems. The results are easily generalized to nonunity feedback systems, as illustrated in Example 17.9.

The closed-loop frequency response function of a unity feedback system may be written in polar form as

$$\frac{C}{R}(\omega) = \left|\frac{C}{R}(\omega)\right| \left/ \arg \frac{C}{R}(\omega) \right. = \frac{G(\omega)}{1 + G(\omega)} = \frac{|G(\omega)| \left/ \phi_G \right.}{1 + |G(\omega)| \left/ \phi_G \right.} \qquad (17.11)$$

where $\phi_G \equiv \arg G(\omega)$.

The locus of points on a db magnitude-phase angle plot for which

$$\left|\frac{C}{R}(\omega)\right| = M = \text{constant}$$

is defined by the equation

$$|G(\omega)|^2 + \frac{2M^2}{M^2 - 1}|G(\omega)|\cos\phi_G + \frac{M^2}{M^2 - 1} = 0 \qquad (17.12)$$

For a fixed value of $M$, this locus can be plotted in three steps: (1) choose numerical values for $|G(\omega)|$; (2) solve the resultant equations for $\phi_G$, excluding values of $|G(\omega)|$ for which $|\cos\phi_G| > 1$; and (3) plot the points obtained on a db magnitude-phase angle plot. Note that for fixed values of $M$ and $|G(\omega)|$, $\phi_G$ is multiple-valued because it appears in the equation as $\cos\phi_G$.

**EXAMPLE 17.5.**  The locus of points for which

$$\left|\frac{C}{R}(\omega)\right| = \sqrt{2}$$

or, equivalently,

$$20 \log_{10} \left| \frac{C}{R}(\omega) \right| = 3 \text{ db}$$

is graphed in Fig. 17-8. A similar curve appears at all odd multiples of 180° along the arg $G(\omega)$ axis.

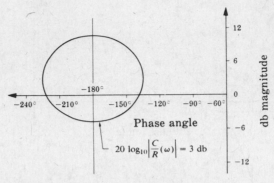

Fig. 17-8

The locus of points on a db magnitude-phase angle plot for which arg$(C/R)(\omega)$ is constant or, equivalently,

$$\tan \left[ \arg \frac{C}{R}(\omega) \right] = N = \text{constant}$$

is defined by the equation

$$|G(\omega)| + \cos \phi_G - \frac{1}{N} \sin \phi_G = 0 \qquad (17.13)$$

For a fixed value of $N$, this locus of points can be plotted in three steps: (1) choose values for $\phi_G$; (2) solve the resultant equations for $G(\omega)$; and (3) plot the points obtained on a db magnitude-phase angle plot.

**EXAMPLE 17.6.** The locus of points for which arg$(C/R)(\omega) = -60°$ or, equivalently,

$$\tan \left[ \arg \frac{C}{R}(\omega) \right] = -\sqrt{3}$$

is graphed in Fig. 17-9. A similar curve appears at all multiples of 180° along the arg $G(\omega)$ axis.

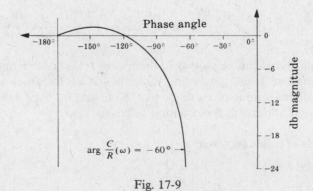

Fig. 17-9

**Definition 17.2:**    A **Nichols chart** is a db magnitude-phase angle plot of the loci of constant db magnitude and phase angle of $(C/R)(\omega)$, graphed as $|G(\omega)|$ versus $\arg G(\omega)$.

**EXAMPLE 17.7.** A Nichols chart is shown in Fig. 17-10. The range of $\arg G(\omega)$ on this chart is well suited to control system analysis.

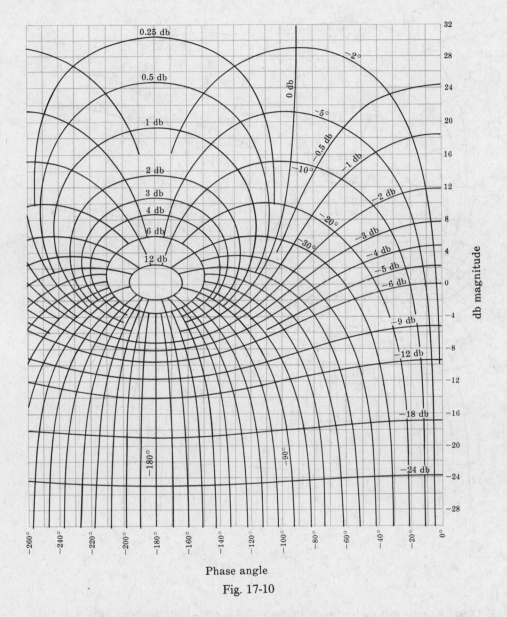

Phase angle

Fig. 17-10

**Definition 17.3:**    A **Nichols chart plot** is a db magnitude-phase angle plot of a frequency response function $P(\omega)$ superimposed on a Nichols chart.

## 17.6  CLOSED-LOOP FREQUENCY RESPONSE FUNCTIONS

The frequency response function $(C/R)(\omega)$ of a unity feedback system can be determined from the Nichols chart plot of $G(\omega)$. Values of $|(C/R)(\omega)|$ in db and $\arg(C/R)(\omega)$ are determined directly from the plot as the points where the graph of $G(\omega)$ intersects the graphs of loci of constant $|(C/R)(\omega)|$ and $\arg(C/R)(\omega)$.

**EXAMPLE 17.8.** The Nichols chart plot of $GH(\omega)$ for the continuous-time system of Example 17.3 is shown in Fig. 17-11. Assuming that it is a unity feedback system ($H = 1$), values for $|(C/R)(\omega)|$ and $\arg(C/R)(\omega)$ are obtained from this graph and plotted as a db magnitude-phase angle plot of $(C/R)(\omega)$ in Fig. 17-12.

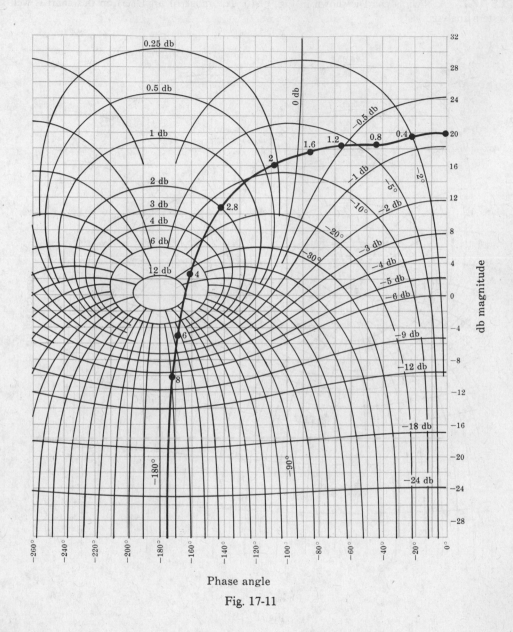

Phase angle

Fig. 17-11

**EXAMPLE 17.9.** Assume that the system in Example 17.3 is not a unity feedback system and that

$$G(\omega) = \frac{10}{(1 + j\omega)\left[1 - (\omega/2)^2 + j\omega/2\right]} \qquad H(\omega) = 1 + j\frac{\omega}{2}$$

Then

$$\frac{C}{R}(\omega) = \frac{1}{H(\omega)}\left[\frac{GH(\omega)}{1 + GH(\omega)}\right] = \frac{1}{H(\omega)}\left[\frac{G'(\omega)}{1 + G'(\omega)}\right]$$

where $G' \equiv GH$. The db magnitude-phase angle plot of $G'(\omega)/(1 + G'(\omega))$ was derived in Example 17.8 and is shown in Fig. 17-12. The db magnitude-phase angle plot of $(C/R)(\omega)$ can be obtained by point-by-point addition of the magnitude and phase angle of the pole $1/(1 + j\omega/2)$ to this graph. The magnitude and phase angle of $1/(1 + j\omega/2)$ can be obtained from Fig. 17-4 for $p = 2$. The result is shown in Fig. 17-13.

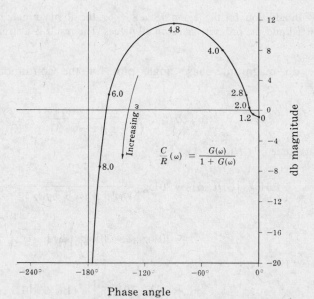

Fig. 17-12

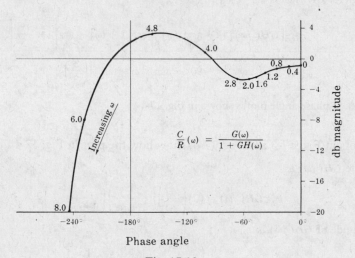

Fig. 17-13

# Solved Problems

## db MAGNITUDE-PHASE ANGLE PLOTS

**17.1.** Show that the db magnitude-phase angle plot for a pole of order $l$ at the origin of the $s$-plane, $1/(j\omega)^l$, is a straight line parallel to the db magnitude axis with an abscissa of $-90l°$ for $\omega \geq 0$.

In polar form, $j\omega = \omega\underline{/90°}$, $\omega \geq 0$. Therefore

$$\frac{1}{(j\omega)^l} = \frac{1}{\omega^l}\underline{/-90l°}, \qquad \omega \geq 0$$

$$20\log_{10}\left|\frac{1}{(j\omega)^l}\right| = 20\log_{10}\frac{1}{\omega^l} = -20\log_{10}\omega^l$$

and $\arg 1/(j\omega)^l = -90l°$. We see that $\arg 1/(j\omega)^l$ is independent of $\omega$; hence the abscissa of the plot is a constant $-90l°$. In addition, for the region $0 \le \omega \le +\infty$, the db magnitude ranges from $+\infty$ to $-\infty$. Thus the abscissa is fixed and the ordinate takes on all values. The result is a straight line as shown in Fig. 17-3.

**17.2.** Construct the db magnitude-phase angle plot for the continuous-time open-loop transfer function

$$GH = \frac{2}{s(1+s)(1+s/3)}$$

The db magnitude of $GH(j\omega)$ is

$$20\log_{10}|GH(j\omega)| = 20\log_{10}\frac{2}{|j\omega||1+j\omega||1+j\omega/3|}$$

$$= 20\log_{10}2 - 20\log_{10}\left[\omega\sqrt{1+\omega^2}\sqrt{1+\frac{\omega^2}{9}}\right]$$

$$= 6.02 - 10\log_{10}\left[\omega^2(1+\omega^2)\left(1+\frac{\omega^2}{9}\right)\right]$$

The phase angle of $GH(j\omega)$ is

$$\arg[GH(j\omega)] = -\arg[j\omega] - \arg[1+j\omega] - \arg\left[1+\frac{j\omega}{3}\right]$$

$$= -90° - \tan^{-1}\omega - \tan^{-1}\left(\frac{\omega}{3}\right)$$

The db magnitude-phase angle plot is shown in Fig. 17-14.

**17.3.** Using the plots in Fig. 17-3 and Fig. 17-4, show how the plot in Fig. 17-14 can be approximated.

We rewrite $GH(j\omega)$ as

$$GH(j\omega) = (2)\left(\frac{1}{j\omega}\right)\left(\frac{1}{1+j\omega}\right)\left(\frac{1}{1+j\omega/3}\right)$$

The db magnitude of $GH(j\omega)$ is

$$20\log_{10}|GH(j\omega)| = 20\log_{10}2 + 20\log_{10}\left|\frac{1}{j\omega}\right| + 20\log_{10}\left|\frac{1}{1+j\omega}\right| + 20\log_{10}\left|\frac{1}{1+j\omega/3}\right|$$

The phase angle is

$$\arg GH(j\omega) = \arg(2) + \arg\left(\frac{1}{j\omega}\right) + \arg\left(\frac{1}{1+j\omega}\right) + \arg\left(\frac{1}{1+j\omega/3}\right)$$

We now construct Table 17.2.

The first row contains the db magnitude and phase angle of the Bode gain $K_B = 2$. The second row contains the db magnitude and phase angle of the term $1/j\omega$ for several values of $\omega$. These are obtained from Fig. 17-3 by letting $l = 1$ and taking values from the curve for the frequencies given. The third row corresponds to the term $1/(1+j\omega)$ and is obtained from Fig. 17-4 for $p = 1$. The fourth row corresponds to the term $1/(1+j\omega/3)$ and is obtained from Fig. 17-4 for $p = 3$. Each pair of entries in the final row is obtained by summing the db magnitudes and phase angles in each column and corresponds to the db magnitude and phase angle of $GH(j\omega)$ for the given value of $\omega$. The values in the last row of this table are then plotted (with the exception of the first) and these points are joined graphically to generate an approximation of Fig. 17-14.

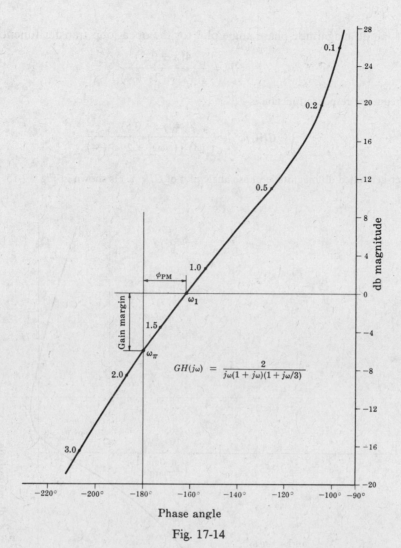

$$GH(j\omega) \; = \; \frac{2}{j\omega(1 + j\omega)(1 + j\omega/3)}$$

Fig. 17-14

**Table 17.2**

| Term \\ Frequency ω | 0 | 0.1 | 0.2 | 0.5 | 1.0 | 1.5 | 2.0 | 3.0 |
|---|---|---|---|---|---|---|---|---|
| $2$ | 6 db 0° | 6 0° | 6 0° | 6 0° | 6 0° | 6 0° | 6 0° | 6 0° |
| $\dfrac{1}{j\omega}$ | ∞ −90° | 20 −90° | 14 −90° | 6 −90° | 0 −90° | −3.6 −90° | −6 −90° | −9.5 −90° |
| $\dfrac{1}{1+j\omega}$ | 0 0° | −0.1 −5.5° | −0.3 −11° | −1.0 −26° | −3.0 −45° | −5.2 −57° | −7.0 −63° | −10 −72° |
| $\dfrac{1}{1+j\omega/3}$ | 0 0° | 0 −2° | −0.1 −4° | −0.2 −9° | −0.5 −17.5° | −1.0 −26° | −1.6 −33° | −3.0 −45° |
| Sum = $GH(j\omega)$ | ∞ −90° | 25.9 −97.5° | 19.6 −105° | 10.8 −125° | 2.5 −152.5° | −3.8 −173° | −8.6 −186° | −16.5 −207° |

**17.4.** Construct the db magnitude-phase angle plot for the open-loop transfer function

$$GH = \frac{4(s + 0.5)}{s^2(s^2 + 2s + 4)}$$

The frequency response function is

$$GH(j\omega) = \frac{4(j\omega + 0.5)}{(j\omega)^2((j\omega)^2 + 2j\omega + 4)}$$

A computer-generated db magnitude-phase angle plot of $GH(j\omega)$ is shown in Fig. 17-15.

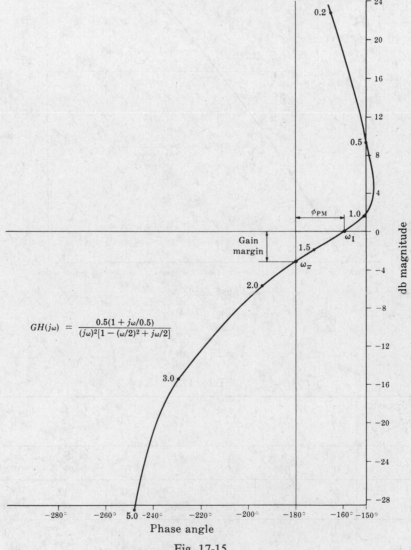

$$GH(j\omega) = \frac{0.5(1 + j\omega/0.5)}{(j\omega)^2[1 - (\omega/2)^2 + j\omega/2]}$$

Fig. 17-15

**17.5.** Construct the db magnitude-phase angle plot for the discrete-time open-loop transfer function

$$GH(z) = \frac{3}{8} \frac{(z + 1)(z + \frac{1}{3})}{(z - 1)(z + \frac{1}{2})}$$

The open-loop frequency response function is

$$GH(e^{j\omega T}) = \frac{3}{8} \frac{(e^{j\omega T} + 1)(e^{j\omega T} + \frac{1}{3})}{(e^{j\omega T} - 1)(e^{j\omega T} + \frac{1}{2})}$$

A computer-generated db magnitude-phase angle plot of $GH$ is shown in Fig. 17-16.

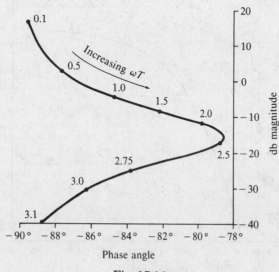

Fig. 17-16

## GAIN AND PHASE MARGINS

**17.6.** Determine the gain and phase margins for the system of Problem 17.2.

The db magnitude-phase angle plot for the open-loop transfer function of this system is given in Fig. 17-14 (Problem 17.2). We see that the curve crosses the 0-db line at a phase angle of $-162°$. Therefore the phase margin is $\phi_{PM} = 180° - 162° = 18°$.

(The gain crossover frequency $\omega_1$ is determined by interpolating along the curve between $\omega = 1.0$ and $\omega = 1.5$ which bound $\omega_1$ below and above, respectively. $\omega_1$ is approximately 1.2 rad/sec.)

The curve crosses the $-180°$ line at a db magnitude of $-6$ db. Hence gain margin $= -(-6) = 6$ db.

(The phase crossover frequency $\omega_\pi$ is determined by interpolating along the curve between $\omega = 1.5$ and $\omega = 2.0$ which bound $\omega_\pi$ below and above. $\omega_\pi$ is approximately 1.75 rad/sec.)

**17.7.** Determine the gain and phase margins for the system of Problem 17.4.

The db magnitude-phase angle plot for the open-loop transfer function of this system is given in Fig. 17-15 (Problem 17.4). We see that the curve crosses the 0-db line at a phase angle of $-159°$. Therefore the phase margin is $\phi_{PM} = 180° - 159° = 21°$.

(The gain crossover frequency $\omega_1$ is found by interpolating along the curve between $\omega = 1.0$ and $\omega = 1.5$ which bound $\omega_1$ below and above, respectively. $\omega_1$ is approximately 1.2 rad/sec.)

The curve crosses the $-180°$ line at a db magnitude of $-3.1$ db. Hence gain margin $= 3.1$ db.

(The phase crossover frequency $\omega_\pi$ is determined by interpolating between $\omega = 1.5$ and $\omega = 2.0$ which bound $\omega_\pi$ below and above, respectively. $\omega_\pi$ is approximately 1.7 rad/sec.)

**17.8.** Determine the gain and phase margins for the system defined by the open-loop frequency response function

$$GH(j\omega) = \frac{1 + j\omega/0.5}{j\omega\left[1 - (\omega/2)^2 + j\omega/2\right]}$$

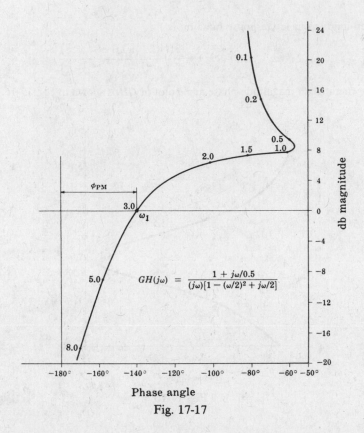

$$GH(j\omega) = \frac{1 + j\omega/0.5}{(j\omega)[1 - (\omega/2)^2 + j\omega/2]}$$

Phase angle

Fig. 17-17

The db magnitude-phase angle plot of $GH(j\omega)$ is given in Fig. 17-17. We see that the curve crosses the 0-db line at a phase angle of $-140°$. Hence the phase margin is $\phi_{PM} = 180° - 140° = 40°$.

The curve does not cross the $-180°$ line for the range of db magnitudes in Fig. 17-17. However, as $\omega \to \infty$,

$$GH(j\omega) \to \frac{j\omega/0.5}{-j\omega(\omega/2)^2} = \frac{8}{\omega^2}\bigg/\!\!-180°$$

The curve approaches the $-180°$ line asymptotically but does not cross it. Therefore the gain margin is indeterminate. This implies that the gain factor can be increased by any amount without producing instability.

**17.9.** Determine the gain and phase margins for the discrete-time system of Problem 17.5.

The db magnitude-phase angle plot for the open-loop transfer function of this system is given in Fig. 17-16 (Problem 17.5). We see that the curve crosses the 0-db line at a phase angle of $-87°$. Therefore the phase margin $\phi_{PM} = 180° - 87° = 93°$.

The gain crossover angle $\omega_1 T$ can be determined by interpolating along the curve between $\omega T = 0.5$ and $\omega T = 1.0$ which bound $\omega_1 T$ below and above, respectively. $\omega_1 T \simeq 0.6$ rad.

The curve never crosses the $-180°$ line, so the gain margin is indeterminate as is the phase crossover angle.

## NICHOLS CHART

**17.10.** Show that the locus of points on a db magnitude-phase angle plot for which the magnitude of the closed-loop frequency response $(C/R)(\omega)$ of either a continuous-time or discrete-time unity feedback system equals a constant $M$ is defined by Equation (17.12).

Using Equation (*17.11*), $|(C/R)(\omega)|$ can be written as

$$\left|\frac{C}{R}(\omega)\right| = \left|\frac{|G(\omega)|\underline{/\phi_G}}{1+|G(\omega)|\underline{/\phi_G}}\right|$$

Since $|G(\omega)|\underline{/\phi_G} = |G(\omega)|\cos\phi_G + j|G(\omega)|\sin\phi_G$, this can be rewritten as

$$\left|\frac{C}{R}(\omega)\right| = \left|\frac{|G(\omega)|\cos\phi_G + j|G(\omega)|\sin\phi_G}{1+|G(\omega)|\cos\phi_G + j|G(\omega)|\sin\phi_G}\right|$$

$$= \sqrt{\frac{|G(\omega)|^2\cos^2\phi_G + |G(\omega)|^2\sin^2\phi_G}{\left[1+|G(\omega)|\cos\phi_G\right]^2 + |G(\omega)|^2\sin^2\phi_G}} = \sqrt{\frac{|G(\omega)|^2}{1+2|G(\omega)|\cos\phi_G + |G(\omega)|^2}}$$

If we set the last expression equal to $M$, square both sides, and clear the fraction, we obtain

$$M^2\left[|G(\omega)|^2 + 2|G(\omega)|\cos\phi_G + 1\right] = |G(\omega)|^2$$

which can be written as

$$(M^2-1)|G(\omega)|^2 + 2M^2|G(\omega)|\cos\phi_G + M^2 = 0$$

Dividing by $(M^2-1)$, we obtain Equation (*17.12*), as required.

**17.11.** Show that the locus of points on a db magnitude-phase angle plot for which the tangent of the argument of the closed-loop frequency response function $(C/R)(\omega)$ of a unity feedback system equals a constant $N$ is defined by Equation (*17.13*).

Using Equation (*17.11*), $\arg(C/R)(\omega)$ can be written as

$$\arg\left[\frac{C}{R}(\omega)\right] = \arg\left[\frac{|G(\omega)|\underline{/\phi_G}}{1+|G(\omega)|\underline{/\phi_G}}\right]$$

Since $|G(\omega)|\underline{/\phi_G} = |G(\omega)|\cos\phi_G + j|G(\omega)|\sin\phi_G$,

$$\arg\left[\frac{C}{R}(\omega)\right] = \arg\left[\frac{|G(\omega)|\cos\phi_G + j|G(\omega)|\sin\phi_G}{1+|G(\omega)|\cos\phi_G + j|G(\omega)|\sin\phi_G}\right]$$

Multiplying numerator and denominator of the term in brackets by the complex conjugate of the denominator yields

$$\arg\left[\frac{C}{R}(\omega)\right] = \arg\left[\frac{(|G(\omega)|\cos\phi_G + j|G(\omega)|\sin\phi_G)(1+|G(\omega)|\cos\phi_G - j|G(\omega)|\sin\phi_G)}{(1+|G(\omega)|\cos\phi_G)^2 + |G(\omega)|^2\sin^2\phi_G}\right]$$

Since the denominator of the term in the last brackets is real, $\arg[(C/R)(\omega)]$ is determined by the numerator only. That is,

$$\arg\left[\frac{C}{R}(\omega)\right] = \arg\left[(|G(\omega)|\cos\phi_G + j|G(\omega)|\sin\phi_G)(1+|G(\omega)|\cos\phi_G - j|G(\omega)|\sin\phi_G)\right]$$

$$= \arg\left[|G(\omega)|\cos\phi_G + |G(\omega)|^2 + j|G(\omega)|\sin\phi_G\right]$$

using $\cos^2\phi_G + \sin^2\phi_G = 1$. Therefore

$$\tan\left[\arg\frac{C}{R}(\omega)\right] = \frac{|G(\omega)|\sin\phi_G}{|G(\omega)|\cos\phi_G + |G(\omega)|^2}$$

Equating this to $N$, cancelling the common $|G(\omega)|$ term and clearing the fraction, we obtain

$$N\left[\cos\phi_G + |G(\omega)|\right] = \sin\phi_G$$

which can be rewritten in the form of Equation (*17.13*), as required.

**17.12.** Construct the db magnitude-phase angle plot of the locus defined by Equation (*17.12*) for db magnitude of $(C/R)(\omega)$ equal to 6 db.

$20 \log_{10}|(C/R)(\omega)| = 6$ db implies that $|(C/R)(\omega)| = 2$. Therefore we let $M = 2$ in Equation (17.12) and obtain

$$|G(\omega)|^2 + \frac{8}{3}|G(\omega)|\cos\phi_G + \frac{4}{3} = 0$$

as the equation defining the locus. Since $|\cos\phi_G| \leq 1$, $|G(\omega)|$ may take on only those values for which this constraint is satisfied. To determine bounds of $|G(\omega)|$, we let $\cos\phi_G$ take on its two extreme values of plus and minus unity. For $\cos\phi_G = 1$, the locus equation becomes

$$|G(\omega)|^2 + \frac{8}{3}|G(\omega)| + \frac{4}{3} = 0$$

with solutions $|G(\omega)| = -2$ and $|GH(\omega)| = -\frac{2}{3}$. Since an absolute value cannot be negative, these solutions are discarded. This implies that the locus does not exist on the 0° line (in general, any line which is a multiple of 360°), which corresponds to $\cos\phi_G = 1$.

For $\cos\phi_G = -1$, the locus equation becomes

$$|G(\omega)|^2 - \frac{8}{3}|G(\omega)| + \frac{4}{3} = 0$$

with solutions $|G(\omega)| = 2$ and $|G(\omega)| = \frac{2}{3}$. These are valid solutions for $|G(\omega)|$ and are the extreme values which $|G(\omega)|$ can assume.

Solving the locus equation for $\cos\phi_G$, we obtain

$$\cos\phi_G = \frac{-\left[\frac{4}{3} + |G(\omega)|^2\right]}{\frac{8}{3}|G(\omega)|}$$

The curves obtained from this relationship are periodic with period 360°. The plot is restricted to a single cycle in the vicinity of the $-180°$ line and is obtained by solving for $\phi_G$ at several values of $|G(\omega)|$ between the bounds 2 and $\frac{2}{3}$. The results are given in Table 17.3

Note that there are two values of $\phi_G$ whenever $|\cos\phi_G| < 1$. The resulting plot is shown in Fig. 17-18.

**Table 17.3**

| $|G(\omega)|$ | $20\log_{10}|G(\omega)|$ | $\cos\phi_G$ | $\phi_G$ | |
|---|---|---|---|---|
| 2.0 | 6 db | $-1$ | $-180°$ | — |
| 1.59 | 4 | $-0.910$ | $-204.5°$ | $-155.5°$ |
| 1.26 | 2 | $-0.867$ | $-209.9°$ | $-150.1°$ |
| 1.0 | 0 | $-0.873$ | $-209.2°$ | $-150.8°$ |
| 0.79 | $-2$ | $-0.928$ | $-201.9°$ | $-158.1°$ |
| 0.67 | $-3.5$ | $-1$ | $-180°$ | — |

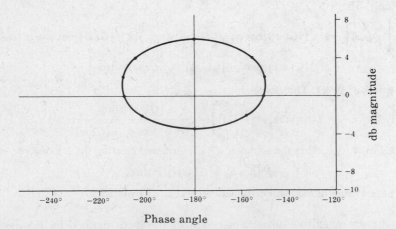

Fig. 17-18

**17.13.** Construct the db magnitude-phase angle plot of the locus defined by Equation (*17.13*) for $\tan[\arg(C/R)(\omega)] = N = -\infty$,

$\tan[\arg(C/R)(\omega)] = -\infty$ implies that $\arg(C/R)(\omega) = -90 + k360°$, $k = 0, \pm 1, \pm 2, \ldots$, or $\arg(C/R)(\omega) = -270° + k360°$, $k = 0, \pm 1, \pm 2, \ldots$. We will plot only the cycle between $-360°$ and $0°$, which corresponds to $k = 0$. Setting $N = -\infty$ in Equation (*17.13*), we obtain the locus equation

$$|G(\omega)| + \cos\phi_G = 0 \qquad \text{or} \qquad \cos\phi_G = -|G(\omega)|$$

Since $|\cos\phi_G| \le 1$, the locus exists only for $0 \le |G(\omega)| \le 1$ or, equivalently,

$$-\infty \le 20\log_{10}|G(\omega)| \le 0$$

To obtain the plot, we use the locus equation to calculate values of db magnitude of $G(\omega)$ corresponding to several values of $\phi_G$. The results of these calculations are given in Table 17.4. The desired plot is shown in Fig. 17-19.

**Table 17.4**

| $\phi_G$ | | $\cos\phi_G$ | $|G(\omega)|$ | $20\log_{10}|G(\omega)|$ |
|---|---|---|---|---|
| $-180°$ | — | $-1$ | 1 | 0 db |
| $-153°$ | $-207°$ | $-0.893$ | 0.893 | $-1.0$ |
| $-135°$ | $-222.5°$ | $-0.707$ | 0.707 | $-3$ |
| $-120°$ | $-240°$ | $-0.5$ | 0.5 | $-6$ |
| $-110.7°$ | $-249.3°$ | $-0.354$ | 0.354 | $-9$ |
| $-104.5°$ | $-255.5°$ | $-0.25$ | 0.25 | $-12$ |
| $-100.3°$ | $-259.8°$ | $-0.178$ | 0.178 | $-15$ |

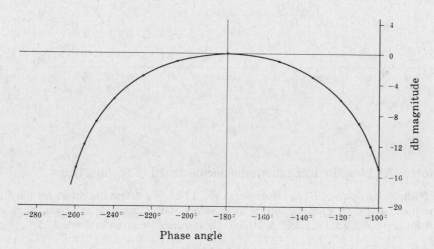

Phase angle

Fig. 17-19

## CLOSED-LOOP FREQUENCY RESPONSE FUNCTIONS

**17.14.** Construct the db magnitude-phase angle plot of the closed-loop frequency response function $(C/R)(j\omega)$ of the unity feedback system whose open-loop transfer function is

$$G = \frac{2}{s(1+s)(1+s/3)}$$

$$\frac{C}{R}(j\omega) = \frac{G(j\omega)}{1+G(j\omega)} = \frac{6}{(j\omega)^3 + 4(j\omega)^2 + 3j\omega + 6} = \frac{6}{(6-4\omega^2)+j(3\omega-\omega^3)}$$

Therefore

$$20\log_{10}\left|\frac{C}{R}(j\omega)\right| = 10\log_{10}\left|\frac{C}{R}(j\omega)\right|^2 = 10\log_{10}\frac{36}{\left(6-4\omega^2\right)^2+\left(3\omega-\omega^3\right)^2}$$

and

$$\arg\left[\frac{C}{R}(j\omega)\right] = -\tan^{-1}\frac{3\omega-\omega^3}{6-4\omega^2}$$

A computer-generated db magnitude-phase angle plot of $(C/R)(j\omega)$ is shown by the solid line in Fig. 17-20.

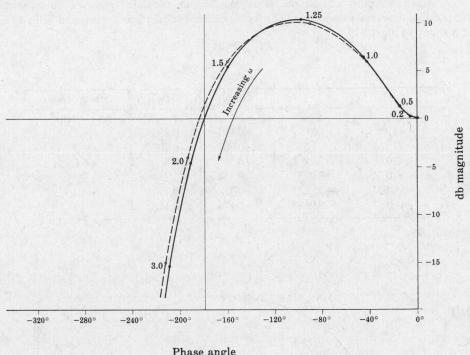

Phase angle

Fig. 17-20

**17.15.** Solve Problem 17.14 again, using the technique discussed in Section 17.6.

Mathcad

The Nichols chart plot of $G(j\omega)$ is shown in Fig. 17-21. We determine values for the db magnitude of $|(C/R)(j\omega)|$ and $\arg[(C/R)(j\omega)]$ by interpolating values of db magnitude and phase angle on the Nichols chart plot for $\omega = 0, 0.2, 0.5, 1.0, 1.25, 1.5, 2.0, 3.0$. These values are given in Table 17.5.

**Table 17.5**

| $\omega$ | $20\log_{10}\left|\dfrac{C}{R}(j\omega)\right|$ | $\arg\left[\dfrac{C}{R}(j\omega)\right]$ |
|:---:|:---:|:---:|
| 0 | 0 db | 0° |
| 0.2 | 0.2 | −6° |
| 0.5 | 1.2 | −15° |
| 1.0 | 6.0 | −42° |
| 1.25 | 10.0 | −90° |
| 1.5 | 6.0 | −155° |
| 2.0 | −4.0 | −194° |
| 3.0 | −15.0 | −212° |

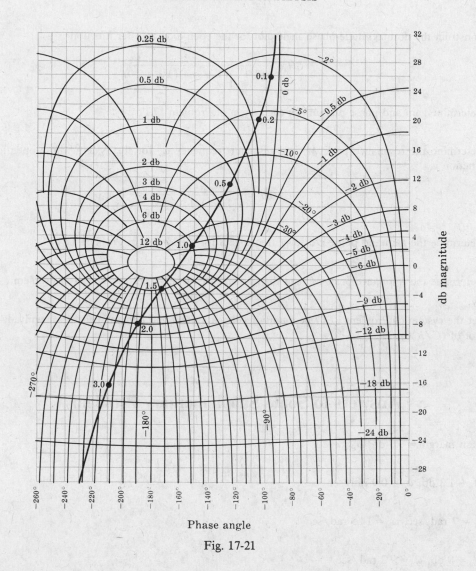

Fig. 17-21

The db magnitude-phase angle plot of $(C/R)(j\omega)$, graphed using the values in the table, is illustrated by the broken line in Fig. 17-20. The differences between the two curves is due to the interpolation necessary to obtain values of db magnitude and phase angle.

# Supplementary Problems

**17.16.** Construct the db magnitude-phase angle plot for the open-loop transfer function

$$GH = \frac{5(s+2)}{s(s+3)(s+5)}$$

**17.17.** Construct the db magnitude-phase angle plot for the open-loop transfer function

$$GH = \frac{10}{s(1+s/5)(1+s/50)}$$

**17.18.** Construct the db magnitude-phase angle plot for the open-loop transfer function

$$GH = \frac{1 + s/2}{s(1+s)(1+s/4)(1+s/20)}$$

**17.19.** Determine gain and phase margins for the system of Problem 17.17.

**17.20.** Determine the resonance peak $M_p$ and resonant frequency $\omega_p$ for the system whose open-loop transfer function is

$$GH = \frac{1}{s(1+s)(1+s/4)}$$

**17.21.** Determine the gain and phase crossover frequencies for the system of Problem 17.17.

**17.22.** Determine the resonance peak $M_p$ and the resonant frequency $\omega_p$ of the system in Problem 17.17.

**17.23.** Let the system of Problem 17.17 be a unity feedback system and construct the db magnitude-phase angle plot of $(C/R)(j\omega)$.

# Answers to Some Supplementary Problems

**17.19.** Gain margin = 9.5 db, $\phi_{PM} = 25°$

**17.20.** $M_p = 1.3$ db, $\omega_p = 0.9$ rad/sec

**17.21.** $\omega_1 = 7$ rad/sec, $\omega_\pi = 14.5$ rad/sec

**17.22.** $M_p = 8$ db, $\omega_p = 7.2$ rad/sec

# Chapter 18

## Nichols Chart Design

### 18.1 DESIGN PHILOSOPHY

Design by analysis in the frequency domain using Nichols chart techniques is performed in the same general manner as the design methods described in previous chapters: appropriate compensation networks are introduced in the forward and/or feedback paths and the behavior of the resulting system is critically analyzed. In this manner, the Nichols chart plot is shaped and reshaped until the performance specifications are met. These specifications are most conveniently expressed in terms of frequency-domain figures of merit such as gain and phase margin for transient performance and the error constants (Chapter 9) for the steady state time-domain response.

The Nichols chart plot is a graph of the open-loop frequency response function $GH(\omega)$, for a continuous-time or discrete-time system, and compensation can be introduced in the forward and/or feedback paths, thus changing $G(\omega)$, $H(\omega)$, or both. We emphasize that no single compensation scheme is universally applicable.

### 18.2 GAIN FACTOR COMPENSATION

We have seen in several previous chapters (5, 12, 13, 16) that an unstable feedback system can sometimes be stabilized, or a stable system destabilized, by adjustment of the gain factor $K$ of $GH$. Nichols chart plots are particularly well suited for determining gain factor adjustments. However, when using Nichols techniques for continuous-time systems, it is more convenient to use the Bode gain $K_B$ (Section 15.3), expressed in decibels (db), than the gain factor $K$. Changes in $K_B$ and $K$, when given in decibels, are equal.

**EXAMPLE 18.1.** The db magnitude-phase angle plot for an unstable continuous-time system, represented by $GH(j\omega)$ with the Bode gain $K_B = 5$, is shown in Fig. 18-1. The instability of this system can be verified by a sketch of the Nyquist plot, or application of the Routh criterion. The Nyquist plot in Example 12.1 chapter 12, illustrates the general shape for all Nyquist plots of systems with one pole at the origin and two real poles in the left-half plane. This graph indicates that positive phase and gain margins guarantee stability and negative phase and gain margins guarantee instability for such a system, which implies that a sufficient decrease in the Bode gain stabilizes the system. If the Bode gain is decreased from $20 \log_{10} 5$ db to $20 \log_{10} 2$ db, the system is stabilized. The db magnitude-phase angle plot for the compensated system is shown in Fig. 18-2. Further decrease in gain does not alter stability.

Note that the curves for $K_B = 5$ and $K_B = 2$ have identical shapes, the only difference being that the ordinates on the $K_B = 5$ curve exceed those on the $K_B = 2$ curve by $20 \log_{10} (5/2)$ db. Therefore changing the gain on a db magnitude-phase angle plot is accomplished by simply shifting the locus of $GH(j\omega)$ up or down by an appropriate number of decibels.

Even though absolute stability can often be altered by gain factor adjustment, this form of compensation is inadequate for most designs because other performance criteria such as those concerned with relative stability cannot usually be met without the inclusion of other types of compensators.

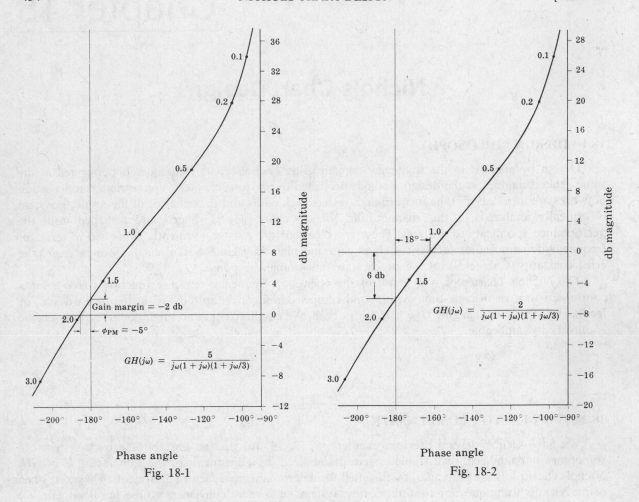

Phase angle
Fig. 18-1

Phase angle
Fig. 18-2

## 18.3 GAIN FACTOR COMPENSATION USING CONSTANT AMPLITUDE CURVES

The Nichols chart may be used to determine the gain factor $K$ (for a *unity feedback* system) for a specified resonant peak $M_p$ (in decibels). The following procedure requires drawing the db magnitude-phase angle plot only once.

**Step 1:** Draw the db magnitude-phase angle plot of $G(\omega)$ for $K = 1$ on tracing paper. The scale of the graph must be the same as that on the Nichols chart.

**Step 2:** Overlay this plot on the Nichols chart so that the magnitude and phase angle scales of each sheet are aligned.

**Step 3:** Fix the Nichols chart and slide the plot up or down until it is just tangent to the constant amplitude curve of $M_p$ db. The amount of shift in decibels is the required value of $K$.

**EXAMPLE 18.2.** In Fig. 18-3(*a*), the db magnitude-phase angle plot of the open-loop frequency response function of a particular unity feedback system with $K = 1$ is shown superimposed on a Nichols chart. The desired $M_p$ is 4 db. We see in Fig. 18-3(*b*) that, if the overlay is shifted upward by 4 db, then the resonant peak $M_p$ of the system is 4 db. Thus the desired $K$ is 4 db.

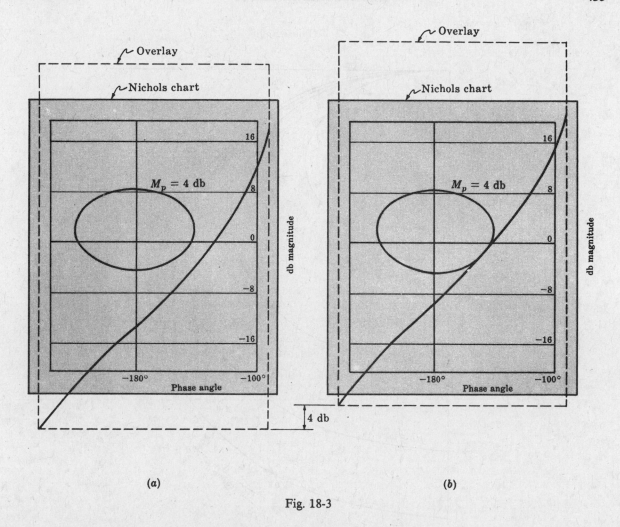

Fig. 18-3

## 18.4  LEAD COMPENSATION FOR CONTINUOUS-TIME SYSTEMS

The Bode form of the transfer function for a lead network is

$$P_{\text{Lead}} = \frac{(a/b)\left(1 + \dfrac{s}{a}\right)}{1 + \dfrac{s}{b}} \qquad (18.1)$$

where $a/b < 1$. The db magnitude-phase angle plots of $P_{\text{Lead}}$ for several values of $b/a$ and with the normalized frequency $\omega/a$ as the parameter are shown in Fig. 18-4.

For some systems in which lead compensation in the forward loop is applicable, appropriate choice of $a$ and $b$ permits an increase in $K_B$, providing greater accuracy and less sensitivity, without adversely affecting transient performance. Conversely, for a given $K_B$, the transient performance can be improved. It is also possible to improve both the steady state and transient responses with lead compensation.

The important properties of a lead network compensator are its phase lead contribution in the low-to-medium-frequency range (the vicinity of the resonant frequency $\omega_p$) and its negligible attenuation at high frequencies. If a very large phase lead is required, several lead networks may be cascaded.

Lead compensation generally increases the bandwidth of a system.

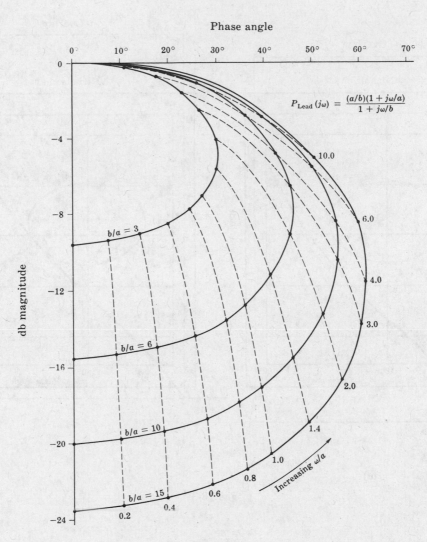

Fig. 18-4

**EXAMPLE 18.3.** The uncompensated continuous-time unity feedback system whose open-loop transfer function is

$$GH = \frac{2}{s(1+s)(1+s/3)}$$

is to be designed to meet the following performance specifications:

1.  When the input is a unit ramp function, the steady state position error must be less than 0.25.
2.  $\phi_{PM} \cong 40°$.
3.  Resonance peak $\cong 4$ db.

Note that the Bode gain is equal to the velocity error constant $K_v$. Therefore the steady state error for the uncompensated system is $e(\infty) = 1/K_v = \frac{1}{2}$ [Equation (9.13)]. From the db magnitude-phase angle plot of $GH$ in Fig. 18-5, we see that $\phi_{PM} = 18°$ and $M_p = 11$ db.

The steady state error is too large by a factor of 2; therefore the Bode gain must be increased by a factor of 2 (6 db). If we increase the Bode gain by 6 db, we obtain the plot labeled $GH_1$ in Fig. 18-5. The phase margin of $GH_1$ is about zero and the resonant peak is near infinity. Therefore the system is on the verge of instability.

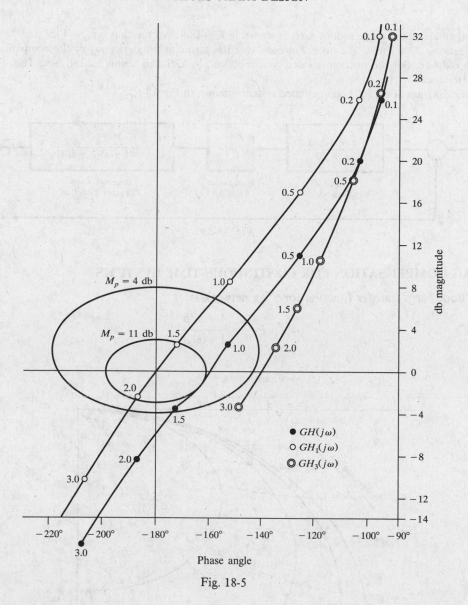

Fig. 18-5

Phase lead compensation can be used to improve the relative stability of the system. The compensated open-loop transfer function is

$$GH_2 = \frac{K_B(a/b)(1+s/a)}{s(1+s)(1+s/3)(1+s/b)} = \frac{4(1+s/a)}{s(1+s)(1+s/3)(1+s/b)}$$

where $K_B = 4(b/a)$ to satisfy the steady state error.

One way of satisfying the requirements on $\phi_{PM}$ and $M_p$ is to add 40° to 50° of phase lead to the $GH_1$ curve in the region $1 \le \omega \le 2.5$ without substantially changing the db magnitude. We have already chosen $K_B = 4(b/a)$ to compensate for $a/b$ in the lead network. Therefore we need concern ourselves only with the effect that the factor $(1+s/a)/(1+s/b)$ has on the $GH_1$ curve. Referring to Fig. 18-4, we see that in order to provide the necessary phase lead we will require $b/a \ge 10$. We note that the curves of Fig. 18-4 include the effect of $a/b$ of the lead network. Since we have already compensated for this, we must add $20 \log_{10}(b/a)$ to the db magnitudes on the curve. In order to keep the db magnitude contribution of the lead network small in the region $1 \le \omega \le 2.5$, we let $b/a = 15$ and choose $a$ so that only the lower portion of the curve ($\omega/a \le 3.0$) contributes in the region of interest $1 \le \omega \le 2.5$. In particular, we let $a = 1.333$. Then the compensated open-loop transfer function is

$$GH_3 = \frac{4(1+s/1.333)}{s(1+s)(1+s/3)(1+s/20)}$$

The db magnitude-phase angle plot of $GH_3$ is shown in Fig. 18-5. We see that $\phi_{PM} = 40.5°$ and $M_p = 4$ db. Thus the specifications are all met. We note, however, that the resonant frequency $\omega_p$ of the compensated system is about 2.25 rad/sec. For the uncompensated system defined by $GH$ it is about 1.2 rad/sec. Thus the bandwidth has been increased.

A block diagram of the fully compensated system is shown in Fig. 18-6.

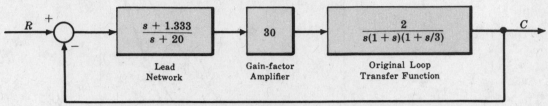

Fig. 18-6

 **18.5 LAG COMPENSATION FOR CONTINUOUS-TIME SYSTEMS**

The Bode form transfer function for a lag network is

$$P_{\text{Lag}} = \frac{1 + s/b}{1 + s/a} \qquad (18.2)$$

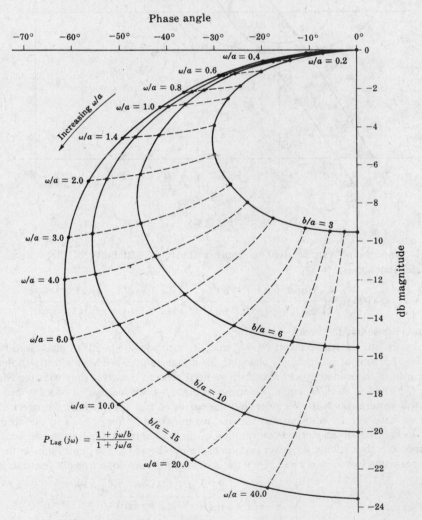

Fig. 18-7

where $a < b$. The db magnitude-phase angle plots of $P_{Lag}$ for several values of $b/a$ and with the normalized frequency $\omega/a$ as the parameter are shown in Fig. 18-7.

The lag network provides compensation by attenuating the high-frequency portion of the db magnitude-phase angle plot. Higher attenuation is provided by cascading several lag networks.

Several general effects of lag compensation are:

1. The bandwidth of the system is usually decreased.

2. The dominant time constant $\tau$ of the system is usually increased, producing a more sluggish system.

3. For a given relative stability, the value of the error constant is increased.

4. For a given error constant, relative stability is improved.

The procedure for using lag compensation is essentially the same as that for lead compensation.

**EXAMPLE 18.4.**    Let us redesign the system of Example 18.3 using gain factor plus lag compensation. The steady state specification is again satisfied by $GH_1$. The db magnitude-phase angle plot of $GH_1$ is repeated in Fig. 18-8. Since $P_{Lag}(j0) = 1$, introduction of the lag network after the steady state specification has been met by gain factor compensation does not require an additional increase in gain factor.

Incorporating the lag network, we get the open-loop transfer function

$$GH_4 = \frac{4(1+s/b)}{s(1+s)(1+s/3)(1+s/a)}$$

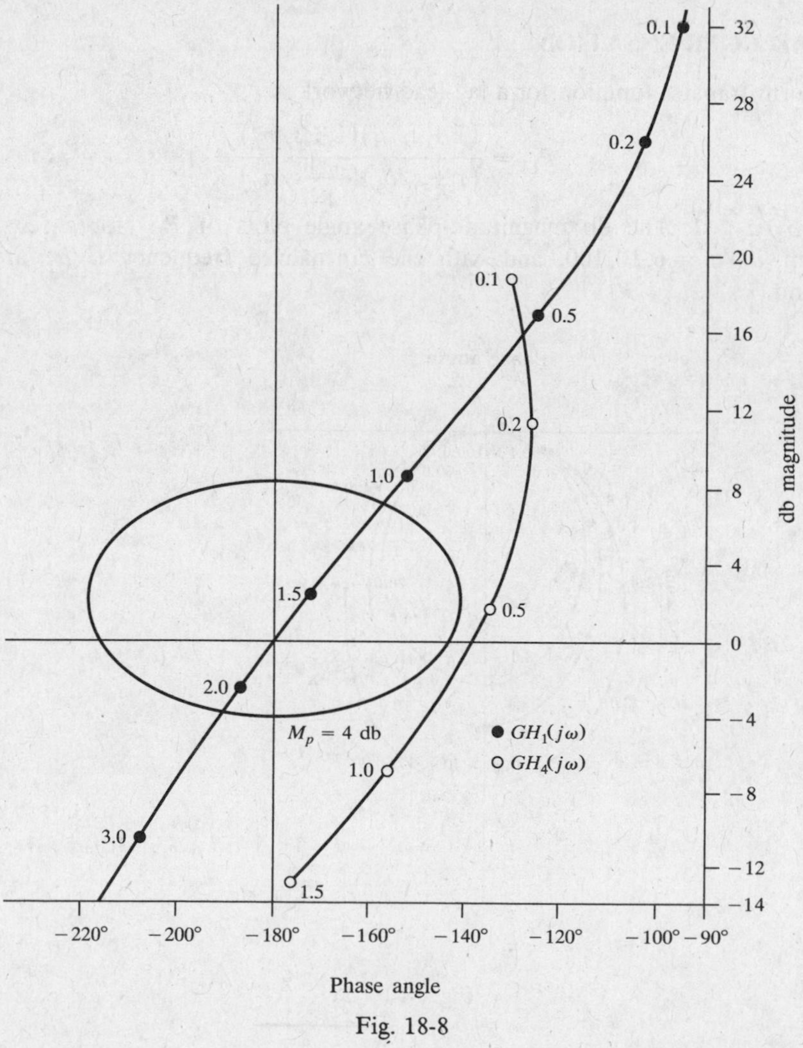

Fig. 18-8

One way of satisfying the requirements on $\phi_{PM}$ and $M_p$ is to choose $a$ and $b$ such that the $GH_1$ curve is attenuated by about 12 db in the region $0.7 \leq \omega \leq 2.0$ without substantial change in the phase angle. Since the lag network introduces some phase lag, it is necessary to attenuate the curve more than 12 db. Referring to Fig. 18-7, we see that if we choose $b/a = 6$, a maximum of 15.5-db attenuation is possible. If we choose $a = 0.015$, then at a frequency $\omega = 0.5$ ($\omega/a = 33.33$) 15.4 db of attenuation is obtained from the lag network, with a phase lag of $-9°$. $GH_4$ can now be written as

$$GH_4 = \frac{4(1 + s/0.09)}{s(1 + s)(1 + s/3)(1 + s/0.015)}$$

where $b = 6a = 0.09$. The db magnitude-phase angle plot of $GH_4$ is given in Fig. 18-8. We see that $\phi_{PM} = 41°$ and $M_p \cong 4$, which satisfy the specifications. We note that the resonant frequency $\omega_p$ of the compensated system is about 0.5 rad/sec. For the uncompensated system defined by $GH$, $\omega_p$ is about 1.2 rad/sec. A block diagram of the fully compensated system is shown in Fig. 18-9.

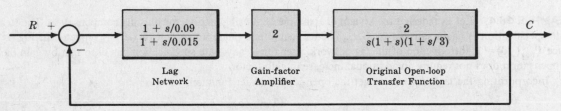

Fig. 18-9

## 18.6 LAG-LEAD COMPENSATION

The Bode form transfer function for a lag-lead network is

$$P_{LL} = \frac{(1 + s/a_1)(1 + s/b_2)}{(1 + s/b_1)(1 + s/a_2)} \tag{18.3}$$

where $b_1/a_1 = b_2/a_2 > 1$. The db magnitude-phase angle plots of $P_{LL}$ for a few values of $b_1/a_1$ ($= b_2/a_2$), when $a_1/a_2 = 6, 10, 100$, and with the normalized frequency $\omega/a_2$ are shown in Fig. 18-10($a$), ($b$), and ($c$)

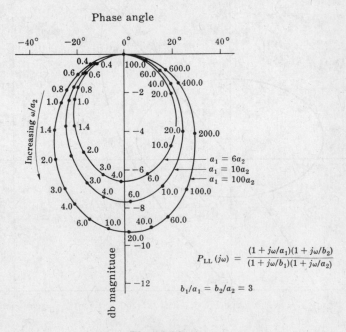

Fig. 18-10($a$)

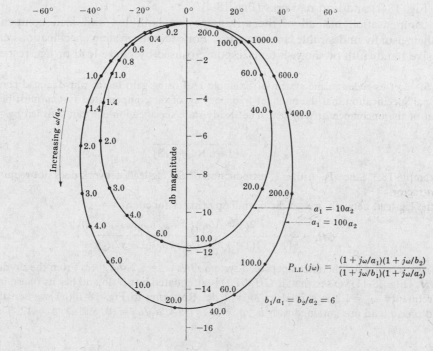

Fig. 18-10($b$)

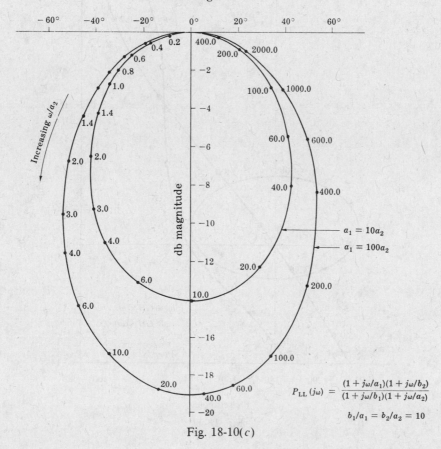

Fig. 18-10($c$)

Additional plots of $P_{LL}$ for other values of $b_1/a_1$ and $a_1/a_2$ can be obtained by combining plots of lag networks (Fig. 18-7) and lead networks (Fig. 18-4).

Lag-lead compensation has all of the advantages of both lag and lead compensation and a minimum of their usually undesirable characteristics. For example, system specifications can be satisfied without excessive bandwidth or sluggish time response caused by phase lead or lag, respectively.

**EXAMPLE 18.5.** Let us redesign the system of Example 18.3 using gain factor plus lag-lead compensation. We add the additional specification that the resonant frequency $\omega_p$ of the compensated system must be approximately the same as that of the uncompensated system. The steady state specification is again satisfied by

$$GH_1 = \frac{4}{s(1+s)(1+s/3)}$$

as shown in Example 18.3. Since $P_{LL}(j0) = 1$, introduction of the lag-lead network does not require an additional increase in gain factor.

Inserting the lag-lead network, we get the open-loop transfer function

$$GH_5 = \frac{4(1+s/a_1)(1+s/b_2)}{s(1+s)(1+s/3)(1+s/b_1)(1+s/a_2)}$$

From Fig. 18-5, we see that for the uncompensated system $GH$, $\omega_p = 1.2$ rad/sec. From the db magnitude-phase angle plot of $GH_1$ (Fig. 18-11) we see that, if $GH_1(j1.2)$ is attenuated by 6.5 db and has its phase increased by 20°, the resonant frequency $\omega_p = 1.2$ is shifted to $M_p = 4$ db. Referring to Fig. 18-10($a$), we see that the desired attenuation and phase lead are obtained with $b_1/a_1 = b_2/a_2 = 3$, $a_1/a_2 = 10$, and $\omega/a_2 = 12$. The constants $a_1$,

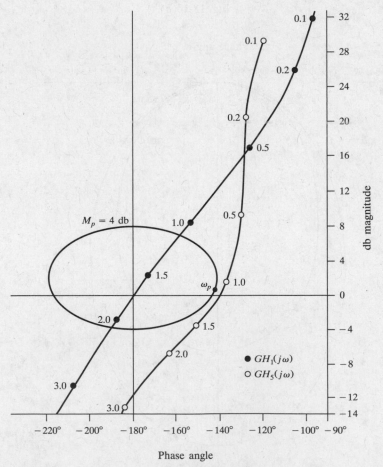

Fig. 18-11

$a_2$, $b_1$, and $b_2$ are determined by noting that

$$a_2 = \frac{\omega_p}{12} = \frac{1.2}{12} = 0.1 \qquad a_1 = 10a_2 = 1 \qquad b_2 = 3a_2 = 0.3 \qquad \text{and} \qquad b_1 = 3a_1 = 3$$

$GH_5$ then becomes

$$GH_5 = \frac{4(1+s)(1+s/0.3)}{s(1+s)(1+s/3)(1+s/3)(1+s/0.1)} = \frac{4(1+s/0.3)}{s(1+s/3)^2(1+s/0.1)}$$

The complete db magnitude-phase angle plot of $GH_5$ is shown in Fig. 18-11. We see that $\phi_{PM} = 40.5°$, $M_p = 4$ db, and the resonant frequency $\omega_p \cong 1.15$. Thus all specifications have been satisfied.

## 8.7  NICHOLS CHART DESIGN OF DISCRETE-TIME SYSTEMS

As with Bode methods (Section 16.6), design of discrete-time systems using Nichols charts is not as straightforward as the design of continuous-time systems using either of these approaches. But, again, the $w$-transform can facilitate the process as it did for Bode design of discrete-time systems. The method is the same as that developed in Section 16.6.

**EXAMPLE 18.6.**  The uncompensated discrete-time unity feedback system with plant transfer function

$$G_2(z) = \frac{9}{4} \frac{(z+1)^3}{z(z+\tfrac{1}{2})^2}$$

is to be designed to yield an overall phase margin of 40° and the same gain crossover frequency $\omega_1$ as the uncompensated system. Since both of these specifications are in the frequency domain, we transform the problem directly into the $w$-domain by substituting $z = (1+w)/(1-w)$, thus forming

$$G_2'(w) = \frac{72}{(w+1)(w+3)^2}$$

The db magnitude-phase angle plot for this system is shown in Fig. 18-12. The gain crossover frequency obtained from this plot is $\omega_{w1} = 3.4$ rad/sec and the phase margin is 10°. A lead compensator with somewhat arbitrary $a$ and $b$ can be chosen as long as the phase lead at $\omega_{w1} = 3.4$ rad/sec is sufficient to raise the phase margin from 10° to 40°. The minimum $b/a$ ratio that yields 30° of phase lead is about 3.3 from Fig. 18-4. We choose $a$ and $b$ so that the maximum phase lead occurs at $\omega_{w1} = 3.4$ rad/sec. From Section 16.3, this occurs when $\omega_{w1} = 3.4 = \sqrt{ab}$. Since $b = 3.3a$, we find $b = 6.27$ and $a = 1.90$. This compensator produces about $20\log_{10}\sqrt{6.27/1.90} = 5$ db of attenuation at $\omega_{w1} = 3.4$ rad/sec. Thus an amplifier with gain of 5.2 db, or gain factor 1.82, is required in addition

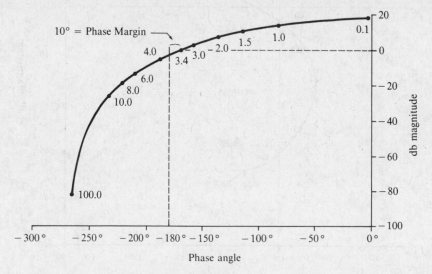

Fig. 18-12

to the lead compensator to maintain $\omega_{w1}$ at 3.4 rad/sec. The $w$-domain transfer function for the compensator is therefore given by

$$G_1(w) = \frac{1.82(w + 1.90)}{w + 6.27}$$

This is transformed back to the $z$-domain by letting $w = (z - 1)/(z + 1)$, thus forming

$$G_1(z) = \frac{0.7229(z + 0.3007)}{z + 0.7222}$$

The compensated control system is shown in Fig. 18-13.

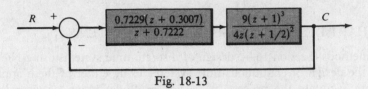

Fig. 18-13

# Solved Problems

## GAIN FACTOR COMPENSATION

**18.1.** The db magnitude-phase angle plot of the open-loop continuous-time frequency response function

$$GH(j\omega) = \frac{K_B\left[1 - (\omega/2)^2 + j\omega/2\right]}{j\omega(1 + j\omega/0.5)^2(1 + j\omega/4)}$$

is shown in Fig. 18-14 for $K_B = 1$. The closed-loop system defined by $GH(j\omega)$ is stable for $K_B = 1$. Determine a value of $K_B$ for which the phase margin is $45°$.

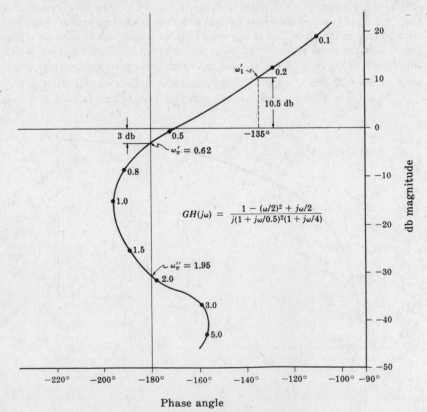

Phase angle

Fig. 18-14

$\phi_{PM} = 180° + \arg GH(j\omega_1)$, where $\omega_1$ is the gain crossover frequency. For $\phi_{PM} = 45°$, $\omega_1$ must be chosen so that $\arg GH(j\omega_1) = -135°$. If we draw a vertical line with abscissa of $-135°$, it intersects the $GH(j\omega)$ curve at a point $\omega_1' \cong 0.25$ rad/sec, where $\arg GH(j\omega_1') = -135°$. The ordinate of this point of intersection is 10.5 db. If we decrease $K_B$ by 10.5 db, the gain crossover frequency becomes $\omega_1'$, and $\phi_{PM} = 45°$. A decrease of 10.5 db implies that $20 \log_{10} K_B = -10.5$, or $K_B = 10^{-10.5/20} = 0.3$. Further decrease in $K_B$ increases $\phi_{PM}$ beyond 45°.

**18.2.** For the system in Problem 18.1, determine the value of $K_B$ for which the system is stable and the gain margin is 10 db.

Gain margin $= -20 \log_{10}|GH(j\omega_\pi)|$ db, where $\omega_\pi$ is the phase crossover frequency. Referring to Fig. 18-14, we see that there are two phase crossover frequencies: $\omega_\pi' \cong 0.62$ rad/sec and $\omega_\pi'' \cong 1.95$ rad/sec. For $\omega_\pi' = 0.62$, we have $20 \log_{10}|GH(j\omega_\pi')| = -3$ db. Therefore the gain margin is 3 db. It can be increased to 10 db by shifting the $GH(j\omega)$ curve downward by 7 db. The phase crossover frequency $\omega_\pi'$ is the same in the new position, but $20 \log_{10}|GH(j\omega_\pi')| = -10$ db. A gain decrease of 7 db implies that $K_B = 10^{-7/20} = 0.447$. Since the system is stable for $K_B = 1$, it remains stable when the $GH(j\omega)$ curve is shifted downward. Absolute stability is not affected unless the $GH(j\omega)$ curve is shifted upward and across the point defined by 0 db and $-180°$, as would be necessary if $-20 \log_{10} GH(j\omega_\pi'') = 10$ db.

**18.3.** For the system of Problem 18.1, determine a value for $K_B$ such that: gain margin $\geq 10$ db, $\phi_{PM} \geq 45°$.

In Problem 18.1, it was shown that $\phi_{PM} \geq 45°$ if $K_B \leq 0.3$; in Problem 18.2, gain margin $\geq 10$ db if $K_B \leq 0.447$. Therefore both requirements can be satisfied by setting $K_B \leq 0.3$. Note that if we had specified gain margin $= 10$ db and $\phi_{PM} = 45°$, then the specifications could not be met by gain factor compensation alone.

**18.4.** Assume that the system of Problem 18.1 is a unity feedback system and determine a value for $K_B$ such that the resonant peak $M_p$ is 5 db.

Mathcad

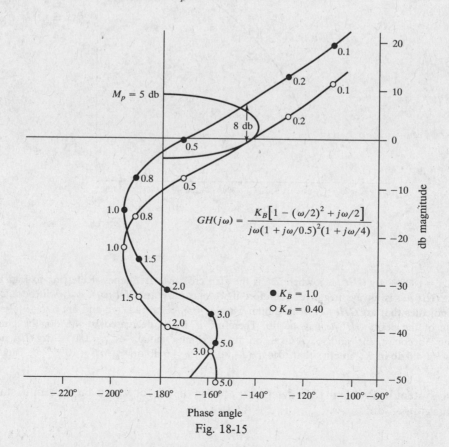

$$GH(j\omega) = \frac{K_B[1 - (\omega/2)^2 + j\omega/2]}{j\omega(1 + j\omega/0.5)^2(1 + j\omega/4)}$$

• $K_B = 1.0$
○ $K_B = 0.40$

Fig. 18-15

The db magnitude-phase angle plot of $GH(j\omega)$ for $K_B = 1$ is shown in Fig. 18-15 along with the locus of points for which $|(C/R)(j\omega)| = 2$ db ($M_p = 2$ db). We see that if $K_B$ is decreased by 8 db, the resulting $GH(j\omega)$ curve is just tangent to the $M_p = 2$ db curve. A decrease of 8 db implies that $K_B = 10^{-8/20} = 0.40$.

**18.5.** The db magnitude-phase angle plot of the open-loop frequency response function

$$GH(j\omega) = \frac{K_B(1 + j\omega/0.5)}{(j\omega)^2\left[1 - (\omega/2)^2 + j\omega/2\right]}$$

is given in Fig. 18-16 for $K_B = 0.5$. The closed-loop system defined by $GH(j\omega)$ is stable for $K_B = 0.5$. Determine the value of $K_B$ which maximizes the phase margin.

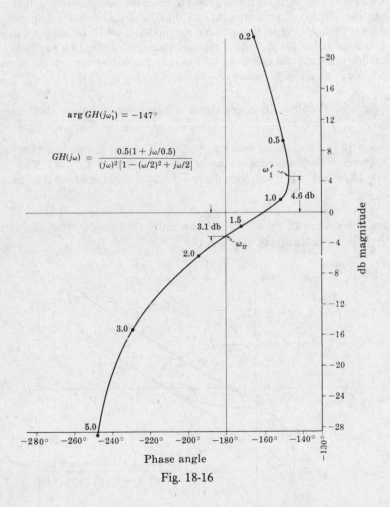

Fig. 18-16

$\phi_{\text{PM}} = 180° + \arg GH(j\omega_1)$, where $\omega_1$ is the gain crossover frequency. Referring to Fig. 18-16, we see that $\arg GH(j\omega)$ is always negative. Therefore if we maximize $\arg GH(j\omega_1)$, $\phi_{\text{PM}}$ will be maximized. Fig. 18-16 indicates that $\arg GH(j\omega)$ is maximum when $\omega = \omega_1' \cong 0.8$ rad/sec and $\arg GH(j\omega_1') = -147°$. The ordinate of the point $GH(j\omega_1')$ is 4.6 db. Therefore if $K_B$ is decreased by 4.6 db, the phase crossover frequency becomes $\omega_1'$; and $\phi_{\text{PM}}$ takes on its maximum value: $\phi_{\text{PM}} = 180° + \arg GH(j\omega_1') = 33°$. A decrease of 4.6 db in $K_B$ implies that $20\log_{10}(K_B/0.5) = -4.6$ db or $K_B/0.5 = 10^{-4.6/20}$. Then $K_B = 0.295$.

**18.6.** For the system in Problem 18.5, determine a value of $K_B$ for which the system is stable and the gain margin is 8 db.

Gain margin = $-20\log_{10}|GH(j\omega_\pi)|$ db. Referring to Fig. 18-16, we see that the gain margin is 3.1 db. This can be increased to 8 db by shifting the curve down by 4.9 db; $\omega_\pi$ remains the same, as it is independent of $K_B$. A decrease of 4.9 db in $K_B$ implies that $20\log_{10}(K_B/0.5) = -4.9$ or $K_B = 0.254$.

## PHASE COMPENSATION

**18.7.** The db magnitude-phase angle plot of the open-loop transfer function $G(j\omega)$ for a particular unity feedback system has been determined experimentally as shown in Fig. 18-17. In addition, the steady state error $e(\infty)$ for a unit ramp function input was measured and found to be $e(\infty) = 0.2$. The open-loop transfer function is known to have a pole at the origin. Determine a combination of phase lead plus gain compensation such that: $M_p \cong 3.5$ db, $\phi_{PM} \cong 40°$, and the steady state error for a unit ramp input is $e(\infty) = 0.1$.

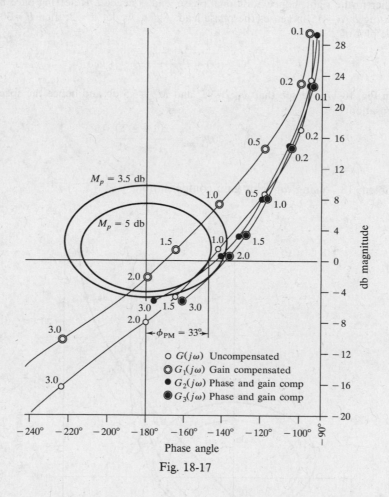

Fig. 18-17

Since $e(\infty) = 1/K_v = 1/K_B$, the steady state requirement can be satisfied by doubling $K_B$. The compensation has the form

$$K'P_{\text{Lead}}(j\omega) = \frac{K'(a/b)(1+s/a)}{1+s/b}$$

Hence $K_B$ is doubled by letting $K'(a/b) = 2$, or $K' = 2(b/a)$.

The db magnitude-phase angle plot for the gain compensated open-loop frequency response function

$$G_1(j\omega) = 2G(j\omega)$$

is shown in Fig. 18-17. $G_1(j\omega)$ satisfies the steady state specification. To satisfy the specifications on $M_p$ and $\phi_{PM}$, the $G_1(j\omega)$ curve must be shifted to the right by about 30° to 40° in the region $1.2 \le \omega \le 2.5$

without substantially changing the db magnitude. This is done by proper choice of $a$ and $b$. Referring to Fig. 18-4, we see that, for $b/a = 10$, 30° phase lead is obtained for $\omega/a \geq 0.65$. Since the lead ratio $a/b$ of the lead network is taken into account by designing for the gain factor $K' = 2(b/a) = 20$, we must add $20\log(b/a) = 20\log_{10}10 = 20$ db to all db magnitudes taken from Fig. 18-4.

To obtain 30° or more phase lead in the frequency range of interest, we let $a = 2$. For this choice we have $\omega = (2)(0.65) = 1.3$ and obtain 30° phase lead. Since $b/a = 10$, then $b = 20$. The compensated open-loop frequency response function is

$$G_2(j\omega) = \frac{2(1 + j\omega/2)}{1 + j\omega/20}G(j\omega)$$

The db magnitude-phase angle plot of $G_2(j\omega)$ is shown in Fig. 18-17. We see that $M_p \cong 4.0$ db and $\phi_{PM} = 36°$; therefore the specifications are not satisfied by this compensation. We need to shift $G_2(j\omega)$ 5° to 10° further to the right; hence additional phase lead is needed. Referring once more to Fig. 18-4, we see that letting $b/a = 15$ increases the phase lead. Again, we let $a = 2$; then $b = 30$. The db magnitude-phase angle plot of

$$G_3(j\omega) = \frac{2(1 + j\omega/2)}{1 + j\omega/30}G(j\omega)$$

is shown in Fig. 18-17. We see that $\phi_{PM} = 41°$ and $M_p \cong 3.5$ db and hence the specifications are met by the compensation

$$30P_{\text{Lead}} = \frac{2(1 + s/2)}{1 + s/30}$$

**18.8.** Solve Problem 18.7 using *lag* plus gain compensation.

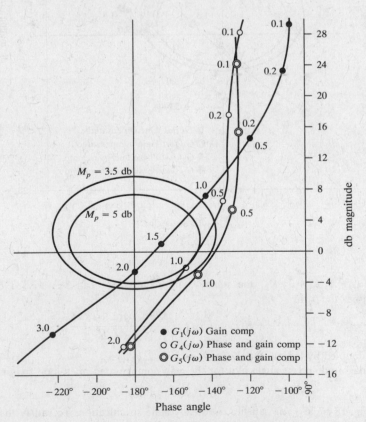

Fig. 18-18

In Problem 18.7 we found that the Bode gain $K_B$ must be increased by a factor of 2 to satisfy the steady state specification. But the Bode gain of a lag network is

$$\lim_{s \to 0} P_{\text{Lag}} = \lim_{s \to 0} \frac{1 + s/b}{1 + s/a} = 1$$

Therefore the compensation required in this problem has the form $2(1 + s/a)/(1 + s/b)$ where the twofold gain factor increase is supplied by an amplifier and $a$ and $b$ for the lag network must be chosen to satisfy the requirements on $M_p$ and $\phi_{\text{PM}}$. The gain-compensated function is shown as $G_1(j\omega) = 2G(j\omega)$ in Fig. 18-18; $G_1(j\omega)$ must be shifted downward by 7 to 10 db in the region $0.7 \leq \omega \leq 2.0$, with no substantial increase in phase lag, to meet the transient specifications.

Referring to Fig. 18-7, we see that, for $b/a = 3$, we can obtain a maximum attenuation of 9.5 db. For $a = 0.1$, the phase lag is $-15°$ at $\omega = 0.7$ ($\omega/a = 7$) and $-6°$ at $\omega = 2.0$ ($\omega/a = 20$), that is, the phase lag is relatively small in the frequency region of interest. The db magnitude-phase angle plot for

$$G_4(j\omega) = \frac{2(1 + j\omega/0.3)}{1 + j\omega/0.1} G(j\omega)$$

is also shown in Fig. 18-18, with $M_p \cong 5$ db and $\phi_{\text{PM}} = 32°$; hence this system does not meet the specifications. To decrease the phase lag introduced in the frequency region $0.7 \leq \omega \leq 2.0$, we change $a$ to 0.05 and $b$ to 0.15. The phase lag is now $9°$ at $\omega = 0.7$ ($\omega/a = 14$). The db magnitude-phase angle plot for

$$G_5(j\omega) = \frac{2(1 + j\omega/0.15)}{1 + j\omega/0.05} G(j\omega)$$

is shown in Fig. 18-18. We see that $M_p \cong 3.5$ db and $\phi_{\text{PM}} = 41°$. Thus the specifications are satisfied. The desired compensation is given by

$$2P_{\text{Lag}} = \frac{2(1 + s/0.15)}{1 + s/0.05}$$

**18.9.** Solve Problem 18.7 using *lag-lead* plus gain compensation. In addition to the previous specifications, we require that the resonant frequency $\omega_p$ of the compensated system be approximately the same as that for the uncompensated system.

In Problems 18.7 and 18.8 we found that the Bode gain $K_B$ must be increased by a factor of 2 to satisfy the steady state specification. The frequency response function of the lag-lead plus gain compensation is therefore given by

$$2P_{\text{LL}}(j\omega) = \frac{2(1 + j\omega/a_1)(1 + j\omega/b_2)}{(1 + j\omega/b_1)(1 + j\omega/a_2)}$$

We must now choose $a_1$, $b_1$, $b_2$, and $a_2$ to satisfy the requirements on $M_p$, $\phi_{\text{PM}}$ and $\omega_p$. Referring to Fig. 18-17, we see that the resonant frequency for the uncompensated system is about 1.1 rad/sec. The db magnitude-phase angle plot of $G_1(j\omega) = 2G(j\omega)$ shown in Fig. 18-19 indicates that, if the $G_1(j\omega)$ curve is attenuated by 6.5 db and $10°$ of phase lead is added at a frequency of $\omega = 1.0$ rad/sec, then the resulting curve will be tangent to the $M_p = 2$ db curve at about 1 rad/sec. Referring to Fig. 18-10, if we let $b_1/a_1 = b_2/a_2 = 3$, $a_1 = 6a_2$, and $\omega/a_2 = 6.0$ for $\omega = 1$, we obtain the desired attenuation and phase lead. Solving for the remaining parameters, we get $a_2 = 1/6 = 0.167$, $b_2 = 3a_2 = 0.50$, $a_1 = 6a_2 = 1.0$, $b_1 = 3a_1 = 3.0$. The db magnitude-phase angle plot for the resulting open-loop frequency response function

$$G_6(j\omega) = \frac{2(1 + j\omega)(1 + j\omega/0.5)}{(1 + j\omega/3)(1 + j\omega/0.167)} G(j\omega)$$

is shown in Fig. 18-19, where $M_p \cong 3.5$ db, $\phi_{\text{PM}} = 44°$, and $\omega_p \cong 1.0$ rad/sec. These values approximately satisfy the specifications.

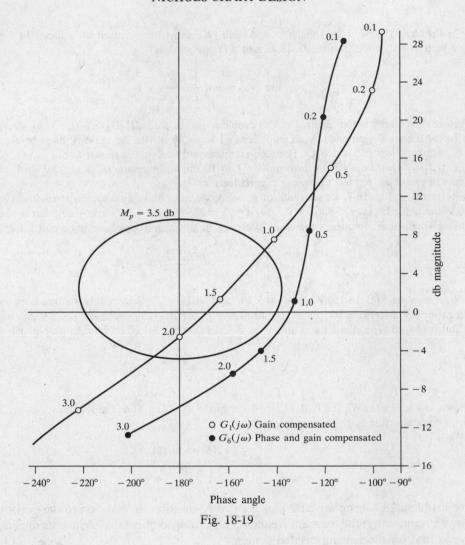

Fig. 18-19

**18.10.** Design compensation for the discrete-time system with open-loop transfer function

$$GH(z) = \frac{K(z+1)^3}{(z-1)\left(z+\frac{2}{3}\right)^2}$$

such that the following performance specifications are satisfied:

1.  gain margin $\geq 6$ db
2.  phase margin $\phi_{PM} \geq 45°$
3.  gain crossover frequency $\omega_1$ such that $\omega_1 T \leq 1.6$ rad
4.  velocity constant $K_v \geq 10$

The Nichols chart plot of $GH$ shown in Fig. 18-20 indicates that $\omega_1 T = 1.6$ rad for $K = -3$ db. The gain and phase margins are met if $K < 4.7$ db; but the steady state specification requires that $K > 10.8$ db (gain factor of 3.47). Substituting $z = (1 + w)/(1 - w)$, we transform the open-loop transfer function from the $z$-domain to the $w$-domain, thus forming

$$GH'(w) = \frac{36}{25} \frac{K}{w(1 + w/5)^2}$$

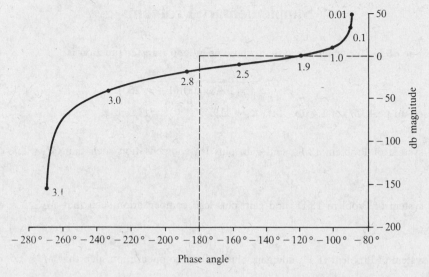

Fig. 18-20

In the $w$-domain the gain crossover frequency specification becomes

$$\omega_{w1} = \tan\left(\frac{\omega_1 T}{2}\right) = 1.02 \text{ rad/sec}$$

A low-frequency cascade lag compensator with $b/a = 3.5$ can be used to increase $K_v$ to 10, while maintaining the gain crossover frequency $\omega_1$ and the gain and phase margins at their previous values. A lag compensator with $b = 0.35$ and $a = 0.1$ satisfies the requirements.

The lag compensator in the $w$-plane is

$$G_1(w) = \frac{3.5(1 + w/0.35)}{1 + w/0.1}$$

This is transformed back into the $z$-domain by substituting $w = (z-1)/z + 1)$, thus forming

$$G_1(z) = 1.2273\left(\frac{z - 0.4815}{z - 0.8182}\right)$$

The db magnitude-phase angle plot for the compensated discrete-time system is shown in Fig. 18-21.

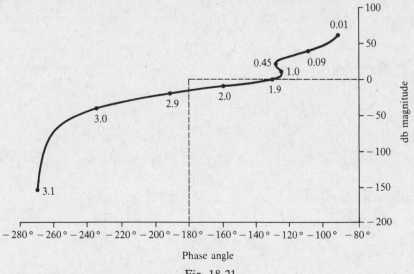

Fig. 18-21

# Supplementary Problems

**18.11.** Find a value of $K_B$ for which the system whose open-loop transfer function is

$$GH = \frac{K_B}{s(1 + s/200)(1 + s/250)}$$

has a resonant peak $M_p$ of 1.4 db. *Ans.* $K_B = 119.4$.

**18.12.** For the system of Problem 18.11, find gain plus lag compensation such that $M_p \leq 1.7$, $\phi_{PM} \geq 35°$, and $K_v \geq 50$.

**18.13.** For the system of Problem 18.11, find gain plus lead compensation such that $M_p \leq 1.7$, $\phi_{PM} \geq 50°$, and $K_v \geq 50$.

**18.14.** For the system of Problem 18.11, find gain plus lag-lead compensation such that $M_p \leq 1.5$, $\phi_{PM} \geq 40°$, and $K_v \geq 100$.

**18.15.** Find gain plus lag compensation for the system whose open-loop transfer function is

$$GH = \frac{K_B}{s(1 + s/10)(1 + s/5)}$$

such that $K_v = 30$ and $\phi_{PM} \geq 40°$.

**18.16.** For the system of Problem 18.15, find gain plus lead compensation such that $K_v \geq 30$ and $\phi_{PM} \geq 45°$. *Hint.* Cadcade two lead compensation networks.

**18.17.** Find gain plus lead compensation for the system whose open-loop transfer function is

$$GH = \frac{K_B}{s(1 + s/2)}$$

such that $K_v = 20$ and $\phi_{PM} = 45°$.

# Chapter 19

# Introduction to Nonlinear Control Systems

## 19.1 INTRODUCTION

We have thus far confined the discussion to systems describable by linear time-invariant ordinary differential or difference equation models or their transfer functions, excited by Laplace or $z$-transformable input functions. The techniques developed for studying these systems are relatively straightforward and usually lead to practical control system designs. While it is probably true that no physical system is *exactly* linear and time-invariant, such models are often adequate approximations and, as a result, the linear system methods developed in this book have broad application. There are many situations, however, for which linear representations are inappropriate and *nonlinear* models are required.

Theories and methods for analysis and design of nonlinear control systems constitute a large body of knowledge, some of it quite complex. The purpose of this chapter is to introduce some of the prevailing classical techniques, utilizing mathematics at about the same level as in earlier chapters.

*Linear* systems are defined in Definition 3.21. Any system that does not satisfy this definition is nonlinear. The major difficulty with nonlinear systems, especially those described by nonlinear ordinary differential or difference equations, is that analytical or closed-form solutions are available only for very few special cases, and these are typically not of practical interest in control system analysis or design. Furthermore, unlike linear systems, for which free and forced responses can be determined separately and the results superimposed to obtain the total response, free and forced responses of nonlinear systems normally *interact* and cannot be studied separately, and superposition does not generally hold for inputs or initial conditions.

In general, the characteristic responses and stability of nonlinear systems depend qualitatively as well as quantitatively on initial condition values, and the magnitude, shape, and form of system inputs. On the other hand, time-domain solutions to nonlinear system equations usually can be obtained, for *specified* inputs, parameters, and initial conditions, by computer simulation techniques. Algorithms and software for simulation, a special topic outside the scope of this book, are widely available and therefore are not developed further here. Instead, we focus on several analytical methods for studying nonlinear control systems.

Nonlinear control system problems arise when the structure or fixed elements of the system are inherently nonlinear, and/or nonlinear compensation is introduced into the system for the purpose of improving its behavior. In either case, stability properties are a central issue.

**EXAMPLE 19.1.** Fig. 19-1($a$) is a block diagram of a nonlinear feedback system containing two blocks. The *linear* block is represented by the transfer function $G_2 = 1/D(D + 1)$, where $D \equiv d/dt$ is the *differential* operator. $D$ is used instead of $s$ in this linear transfer function because the Laplace transform and its inverse are generally not strictly applicable for nonlinear analysis of systems with both linear and nonlinear elements. Alternatively, when using the describing function method (Section 19.5), an approximate frequency response technique, we

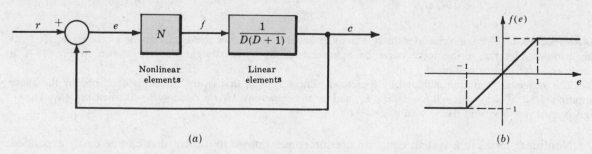

$(a)$        $(b)$

Fig. 19-1

usually write

$$G_2(j\omega) = \frac{1}{j\omega(j\omega + 1)}$$

The *nonlinear block* $N$ has the transfer characteristic $f(e)$ defined in Figure 19-1(b). Such nonlinearities are called (piecewise-linear) **saturation** functions, described further in the next section.

**EXAMPLE 19.2.** If the earth is assumed spherical and all external forces other than gravity are negligible, then the motion of an earth satellite lies in a plane called the *orbit plane*. This motion is defined by the following set of nonlinear differential equations (see Problem 3.3):

$$r\frac{d^2\theta}{dt^2} + 2\frac{dr}{dt}\frac{d\theta}{dt} = 0 \qquad \text{(transverse force equation)}$$

$$\frac{d^2r}{dt^2} - r\left(\frac{d\theta}{dt}\right)^2 = -\frac{k^2}{pr^2} \qquad \text{(radial force equation)}$$

The satellite, together with any controller designed to modify its motion, constitutes a nonlinear control system.

Several popular methods for nonlinear analysis are summarized below.

## 19.2   LINEARIZED AND PIECEWISE-LINEARIZED APPROXIMATIONS OF NONLINEAR SYSTEMS

Nonlinear terms in differential or difference equations can sometimes be approximated by linear terms or zero-order (constant) terms, over limited ranges of the system response or system forcing function. In either case, one or more linear differential or difference equations can be obtained as approximations of the nonlinear system, valid over the same limited operating ranges.

**EXAMPLE 19.3.** Consider the spring-mass system of Fig. 19-2, where the spring force $f_s(x)$ is a nonlinear function of the displacement $x$ measured from the rest position, as shown in Fig. 19-3.

The equation of motion of the mass is $M(d^2x/dt^2) + f_s(x) = 0$. However, if the absolute magnitude of the displacement does not exceed $x_0$, then $f_s(x) = kx$, where $k$ is a constant. In this case, the equation of motion is a constant-coefficient linear equation given by $M(d^2x/dt^2) + kx = 0$, valid for $|x| \leq x_0$.

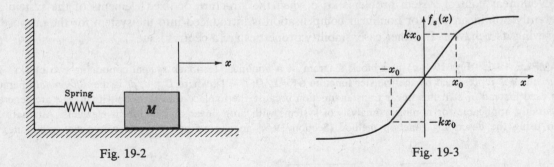

Fig. 19-2                                   Fig. 19-3

**EXAMPLE 19.4.** We again consider the system of Example 19.3, but now the displacement $x$ exceeds $x_0$. To treat this problem, let the spring force curve be approximated by three straight lines as shown in Fig. 19-4, a *piecewise-linear* approximation of $f_s(x)$.

The *system* is then approximated by a piecewise-linear system; that is, the system is described by the linear equation $M(d^2x/dt^2) + kx = 0$ when $|x| \leq x_1$, and by the equations $M(d^2x/dt^2) \pm F_1 = 0$ when $|x| > x_1$. The $+$ sign is used if $x > x_1$ and the $-$ sign if $x < -x_1$.

Nonlinear terms in a system equation are sometimes known in a form that can be easily expanded in a series, for example, a Taylor or a Maclaurin series. In this manner, a nonlinear term can be approximated by the first few terms of the series, excluding terms higher than first degree.

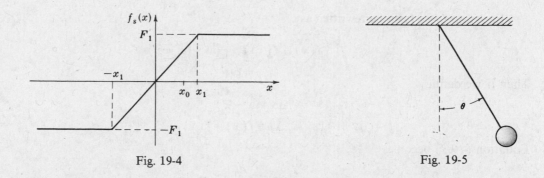

Fig. 19-4                                                    Fig. 19-5

**EXAMPLE 19.5.**   Consider the nonlinear equation describing the motion of a pendulum (see Fig. 19-5):

$$\frac{d^2\theta}{dt^2} + \frac{g}{l}\sin\theta = 0$$

where $l$ is the length of the pendulum bob and $g$ is the acceleration of gravity. If small motions of the pendulum about the "operating point" $\theta = 0$ are of interest, then the equation of motion can be linearized about this operating point. This is done by forming a Taylor series expansion of the nonlinear term $(g/l)\sin\theta$ about the point $\theta = 0$ and retaining only the first degree terms. The nonlinear equation is

$$\frac{d^2\theta}{dt^2} + \frac{g}{l}\sin\theta = \frac{d^2\theta}{dt^2} + \frac{g}{l}\sum_{k=0}^{\infty}\frac{\theta^k}{k!}\left(\frac{d^k}{d\theta^k}(\sin\theta)\bigg|_{\theta=0}\right)$$

$$= \frac{d^2\theta}{dt^2} + \frac{g}{l}\left[\theta - \frac{\theta^3}{3!} + \cdots\right] = 0$$

The linear equation is $d^2\theta/dt^2 + (g/l)\theta = 0$, valid for small variations in $\theta$.

It is instructive to express the linearization process more formally for Taylor series applications, to better establish its applicability and limitations.

### Taylor Series

The infinite series expansion of a general nonlinear function $f(x)$ can be quite useful in nonlinear systems analysis. The function $f(x)$ can be written as the following infinite series, expanded about the point $\bar{x}$:

$$f(x) = f(\bar{x}) + \frac{df}{dx}\bigg|_{x=\bar{x}}(x-\bar{x}) + \frac{1}{2!}\frac{d^2f}{dx^2}\bigg|_{x=\bar{x}}(x-\bar{x})^2 + \cdots$$

$$= \sum_{k=0}^{\infty}\frac{(x-\bar{x})^k}{k!}\frac{d^kf}{dx^k}\bigg|_{x=\bar{x}} \tag{19.1}$$

where $(d^kf/dx^k)|_{x=\bar{x}}$ is the value of the $k$th derivative of $f$ with respect to $x$ evaluated at the point $x = \bar{x}$. Clearly, this expansion exists (is feasible) only if all the required derivatives exist.

If the sum of the terms of Equation (19.1) second-degree and higher-degree in $(x - \bar{x})$ are negligible compared with the sum of the first two terms, then we can write

$$f(x) \cong f(\bar{x}) + \frac{df}{dx}\bigg|_{x=\bar{x}}(x-\bar{x}) \tag{19.2}$$

This approximation usually works if $x$ is "close enough" to $\bar{x}$, or, equivalently, if $x - \bar{x}$ is "small enough," in which case higher-degree terms are *relatively* small.

Equation ($19.2$) can be rewritten as

$$f(x) - f(\bar{x}) \cong \left.\frac{df}{dx}\right|_{x=\bar{x}} (x - \bar{x}) \tag{19.3}$$

Then if we define

$$\Delta x \equiv x - \bar{x} \tag{19.4}$$

$$\Delta f \equiv f(x) - f(\bar{x}) \tag{19.5}$$

Equation ($19.3$) becomes

$$\Delta f \cong \left.\frac{df}{dx}\right|_{x=\bar{x}} \Delta x \tag{19.6}$$

If $x = x(t)$ is a function of time $t$, or any other independent variable, then in most applications $t$ can be treated as a fixed parameter when performing the linearization computations above, and $\Delta x = \Delta x(t) \equiv x(t) - \bar{x}(t)$, etc.

**EXAMPLE 19.6.**  Suppose $y(t) = f[u(t)]$ represents a nonlinear system with input $u(t)$ and output $y(t)$, where $t \geq t_0$ for some $t_0$, and $df/du$ exists for all $u$. If the normal operating conditions for this system are defined by the input $u = \bar{u}$ and output $y = \bar{y}$, then small changes $\Delta y(t) = y(t) - \bar{y}(t)$ in output operation in response to small changes in the input $\Delta u(t) = u(t) - \bar{u}(t)$ can be expressed by the approximate linear relation

$$\Delta y(t) \cong \left.\frac{df}{du}\right|_{u=\bar{u}(t)} \Delta u(t) \tag{19.7}$$

for $t \geq t_0$.

### Taylor Series for Vector Processes

Equations ($19.1$) through ($19.7$) are readily generalized for nonlinear $m$-vector functions of $n$-vector arguments, $\mathbf{f}(\mathbf{x})$, where

$$\mathbf{f} \equiv \begin{bmatrix} f_1 \\ f_2 \\ \vdots \\ f_m \end{bmatrix} \qquad \mathbf{x} \equiv \begin{bmatrix} x_1 \\ x_2 \\ \vdots \\ x_n \end{bmatrix}$$

and $m$ and $n$ are arbitrary. In this case, $\Delta\mathbf{x} \equiv \mathbf{x} - \bar{\mathbf{x}}$, $\Delta\mathbf{f} \equiv \mathbf{f}(\mathbf{x}) - \mathbf{f}(\bar{\mathbf{x}})$, and Equation ($19.6$) becomes

$$\Delta\mathbf{f} \cong \left.\frac{d\mathbf{f}}{d\mathbf{x}}\right|_{x=\bar{x}} \Delta\mathbf{x} \tag{19.8}$$

where $d\mathbf{f}/d\mathbf{x}$ is a matrix defined as

$$\frac{d\mathbf{f}}{d\mathbf{x}} = \begin{bmatrix} \dfrac{\partial f_1}{\partial x_1} & \dfrac{\partial f_1}{\partial x_2} & \cdots & \dfrac{\partial f_1}{\partial x_n} \\ \vdots & \vdots & \ddots & \vdots \\ \dfrac{\partial f_m}{\partial x_1} & \dfrac{\partial f_m}{\partial x_2} & \cdots & \dfrac{\partial f_m}{\partial x_n} \end{bmatrix} \tag{19.9}$$

**EXAMPLE 19.7.**  For $m = 1$ and $n = 2$, Equation ($19.9$) reduces to

$$\frac{df}{d\mathbf{x}} = \begin{bmatrix} \dfrac{\partial f}{\partial x_1} & \dfrac{\partial f}{\partial x_2} \end{bmatrix}$$

and Equation ($19.8$) is

$$\Delta f \cong \left[ \frac{\partial f}{\partial x_1} \quad \frac{\partial f}{\partial x_2} \right] \left[ \begin{matrix} \Delta x_1 \\ \Delta x_2 \end{matrix} \right] = \frac{\partial f}{\partial x_1} \Delta x_1 + \frac{\partial f}{\partial x_2} \Delta x_2 \qquad (19.10)$$

Equation ($19.10$) represents the common case where a nonlinear scalar function $f$ of two variables, say $x_1 \equiv x$ and $x_2 \equiv y$, are linearized about a point $\{\bar{x}, \bar{y}\}$ in the plane.

### Linearization of Nonlinear Differential Equations

We follow the same procedure to linearize differential equations as we did above in linearizing functions $\mathbf{f}(x)$. Consider a **nonlinear differential system** written in state variable form:

$$\frac{d\mathbf{x}}{dt} = \mathbf{f}[\mathbf{x}(t), \mathbf{u}(t)] \qquad (19.11)$$

where the vector of $n$ state variables $\mathbf{x}(t)$ and the $r$-input vector $\mathbf{u}(t)$ are defined as in Chapter 3, Equations ($3.24$) and ($3.25$), and $t \geq t_0$. In Equation ($19.11$), $\mathbf{f}$ is an $n$-vector of nonlinear functions of $\mathbf{x}(t)$ and $\mathbf{u}(t)$.

Similarly, **nonlinear output equations** may be written in vector form:

$$\mathbf{y}(t) = \mathbf{g}[\mathbf{x}(t)] \qquad (19.12)$$

where $\mathbf{y}(t)$ is an $m$-vector of outputs and $\mathbf{g}$ is an $m$-vector of nonlinear functions of $\mathbf{x}(t)$.

**EXAMPLE 19.8.** One example of a nonlinear SISO differential system of the form of Equations ($19.11$) and ($19.12$) is

$$\frac{dx_1}{dt} = f_1(\mathbf{x}, u) = c_1 u x_2 - c_2 x_1^2$$

$$\frac{dx_2}{dt} = f_2(\mathbf{x}, u) = \frac{c_3 x_1}{c_4 + x_1}$$

$$y = g(\mathbf{x}) = c_5 x_1^2$$

The *linearized versions* of Equations ($19.11$) and ($19.12$) are given by

$$\frac{d(\Delta \mathbf{x})}{dt} \cong \frac{\partial \mathbf{f}}{\partial \mathbf{x}} \bigg|_{\substack{\mathbf{x}=\bar{\mathbf{x}}(t) \\ \mathbf{u}=\bar{\mathbf{u}}(t)}} \Delta \mathbf{x} + \frac{\partial \mathbf{f}}{\partial \mathbf{u}} \bigg|_{\substack{\mathbf{x}=\bar{\mathbf{x}}(t) \\ \mathbf{u}=\bar{\mathbf{u}}(t)}} \Delta \mathbf{u} \qquad (19.13)$$

$$\Delta y(t) \cong \frac{\partial \mathbf{g}}{\partial \mathbf{x}} \bigg|_{\mathbf{x}=\bar{\mathbf{x}}(t)} \Delta \mathbf{x} \qquad (19.14)$$

where the partial derivative matrices in these equations are defined as in Equations ($19.9$) and ($19.10$), each evaluated at the "point" $\{\bar{\mathbf{x}}, \bar{\mathbf{u}}\}$. The pair $\bar{\mathbf{x}} \equiv \bar{\mathbf{x}}(t)$ and $\bar{\mathbf{u}} \equiv \bar{\mathbf{u}}(t)$ are actually functions of time, but they are treated like "points" in the indicated computations.

Linearized equations ($19.13$) and ($19.14$) are usually interpreted as follows. If the input is perturbed or deviates from an "operating point" $\bar{\mathbf{u}}(t)$ by a small enough amount $\Delta \mathbf{u}(t)$, generating small enough perturbations $\Delta \mathbf{x}(t)$ in the state and small enough perturbations in the output $\Delta \mathbf{y}(t)$ about their operating points, then the *linear* equations ($19.13$) and ($19.14$) are reasonable approximation equations for the perturbed states $\Delta \mathbf{x}(t)$ and perturbed outputs $\Delta \mathbf{y}(t)$.

Linearized equations ($19.13$) and ($19.14$) are often called the (*small*) **perturbation equations** for the nonlinear differential system. They are *linear* in $\Delta \mathbf{x}$ and ($\Delta \mathbf{u}$), because the coefficient matrices:

$$\frac{\partial \mathbf{f}}{\partial \mathbf{x}} \bigg|_{\substack{\mathbf{x}=\bar{\mathbf{x}}(t) \\ \mathbf{u}=\bar{\mathbf{u}}(t)}} \qquad \frac{\partial \mathbf{f}}{\partial \mathbf{u}} \bigg|_{\substack{\mathbf{x}=\bar{\mathbf{x}}(t) \\ \mathbf{u}=\bar{\mathbf{u}}(t)}} \qquad \frac{\partial \mathbf{g}}{\partial \mathbf{x}} \bigg|_{\mathbf{x}=\bar{\mathbf{x}}(t)}$$

having been evaluated at $\bar{\mathbf{x}}(t)$ and/or $\bar{\mathbf{u}}(t)$, are *not* functions of $\Delta \mathbf{x}(t)$ [or $\Delta \mathbf{u}(t)$].

Linearized equations ($19.13$) and ($19.14$) are also *time-invariant* if $\bar{\mathbf{u}}(t) = \bar{\mathbf{u}} = $ constant and $\bar{\mathbf{x}}(t) = \bar{\mathbf{x}}$ = constant. In this case, all of the methods developed in this book for time-invariant ordinary differential systems can be applied. Nevertheless, the results must be interpreted judiciously because, again, the linearized model is an approximation, valid only for "small enough" perturbations about an operating point and, generally speaking, "small enough" perturbations are not always easy to ascertain.

**EXAMPLE 19.9.**  The linearized (perturbation) equations for the system given in Example 19.8 are determined as follows from Equations ($19.13$) and ($19.14$). For convenience, we first define

$$\left. \frac{\partial f}{\partial x} \right|_{\substack{x = \bar{x}(t) \\ u = \bar{u}(t)}} \equiv \frac{\partial \bar{f}}{\partial x}$$

etc., to simplify the notation. Then

$$\frac{d(\Delta x_1)}{dt} = \frac{\partial \bar{f}_1}{\partial x_1} \Delta x_1 + \frac{\partial \bar{f}_1}{\partial x_2} \Delta x_2 + \frac{\partial \bar{f}_1}{\partial u} \Delta u = -2 c_2 \bar{x}_1 \Delta x_1 + c_1 \bar{u} \Delta x_2 + c_1 \bar{x}_2 \Delta u$$

Similarly,

$$\frac{d(\Delta x_2)}{dt} \cong \frac{\partial \bar{f}_2}{\partial x_1} \Delta x_1 + \frac{\partial \bar{f}_2}{\partial x_2} \Delta x_2 + \frac{\partial \bar{f}_2}{\partial u} \Delta u$$

$$= \frac{c_3 c_4}{\left( c_4 + \bar{x}_1 \right)^2} \Delta x_1 + 0 + 0 = \frac{c_3 c_4 \, \Delta x_1}{\left( c_4 + \bar{x}_1 \right)^2}$$

and the output perturbation equation is

$$\Delta y \cong \frac{\partial \bar{g}}{\partial x_1} \Delta x_1 + \frac{\partial \bar{g}}{\partial x_2} \Delta x_2 = 2 c_5 \bar{x}_1 \Delta x_1$$

### Linearization of Nonlinear Discrete-Time Equations

The Taylor series linearization procedure can be applied to many discrete-time system problems, but sufficient care must be taken to justify the existence of the series. The application is often justified if the discrete-time equations represent reasonably well-behaved nonlinear processes, such as discrete-time representations of continuous systems with state variables expressed only at discrete-time instants.

**EXAMPLE 19.10.**  The time-invariant discrete-time system represented by the nonlinear difference equation $x(k + 1) = a x^2(k)$, with $a < 0$ and $x(0) \neq 0$, is easily linearized, because the nonlinear term $a x^2(k)$ is a smooth function of $x$. We have

$$x(k + 1) = a x^2(k) \equiv f(x)$$

$$\Delta f = f(x) - f(\bar{x})$$

$$\left. \frac{\partial f}{\partial x} \right|_{x = \bar{x}} = 2 a \bar{x}$$

$$x(k) = \bar{x}(k) + \Delta x(k)$$

$$\bar{x}(k + 1) = a \bar{x}^2(k)$$

Substitution of these equations into Equation ($19.6$) and rearranging terms yields

$$\Delta x(k + 1) \cong 2 a \bar{x}(k) \, \Delta x(k)$$

which is linear in $\Delta x$, but time-varying in general.

## 19.3   PHASE PLANE METHODS

In Sections 3.15 and 4.6, the state variable form of linear differential equations was introduced and shown to be a useful tool for analysis of linear systems. In Section 19.2, this representation was applied to nonlinear systems via the concept of linearization. In this section, **phase plane** methods are developed for analyzing nonlinear differential equations in state variable form, without the need for linearization.

A second-order differential equation of the form:

$$\frac{d^2x}{dt^2} = f\left(x, \frac{dx}{dt}\right) \qquad (19.15)$$

can be rewritten as a pair of first-order differential equations, as in Section 3.15, by making the change of variables $x = x_1$ and $dx/dt = x_2$, yielding

$$\frac{dx_1}{dt} = x_2 \qquad (19.16)$$

$$\frac{dx_2}{dt} = f(x_1, x_2) \qquad (19.17)$$

The two-tuple, or pair of state variables $(x_1, x_2)$, may be considered as a point in the plane. Since $x_1$ and $x_2$ are functions of time, then as $t$ increases, $(x_1(t), (x_2(t))$ describes a *path* or *trajectory* in the plane. This plane is called the **phase plane**, and the trajectory is a parametric plot of $x_2$ versus $x_1$, parametrized by $t$.

If we eliminate time as the independent variable in Equations ($19.16$) and ($19.17$), we obtain the first-order differential equation

$$\frac{dx_1}{dx_2} = \frac{x_2}{f(x_1, x_2)} \qquad (19.18)$$

Solution of Equation ($19.18$) for $x_1$ as a function of $x_2$ (or vice versa) defines a trajectory in the phase plane. By solving this equation for various initial conditions on $x_1$ and $x_2$ and examining the resulting phase plane trajectories, we can determine the behavior of the second-order system.

**EXAMPLE 19.11.**   The differential equation

$$\frac{d^2x}{dt^2} + \left(\frac{dx}{dt}\right)^2 = 0$$

with the initial conditions $x(0) = 0$ and $(dx/dt)|_{t=0} = 1$, can be replaced by the two first-order equations

$$\frac{dx_1}{dt} = x_2 \qquad x_1(0) = 0$$

$$\frac{dx_2}{dt} = -x_2^2 \qquad x_2(0) = 1$$

where $x \equiv x_1$ and $dx/dt \equiv x_2$. Eliminating time as the independent variable, we obtain

$$\frac{dx_1}{dx_2} = -\frac{x_2}{x_2^2} = -\frac{1}{x_2} \qquad \text{or} \qquad dx_1 = -\frac{dx_2}{x_2}$$

Integration of this equation for the given initial conditions yields

$$\int_{x_1(0)=0}^{x_1} dx_1' = x_1 = -\int_{x_2(0)=1}^{x_2} \frac{dx_2'}{x_2'} = -\ln x_2 \qquad \text{or} \qquad x_2 = e^{-x_1}$$

The phase plane trajectory defined by this equation is plotted in Fig. 19-6. Its direction in the phase plane is

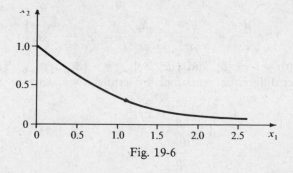

Fig. 19-6

determined by noting that $dx_2/dt = -x_2^2 < 0$ for all $x_2 \neq 0$. Therefore $x_2$ always decreases and we obtain the trajectory shown.

## On-Off Control Systems

A particularly useful application of phase plane methods is designing *on-off controllers* (Definition 2.25), for the special class of feedback control systems with linear continuous-time second-order plants, as in Fig. 19-7 and Equation (*19.19*).

$$\frac{d^2c}{dt^2} + a\frac{dc}{dt} = u \qquad a \geq 0 \tag{19.19}$$

The initial conditions $c(0)$ and $(dc/dt)|_{t=0}$ for Equation (*19.19*) are arbitrary. The on-off controller with input $e = r - c$ generates the control signal $u$ which attains only two values, $u = \pm 1$.

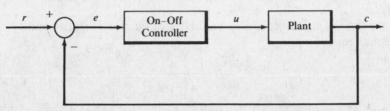

Fig. 19-7

## On-Off Controller Design Specifications

If the reference input $r$ is a unit step function applied at time zero, typical design specifications for the system of Fig. 19-7 are the following. The control input $u$ to the plant must drive the plant output $c(t)$ to $c(t') = 1$, and its derivative $dc/dt$ to $(dc/dt)|_{t=t'} = 0$, simultaneously, *and* in the minimum possible time $t'$. The steady state error becomes zero at $t'$ and remains zero if the control signal is turned off ($u = 0$).

Since $t'$ is required to be minimum, this is an *optimal control problem* (see Section 20.5). It can be shown that $t'$ is minimized only if the control signal $u$ switches values, from $+1$ to $-1$ or from $-1$ to $+1$, at most once during the time interval $0 \leq t \leq t'$.

## On-Off Controller Design

In solving this design problem, it is convenient to use the error $e = r - c$, where $r = \mathbf{1}(t)$, as the variable of interest, rather than the controlled output $c$, because $e = 0$ and $de/dt = 0$ when $c = 1$ and $dc/dt = 0$. Therefore requiring that the error $e$ and its derivative go to zero in minimum time is equivalent to our original problem.

To solve the problem, we first generate a differential equation for $e$:

$$\frac{de}{dt} = \frac{d}{dt}(r - c) = -\frac{dc}{dt}$$

$$\frac{d^2e}{dt^2} = -\frac{d^2c}{dt^2} = a\frac{dc}{dt} - u = -a\frac{de}{dt} - u \tag{19.20}$$

with initial conditions $e(0) = 1 - c(0)$ and $(de/dt)|_{t=0} = -(dc/dt)|_{t=0}$. Then we replace Equation (*19.20*) with two first-order differential equations, by letting $e \equiv x_1$ and $de/dt \equiv x_2$:

$$\frac{dx_1}{dt} = x_2 \tag{19.21}$$

$$\frac{dx_2}{dt} = -ax_2 - u \tag{19.22}$$

with initial conditions $x_1(0) = e(0) = 1 - c(0)$ and $x_2(0) = (de/dt)|_{t=0} = -(dc/dt)|_{t=0}$. Eliminating time as the independent variable, we obtain

$$\frac{dx_2}{dx_1} = -\frac{ax_2 + u}{x_2} \qquad \text{or} \qquad dx_1 = -\frac{x_2\, dx_2}{ax_2 + u} \qquad (19.23)$$

This equation plus the initial conditions on $x_1(0)$ and $x_2(0)$ define a trajectory in the phase plane.

Since the control signal $u$ switches ($+1$ to $-1$ or $-1$ to $+1$) no more than once, we can separate the trajectory into two parts, the first prior to the switching time and the second after switching. We consider the second part first, as it terminates at the origin of the phase plane, $x_1 = x_2 = 0$. We set $u = \pm 1$ in Equation $(19.23)$ and then integrate between a general set of initial conditions $x_1(t)$ and $x_2(t)$ and the terminal conditions $x_1 = x_2 = 0$. To perform the integration, we consider four different sets of initial conditions, each corresponding to one of the quadrants of the phase plane.

In the first quadrant, $x_1 > 0$ and $x_2 > 0$. Note that $dx_1/dt = x_2 > 0$. Thus $x_1$ increases when $x_2$ is in the first quadrant, and when $x_2$ goes to zero, $x_1$ cannot be zero. Therefore trajectories which start in the first quadrant cannot terminate at the origin of the phase plane if $u$ does not switch.

An identical argument holds when the initial conditions are in the third quadrant, that is, if $x_1 < 0$ and $x_2 < 0$, the trajectory cannot terminate at the origin if $u$ does not switch.

In the second quadrant, $x_1 < 0$ and $x_2 > 0$. Since $dx_1/dt = x_2 > 0$, $x_1$ will increase as long as $x_2 > 0$. Since $a > 0$, then $-ax_2 < 0$ and thus $dx_2/dt < 0$ for $u = +1$ whenever $x_2 > 0$. Integration of Equation $(19.23)$ with $u = +1$, initial conditions in the second quadrant, and terminal conditions $x_1 = x_2 = 0$, yields

$$\int_{x_1(t)}^{0} dx_1 = -x_1(t) = -\int_{x_2(t)}^{0} \frac{x_2\, dx_2}{ax_2 + 1}$$

or
$$x_1(t) = \frac{1}{a^2}\left[ ax_2 + 1 - \ln(ax_2 + 1) \right]\Bigg|_{x_2(t)}^{0} = -\frac{x_2(t)}{a} + \frac{1}{a^2}\ln\left[ ax_2(t) + 1 \right] \qquad (19.24)$$

where $x_1(t) \le 0$, $x_2(t) \ge 0$. This equation defines a curve in the second quadrant of the phase plane such that, for any point on this curve, the trajectory terminates at the origin if $u = +1$. That is, the control signal $u = +1$ drives $x_1$ and $x_2$ to zero simultaneously.

By an identical argument, there exists a curve in the fourth quadrant defined by

$$x_1(t) = -\frac{x_2(t)}{a} - \frac{1}{a^2}\ln\left[ -ax_2(t) + 1 \right] \qquad (19.25)$$

where $x_1(t) \ge 0$, $x_2(t) \le 0$ such that for any $(x_1(t), x_2(t))$ on this curve the control signal $u = -1$ drives $x_1$ and $x_2$ to zero simultaneously.

The curves defined by Equations $(19.24)$ and $(19.25)$ join at $x_1 = x_2 = 0$ and together define the **switching curve** for the on-off controller. The switching curve divides the entire phase plane into two regions, as indicated in Fig. 19-8. The part of any trajectory after switching always starts on this curve, moves along the curve, and terminates at $x_1 = x_2 = 0$.

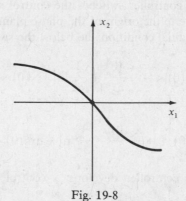

Fig. 19-8

Now we consider the part of the trajectory prior to switching. First, we explore a monotone property of the switching curve. In the second quadrant, where $u = +1$, $x_2 > 0$, and the slope of the curve is negative:

$$\frac{dx_2}{dx_1} = -\left(a + \frac{1}{x_2}\right) < 0$$

In the fourth quadrant, where $u = -1$, $x_2 < 0$, and

$$\frac{dx_2}{dx_1} = -\left(a - \frac{1}{x_2}\right) < 0$$

Therefore the slope of the entire switching curve is negative for all $(x_1, x_2)$ on the curve, that is, the switching curve is *monotone decreasing*. Thus, corresponding to any specific value of $x_1$, there is one and only one corresponding value of $x_2$. Because of the monotone property of the switching curve, the region above the switching curve is the same as the region to the right of the switching curve, that is, it consists of the set of points $(x_1, x_2)$ such that

$$x_1 > -\frac{x_2}{a} + \frac{1}{a^2}\ln(ax_2 + 1) \tag{19.26}$$

when $x_2 \geq 0$ and

$$x_1 > -\frac{x_2}{a} - \frac{1}{a^2}\ln(-ax_2 + 1) \tag{19.27}$$

when $x_2 \leq 0$.

We consider the part of the trajectory prior to switching, when the conditions $(x_1(0), x_2(0))$ lie above the switching curve. For this case, $u = +1$ and the first part of the trajectory is obtained by integration of Equation (19.23) with $u = +1$ between the initial conditions $(x_1(0), x_2(0))$ and an arbitrary pair of points $(x_1(t), x_2(t))$ which satisfy the inequalities (19.26) and (19.27). We obtain the trajectory by integrating Equation (19.23), which yields

$$\int_{x_1(0)}^{x_1(t)} dx_1 = x_1(t) - x_1(0) = -\int_{x_2(0)}^{x_2(t)} \frac{x_2 \, dx_2}{ax_2 + 1} = -\frac{1}{a^2}\left[ax_2 + 1 - \ln(ax_2 + 1)\right]\Bigg|_{x_2(0)}^{x_2(t)}$$

or
$$x_1(t) = x_1(0) + \frac{x_2(0)}{a} - \frac{1}{a^2}\ln\left[ax_2(0) + 1\right] - \frac{x_2(t)}{a} + \frac{1}{a^2}\ln\left[ax_2(t) + 1\right] \tag{19.28}$$

Note that this part of the trajectory has the same shape as that in Equation (19.24), but that it is shifted to the right. So, when $x_2(t) = 0$, $x_1(t) = x_1(0) + (1/a)[x_2(0) - (1/a)\ln(ax_2(0) + 1)]$, which is greater than 0 because of inequality (19.26).

Thus, when $(x_1(0), x_2(0))$ lies above the switching curve, the on-off controller develops a control signal $u = +1$ and the resultant trajectory $(x_1(t), x_2(t))$ is defined by Equation (19.28). When this trajectory intersects the switching curve, that is, when $(x_1(t), x_2(t))$ satisfies Equations (19.25) and (19.28) simultaneously, the on-off controller switches the control signal to $u = -1$ and the trajectory continues along the switching curve to the origin of the phase plane.

By identical reasoning, if the initial conditions lie below the switching curve, that is,

$$x_1(0) < -\frac{x_2(0)}{a} + \frac{1}{a^2}\ln\left[ax_2(0) + 1\right]$$

when $x_2(0) \geq 0$, or

$$x_1(0) < -\frac{x_2(0)}{a} - \frac{1}{a^2}\ln\left[-ax_2(0) + 1\right]$$

when $x_2(0) \leq 0$, then the on-off controller develops a control signal $u = -1$ and the trajectory

$(x_1(t), x_2(t))$ satisfies

$$x_1(t) = x_1(0) + \frac{x_2(0)}{a} + \frac{1}{a^2}\ln[-ax_2(0) + 1] - \frac{x_2(t)}{a} - \frac{1}{a^2}\ln[-ax_2(t) + 1] \qquad (19.29)$$

When this trajectory intersects the switching curve, that is, when $(x_1(t), x_2(t))$ satisfies Equations (19.24) and (19.29) simultaneously, the on-off controller switches the control signal to $u = +1$ and the trajectory moves along the switching curve in the second quadrant and terminates at the origin of the phase plane.

Recalling that $x_1 \equiv e$ and $x_2 \equiv \dot{e}$, the switching logic of the on-off controller is as follows:

(a)   When $\dot{e} > 0$ and $e + \dfrac{\dot{e}}{a} - \dfrac{1}{a^2}\ln(a\dot{e} + 1) > 0$, then $u = +1$

(b)   When $\dot{e} < 0$ and $e + \dfrac{\dot{e}}{a} + \dfrac{1}{a^2}\ln(-a\dot{e} + 1) > 0$, then $u = +1$

(c)   When $\dot{e} > 0$ and $e + \dfrac{\dot{e}}{a} - \dfrac{1}{a^2}\ln(a\dot{e} + 1) < 0$, then $u = -1$

(d)   When $\dot{e} < 0$ and $e + \dfrac{\dot{e}}{a} + \dfrac{1}{a^2}\ln(-a\dot{e} + 1) < 0$, then $u = -1$

**EXAMPLE 19.12.**   For the feedback control system depicted in Fig. 19-7 and plant defined by Equation (19.19) with parameter $a = 1$, the switching curve is defined by

$$e = -\dot{e} + \ln(\dot{e} + 1) \qquad \text{for} \quad \dot{e} > 0$$
$$e = -\dot{e} - \ln(-\dot{e} + 1) \qquad \text{for} \quad \dot{e} < 0$$

and the switching logic for the on-off controller is given in Table 19.1.

**Table 19.1**

| $\dot{e} > 0$ | $f_1(e) = e + \dot{e} - \ln(\dot{e} + 1) > 0$ | $f_2(e) = e + \dot{e} + \ln(-\dot{e} + 1) > 0$ | $u$ |
|---|---|---|---|
| No | No | No | $-1$ |
| No | No | Yes | $+1$ |
| No | Yes | No | $-1$ |
| No | Yes | Yes | $+1$ |
| Yes | No | No | $-1$ |
| Yes | No | Yes | $-1$ |
| Yes | Yes | No | $+1$ |
| Yes | Yes | Yes | $+1$ |

### Generalization

Phase plane methods apply to second-order systems. The approach has been generalized to third- and higher-order systems, but the analysis is typically much more complex. For example, to design on-off controllers in this way for third-order systems, switching curves are replaced by switching *surfaces* and the switching logic becomes far more extensive than that given in Table 19.1 for second-order systems.

## 19.4   LYAPUNOV'S STABILITY CRITERION

The stability criteria presented in Chapter 5 cannot be applied to nonlinear systems in general, although they may be applicable if the system is linearized, as in Section 19.2, if the perturbations $\Delta \mathbf{x}$ are small enough, and if $\bar{\mathbf{u}}(t)$ and $\bar{\mathbf{x}}(t)$ are constant, that is, if the linearized equations are time-invariant. A more general method is provided by the Lyapunov theory, for exploring the stability of system states $\mathbf{x}(t)$ and outputs $\mathbf{y}(t)$ in the time domain, for any size perturbations $\Delta\mathbf{x}(t)$. It can be used for both linear and nonlinear systems described by sets of simultaneous first-order ordinary differential or difference equations, which we write concisely here in state variable form:

$$\dot{\mathbf{x}} = \mathbf{f}(\mathbf{x}, \mathbf{u}) \qquad (19.30)$$

or

$$\mathbf{x}(k+1) = \mathbf{f}[\mathbf{x}(k), \mathbf{u}(k)] \qquad (19.31)$$

The following stability definitions are for unforced systems, that is, for $\mathbf{u} = \mathbf{0}$, and for simplicity we write $\dot{\mathbf{x}} = \mathbf{f}(\mathbf{x})$ or $\mathbf{x}(k+1) = \mathbf{f}[\mathbf{x}(k)]$.

A point $\mathbf{x}_s$ for which $\mathbf{f}(\mathbf{x}_s) = \mathbf{0}$ is called a **singular point**. A singular point $\mathbf{x}_s$ is said to be **stable** if, for any hyperspherical region $S_R$ (e.g., a circle in two dimensions) of radius $R$ centered at $\mathbf{x}_s$, there exists a hyperspherical region $S_r$ of radius $r \le R$ also centered at $\mathbf{x}_s$ in which any motion $\mathbf{x}(t)$ of the system beginning in $S_r$ remains in $S_R$ ever after.

A singular point $\mathbf{x}_s$ is **asymptotically stable** if it is stable and all trajectories (motions) $\mathbf{x}(t)$ tend toward $\mathbf{x}_s$ as time goes to infinity.

The **Lyapunov stability criterion** states that, if the origin is a singular point, then it is stable if a **Lyapunov function** $V(\mathbf{x})$ can be found with the following properties:

(a) $V(\mathbf{x}) > 0$ for all values of $\mathbf{x} \ne \mathbf{0}$ \hfill $(19.32)$

(b) $dV/dt \le 0$ for all $\mathbf{x}$, for continuous systems, or $\Delta V[\mathbf{x}(k)] \equiv V[\mathbf{x}(k+1)] - V[\mathbf{x}(k)] \le 0$, for all $\mathbf{x}$, for discrete-time systems \hfill $(19.33)$

Furthermore, if $dV/dt$ (or $\Delta V$) is never zero except at the origin, the origin is *asymptotically stable*.

**EXAMPLE 19.13.** A nonlinear continuous system represented by

$$\frac{d^2x}{dt^2} + \frac{dx}{dt} + \left(\frac{dx}{dt}\right)^3 + x = 0$$

or, equivalently, the pair of equations

$$\frac{dx_1}{dt} = x_2 \qquad \frac{dx_2}{dt} = -x_2 - x_2^3 - x_1$$

where $x_1 \equiv x$, has a singular point at $x_1 = x_2 = 0$. The function $V = x_1^2 + x_2^2$ is positive for all $x_1$ and $x_2$, except $x_1 = x_2 = 0$ where $V = 0$. The derivative

$$\frac{dV}{dt} = 2x_1\frac{dx_1}{dt} + 2x_2\frac{dx_2}{dt} = 2x_1x_2 + 2x_2(-x_2 - x_2^3 - x_1) = -2x_2^2 - 2x_2^4$$

is never positive. Therefore the origin is stable.

**EXAMPLE 19.14.** The nonlinear system shown in Fig. 19-9 is represented by the differential equations [with $x_1(t) \equiv -c(t)$]:

$$\dot{x}_1 = -x_1 + x_2$$
$$\dot{x}_2 = -f(x_1 + r)$$

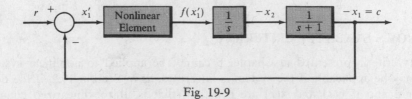

Fig. 19-9

Also, $f(0) \equiv 0$ for this particular nonlinear element. If $r$ is constant, we can make the changes of variables $x_1' \equiv x_1 + r$, $x_2' \equiv x_2 + r$, and the state equations become

$$\dot{x}_1' = -x_1' + x_2'$$
$$\dot{x}_2' = -f(x_1')$$

The origin $x_1' = x_2' = 0$ is a singular point since $\dot{x}_1' = \dot{x}_2' = 0$ at the origin. The Lyapunov function is defined by $V \equiv 2\int_0^{x_1'} f(e)\, de + x_2'^2 > 0$ for all $x_1'$, $x_2' \neq 0$, if $x_1' f(x_1') > 0$ for all $x_1' \neq 0$. Differentiating $V$,

$$\dot{V} = 2f(x_1')\dot{x}_1' + 2x_2'\dot{x}_2' = 2f(x_1')(-x_1' + x_2') - 2x_2'f(x_1') = -2x_1'f(x_1')$$

Thus, if we restrict $x_1' f(x_1') > 0$, to maintain $V > 0$, $\dot{V} \leq 0$ for $x_1' \neq 0$. Therefore the system is stable for any nonlinear element satisfying the conditions

$$f(0) = 0$$
$$x_1' f(x_1') > 0 \qquad \text{for} \quad x_1' \neq 0$$

Note that this result is very general in that only the conditions above are required to assure stability.

If $r$ is not constant, the solution for $x_1(t)$ and $x_2(t)$ corresponding to $r(t)$ is in general not constant. But, if the solution were known, the stability of the solution could be analyzed in a similar way.

**EXAMPLE 19.15.**   For the discrete-time system

$$x_1(k+1) = x_2(k)$$
$$x_2(k+1) = -f[x_1(k)]$$

where $f(x_1)$ is the saturation nonlinearity in Fig. 19-1($b$), the origin is a singular point because $x_1(k) = x_2(k) = 0$ implies $x_1(k+1) = x_2(k+1) = 0$. Let $V \equiv x_1^2 + x_2^2$, which is greater than zero for all $x_1$, $x_2 \neq 0$. Then

$$\Delta V = x_1^2(k+1) + x_2^2(k+1) - x_1^2(k) - x_2^2(k)$$
$$= x_2^2(k) + f^2[x_1(k)] - x_1^2(k) - x_2^2(k)$$
$$= -x_1^2(k) + f^2[x_1(k)]$$

Since $f^2(x_1) \leq x_1^2$ for all $x_1$, $\Delta V \leq 0$ for all $x_1$, $x_2$ and therefore the origin is stable.

### Choosing Lyapunov Functions

For many problems, a convenient choice for the Lyapunov function $V(\mathbf{x})$ is the scalar quadratic form function $V(\mathbf{x}) = \mathbf{x}^T P \mathbf{x}$, where $\mathbf{x}^T$ is the transpose of the column vector $\mathbf{x}$ and $P$ is a real symmetric matrix. To render $V > 0$, the matrix $P$ must be *positive definite*. By Sylvester's theorem [7], $P$ is **positive definite** if and only if all its discriminants are positive, that is,

$$P_{11} > 0$$

$$\begin{vmatrix} P_{11} & P_{12} \\ P_{21} & P_{22} \end{vmatrix} > 0$$

$$\vdots$$

$$\begin{vmatrix} P_{11} & \cdots & P_{1n} \\ \vdots & \vdots & \vdots \\ P_{n1} & \cdots & P_{nn} \end{vmatrix} > 0 \qquad\qquad (19.34)$$

For continuous systems $\dot{\mathbf{x}} = \mathbf{f}(\mathbf{x})$, the derivative of $V(\mathbf{x}) = \mathbf{x}^T P \mathbf{x}$ is given by

$$\dot{V}(\mathbf{x}) = \dot{\mathbf{x}}^T P \mathbf{x} + \mathbf{x}^T P \dot{\mathbf{x}} = \mathbf{f}^T(\mathbf{x}) P \mathbf{x} + \mathbf{x}^T P \mathbf{f}(\mathbf{x})$$

For discrete systems, $\mathbf{x}(k+1) = \mathbf{f}[\mathbf{x}(k)]$ and

$$\Delta V(k) = V(k+1) - V(k) = \mathbf{x}^T(k+1) P \mathbf{x}(k+1) - \mathbf{x}^T(k) P \mathbf{x}(k)$$
$$= \mathbf{f}^T[\mathbf{x}(k)] P \mathbf{f}[\mathbf{x}(k)] - \mathbf{x}^T(k) P \mathbf{x}(k)$$

**EXAMPLE 19.16.** For the system represented by $\dot{\mathbf{x}} = A\mathbf{x}$ with $A = \begin{bmatrix} -2 & 1 \\ 2 & -3 \end{bmatrix}$, let $V = \mathbf{x}^T P \mathbf{x}$ with $P = \begin{bmatrix} 1 & 0 \\ 0 & 1 \end{bmatrix}$. Then

$$\dot{V} = \mathbf{x}^T[A^T P + PA]\mathbf{x} = \mathbf{x}^T\left[\begin{bmatrix} -2 & 2 \\ 1 & -3 \end{bmatrix} + \begin{bmatrix} -2 & 1 \\ 2 & -3 \end{bmatrix}\right]\mathbf{x}$$

$$\dot{V} = \mathbf{x}^T\begin{bmatrix} -4 & 3 \\ 3 & -6 \end{bmatrix}\mathbf{x} = -\mathbf{x}^T Q \mathbf{x}$$

where

$$Q = \begin{bmatrix} 4 & -3 \\ -3 & 6 \end{bmatrix}$$

Since $P$ is positive definite, $V > 0$ for all $\mathbf{x} \neq 0$. The discriminants of $Q$ are 4 and $(24 - 9) = 15$. Therefore $Q$ is positive definite and $-Q$ is negative definite, which guarantees that $\dot{V} < 0$ for all $\mathbf{x} \neq 0$. The origin is therefore asymptotically stable for this system.

## 19.5  FREQUENCY RESPONSE METHODS

### Describing Functions

Describing functions are approximate frequency response functions for the nonlinear elements of a system, which can be used to analyze the overall system using frequency response techniques developed in earlier chapters.

A describing function is developed for a nonlinear element by analyzing its response to a sinusoidal input $A \sin \omega t$, which can be written as a Fourier series:

$$\sum_{n=1}^{\infty} B_n \sin(n\omega t + \phi_n) \tag{19.35}$$

The **describing function** is the ratio of the complex Fourier coefficient $B_1 e^{j\phi_1}$ of the fundamental frequency of this output, to the amplitude $A$ of the input. That is, the *describing function* is the complex function of $\omega$, $(B_1/A)e^{j\phi_1}$, a frequency response function of an approximation of the nonlinear element. Thus the describing function represents the effective gain of the nonlinear element at the frequency of the input sinusoid.

In general, $B_1$ and $\phi_1$ are functions of both the input frequency $\omega = 2\pi/T$ and the input amplitude $A$. Therefore we may write $B_1 = B_1(A, \omega)$, $\phi_1 = \phi_1(A, \omega)$ and the describing function as

$$\overline{N}(A, \omega) = \frac{B_1 e^{j\phi_1}}{A} = \frac{B_1(A, \omega) e^{j\phi_1(A, \omega)}}{A} \tag{19.36}$$

To apply the method, we replace system nonlinearities by describing functions and then apply the frequency domain techniques of Chapters 11, 12, and 15 through 18, with some modifications to account for the dependence of $B_1$ and $\phi_1$ on $A$.

**EXAMPLE 19.17.** The output of the nonlinear function $f(e) = e^3$ in response to an input $e = A \sin \omega t$ is

$$f(e) = A^3 \sin^3 \omega t = \frac{A^3}{4}(3 \sin \omega t - \sin^3 \omega t)$$

From Equation (19.36), the describing function for $f(e)$ is

$$\overline{N}(A) = \frac{3A^2}{4}$$

Note that this nonlinearity produces no phase shift, so that $\phi_1(A, \omega) = 0$.

**Hysteresis**

A common type of nonlinearity called **hysteresis** or **backlash** is shown in Fig. 19-10. In electrical systems, it may occur due to nonlinear electromagnetic properties and, in mechanical systems, it may result from backlash in gear trains or mechanical linkages. For another example, see Problem 2.16.

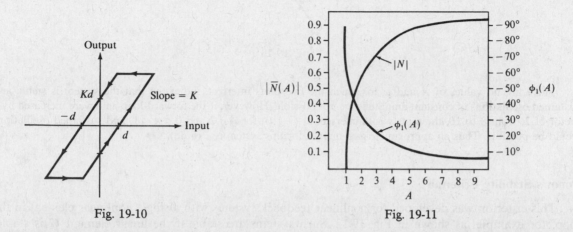

Fig. 19-10                    Fig. 19-11

The describing function characteristic for hysteresis, normalized to **dead zone** parameter $d = 1$ and slope $K = 1$, is shown in Fig. 19-11. The phase lag $\phi_1(A)$ of this describing function is a function of the input amplitude $A$, but is independent of the input frequency $\omega$.

The describing function technique is particularly well suited for analysis of continuous or discrete-time systems containing a single nonlinear element, as illustrated in Fig. 19-12, with open-loop transfer function $GH = \overline{N}(A, \omega)G(\omega)$. Frequency response analysis of such systems typically entails first determining whether there exist values of $A$ and $\omega$ that satisfy the characteristic equation, $1 + \overline{N}(A, \omega)G(\omega) = 0$, or

$$G(\omega) = -\frac{1}{\overline{N}(A, \omega)}$$

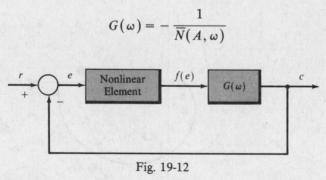

Fig. 19-12

that is, values of $A$ and $\omega$ permitting oscillations. Nyquist, Bode, or Nichols chart plots of $G$ and $-1/\overline{N}$ separately can be used to resolve this problem, because the plots must intersect if such $A$ and $\omega$ exist. Relative stability can also be evaluated from such plots, by determining the additional gain (gain margin) and/or phase shift (phase margin) required to have the curves intersect.

It must be kept in mind that the describing function is only an approximation for the nonlinearity. The accuracy of describing function methods, using frequency response analysis based on linear system methods, depends upon the effective filtering by the plant $G(\omega)$ of the (neglected) higher than first-order harmonics produced by the nonlinearity. Since most plants have more poles than zeros, it is often a reasonable approximation.

**EXAMPLE 19.18.**   Consider the system of Fig. 19-12 with $G(\omega) = 8/j\omega(j\omega + 2)^2$ and the saturation nonlinearity of Problem 19.17. Polar plots of $G(\omega)$ and $-1/\overline{N}(A)$ are shown in Fig. 19-13.

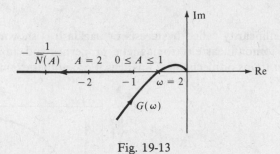

Fig. 19-13

There are no values of $A$ and $\omega$ for which the two plots intersect, indicating that the system is stable and sustained oscillations of constant amplitude are not possible. However, if the forward-loop gain were increased by a factor of 2, from 8 to 16, the plots would intersect at $(-1, 0)$ for $\omega = 2$ and $0 < A < 1$, and sustained oscillations would be possible. Thus an approximate gain margin for this system is 2 (6 db).

## Popov's Stability Criterion

This criterion was developed for nonlinear feedback systems with a single nonlinear element in the loop, for example, as shown in Fig. 19-12. Such systems are stable if the linear element $G$ is stable, $\operatorname{Re} G(\omega) > -1/K$, and the nonlinear element $f(e)$ satisfies the conditions: $f(0) = 0$ and $0 < f(e)/e < K$ for $e \neq 0$. Note that this criterion does not involve any approximations. Nyquist analysis is particularly well suited for its application.

**EXAMPLE 19.19.** For the system of Fig. 19-12, with $G = 1/(j\omega + 1)^3$, the Polar Plot is shown in Fig. 19-14. For all $\omega$, $\operatorname{Re} G \geq -1/4$. Therefore the nonlinear system is stable if $K < 4$, $f(0) = 0$, and $0 < f(e)/e < K$ for $e \neq 0$.

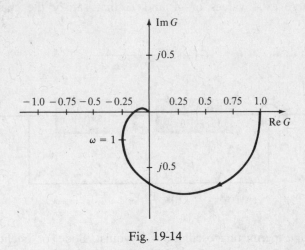

Fig. 19-14

**EXAMPLE 19.20.** For the nonlinear system in Fig. 19-12, with a stable discrete-time plant $G = 1/z$,

$$G(e^{j\omega T}) = e^{-j\omega T} = \cos \omega T - j \sin \omega T$$

The circular Polar Plot of $G$ is shown in Fig. 19-15, and

$$\operatorname{Re} G(e^{j\omega T}) > \frac{-1}{K} \quad \text{for} \quad K < 1$$

Thus the system is stable if $f(0) = 0$ and $0 < f(e)/e < K < 1$ for $e \neq 0$.

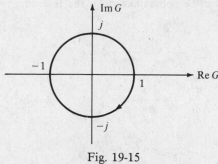

Fig. 19-15

# Solved Problems

## NONLINEAR CONTROL SYSTEMS

**19.1.** Several types of *control laws* or *control algorithms* were presented in Definitions 2.25 through 2.29. Which of these are nonlinear and which are linear, from the viewpoint of their input-output characteristics?

The on-off (binary) controller of Definition 2.25 is clearly nonlinear, its output being a discontinuous function of its input. The remaining controllers, that is, the proportional ($P$), derivative ($D$), integral ($I$) and PD, PI, DI, and PID controllers given in Definitions 2.26 through 2.29, are all linear. Each of their outputs are defined by linear operations, or linear combinations of linear operations, on each of their inputs.

**19.2.** Why is the thermostatically controlled heating system described in Problem 2.16 nonlinear?

The thermostat controller in this system is a nonlinear binary device, with a hysteresis input-output characteristic, as described in Problem 2.16. This controller regulates the room temperature output of this control system in an oscillatory manner between upper and lower limits bracketing the desired temperature setting. This type of behavior is characteristic of many nonlinear control systems.

## LINEARIZED AND PIECEWISE-LINEAR SYSTEM APPROXIMATIONS

**19.3.** The differential equation of a certain physical system is given by

$$\frac{d^3y}{dt^3} + 4\frac{d^2y}{dt^2} + f(y) = 0$$

The function $f(y)$ is nonlinear, but it can be approximated by the piecewise-linear graph illustrated in Fig. 19-16. Determine a piecewise-linear approximation for the nonlinear system differential equation.

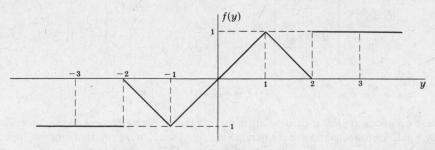

Fig. 19-16

The nonlinear system can be approximated by the following set of five linear equations over the indicated ranges of $y$:

$$\frac{d^3y}{dt^3} + 4\frac{d^2y}{dt^2} - 1 = 0 \qquad y < -2$$

$$\frac{d^3y}{dt^3} + 4\frac{d^2y}{dt^2} - y - 2 = 0 \qquad -2 \leq y < -1$$

$$\frac{d^3y}{dt^3} + 4\frac{d^2y}{dt^2} + y = 0 \qquad -1 \leq y \leq 1$$

$$\frac{d^3y}{dt^3} + 4\frac{d^2y}{dt^2} - y + 2 = 0 \qquad 1 < y \leq 2$$

$$\frac{d^3y}{dt^3} + 4\frac{d^2y}{dt^2} + 1 = 0 \qquad 2 < y$$

**19.4.** A solution of the nonlinear differential equation

Mathcad

$$\frac{d^2y}{dt^2} + y\cos y = u$$

with input $u = 0$, is $y = 0$. Linearize the differential equation about this input and output using a Taylor series expansion of the function $d^2y/dt^2 + y\cos y - u$ about the point $u = y = 0$.

The Taylor series expansion of $\cos y$ about $y = 0$ is

$$\cos y = \sum_{k=0}^{\infty} \frac{y^k}{k!}\left[\frac{d^k}{dy^k}(\cos y)\bigg|_{y=0}\right] = 1 - \frac{1}{2!}y^2 + \cdots$$

Therefore
$$\frac{d^2y}{dt^2} + y\cos y - u = \frac{d^2y}{dt^2} + y\left(1 - \frac{y^2}{2!} + \cdots\right) - u$$

Keeping only first-degree terms, the linearized equation is $d^2y/dt^2 + y = u$. This equation is valid only for small deviations (perturbations) about the operating point $u = y = 0$.

**19.5.** Write the perturbation equations determined in Example 19.9 in vector-matrix form. Why are they linear? Under what conditions would they be time-invariant?

$$\frac{d(\Delta\mathbf{x})}{dt} \equiv \begin{bmatrix} \dfrac{d(\Delta x_1)}{dt} \\[2mm] \dfrac{d(\Delta x_2)}{dt} \end{bmatrix} \cong \begin{bmatrix} -2c_2\bar{x}_1(t) & c_1\bar{u}(t) \\[2mm] \dfrac{c_3c_4}{[c_4 + \bar{x}_1(t)]^2} & 0 \end{bmatrix} \Delta\mathbf{x} + \begin{bmatrix} c_1\bar{x}_2(t) \\ 0 \end{bmatrix}\Delta u$$

$$\Delta y \cong \begin{bmatrix} 2c_5\bar{x}_1(t) & 0 \end{bmatrix}\Delta\mathbf{x}$$

These equations are linear because the matrices premultiplying $\Delta\mathbf{x}$ and $\Delta u$ are independent of $\Delta\mathbf{x}$ and $\Delta\mathbf{u}$. They would be time-invariant if the parameters $c_1, c_2, \ldots, c_5$ were constant and the "operating point" of the system, for $u = \bar{u}(t)$ and $x = \bar{x}(t)$, were also constant. This would be the case if $\bar{u} = $ constant.

**19.6.** Derive the linearized Equations (*19.13*) and (*19.14*) for the nonlinear differential system given by (*19.11*) and (*19.12*).

We consider changes $\Delta\mathbf{x}$ in $\mathbf{x}$ as a result of changes $\Delta\mathbf{u}$ in $\mathbf{u}$, each about operating points $\bar{\mathbf{x}}$ and $\bar{\mathbf{u}}$, respectively, that is,

$$\mathbf{x}(t) = \bar{\mathbf{x}}(t) + \Delta\mathbf{x}(t)$$

$$\mathbf{u}(t) = \bar{\mathbf{u}}(t) + \Delta\mathbf{u}(t)$$

In these equations, $t$ is considered a parameter, held constant in the derivation. We therefore suppress $t$, for convenience. Substitution of $\bar{\mathbf{x}} + \Delta\mathbf{x}$ for $\mathbf{x}$ and $\bar{\mathbf{u}} + \Delta\mathbf{u}$ for $\mathbf{u}$ in (*19.11*) gives

$$\frac{d\mathbf{x}}{dt} = \frac{d\bar{\mathbf{x}}}{dt} + \frac{d(\Delta\mathbf{x})}{dt} = \mathbf{f}(\bar{\mathbf{x}} + \Delta\mathbf{x}, \bar{\mathbf{u}} + \Delta\mathbf{u})$$

Now we expand this equation in a Taylor series about $\{\bar{\mathbf{x}}, \bar{\mathbf{u}}\}$, retaining only first-order terms:

$$\frac{d\mathbf{x}}{dt} + \frac{d(\Delta\mathbf{x})}{dt} \cong \mathbf{f}(\bar{\mathbf{x}}, \bar{\mathbf{u}}) + \left.\frac{d\mathbf{f}}{\partial\mathbf{x}}\right|_{\substack{\mathbf{x}=\bar{\mathbf{x}}(t)\\\mathbf{u}=\bar{\mathbf{u}}(t)}} \Delta\mathbf{x} + \left.\frac{\partial\mathbf{f}}{\partial\mathbf{u}}\right|_{\substack{\mathbf{x}=\bar{\mathbf{x}}(t)\\\mathbf{u}=\bar{\mathbf{u}}(t)}} \Delta\mathbf{u}$$

Then, since $d\bar{\mathbf{x}}/dt = f(\bar{\mathbf{x}}, \bar{\mathbf{u}})$, Equation (*19.11*) follows immediately after subtracting these corresponding terms from both sides of the equation above. Similarly, for

$$\mathbf{y} = \mathbf{g}(\mathbf{x})$$

$$\mathbf{y} = \bar{\mathbf{y}} + \Delta\mathbf{y} = \mathbf{g}(\bar{\mathbf{x}} + \Delta\mathbf{x}) \cong \mathbf{g}(\bar{\mathbf{x}}) + \left.\frac{\partial\mathbf{g}}{\partial\mathbf{x}}\right|_{\mathbf{x}=\mathbf{x}} \Delta\mathbf{x} = \bar{\mathbf{y}} + \left.\frac{\partial\mathbf{g}}{\partial\mathbf{x}}\right|_{\mathbf{x}=\bar{\mathbf{x}}} \Delta\mathbf{x}$$

Subtracting $\bar{\mathbf{y}}$ from both sides finally gives

$$\Delta\mathbf{y} \cong \left.\frac{\partial\mathbf{g}}{\partial\mathbf{x}}\right|_{\mathbf{x}=\bar{\mathbf{x}}} \Delta\mathbf{x}$$

**19.7.** The equations describing the motion of an earth satellite in the orbit plane are

$$r\frac{d^2\theta}{dt^2} + 2\frac{dr}{dt}\frac{d\theta}{dt} = 0 \qquad \frac{d^2r}{dt^2} - r\left(\frac{d\theta}{dt}\right)^2 = -\frac{k^2}{pr^2}$$

(See Problem 3.3 and Example 19.2 for more details.) A satellite is in a nearly circular orbit determined by $r$ and $d\theta/dt \equiv \omega$. An exactly circular orbit is defined by

$$r = r_0 = \text{constant} \qquad \omega = \omega_0 = \text{constant}$$

Since $dr_0/dt = 0$ and $d\omega_0/dt = 0$, the first differential equation is eliminated for a circular orbit. The second equation reduces to $r_0\omega_0^2 = k^2/pr_0^2$. Find a set of linear equations which approximately describes the differences

$$\delta r \equiv r - r_0 \qquad \delta\omega \equiv \omega - \omega_0$$

In the equations of motion we make the substitutions

$$r = r_0 + \delta r \qquad \omega = \omega_0 + \delta\omega$$

and obtain the equations

$$(r_0 + \delta r)\frac{d(\omega_0 + \delta\omega)}{dt} + 2\frac{d(r_0 + \delta r)}{dt}(\omega_0 + \delta\omega) = 0$$

$$\frac{d^2(r_0 + \delta r)}{dt^2} - (r_0 + \delta r)(\omega_0 + \delta\omega)^2 = -\frac{k}{p(r_0 + \delta r)^2}$$

We note that

$$\frac{d(r_0 + \delta r)}{dt} = \frac{d(\delta r)}{dt} \qquad \frac{d^2(r_0 + \delta r)}{dt^2} = \frac{d^2(\delta r)}{dt^2} \qquad \frac{d(\omega_0 + \delta\omega)}{dt} = \frac{d(\delta\omega)}{dt}$$

since both $r_0$ and $\omega_0$ are constant. The first differential equation then becomes

$$r_0\frac{d(\delta\omega)}{dt} + (\delta r)\frac{d(\delta\omega)}{dt} + 2\omega_0\frac{d(\delta r)}{dt} + 2\frac{d(\delta r)}{dt}\delta\omega = 0$$

Since the differences $\delta r$, $\delta\omega$ and their derivatives are small, the second-order terms $(\delta r)(d(\delta\omega)/dt)$ and $2(d(\delta r)/dt)\delta\omega$ can be assumed negligible and eliminated. The resulting linear equation is

$$r_0\frac{d(\delta\omega)}{dt} + 2\omega_0\frac{d(\delta r)}{dt} = 0$$

which is one of the two desired equations. The second differential equation can be rewritten as

$$\frac{d^2(\delta r)}{dt^2} - r_0\omega_0^2 - 2r_0\omega_0\delta\omega - r_0(\delta\omega)^2 - \omega_0^2\delta r - 2\omega_0(\delta r)(\delta\omega) - (\delta\omega)^2\delta r$$

$$= -\frac{k}{pr_0^2} - \frac{2k\delta r}{r_0^3} + \text{higher-order terms in } \delta r \text{ and } \delta\omega$$

where the right-hand side is the Taylor series expansion of $-k/pr^2$ about $r_0$. All terms of order 2 and greater in $\delta r$ and $\delta\omega$ may again be assumed negligible and eliminated leaving the linear equation

$$\frac{d^2(\delta r)}{dt^2} - r_0\omega_0^2 - 2r_0\omega_0\delta\omega - \omega_0^2\delta\omega - \omega_0^2\delta r = -\frac{k}{pr_0^2} - \frac{2k\delta r}{pr_0^3}$$

In the problem statement we saw that $r_0\omega_0^2 = k/pr_0^2$. Hence the final equation is

$$\frac{d^2(\delta r)}{dt^2} - 2r_0\omega_0\delta\omega - \omega_0^2\delta r = -\frac{2k\delta r}{pr_0^3}$$

which is the second of the two desired linearized equations.

## PHASE PLANE METHODS

**19.8.** Show the equation $d^2x/dt^2 = f(x, dx/dt)$ can be equivalently described by a pair of first-order differential equations.

We define a set of new variables: $x_1 \equiv x$ and $x_2 \equiv dx_1/dt = dx/dt$.

$$\frac{d^2x}{dt^2} = \frac{d^2x_1}{dt} = \frac{dx_2}{dt} = f\left(x, \frac{dx}{dt}\right) = f\left(x_1, \frac{dx_1}{dt}\right) = f(x_1, x_2)$$

The two desired equations are therefore

$$\frac{dx_1}{dt} = x_2 \qquad \frac{dx_2}{dt} = f(x_1, x_2)$$

**19.9.** Show that the phase plane trajectory of the solution of the differential equation

Mathcad

$$\frac{d^2x}{dt^2} + x = 0$$

with initial conditions $x(0) = 0$ and $(dx/dt)|_{t=0} = 1$ is a circle of unit radius centered at the origin.

Letting $x \equiv x_1$ and $x_2 \equiv dx_1/dt$, we obtain the pair of equations

$$\frac{dx_1}{dt} = x_2 \qquad x_1(0) = 0$$

$$\frac{dx_2}{dt} = -x_1 \qquad x_2(0) = 1$$

We eliminate time as the independent variable by writing

$$\frac{dx_1}{dx_2} = -\frac{x_2}{x_1} \qquad \text{or} \qquad x_1\, dx_1 + x_2\, dx_2 = 0$$

Integrating this equation for the given initial conditions, we obtain

$$\int_0^{x_1} x_1'\, dx_1' + \int_1^{x_2} x_2'\, dx_2' = \tfrac{1}{2}x_1^2 + \tfrac{1}{2}x_2^2 - \tfrac{1}{2} = 0 \qquad \text{or} \qquad x_1^2 + x_2^2 = 1$$

which is the equation of a circle of unit radius centered at the origin.

**19.10.** Determine the equation of the phase plane trajectory of the equation

$$\frac{d^2x}{dt^2} + \frac{dx}{dt} = 0$$

with the initial conditions $x(0) = 0$ and $(dx/dt)|_{t=0} = 1$.

With $x_1 \equiv x$ and $x_2 \equiv dx_1/dt$ we obtain the pair of first-order equations

$$\frac{dx_1}{dt} = x_2 \qquad x_1(0) = 0$$

$$\frac{dx_2}{dt} = -x_2 \qquad x_2(0) = 1$$

We eliminate time as the independent variable by writing

$$\frac{dx_1}{dx_2} = -\frac{x_2}{x_2} = -1 \qquad \text{or} \qquad dx_1 + dx_2 = 0$$

Then

$$\int_0^{x_1} dx_1' + \int_1^{x_2} dx_2' = x_1 + x_2 - 1 = 0 \qquad \text{or} \qquad x_1 + x_2 = 1$$

which is the equation of a straight line, as shown in Fig. 19-17. The direction of the motion in the phase plane is indicated by the arrow and is determined by noting that, initially, $x_2(0) = 1$; therefore $dx_1/dt > 0$ and $x_1$ is increasing, and $dx_2/dt < 0$ and $x_2$ is decreasing. The trajectory ends at the point $(x_1, x_2) = (1, 0)$, where $dx_1/dt = dx_2/dt = 0$ and thus motion terminates.

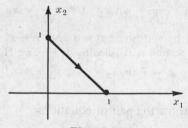

Fig. 19-17

**19.11.** Design an on-off controller for the system given by Equation (*19.19*) and Fig. 19-7, with $a = 0$.

For $a = 0$ in Equation (*19.19*), Equation (*19.23*) becomes

$$dx_1 = \frac{x_2 \, dx_2}{u}$$

The switching curve is generated by integrating this equation in the second quadrant with $u = +1$ and terminating at the origin, yielding

$$x_1(t) = -\frac{x_2^2(t)}{2} \quad \text{or} \quad e = -\frac{\dot{e}^2}{2}$$

and integrating in the fourth quadrant with $u = -1$ and terminating at the origin, yielding

$$x_1(t) = \frac{x_2^2(t)}{2} \quad \text{or} \quad e = \frac{\dot{e}^2}{2}$$

The switching curve is sketched in Fig. 19-18. The switching logic of this on-off controller is given in Table 19.2.

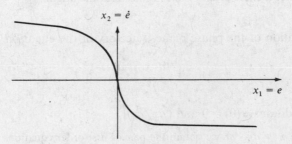

Fig. 19-18

**Table 19.2**

| $\dot{e} > 0$ | $e + \dot{e}^2/2 > 0$ | $e - \dot{e}^2/2 > 0$ | $u$ |
|---|---|---|---|
| No | No | No | $-1$ |
| No | No | Yes | $+1$ |
| No | Yes | No | $-1$ |
| No | Yes | Yes | $+1$ |
| Yes | No | No | $-1$ |
| Yes | No | Yes | $-1$ |
| Yes | Yes | No | $+1$ |
| Yes | Yes | Yes | $+1$ |

## LYAPUNOV'S STABILITY CRITERION

**19.12.** Find the singular points of the pair of equations

$$\frac{dx_1}{dt} = \sin x_2 \qquad \frac{dx_2}{dt} = x_1 + x_2$$

Singular points are found by setting $\sin x_2 = 0$ and $x_1 + x_2 = 0$. The first equation is satisfied by $x_2 = \pm n\pi$, $n = 0, 1, 2, \ldots$. The second is satisfied by $x_1 = -x_2$. Hence the singular points are defined by

$$x_1 = \mp n\pi, \, x_2 = \pm n\pi \qquad n = 0, 1, 2, \ldots$$

**19.13.** The origin is a singular point for the pair of equations

$$\frac{dx_1}{dt} = ax_1 + bx_2 \qquad \frac{dx_2}{dt} = cx_1 + dx_2$$

Using Lyapunov theory, find sufficient conditions on $a$, $b$, $c$, and $d$ such that the origin is asymptotically stable.

We choose a function

$$V = x_1^2 + x_2^2$$

which is positive for all $x_1, x_2$ except $x_1 = x_2 = 0$. The time derivative of $V$ is

$$\frac{dV}{dt} = 2x_1 \frac{dx_1}{dt} + 2x_2 \frac{dx_2}{dt} = 2ax_1^2 + 2bx_1x_2 + 2cx_1x_2 + 2dx_2^2$$

To make $dV/dt$ negative for all $x_1, x_2$, we might choose $a < 0$, $d < 0$, and $b = -c$. In this case,

$$\frac{dV}{dt} = 2ax_1^2 + 2dx_2^2 < 0$$

except when $x_1 = x_2 = 0$. Hence one set of sufficient conditions for asymptotic stability are $a < 0$, $d < 0$, and $b = -c$. There are other possible solutions to this problem.

**19.14.** Determine sufficient conditions for the stability of the origin of the nonlinear discrete-time system described by

$$x_1(k+1) = x_1(k) - f[x_1(k)]$$

Let $V[x(k)] = [x_1(k)]^2$, which is greater than 0 for all $x \neq 0$. Then

$$\Delta V = x_1^2(k+1) - x_1^2(k) = (x_1(k) - f[x_1(k)])^2 - x_1^2(k)$$

$$= -x_1(k)f[x_1(k)]\left(2 - \frac{f[x_1(k)]}{x_1}\right)$$

Therefore sufficient conditions for $\Delta V \leq 0$ and thus stability of the system are

$$x_1 f(x_1) \geq 0$$

$$\frac{f(x_1)}{x_1} \leq 2 \qquad \text{for all} \quad x_1$$

**19.15.** Determine sufficient conditions for the stability of the system

$$\dot{\mathbf{x}} = A\mathbf{x} + \mathbf{b}f(x_1) \qquad \text{where} \quad A = \begin{bmatrix} -2 & -1 \\ 0 & -2 \end{bmatrix}, \mathbf{b} = \begin{bmatrix} 1 \\ 2 \end{bmatrix}$$

Let $V = \mathbf{x}^T P \mathbf{x}$ and $P = \begin{bmatrix} a & c \\ c & 1 \end{bmatrix}$. Then

$$\dot{V} = \mathbf{x}^T(PA + A^T P)\mathbf{x} + \mathbf{x}^T P\mathbf{b}f(x_1) + f(x_1)\mathbf{b}^T P\mathbf{x}$$

$$= \mathbf{x}^T \begin{bmatrix} -4a & -a - 4c \\ -a - 4c & -2c - 4 \end{bmatrix} \mathbf{x} + 2(a + 2c)x_1 f(x_1) + 2(c + 2)x_2 f(x_1)$$

To eliminate the cross-product term $x_2 f(x_1)$, set $c = -2$. Then

$$\dot{V} = -\mathbf{x}^T Q\mathbf{x} + 2(a - 4)x_1 f(x_1)$$

where $Q \equiv \begin{bmatrix} 4a & a - 8 \\ a - 8 & 0 \end{bmatrix}$. For $Q \geq 0$, $a = 8$. The resulting $\dot{V}$ is

$$\dot{V} = -32x_1^2 + 8x_1 f(x_1) = -8x_1^2\left(4 - \frac{f(x_1)}{x_1}\right)$$

Then $\dot{V} \leq 0$ and the system is stable if $f(x_1)/x_1 \leq 4$ for all $x_1 \neq 0$.

**19.16.** Determine sufficient conditions for stability of the nonlinear discrete-time system

$$\mathbf{x}(k+1) = A\mathbf{x}(k) + \mathbf{b}f\big[x_1(k)\big]$$

where $A = \begin{bmatrix} 1 & 1 \\ 0 & -1 \end{bmatrix}$ and $\mathbf{b} = \begin{bmatrix} 0 \\ -1 \end{bmatrix}$.

Let $V = \mathbf{x}^T P \mathbf{x}$, where $P = \begin{bmatrix} a & c \\ c & 1 \end{bmatrix}$. Then

$$\Delta V = V[\mathbf{x}(k+1)] - V[\mathbf{x}(k)] = \mathbf{x}(k+1)^T P \mathbf{x}(k+1) - \mathbf{x}(k)^T P \mathbf{x}(k)$$

$$= \big[ f[x_1(k)]\mathbf{b}^T + \mathbf{x}(k)^T A^T \big] P \big[ A\mathbf{x}(k) + \mathbf{b}f[x_1(k)] \big] - \mathbf{x}(k)^T P \mathbf{x}(k)$$

$$= \mathbf{x}^T (A^T PA - P)\mathbf{x} + f(x_1)\mathbf{b}^T P \mathbf{b}f(x_1) + f(x_1)\mathbf{b}^T PA\mathbf{x} + \mathbf{x}^T A^T P \mathbf{b}f(x_1)$$

where

$$A^T PA - P = \begin{bmatrix} 0 & a-2c \\ a-2c & a-2c \end{bmatrix} \quad \text{and} \quad \mathbf{b}^T PA = [\, -c \quad 1-c \,]$$

Now, in order for $A^T PA - P \le 0$, we set $a = 2c$ and, to eliminate the cross-product term $x_2 f(x_1)$, we set $c = 1$. Then $A^T PA - P = 0$ and

$$\Delta V = [f(x_1)]^2 - 2x_1 f(x_1) = -x_1 f(x_1)\left(2 - \frac{f(x_1)}{x_1}\right)$$

Sufficient conditions for $\Delta V \le 0$ and stability of the origin are then

$$x_1 f(x_1) \ge 0 \quad \text{and} \quad \frac{f(x_1)}{x_1} \le 2 \quad \text{for all} \quad x_1.$$

## FREQUENCY RESPONSE METHODS

**19.17.** Show that the describing function for the piecewise-linear saturation element in Example 19.1 is given by

$$\frac{B_1}{A} e^{j\phi_1} = \frac{2}{\pi}\left[ \sin^{-1}\frac{1}{A} + \frac{1}{A}\cos\sin^{-1}\frac{1}{A} \right]$$

We see from Fig. 19-1($b$) that, when the magnitude of the input is less than 1.0, the output equals the input. When the input exceeds 1.0, then the output equals 1.0. Using the notation of Example 19.1, if

$$e(t) = A \sin \omega t \qquad A > 1$$

then $f(t)$ is as shown in Fig. 19-19 and can be written as

$$f(t) = \begin{cases} A\sin\omega t & \begin{cases} 0 \le t \le t_1 \\ t_2 \le t \le t_3 \\ t_4 \le t \le 2\pi/\omega \end{cases} \\ 1 & t_1 \le t \le t_2 \\ -1 & t_3 \le t \le t_4 \end{cases}$$

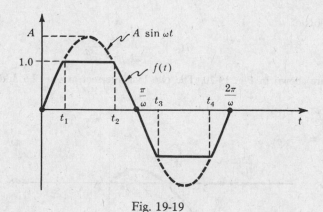

Fig. 19-19

The time $t_1$ is obtained by noting that

$$A \sin \omega t_1 = 1 \qquad \text{or} \qquad t_1 = \frac{1}{\omega} \sin^{-1} \frac{1}{A}$$

Similarly,

$$t_2 = \frac{\pi}{\omega} - \frac{1}{\omega} \sin^{-1} \frac{1}{A} \qquad t_3 = \frac{\pi}{\omega} + \frac{1}{\omega} \sin^{-1} \frac{1}{A} \qquad t_4 = \frac{2\pi}{\omega} - \frac{1}{\omega} \sin^{-1} \frac{1}{A}$$

The magnitude $B_1$ and phase angle $\phi_1$ of the describing function are determined from the expression for the first Fourier coefficient:

$$B_1 = \frac{\omega}{\pi} \int_0^{2\pi/\omega} f(t) \sin \omega t \, dt$$

Since $f(t)$ is an odd function, the phase angle $\phi_1$ is zero. The integral defining $B_1$ can be rewritten as

$$B_1 = \frac{\omega}{\pi} \int_0^{t_1} A \sin^2 \omega t \, dt + \frac{\omega}{\pi} \int_{t_1}^{t_2} \sin \omega t \, dt$$

$$+ \frac{\omega}{\pi} \int_{t_2}^{t_3} A \sin^2 \omega t \, dt - \frac{\omega}{\pi} \int_{t_3}^{t_4} \sin \omega t \, dt + \frac{\omega}{\pi} \int_{t_4}^{2\pi/\omega} A \sin^2 \omega t \, dt$$

But

$$\int_0^{t_1} A \sin^2 \omega t \, dt = \int_{t_4}^{2\pi/\omega} A \sin^2 \omega t \, dt = \frac{1}{2} \int_{t_2}^{t_3} A \sin^2 \omega t \, dt$$

and

$$\int_{t_1}^{t_2} \sin \omega t \, dt = - \int_{t_3}^{t_4} \sin \omega t \, dt = 2 \int_{t_1}^{\pi/2\omega} \sin \omega t \, dt$$

We can thus write $B_1$ as

$$B_1 = \frac{4\omega}{\pi} \int_0^{t_1} A \sin^2 \omega t \, dt + \frac{4\omega}{\pi} \int_{t_1}^{\pi/2\omega} \sin \omega t \, dt = \frac{2}{\pi} \left[ A \omega t_1 - \frac{A}{2} \sin 2 \omega t_1 + 2 \cos \omega t_1 \right]$$

Substituting $t_1 = (1/\omega)\sin^{-1}(1/A)$ and simplifying, we obtain

$$B_1 = \frac{2}{\pi} \left[ A \sin^{-1} \frac{1}{A} + \cos \sin^{-1} \frac{1}{A} \right]$$

Finally, the describing function is

$$\frac{B_1}{A} = \frac{2}{\pi} \left[ \sin^{-1} \frac{1}{A} + \frac{1}{A} \cos \sin^{-1} \frac{1}{A} \right]$$

**19.18.** Determine the amplitude $A$ and frequency $\omega$ for which oscillations could be maintained in the system of Example 19.18 with the forward-loop gain increased to 32 from 8.

The Polar Plots of

$$G(\omega) = \frac{32}{j\omega(j\omega + 2)^2}$$

and $-1/\overline{N}(A)$ are shown in Fig. 19-20. The two loci intersect at $A = 2.5$ and $\omega = 2$, the conditions for oscillation.

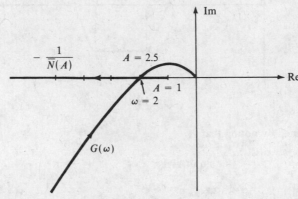

Fig. 19-20

**19.19.** Determine the amplitude and frequency of possible oscillations for the system of Fig. 19-12 with  $f(e) = e^3$ and

$$G(\omega) = \frac{1}{(j\omega + 1)^3}$$

From Example 19.17, the describing function for this nonlinearity is

$$\overline{N}(A) = \frac{3A^2}{4} \quad \text{and} \quad -\frac{1}{\overline{N}} = -\frac{4}{3A^2}$$

From the Polar Plots shown in Fig. 19-21, $G(\omega)$ and $-1/\overline{N}$ intersect for $\omega = 1.732$ and $A = 3.27$, the conditions for oscillation.

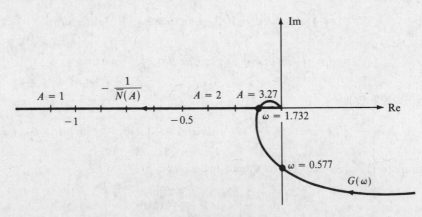

Fig. 19-21

**19.20.** Determine the amplitude and frequency of possible oscillations for the system of Fig. 19-12, with the hysteresis nonlinearity shown in Fig. 19-22, and $G(\omega) = 2/j\omega(j\omega + 1)$.

The system block diagram can be manipulated as shown in Fig. 19-23, so that the hysteresis element is normalized, with a dead zone of 1 and a slope of 1. Figure 19-11 can then be used to construct the Polar Plot of $-1/N$, shown in Fig. 19-24 with the Polar Plot of $2G(\omega)$, rather than $G(\omega)$, because the loop transfer function excluding the nonlinearity is $4G(\omega)/2 = 2G(\omega)$.

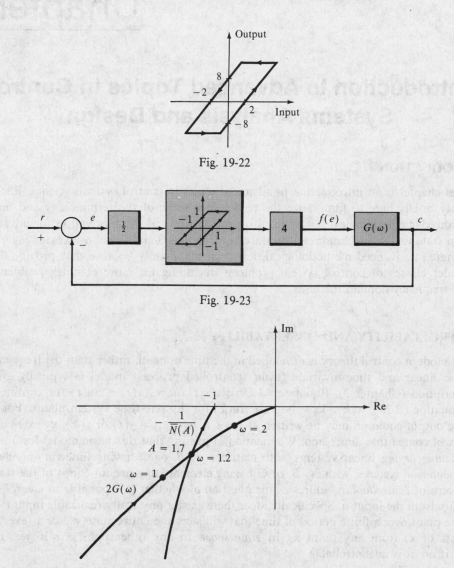

Fig. 19-22

Fig. 19-23

Fig. 19-24

The two curves intersect for $\omega = 1.2$ rad/sec and $A = 1.7$, the conditions for oscillation of the system. Note that $A$ is the amplitude of the input to the normalized nonlinearity. Therefore the amplitude for oscillations is 3.4, in terms of $e$.

## Supplementary Problems

**19.21.** Determine the phase plane trajectory of the solution of the differential equation

$$\frac{d^2x}{dt^2} + 2\frac{dx}{dt} + 4x = 0$$

**19.22.** Using Lyapunov theory, find sufficient conditions on $a_1$ and $a_0$ which guarantee that the point $x = 0$, $dx/dt = 0$ is stable for the equation

$$\frac{d^2x}{dt^2} + a_1\frac{dx}{dt} + a_0x = 0$$

# Chapter 20

# Introduction to Advanced Topics in Control Systems Analysis and Design

## 20.1 INTRODUCTION

This final chapter is an introduction to advanced topics in control systems science. Each subject is discussed only briefly here to familiarize the reader with some of the terminology and mathematical level of advanced methodologies. It should also provide some of the motivation for advanced study. Time-domain state variable techniques, introduced in Chapters 3 and 4 and used extensively in Chapter 19, predominate in advanced methodological developments, mainly because they provide the basis for solving broader classes of control system problems, including far more complex problems than are amenable to frequency-domain methods.

## 20.2 CONTROLLABILITY AND OBSERVABILITY

Much of modern control theory is developed in the time domain, rather than the frequency domain, and the basic linear and time-invariant *plant* (controlled process) model is typically given a *state variable* description (Chapter 3), Equation (*3.25b*): $d\mathbf{x}(t)/dt = A\mathbf{x}(t) + B\mathbf{u}(t)$ for continuous system plants, or Equation (*3.36*): $\mathbf{x}(k+1) = A\mathbf{x}(k) + B\mathbf{u}(k)$ for discrete-time system plants. For either type of model, the output equation may be written as $\mathbf{y} = C\mathbf{x}$, where $\mathbf{y} = \mathbf{y}(t)$ or $\mathbf{y}(k)$, $\mathbf{x} = \mathbf{x}(t)$ or $\mathbf{x}(k)$, and $C$ is a matrix of compatible dimension. We mention in passing that this basic model form is often used to represent *time-varying* linear systems, with matrices $A$, $B$, or $C$ having time-varying elements, and (less often) *nonlinear* systems, with $A$, $B$, or $C$ having elements that are functions of the state vector $\mathbf{x}$.

The concept of *controllability* addresses the question of whether it is possible to *control* or *steer* the state (vector) $\mathbf{x}$ from the input $\mathbf{u}$. Specifically, does there exist a physically realizable input $\mathbf{u}$ that can be applied to the plant over a finite period of time that will steer the entire state vector $\mathbf{x}$ (every one of the $n$ components of $\mathbf{x}$) from any point $\mathbf{x}_0$ in *state space* to any other point $\mathbf{x}_1$? If yes, the plant is **controllable**; if no, it is **uncontrollable**.

The concept of *observability* is complementary to that of controllability. It addresses the question of whether it is possible to determine all of the $n$ components of the state vector $\mathbf{x}$ by measurement of the output $\mathbf{y}$ over a finite period of time. If yes, the system is **observable**; if no, it is **unobservable**. Obviously, if $\mathbf{y} = \mathbf{x}$, that is, if all state variables are measured, the system is observable. However, if $\mathbf{y} \neq \mathbf{x}$ and $C$ is not a square matrix, the plant may still be observable.

The controllability and observability properties of the plant have important practical consequences in analysis and, more importantly, design of modern feedback control systems. Intuitively, uncontrollable plants cannot be steered arbitrarily; and it is impossible to know all of the state variables of unobservable plants. These problems are clearly related, because together this means that unobservable states (or state variables) cannot be individually controlled if the control variable $\mathbf{u}$ is required to be a function of $\mathbf{x}$, that is, if *feedback* control is needed.

Linear, time-invariant plant models in state variable form [Equations (*3.25b*) or (*3.36*)] are **controllable** if and only if the following **controllability matrix** has rank $n$ ($n$ linearly independent columns), where $n$ is the number of state variables in the state vector $\mathbf{x}$:

$$\begin{bmatrix} B & AB & A^2B & \cdots & A^{n-1}B \end{bmatrix} \qquad (20.1)$$

Similarly, the plant model is **observable** if and only if the following **observability matrix** has

480

rank $n$ ($n$ linearly independent rows):

$$\begin{bmatrix} C \\ CA \\ CA^2 \\ \vdots \\ CA^{n-1} \end{bmatrix} \qquad (20.2)$$

**EXAMPLE 20.1.**  Consider the following single-input single-output (SISO) plant model, with $\mathbf{x} \equiv \begin{bmatrix} x_1 \\ x_2 \end{bmatrix}$ and $a_{11}, a_{12}, a_{22}$ each nonzero:

$$\frac{d\mathbf{x}}{dt} = \begin{bmatrix} a_{11} & a_{12} \\ 0 & a_{22} \end{bmatrix}\mathbf{x} + \begin{bmatrix} 1 \\ 0 \end{bmatrix}u \qquad y = C\mathbf{x} = \begin{bmatrix} 1 & 0 \end{bmatrix}\mathbf{x}$$

To test if this model is controllable, we first evaluate the matrix given by Equation (20.1):

$$\begin{bmatrix} B & AB \end{bmatrix} = \begin{bmatrix} 1 & a_{11} \\ 0 & 0 \end{bmatrix}$$

By Definition 3.11, the two columns $\begin{bmatrix} 1 \\ 0 \end{bmatrix}$ and $\begin{bmatrix} a_{11} \\ 0 \end{bmatrix}$ would be linearly independent if the only constants $\alpha$ and $\beta$ for which

$$\alpha \begin{bmatrix} 1 \\ 0 \end{bmatrix} + \beta \begin{bmatrix} a_{11} \\ 0 \end{bmatrix} = \begin{bmatrix} 0 \\ 0 \end{bmatrix}$$

where $\alpha \equiv \beta \equiv 0$. This is clearly *not* the case, because $\alpha = 1$ and $\beta = -1/a_{11}$ satisfies this equation. Therefore the two columns of $\begin{bmatrix} B & AB \end{bmatrix}$ are linearly *dependent*, the rank of $\begin{bmatrix} B & AB \end{bmatrix} - 1 \neq 2 = n$, and this plant is therefore *uncontrollable*.

Similarly, from Equation (20.2),

$$\begin{bmatrix} C \\ CA \end{bmatrix} = \begin{bmatrix} 1 & 0 \\ a_{11} & a_{12} \end{bmatrix}$$

For this matrix, the only $\alpha$ and $\beta$ for which $\alpha[1 \quad 0] + \beta[a_{11} \quad a_{12}] = [0 \quad 0]$ are $\alpha \equiv \beta \equiv 0$, because $a_{12} \neq 0$. Therefore the rank of $\begin{bmatrix} C \\ CA \end{bmatrix}$ is $n = 2$ and this plant is *observable*.

## 20.3   TIME-DOMAIN DESIGN OF FEEDBACK SYSTEMS (STATE FEEDBACK)

Design of many feedback control systems may be accomplished using time-domain representations and the concepts of controllability and observability discussed above. As noted in earlier chapters, particularly Chapter 14, Root-Locus Design, linear control system design is often performed by manipulating the locations of the poles of the closed-loop transfer function (the roots of the characteristic equation), using appropriate compensators in the feedforward or feedback path to meet performance specifications. This approach is satisfactory in many circumstances, but it has certain limitations that can be overcome using a different design philosophy, called *state feedback* design, that permits *arbitrary* pole placement, thereby providing substantially more flexibility in design.

The basic idea underlying state feedback control system design is as follows for single-input continuous plants $d\mathbf{x}/dt = A\mathbf{x} + Bu$. The procedure is the same for discrete-time systems.

With reference to Fig. 2-1, we seek a state feedback control:

$$u = -G\mathbf{x} + r \qquad (20.3)$$

where $G$ is a $1 \times n$ feedback matrix of constant gains (to be designed) and $r$ is the reference input. Combining these equations, the closed-loop system is given by

$$\frac{d\mathbf{x}}{dt} = (A - BG)\mathbf{x} + Br \qquad (20.4)$$

If the plant is *controllable*, the matrix $G$ exists that can yield any (arbitrary) set of desired roots for the characteristic equation of this closed-loop system, represented by $|\lambda I - A + BG| = 0$, where the $\lambda$ solutions of this determinant equation are the roots. This is the basic result.

**EXAMPLE 20.2.**  A block diagram of the state feedback system given by Equations (*20.3*) and (*20.4*) is shown in Fig. 20-1.

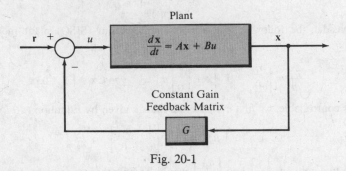

Fig. 20-1

To implement a state feedback design, the entire state vector $\mathbf{x}$ must somehow be made available, either as $\mathbf{x}$ exactly, or as an adequate approximation, denoted $\hat{\mathbf{x}}$. If the output is $\mathbf{y} = \mathbf{x}$, as in Fig. 20-1, there obviously is no problem. But, if all states are not available as outputs, which is more common, then *observability* of the plant model differential and output equations ($d\mathbf{x}/dt = A\mathbf{x} + B\mathbf{u}$ and $\mathbf{y} = C\mathbf{x}$) is required to obtain the needed state *estimate* or *observer* $\hat{\mathbf{x}}$. The equations for a typical state observer system are given by

$$\frac{d\hat{\mathbf{x}}}{dt} = (A - LC)\hat{\mathbf{x}} + L\mathbf{y} + B\mathbf{u} \qquad (20.5)$$

where $A$, $B$, and $C$ are matrices of the plant and output measurement systems and $L$ is an *observer design matrix* to be determined in a particular problem.

**EXAMPLE 20.3.**  A detailed block diagram of the state observer system given by Equation (*20.5*) is shown in Fig. 20-2, along with the plant and measurement system block diagram (upper portion) for generating the needed input signals for the observer system (lower portion).

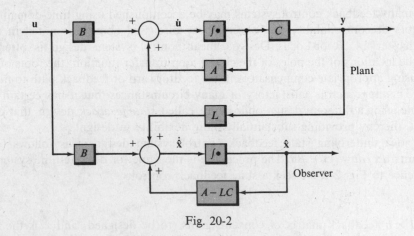

Fig. 20-2

**EXAMPLE 20.4.**  Under suitable conditions, which include controllability and observability of the plant to be controlled, a *separation principle* applies and the state feedback portion (matrix $G$) and observer portion (matrix $L$) of a state feedback control system (with $\mathbf{y} \neq \mathbf{x}$) can be designed independently. A block diagram of the combined systems is shown in Fig. 20-3.

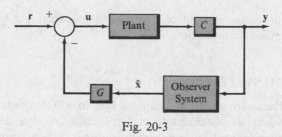

Fig. 20-3

We have omitted many details in this introductory material, and state feedback control systems are often more complex than described above.

## 20.4  CONTROL SYSTEMS WITH RANDOM INPUTS

System stimuli often include random or otherwise "unknown" components. This means that input functions may sometimes be more appropriately described probabilistically than deterministically. Such excitations are called **random processes.** System disturbances $n$ (Definition 2.21), illustrated in several previous chapters, are sometimes represented by random process models in modern control theory and practice.

A random process can be viewed as a function of two variables, $t$ and $\eta$, where $t$ represents time and $\eta$ a *random event*. The value of $\eta$ is determined by chance.

**EXAMPLE 20.5.**  A particular random process is denoted by $x(t, \eta)$. The random event $\eta$ is the result of tossing an unbiased coin; heads or tails appears with equal probability. We define

$$x(t, \eta) \equiv \begin{cases} \text{a unit step function if } \eta = \text{heads} \\ \text{a unit ramp function if } \eta = \text{tails} \end{cases}$$

Thus $x(t, \eta)$ consists of two simple functions but is a random process because chance dictates which function occurs.

In practice, random processes consist of an infinity of possible time functions, called *realizations*, and we usually cannot describe them as explicitly as the one in Example 20.5. Instead, they must be described, in a statistical sense, by averages over all possible functions of time. The performance criteria discussed previously have all been related to specific inputs (e.g., $K_P$ is defined for a unit step input, $M_P$ and $\phi_{\text{PM}}$ for sine waves). But satisfaction of performance specifications defined for one input signal does not necessarily guarantee satisfaction for others. Therefore, for a random input, we cannot design for a *particular* signal, such as step function, but must design for the statistical average of random input signals.

**EXAMPLE 20.6.**  The unit feedback system in Fig. 20-4 is excited by a random process input $r$ having an infinity of possibilities. We want to determine compensation so that the error $e$ is not excessive. There are an infinity of possibilities for $r$ and, therefore, for $e$. Hence we cannot ask that each possible error satisfy given performance criteria but only that average errors be small. For instance, we might ask that $G_1$ be chosen from the set of all causal systems such that, as time goes to infinity, the statistical average of $e^2(t)$ does not exceed some constant, or is minimized.

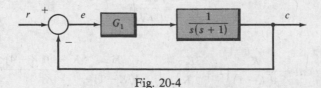

Fig. 20-4

The study of random processes in control systems, often called *stochastic control theory*, is an advanced level subject in applied mathematics.

## 20.5  OPTIMAL CONTROL SYSTEMS

The design problems discussed in earlier chapters are, in an elementary sense, optimal control problems. The classical measures of system performance such as steady state error, gain margin, and phase margin are essentially criteria of optimality, and control system compensators are designed to meet these requirements. In more general optimal control problems, the system measure of performance, or *performance index*, is not fixed beforehand. Instead, compensation is chosen so that the performance index is *maximized* or *minimized*. The value of the performance index is unknown until the completion of the optimization process.

In many problems, the performance index is a measure or function of the error $e(t)$ between the actual and ideal responses. It is formulated in terms of the design parameters chosen, to optimize the performance index, subject to existing physical constraints.

**EXAMPLE 20.7.**   For the system illustrated in Fig. 20-5 we want to find a $K \geq 0$ such that the integral of the square of the error $e$ is minimized when the input is a unit step function. Since $e \equiv e(t)$ is not constant, but a function of time, we can formulate this problem as follows: Choose $K \geq 0$ such that $\int_0^\infty e^2(t)\,dt$ is minimized, where

$$e(t) = \mathscr{L}^{-1}\left[\frac{s+2}{s^2+2s+K}\right] = \sqrt{\frac{K}{K-1}}\, e^{-t}\sin(\sqrt{K-1}\,t + \tan^{-1}\sqrt{K-1})$$

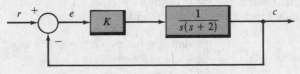

Fig. 20-5

The solution may be obtained for $K > 1$ using conventional minimization techniques of integral calculus, as follows:

$$\int_0^\infty e^2(t)\,dt = \frac{K}{K-1}\int_0^\infty \left[e^{-t}\sin(\sqrt{K-1}\,t + \tan^{-1}\sqrt{K-1})\right]^2 dt$$

Integration yields

$$\int_0^\infty e^2(t)\,dt = \left(\frac{K}{K-1}\right)\left(\frac{e^{-2t}}{4}\right)\left[-1 - \frac{\cos\left(2\sqrt{K-1}\,t + 2\tan^{-1}\sqrt{K-1} - \tan^{-1}(-\sqrt{K-1})\right)}{\sqrt{K}}\right]\Bigg|_0^\infty$$

$$= \frac{K}{4(K-1)}\left[1 + \frac{\cos\left(2\tan^{-1}\sqrt{K-1} - \tan^{-1}(-\sqrt{K-1})\right)}{K}\right]$$

But

$$\cos\left(2\tan^{-1}\sqrt{K-1} - \tan^{-1}(-\sqrt{K-1})\right) = -\cos 3\sqrt{K-1} = 3\cos\sqrt{K-1} - 4\cos^3\sqrt{K-1}$$

$$= \frac{3K-4}{K\sqrt{K}}$$

Therefore

$$\int_0^\infty e^2(t)\,dt = \frac{K}{4(K-1)}\left(1 + \frac{3K-4}{K^2}\right) = \frac{K}{4(K-1)}\,\frac{(K-1)(K+4)}{K^2} = \frac{K+4}{4K}$$

The first derivative of $\int_0^\infty e^2(t)\,dt$ with respect to $K$ is given by

$$\frac{d}{dK}\left(\frac{K+4}{4K}\right) = -\frac{1}{K^2}$$

Apparently, $\int_0^\infty e^2(t)\,dt$ decreases monotonically as $K$ increases. Therefore the optimal value of $K$ is $K = \infty$, which is of course unrealizable. For this value of $K$,

$$\lim_{K \to \infty}\int_0^\infty e^2(t)\,dt = \lim_{K \to \infty}\left(\frac{K+4}{4K}\right) = \frac{1}{4}$$

Note also that the natural frequency $\omega_n$ of the optimal system is $\omega_n = \sqrt{K} = \infty$ and the damping ratio $\xi = 1/\omega_n = 0$, making it marginally stable. Therefore only a *suboptimal* (less than optimal) system can be practically realized and its design depends on the specific application.

Typical optimal control problems, however, are much more complex than this simple example and they require more sophisticated mathematical techniques for their solution. We do little more here than mention their existence.

## 20.6  ADAPTIVE CONTROL SYSTEMS

In some control systems, certain parameters are either not constant, or they vary in an unknown manner. In Chapter 9 we illustrated one way of minimizing the effects of such contingencies by designing for minimum sensitivity. If, however, parameter variations are large or very rapid, it may be desirable to design for the capability of continuously measuring them and changing the compensation so that system performance criteria are always satisfied. This is called *adaptive control* design.

**EXAMPLE 20.8.**   Figure 20-6 depicts an example block diagram of an adaptive control system. The parameters $A$ and $B$ of the plant are known to vary with time. The block labeled "Identification and Parameter Adjustment" continuously measures the input $u(t)$ and output $c(t)$ of the plant to *identify* (quantify) the parameters $A$ and $B$. In this manner, $a$ and $b$ of the lead compensator are modified by the output of this element to satisfy system specifications. The design of the Identification and Parameter Adjustment block is the major problem of adaptive control, another subject requiring advanced knowledge of applied mathematics.

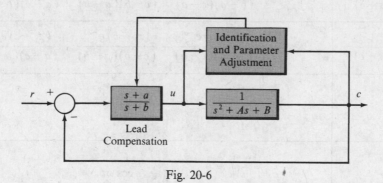

Fig. 20-6

# Appendix A

**Some Laplace Transform Pairs Useful for Control Systems Analysis**

| $F(s)$ | $f(t) \qquad t > 0$ |
|---|---|
| 1 | $\delta(t)$ $\qquad$ unit impulse |
| $e^{-Ts}$ | $\delta(t - T)$ $\qquad$ delayed impulse |
| $\dfrac{1}{s + a}$ | $e^{-at}$ |
| $\dfrac{1}{(s + a)^n}$ | $\dfrac{1}{(n-1)!} t^{n-1} e^{-at} \qquad n = 1, 2, 3, \ldots$ |
| $\dfrac{1}{(s + a)(s + b)}$ | $\dfrac{1}{b - a}(e^{-at} - e^{-bt})$ |
| $\dfrac{s}{(s + a)(s + b)}$ | $\dfrac{1}{a - b}(ae^{-at} - be^{-bt})$ |
| $\dfrac{s + z_1}{(s + a)(s + b)}$ | $\dfrac{1}{b - a}[(z_1 - a)e^{-at} - (z_1 - b)e^{-bt}]$ |
| $\dfrac{1}{(s + a)(s + b)(s + c)}$ | $\dfrac{e^{-at}}{(b-a)(c-a)} + \dfrac{e^{-bt}}{(c-b)(a-b)} + \dfrac{e^{-ct}}{(a-c)(b-c)}$ |
| $\dfrac{s + z_1}{(s + a)(s + b)(s + c)}$ | $\dfrac{(z_1 - a)e^{-at}}{(b-a)(c-a)} + \dfrac{(z_1 - b)e^{-bt}}{(c-b)(a-b)} + \dfrac{(z_1 - c)e^{-ct}}{(a-c)(b-c)}$ |
| $\dfrac{\omega}{s^2 + \omega^2}$ | $\sin \omega t$ |
| $\dfrac{s}{s^2 + \omega^2}$ | $\cos \omega t$ |
| $\dfrac{s + z_1}{s^2 + \omega^2}$ | $\sqrt{\dfrac{z_1^2 + \omega^2}{\omega^2}} \sin(\omega t + \phi) \qquad \phi \equiv \tan^{-1}(\omega / z_1)$ |
| $\dfrac{s \sin\phi + \omega \cos\phi}{s^2 + \omega^2}$ | $\sin(\omega t + \phi)$ |
| $\dfrac{1}{(s + a)^2 + \omega^2}$ | $\dfrac{1}{\omega} e^{-at} \sin \omega t$ |

| $F(s)$ | $f(t) \qquad t > 0$ |
|---|---|
| $\dfrac{1}{s^2 + 2\zeta\omega_n s + \omega_n^2}$ | $\dfrac{1}{\omega_d} e^{-\zeta\omega_n t} \sin \omega_d t \qquad \omega_d \equiv \omega_n \sqrt{1 - \zeta^2}$ |
| $\dfrac{s + a}{(s + a)^2 + \omega^2}$ | $e^{-at} \cos \omega t$ |
| $\dfrac{s + z_1}{(s + a)^2 + \omega^2}$ | $\sqrt{\dfrac{(z_1 - a)^2 + \omega^2}{\omega^2}} \, e^{-at} \sin(\omega t + \phi) \qquad \phi \equiv \tan^{-1}\left(\dfrac{\omega}{z_1 - a}\right)$ |
| $\dfrac{1}{s}$ | $\mathbf{1}(t) \qquad\qquad \text{unit step}$ |
| $\dfrac{1}{s} e^{-Ts}$ | $\mathbf{1}(t - T) \qquad\qquad \text{delayed step}$ |
| $\dfrac{1}{s}(1 - e^{-Ts})$ | $\mathbf{1}(t) - \mathbf{1}(t - T) \qquad\qquad \text{rectangular pulse}$ |
| $\dfrac{1}{s(s + a)}$ | $\dfrac{1}{a}(1 - e^{-at})$ |
| $\dfrac{1}{s(s + a)(s + b)}$ | $\dfrac{1}{ab}\left(1 - \dfrac{be^{-at}}{b - a} + \dfrac{ae^{-bt}}{b - a}\right)$ |
| $\dfrac{s + z_1}{s(s + a)(s + b)}$ | $\dfrac{1}{ab}\left(z_1 - \dfrac{b(z_1 - a)e^{-at}}{b - a} + \dfrac{a(z_1 - b)e^{-bt}}{b - a}\right)$ |
| $\dfrac{1}{s(s^2 + \omega^2)}$ | $\dfrac{1}{\omega^2}(1 - \cos \omega t)$ |
| $\dfrac{s + z_1}{s(s^2 + \omega^2)}$ | $\dfrac{z_1}{\omega^2} - \sqrt{\dfrac{z_1^2 + \omega^2}{\omega^4}} \cos(\omega t + \phi) \qquad \phi \equiv \tan^{-1}(\omega / z_1)$ |
| $\dfrac{1}{s(s^2 + 2\zeta\omega_n s + \omega_n^2)}$ | $\dfrac{1}{\omega_n^2} - \dfrac{1}{\omega_n \omega_d} e^{-\zeta\omega_n t} \sin(\omega_d t + \phi)$ <br><br> $\omega_d \equiv \omega_n \sqrt{1 - \zeta^2} \qquad \phi \equiv \cos^{-1}\zeta$ |
| $\dfrac{1}{s(s + a)^2}$ | $\dfrac{1}{a^2}(1 - e^{-at} - ate^{-at})$ |
| $\dfrac{s + z_1}{s(s + a)^2}$ | $\dfrac{1}{a^2}[z_1 - z_1 e^{-at} + a(a - z_1)te^{-at}]$ |
| $\dfrac{1}{s^2}$ | $t \qquad\qquad \text{unit ramp}$ |
| $\dfrac{1}{s^2(s + a)}$ | $\dfrac{1}{a^2}(at - 1 + e^{-at})$ |
| $\dfrac{1}{s^n} \qquad n = 1, 2, 3, \ldots$ | $\dfrac{t^{n-1}}{(n-1)!} \qquad 0! = 1$ |

## Some z-Transform Pairs Useful for Control Systems Analysis

| $F(z)$ | $k$th term of time sequence $f(k)$, $\quad k = 0, 1, 2, \ldots$ |
|---|---|
| $z^{-k}$ | 1 at $k$, 0 elsewhere<br>(Kronecker delta sequence) |
| $\dfrac{z}{z - e^{-aT}}$ | $e^{-akT}$ |
| $\dfrac{Te^{-aT}z}{\left(z - e^{-aT}\right)^2}$ | $kTe^{-akT}$ |
| $\dfrac{T^2 e^{-aT} z\left(z + e^{-aT}\right)}{\left(z - e^{-aT}\right)^3}$ | $(kT)^2 e^{-akT}$ |
| $\dfrac{z^n}{\left(z - A\right)^n}$ | $\dfrac{(k+1)(k+2)\cdots(k+n-1)}{(n-1)!} A^k$<br>($A$ is any complex number) |
| $\dfrac{z}{z - 1}$ | 1 (unit step sequence) |
| $\dfrac{Tz}{(z-1)^2}$ | $kT$ (unit ramp sequence) |
| $\dfrac{T^2 z(z+1)}{(z-1)^3}$ | $(kT)^2$ |
| $\dfrac{z^n}{(z-1)^n}$ | $\dfrac{(k+1)(k+2)\cdots(k+n-1)}{(n-1)!}$ |
| $\dfrac{z \sin \omega T}{z^2 - 2z \cos \omega T + 1}$ | $\sin \omega kT$ |
| $\dfrac{z(z - \cos \omega T)}{z^2 - 2z \cos \omega T + 1}$ | $\cos \omega kT$ |
| $\dfrac{z e^{-aT} \sin \omega T}{z^2 - 2z e^{-aT} \cos \omega T + e^{-2aT}}$ | $e^{-akT} \sin \omega kT$ |
| $\dfrac{z(z - e^{-aT} \cos \omega T)}{z^2 - 2z e^{-aT} \cos \omega T + e^{-2aT}}$ | $e^{-akT} \cos \omega kT$ |
| $\dfrac{1}{(z-a)(z-b)}$ | 0 for $k = 0$<br>$\dfrac{1}{a - b}\left(a^{k-1} - b^{k-1}\right)$ for $k > 0$ |
| $\dfrac{z}{(z-a)(z-b)}$ | $\dfrac{1}{a - b}\left(a^k - b^k\right)$ |
| $\dfrac{z(1 - a)}{(z-1)(z-a)}$ | $1 - a^k$ |

# References and Bibliography

1. Churchill, R. V. and Brown, J. W., *Complex Variables and Applications*, Fourth Edition, McGraw-Hill, New York, 1984.
2. Hartline, H. K. and Ratliff, F., "Inhibitory Interaction of Receptor Units in the Eye of the Limulus," *J. Gen. Physiol.*, 40:357, 1957.
3. Bliss, J. C. and Macurdy, W. B., "Linear Models for Contrast Phenomena," *J. Optical Soc. America*, 51:1375, 1961.
4. Reichardt, W. and MacGinitie, "On the Theory of Lateral Inhibition," *Kybernetic* (German), 1:155, 1962.
5. Desoer, C. A., "A General Formulation of the Nyquist Criterion," *IEEE Transactions on Circuit Theory*, Vol. CT-12, No. 2, June 1965.
6. Krall, A. M., "An Extension and Proof of the Root-Locus Method," *Journal of the Society for Industrial and Applied Mathematics*, Vol. 9, No. 4, December 1961, pp. 644–653.
7. Wiberg, D. M., *State Space and Linear Systems*, Schaum Outline Series, McGraw-Hill, New York, 1971.
8. LaSalle, J. and Lefschetz, S., *Stability by Liapunov's Direct Method, with Applications*, Academic Press, New York, 1958.
9. Lindorff, D. P., *Theory of Sampled-Data Control Systems*, John Wiley & Sons, New York, 1965.
10. Åström, K. J. and Wittenmark, B., *Computer Controlled Systems*, Prentice-Hall, Englewood Cliffs, New Jersey, 1984.
11. Leigh, J. R., *Applied Digital Control*, Prentice-Hall, Englewood Cliffs, New Jersey, 1985.
12. Chen, C. T., *Introduction to Linear System Theory*, Second Edition, Holt, Rinehart and Winston, New York, 1985.
13. Truxal, J. G., *Automatic Feedback Control System Synthesis*, McGraw-Hill, New York, 1955.
14. Aizerman, M. A., *Theory of Automatic Control*, Addison-Wesley, Reading, Massachusetts, 1963.
15. Bode, H. W., *Network Analysis and Feedback Amplifier Design*, Van Nostrand, Princeton, New Jersey, 1945.
16. Brown, G. S. and Campbell, D. P., *Principles of Servomechanisms*, John Wiley, New York, 1948.
17. James, H. M., Nichols, N. B. and Phillips, R. S., *Theory of Servomechanisms*, McGraw-Hill, New York, 1947.
18. Kuo, B. C., *Automatic Control Systems*, Fifth Edition, Prentice-Hall, Englewood Cliffs, New Jersey 1987.

# Appendix C

## SAMPLE Screens From
## The Companion *Interactive Outline*

As described on the back cover, this book has a companion *Interactive Schaum's Outline* using Mathcad® which is designed to help you learn the subject matter more quickly and effectively. The *Interactive Outline* uses the LIVE-MATH environment of Mathcad technical calculation software to give you on-screen access to approximately 100 representative solved problems from this book, along with summaries of key theoretical points and electronic cross-referencing and hyperlinking. The following pages reproduce a representative sample of screens from the *Interactive Outline* and will help you understand the powerful capabilities of this electronic learning tool. Compare these screens with the associated solved problems from this book (the corresponding page numbers are listed at the start of each problem) to see how one complements the other.

In the *Interactive Schaum's Outline*, you'll find all related text, diagrams, and equations for a particular solved problem together on your computer screen. As you can see on the following pages, all the math appears in familiar notation, including units. The format differences you may notice between the printed *Schaum's Outline* and the *Interactive Outline* are designed to encourage your interaction with the material or show you alternate ways to solve challenging problems.

As you view the following pages, keep in mind that every number, formula, and graph shown *is completely interactive when viewed on the computer screen.* You can change the starting parameters of a problem and watch as new output graphs are calculated before your eyes; you can change any equation and immediately see the effect of the numerical calculations on the solution. Every equation, graph, and number you see is available for experimentation. Each adapted solved problem becomes a worksheet you can modify to solve dozens of related problems. The companion *Interactive Outline* thus will help you to learn and retain the material taught in this book more effectively and can also serve as a working problem-solving tool.

The Mathcad icon shown on the right is printed throughout this *Schaum's Outline*, indicating which problems are included in the *Interactive Outline*.

For more information about system requirements and the availability of titles in *Schaum's Interactive Outline Series*, please see the back cover.

Mathcad is a registered trademark of MathSoft, Inc.

## Stability of Discrete-Time Systems

(Schaum's Feedback and Control Systems, 2nd ed., Solved Problem 5.22, pp. 124 - 125)

**Statement**

Is the system with the following characteristic equation stable?

$$z^4 + 2 \cdot z^3 + 3 \cdot z^2 + z + 1.0 = 0$$

**System Parameters**

In this problem, a numerical root-finding method is used which is justified in detail in **Appendix D**. For now, you may want to just follow along, and concentrate on the stability question. Create a vector of the polynomial coefficients, up to the **n − 1** power in the equation, starting with the zeroth-order term.

$$\text{coeff} := \begin{bmatrix} 1.0 \\ 1 \\ 3 \\ 2 \end{bmatrix} \quad \begin{matrix} \mathbf{1.0} \\ \mathbf{z} \\ \mathbf{z^2} \\ \mathbf{z^3} \end{matrix}$$

Coefficient of the **n**th power term:          $A := 1$

**Solution**

Find the number of coefficients in the vector, and create a subdiagonal matrix of ones.

$$n := 0 .. \text{length}(\text{coeff}) - 2 \qquad\qquad C_{n+1, n} := 1$$

$$C = \begin{bmatrix} 0 & 0 & 0 \\ 1 & 0 & 0 \\ 0 & 1 & 0 \\ 0 & 0 & 1 \end{bmatrix}$$

This is the subdiagonal matrix. For more information on the range variables and matrix functions used here, see **A Mathcad Tutorial**.

Solve for the eigenvalues of the matrix constructed from **C** and the coefficient vector. These are the roots of the equation.

$$Z := \text{eigenvals}\left( \text{augment}\left( C, -\frac{\text{coeff}}{A} \right) \right)$$

The roots of the equation are

$$Z = \begin{bmatrix} -0.043 + 0.641i \\ -0.043 - 0.641i \\ -0.957 + 1.227i \\ -0.957 - 1.227i \end{bmatrix}$$

In order to graph these solutions, index them with a range variable:

$$i := 0 .. \text{length}(Z) - 1$$

Since this is a discrete-time system, the stability requirement is that the roots lie inside the unit circle. This will be graphed parametrically using sines and cosines, so define the range for $\theta$.

$$\theta := 0, 0.1 \cdot \pi .. 2 \cdot \pi$$

The **z**-plane diagram for this system is

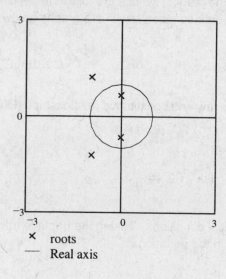

× roots
— Real axis

Because not all the roots are inside the unit circle, the system is unstable. You should take time to carefully examine the numerical root-finding technique shown here; it will be used throughout this Electronic Book.

Also, try changing the numbers in the vector of coefficients, **coeff**. See what sorts of discrete-time systems are stable. Can you find one? Can you find one that's marginally stable? What happens when you change the coefficient of the **nth** term, **A**?

## Lag Compensator

(Schaum's Feedback and Control Systems, 2nd ed., Solved Problems 6.13 and 6.16,
pp. 138 and 139)

**Statement**

(a) Derive the transfer function of the R-C network implementation of the lag compensator shown in the figure below. (b) Derive the transfer function of two simple lag networks connected in cascade.

**System Parameters**

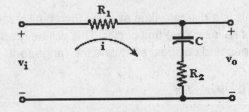

$\Omega \equiv \text{ohm}$

$\mu F \equiv 10^{-6} \cdot \text{farad}$

In order to graph the results, we use the following specific circuit element values, which are defined globally with the graphs below.

$$R_1 = 200 \cdot \Omega \qquad R_2 = 50 \cdot \Omega \qquad C = 25 \cdot \mu F$$

**Solution**

Kirchhoff's voltage law and constitutive relationships (Ohm's law) for the loop yield the equation

$$i \cdot R_1 + \frac{1}{C} \cdot \int_0^t i\, dt + i \cdot R_2 = v_i$$

**(a)**

assuming zero initial conditions. Taking the Laplace transform of these two equations results in the equation

$$\left(R_1 + R_2 + \frac{1}{C \cdot s}\right) \cdot I(s) = V_i(s)$$

Notice that this is the same expression which would have resulted if we used the expression 1/(C*s) for the impedance of the capacitor in Kirchhoff's voltage law. Since the transfer function is the ratio of the output to the input, find $P_{LAG}(s) = V_o(s)/V_i(s)$:

$$V_o = \left(R_2 + \frac{1}{C \cdot s}\right) \cdot I(s)$$

This gives

$$P_{LAG}(s) := \frac{R_2 + \dfrac{1}{C \cdot s}}{R_1 + R_2 + \dfrac{1}{C \cdot s}}$$

Compare this to the definition of a lag compensator given in **Chapter 6**. Here,

$$a := \frac{1}{(R_1 + R_2) \cdot C} \qquad\qquad b := \frac{1}{R_2 \cdot C}$$

where **-a** is the pole of the system. To graph the frequency response, define a suitable range for ω.

$$\omega := 1 \cdot \frac{rad}{sec}, 5 \cdot \frac{rad}{sec} .. 10^4 \cdot \frac{rad}{sec}$$

By changing the values of circuit elements below, examine their effect on the characteristics of the frequency response curves shown in the figures.

$$R_1 \equiv 200 \cdot \Omega \qquad\qquad R_2 \equiv 50 \cdot \Omega \qquad\qquad C \equiv 25 \cdot \mu F$$

Magnitude                                                              Phase

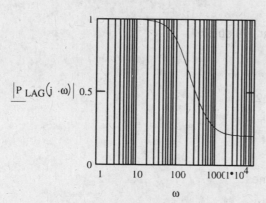

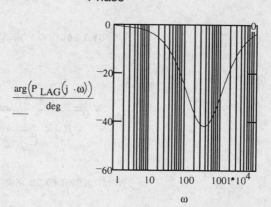

Notice that these graphs are in semilog scale. We can see that this is a lag compensator from the graph of the phase: the response lags the input for all frequencies. It is possible to examine the simple lag network by setting **R₂** equal to zero.

# MATHCAD SAMPLES

**(b)**    Suppose we examine the situation in which two simple lag networks are connected in cascade.

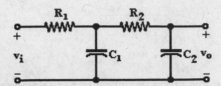

Using a voltage divider and the Laplace transform expression for the impedances,

$$V_2 = \frac{V_i \cdot \left[ \dfrac{1}{R_2 + \dfrac{1}{C_2 \cdot s}} + \dfrac{1}{\left( \dfrac{1}{C_1 \cdot s} \right)} \right]^{-1}}{R_1 + \left[ \dfrac{1}{R_2 + \dfrac{1}{C_2 \cdot s}} + \dfrac{1}{\left( \dfrac{1}{C_1 \cdot s} \right)} \right]^{-1}}$$

simplifies to

$$V_2 = \frac{V_i \cdot (R_2 \cdot C_2 \cdot s + 1)}{s^2 \cdot R_1 \cdot C_1 \cdot R_2 \cdot C_2 + (R_1 \cdot C_1 + R_2 \cdot C_2 + R_1 \cdot C_2) \cdot s + 1}$$

Using a second voltage divider, we obtain

$$V_0 = \frac{V_2 \cdot \dfrac{1}{C_2 \cdot s}}{R_2 + \dfrac{1}{C_2 \cdot s}} = \frac{V_2}{(R_2 \cdot C_2 \cdot s + 1)}$$

which, after expansion, becomes

$$V_0 = \frac{V_i}{s^2 \cdot R_1 \cdot C_1 \cdot R_2 \cdot C_2 + (R_1 \cdot C_1 + R_2 \cdot C_2 + R_1 \cdot C_2) \cdot s + 1}$$

The expression for $V_0$ results in the transfer function

$$P(s) := \frac{1}{s^2 \cdot R_1 \cdot C_1 \cdot R_2 \cdot C_2 + (R_1 \cdot C_1 + R_2 \cdot C_2 + R_1 \cdot C_2) \cdot s + 1}$$

Experiment with the two capacitor values to see their effect on the lag compensator output.

$$C_1 \equiv 25 \cdot \mu F \qquad\qquad C_2 \equiv 20 \cdot \mu F$$

Magnitude

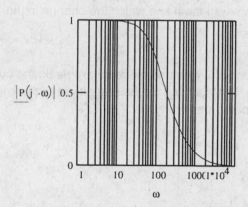

Phase

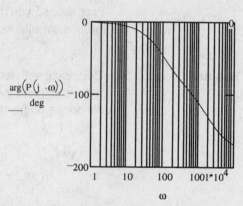

Compare the single-stage and two-stage simple lag compensators:

$$P_{single}(s) := \frac{1}{(R_1 \cdot C \cdot s + 1)}$$

$$P_{double}(s) := \frac{1}{s^2 \cdot R_1 \cdot C_1 \cdot R_2 \cdot C_2 + (R_1 \cdot C_1 + R_2 \cdot C_2 + R_1 \cdot C_2) \cdot s + 1}$$

The second-order pole on the two-stage compensator greatly increases the amount of lag achieved in phase. Think about how you would use this information to best implement a compensator. Is the two-stage system stable? What would you do if you wished to add lag to a circuit operating at higher frequencies (notice that the response is almost zero at 1000 rad/sec)?

# Frequency vs. Time-Domain Specifications

(Schaum's Feedback and Control Systems, 2nd ed., Example Problems 10.2 and 10.3, pp. 233 - 235)

**Statement**

Using the second order system shown first in **Chapter 3**, compare the frequency and time-domain specifications and plots.

**System Parameters**

$$\omega_n = 100 \cdot \frac{rad}{sec} \qquad\qquad \zeta = 0.2 \qquad\qquad dB := 1$$

(These parameters are defined globally next to the graphs at the end of the problem, so you may experiment with them and watch the change in the graphs simultaneously.)

**Solution**

Beginning with the frequency-domain, examine the resonant peak, the cutoff frequency, and the bandwidth. The equation for the magnitude of the impulse response of the canonical second-order system is

$$Y(s) := \frac{\omega_n^2}{s^2 + 2 \cdot \zeta \cdot \omega_n \cdot s + \omega_n^2}$$

The magnitude of the response, in dB, is

$$MAG(\omega) := 20 \cdot \log(\,|Y(j \cdot \omega)|\,)$$

To find the peak value, take the derivative, as was done in **Chapter 10**.

$$D(\omega) := \frac{d}{d\omega}|Y(j \cdot \omega)| \qquad \text{Guess:} \quad \omega := if\!\left(|\zeta| > .5, \frac{\omega_n}{2}, \omega_n\right)$$

Find the frequency at which the derivative is zero.

$$\omega_p := |root(D(\omega), \omega)| \qquad\qquad \omega_p = 95.915 \cdot \frac{rad}{sec}$$

Check: $D\!\left(\omega_p\right) = 8.938 \cdot 10^{-6} \cdot sec$

This is very close to zero, so $\omega_p$ is a good approximation of the resonant frequency.

The magnitude of the resonance peak is given by

$$M_p := \left| Y(j \cdot \omega_p) \right| \qquad\qquad M_p = 2.552$$

The magnitude of the peak could be used to calculate the bandwidth, but since this is a lowpass system, it's probably best to base the bandwidth calculation on the value of the transfer function at dc.

In this case,

$$\left| Y\left(0 \cdot \frac{rad}{sec}\right) \right| = 1 \qquad\qquad \text{or, in decibels,} \qquad\qquad MAG\left(0 \cdot \frac{rad}{sec}\right) = 0 \cdot dB$$

$$\omega_c := \left| \text{root} \left[ \left| Y(j \cdot \omega) \right| - \frac{\left| Y\left(0 \cdot \frac{rad}{sec}\right) \right|}{\sqrt{2}}, \omega \right] \right| \qquad\qquad \omega_c = 150.958 \cdot \frac{rad}{sec}$$

Check:   $MAG\left(\omega_c\right) = -3.01$

which corresponds, as we expect, to a 3 decibel drop.  The bandwidth is equal to the cutoff frequency, in this case, since the first cutoff frequency is zero.

The time-domain output of the system is

$$\omega_d := \omega_n \cdot \sqrt{1 - \zeta^2} \qquad\qquad \text{envelope}(t) := \frac{\omega_n \cdot e^{-\zeta \cdot \omega_n \cdot t}}{\omega_d}$$

$$y(t) := 1 - \text{envelope}(t) \cdot \sin\left(\omega_d \cdot t + \text{atan}\left(\frac{\omega_d}{\zeta \cdot \omega_n}\right)\right)$$

In the time-domain, examine the overshoot and the dominant time constant. The dominant time constant is given by inspection of the solution, from which you can see that the transient response is the decaying exponential.  The time constant is the multiplier in this exponential, described as the function **envelope(t)** above.

$$\tau := \frac{1}{\zeta \cdot \omega_n} \qquad\qquad\qquad \tau = 0.05 \cdot sec$$

The overshoot, as defined in **Chapter 10**, is the maximum difference between the transient and steady state solutions for a unit step input. We can find this value using derivatives and the **root** function, as above:

$$D(t) := \frac{d}{dt} y(t) \qquad\qquad \text{Guess:} \qquad t := \frac{\pi}{\omega_d}$$

Find the time at which the derivative is zero.

$$t_{OS} := \text{root}(D(t), t) \qquad\qquad t_{OS} = 0.032 \cdot \sec$$

The value at this point is

$$\text{value} := y(t_{OS}) \qquad\qquad \text{value} = 1.527$$

The steady-state value is approximately the value after 5 time constants:

$$F := y(5 \cdot \tau) \qquad\qquad F = 0.995$$

So the overshoot is

$$\text{overshoot} := F - \text{value}$$

Now plot both the time and the frequency response, and display the various specifications on the graphs with markers.

Create a time scale: $\qquad t := 0 \cdot \sec, .1 \cdot \tau .. 4 \cdot \tau$

To evenly space points on a logarithmic scale, use the following definitions to create the frequency range.

number of points: $\qquad N := 100 \qquad\qquad\qquad i := 0 .. N - 1$

step size: $\qquad r := \log\!\left(\frac{.01 \cdot \omega_n}{2 \cdot \omega_n}\right) \cdot \frac{1}{N}$

range variable: $\qquad \omega_i := 2 \cdot \omega_n \cdot 10^{i \cdot r}$

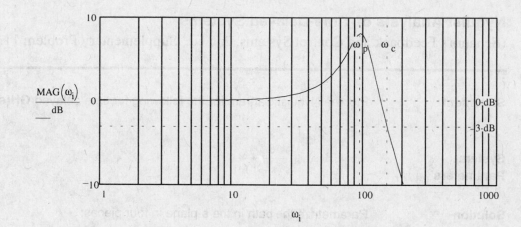

**Change these:**

$$\omega_n \equiv 100 \cdot \frac{rad}{sec}$$

$$\zeta \equiv .2$$

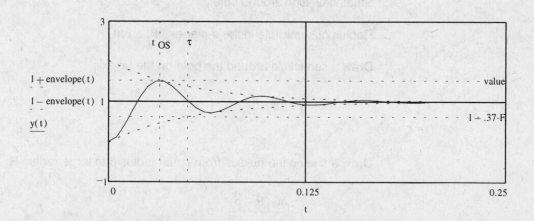

Experiment with the values of the natural frequency and the damping ratio defined next to the graphs. As always, the accuracy of the answers you get will depend somewhat on the guess value you choose for the root-finding routines. An effort has been made to build a guess which works for most values, but be careful to check that answers make physical sense. You may need to adjust the guess in some extreme cases.

What happens to the various specifications as the damping ratio changes? What about the natural frequency? What does this tell you in terms of system design?

---

## Nyquist Analysis of Time-Delayed Systems
(Schaum's Feedback and Control Systems, 2nd ed., Supplementary Problem 11.80, p. 296)

---

**Statement**

Plot the Nyquist diagram for the following for time delayed **GH(s)** shown below

**System Parameters**

$$GH(s) := \frac{e^{-s}}{s \cdot (s + 1)}$$

**Solution**

Parametrize the path in the **s**-plane in four pieces:

Number of points per segment:      $n := 500$          $m := 0 .. n$

Small deviation around pole:       $\rho := .2$

Radius of semicircle in the **s**-plane:   $R := 100$

Draw a semicircle around the pole on the **j$\omega$**-axis.

$$s_m := \rho \cdot e^{j \cdot \left( \frac{\pi \cdot m}{n} - \frac{\pi}{2} \right)}$$

Draw a line on the **j$\omega$**-axis from small radius $\rho$ to large radius **R**.

$$s_{n+m} := j \cdot \left[ \frac{m \cdot (R - \rho)}{n} + \rho \right]$$

Draw a semicircle of radius **R**.

$$s_{2 \cdot n + m} := R \cdot e^{j \cdot \left( \frac{-\pi \cdot m}{n} + \frac{\pi}{2} \right)}$$

Draw a line on the **j$\omega$**-axis from large radius **R** to small radius $\rho$.

$$s_{3 \cdot n + m} := -j \cdot \left[ R - \frac{m \cdot (R - \rho)}{n} \right]$$

Close the path and index it:      $s_0 := s_{4 \cdot n}$          $k := 0 .. 4 \cdot n$

Here is the Nyquist diagram for this system.  Each part of the path above is mapped with a different line type (solid, dashed, etc.).

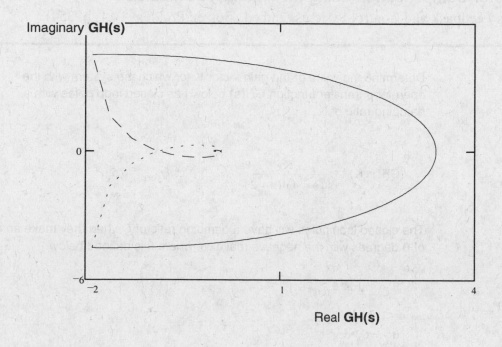

Imaginary **GH(s)**

Real **GH(s)**

Here's an expanded view of the central structure:

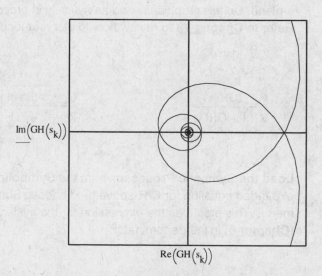

$\underline{\text{Im}}\left(\text{GH}\left(s_k\right)\right)$

$\text{Re}\left(\text{GH}\left(s_k\right)\right)$

The time delay introduces a diminishing spiral to the Nyquist plot of the open-loop transfer function, which spirals in with increasing frequency along the Nyquist path, and back out as the Nyquist path frequency returns to zero.  This spiral is superimposed upon the familiar structure you've seen before for a type 1 system in **Chapter 11**.

## Gain Factor Compensation Using the Root Locus Method
(Schaum's Feedback and Control Systems, 2nd ed., Solved Problem 14.1, p. 354)

**Statement**

Determine the value of the gain factor **K** for which the system with the open-loop transfer function **GH(s)** below has closed loop poles with a damping ratio of ζ.

**System Parameters**

$$GH(s, K) := \frac{K}{s \cdot (s + 4) \cdot (s + 2)} \qquad \zeta_{req} := 0.5$$

**Solution**

The closed loop poles will have a damping ratio of ζ when they make an angle of θ degrees with the negative real axis, where θ is defined below.

$$\theta_\zeta := acos(\zeta_{req})$$

$$\theta_\zeta = 60 \cdot deg$$

We need the value of **K** at which the root-locus crosses the ζ line in the **s**-plane. Do this graphically and analytically in order to verify the answer. Refer to **Chapter 13** to review how to plot root-loci in Mathcad.

$$\frac{C}{R} = \frac{GH(s)}{1 + GH(s)} \qquad or, \qquad \frac{C}{R} = \frac{\dfrac{K}{s \cdot (s+4) \cdot (s+2)}}{1 + \dfrac{K}{s \cdot (s+4) \cdot (s+2)}}$$

**Load the Symbolic Processor** from the **Symbolic** menu.  Then, select the expanded equation for **C/R** above, and choose **Simplify** from the **Symbolic** menu.  This produces the expression for the system characteristic equation (**Chapter 6**) in the denominator:

$$\frac{C}{R} = \frac{K}{\left(s^3 + 6 \cdot s^2 + 8 \cdot s + K\right)}$$

$$s^3 + 6 \cdot s^2 + 8 \cdot s + K_i = 0 \qquad\qquad num_{roots} := 3$$

Solving for the roots of this equation, as shown in **Appendix D**,

$$j := 0 .. num_{roots} - 2 \qquad\qquad C_{j+1,j} := 1$$

$$coeff(K) := \begin{pmatrix} -K \\ -8 \\ -6 \end{pmatrix} \qquad\qquad k := 0 .. num_{roots} - 1$$

$$i := 0 .. 500 \qquad\qquad K_i := \frac{i}{10}$$

$$R^{<i>} := eigenvals\left(augment\left(C, coeff\left(K_i\right)\right)\right)$$

The graph of the ζ line is simply a graph of a line with a slope of θ degrees, where the angle was found above. Plot that line by defining **x** and **y(x)** and including them on the root-locus plot.

$$x := -2.5, -2.4 .. 0 \qquad\qquad y(x) := \tan\left(-\theta_\zeta\right) \cdot x$$

$$p := 75 \qquad\qquad K_p = 7.5$$

Change **p** to see the direction in which the root locus moves with change in gain. This moves the boxes on the trace.

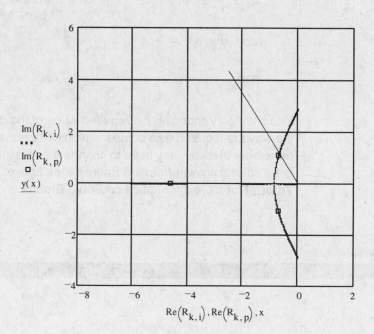

If you change the value of **p** so that one of the boxes moves onto the intersection point of the loci and the damping line, you'll find an approximate value for the desired gain factor, $K_p$. You can graphically read the value of **s** at which the intersection occurs. Use these values as starting guesses for a **Solve Block**:

$$s := -0.5 + 0.8j \qquad\qquad K := 7.5$$

Use the three constraints on the values of **s** and **K**:

Given

$$\arg(s) = \pi - \theta_\zeta \qquad\qquad \text{damping ratio constraint}$$

$$\arg(GH(s, K)) = -1 \cdot \pi \qquad \text{angle constraint (\underline{Chapter 13})}$$

$$|GH(s, K)| = 1 \qquad\qquad \text{magnitude constraint (\underline{Chapter 13})}$$

$$\begin{pmatrix} s \\ K \end{pmatrix} := \text{Find}(s, K) \qquad\qquad \begin{pmatrix} s \\ K \end{pmatrix} = \begin{pmatrix} -0.667 + 1.155i \\ 8.296 \end{pmatrix}$$

Check the solution:

$$\arg(s) = 120 \cdot \deg \qquad\qquad \pi - \theta_\zeta = 120 \cdot \deg$$

$$\arg(GH(s, K)) = -1 \cdot \pi$$

$$|GH(s, K)| = 1$$

You should try changing the required value of the damping ratio to see the way the required gain compensation changes. If you do this, remember that you may have to change the guess values for **s** and **K** to get a correct answer from the **Solve Block** above. See **A Mathcad Tutorial** for more information on **Solve Blocks**.

# Index

507

**ASK FOR THE *SCHAUM'S* SOLVED PROBLEMS SERIES AT YOUR LOCAL BOOKSTORE OR CHECK THE APPROPRIATE BOX(ES) ON THE PRECEDING PAGE AND MAIL WITH THIS COUPON TO:**

McGRAW-HILL, INC.
ORDER PROCESSING S-1
PRINCETON ROAD
HIGHTSTOWN, NJ 08520

OR CALL
1-800-338-3987

**NAME** (PLEASE PRINT LEGIBLY OR TYPE)

_____

**ADDRESS** (NO P.O. BOXES)

_____

**CITY**                                    **STATE**            **ZIP**

**ENCLOSED IS**  ☐ **A CHECK**  ☐ **MASTERCARD**  ☐ **VISA**  ☐ **AMEX**  (✓ ONE)

**ACCOUNT #** _____  **EXP. DATE** _____

**SIGNATURE** _____

MAKE CHECKS PAYABLE TO MCGRAW-HILL, INC.  <u>PLEASE INCLUDE LOCAL SALES TAX AND **$1.25** SHIPPING/HANDLING.</u>
PRICES SUBJECT TO CHANGE WITHOUT NOTICE AND MAY VARY OUTSIDE THE U.S.   FOR THIS INFORMATION, WRITE TO
THE ADDRESS ABOVE OR CALL THE **800** NUMBER.